Fluidized Bed Combustion

MECHANICAL ENGINEERING
A Series of Textbooks and Reference Books

Founding Editor

L. L. Faulkner

*Columbus Division, Battelle Memorial Institute
and Department of Mechanical Engineering
The Ohio State University
Columbus, Ohio*

Additional Volumes in Preparation

Mechanical Engineering Software

Fluidized Bed Combustion

Simeon N. Oka

*Laboratory for Thermal Engineering
and Energy
Institute VINČA
Belgrade, Serbia and Montenegro*

Technical Editor
E. J. Anthony
*CANMET Energy Technology Centre (CETC)
Natural Resources Canada
Ottawa, Ontario, Canada*

CRC Press
Taylor & Francis Group
Boca Raton London New York

CRC Press is an imprint of the
Taylor & Francis Group, an **informa** business

CRC Press
Taylor & Francis Group
6000 Broken Sound Parkway NW, Suite 300
Boca Raton, FL 33487-2742

First issued in paperback 2019

ISBN-13: 978-0-8247-4699-5 (hbk)
ISBN-13: 978-0-367-39501-8 (pbk)

Visit the Taylor & Francis Web site at
http://www.taylorandfrancis.com

and the CRC Press Web site at
http://www.crcpress.com

Foreword

Fluidized bed combustion in both of its major forms—bubbling and circulating FBC—is an important and rapidly maturing technology, employed throughout the world. Bubbling FBC technology has become the standard technology for drying, heat and steam production and power generation for smaller applications (less than 25 MW$_e$). It is widely used in Europe, North America, and China among other places to burn an enormous range of fuels, from various grades of coal, biomass, and industrial and refuse-derived waste. Given the fact that several hundred bubbling FBC boilers exist worldwide it is somewhat surprising that there are relatively few volumes dealing with the subject of fluidized bed combustion in the English language.

Most of the books actually available either deal with peripheral subjects such as fluidization, heat transfer or corrosion, or are dedicated to the newer version of this technology: the circulating fluidized bed. The subject of bubbling fluidized bed combustion has been treated in a number of early volumes such as the 1970 book on fluidized combustion of coal from the British Coal Board, or the volume edited by J. R. Howard in 1983. There is also an excellent short monograph by J. R. Howard, published in 1989, on fluidized bed technology, whose primary objective is "to help beginners and students gain insights into the subject of fluidized bed technology." Nonetheless, there is effectively a complete lack of books dealing with the overall subject of bubbling fluidized bed combustion for the professional engineer or scientist or users of this technology.

This volume provides the reader with a critical overview of the voluminous literature that exists in reports and conference and journal publications. It aims to provide the reader with a balance between the theoretical aspects of this subject and the practical applications of bubbling fluidized bed combustion technology. As such it is an important contribution to the literature in this dynamic area.

Finally, I should say a few words on the editing process followed. Given that this volume is a translation of a text produced in the former Yugoslavia in 1993, there were two choices: either to attempt to completely rewrite the volume, taking into account all current developments, or to respect the format and structure of the original volume, which presents much of the earlier literature, in particular, details on the major programs of R&D undertaken in Yugoslavia on FBC. In such circumstances no choice is entirely adequate, but the path I have followed is to leave most of the book unchanged, while correcting and modifying the text where there have been significant changes, such as the decision of most users to employ circulating fluidized bed combustion technology for larger boilers (greater than 25–50 MW$_e$). However, it is my conviction that this book represents an important contribution to a technology that will continue to be used in a wide range of countries to burn local fuels, biomass and wastes for the foreseeable future.

E. J. Anthony

Preface

It is a great pleasure and honor for a scientist from a small country to present his results to the international scientific and professional community. In the case of this book, this would not have been possible without the significant efforts that Dr. Edward J. Anthony invested in this project. From the first idea and discussions in Vienna at the IEA-FBC Meeting in 1998, he supported my wish to publish a book on fluidized bed combustion (FBC) in English, and made exceptional efforts in helping me present this book project to the publishers Marcel Dekker, Inc.

Dr. Anthony also accepted the task of being editor for this book and agreed to update Chapter 5, to add new information on bubbling fluidized bed boilers as well as correcting the translation. His contribution has helped to increase the value of this book. As such, I gratefully acknowledge his generous efforts, which confirm my optimistic views on the willingness of members of the international scientific and engineering FBC community to work together.

This book is written with the strong belief that the FBC scientific and engineering community needs insights into the state of the art in fluidized bed combustion research, and should have ready access to the data available on the behavior of full-scale bubbling fluidized bed combustion boilers. While fluidized bed combustion technology is only about 20 years old, it has quickly become competitive with conventional coal combustion technologies, and in some aspects shows significant advantages over conventional technology. Moreover, it is the only coal combustion technology originating from the beginning of the first World Energy Crisis that is actually available in the commercial market as an economic, efficient, and ecologically acceptable technology which is also fully competitive with conventional oil and gas burning technologies.

The situation for power production in Yugoslavia does not justify the large-scale importation of oil or gas, and that, together with the need for environmentally benign production of heat and power from thermal power plants, provides the driving force for the implementation of the new, so called, "clean" combustion technologies. Given the economic situation in Eastern Europe and many other countries, fluidized bed combustion boilers represent an optimal solution for using local fuel supplies.

One of the important motives for writing this book was the fact that in 1973, the VINČA Institute of Nuclear Sciences, in the Laboratory of Thermal Engineering and Energy, started a large, long-term program aimed at investigating and developing fluidized bed combustion technology. This program provided many of the important scientific and practical results that are included in this volume.

While this book is intended primarily for researchers, it should also be useful for engineers and students. In order to be of value for these different categories of readers, the book necessarily covers a wide range of issues from the strictly practical to the theoretical. First, it provides a review and critical analysis of the various fundamental investigations on fluidized bed hydrodynamics, heat transfer and combustion processes provided in the research literature focusing in particular on the experimental evidence available to support various ideas. At the same time, it must be pointed out that many processes are not yet fully investigated, and that the present level of knowledge is still inadequate, and likely to be supplemented by new developments. It is hoped that in this context the present volume will allow the interested reader to use the information provided here as a starting point for further investigations on the problems and processes of interest in the FBC field.

The book also provides data on the extensive operational experience gained on commercial-scale FBC boilers and makes numerous recommendations on the choice of boiler concept, analysis and methods to be used to determine both the operation parameters and boiler features. However, the engineer-designer will also find highlighted here the strong connection between FBC boiler characteristics and the physical processes taking place in such boilers with a view to ensuring that FBC boilers are designed and operated to achieve the primary goals of this type of combustion, namely high combustion efficiency, high boiler efficiency and low emissions.

Students, especially postgraduate students, may also use this book as supporting material for lectures on this subject, since the processes are explained systematically and efforts have been made to ensure that the text is not "loaded" with either unnecessary data or mathematical development. Major attention has also been paid to providing explanations of the physical essence of the processes taking place in fluidized bed combustion boilers. The main idea presented here is that bed hydrodynamics defines the conditions under which both heat transfer and the combustion processes take place.

Fluidized bed technology is used extensively for a vast range of mechanical, physical, and chemical processes. This is the reason this book—in particular, Part II, with its chapters on hydrodynamics of the gas-solids fluidization and heat and mass transfer in fluidized beds—ought to be useful for researchers, engineers and students in other engineering fields besides those dealing with combustion or power production.

The book is written so that each section can be read independently, although for readers wishing to gain a better background on the processes taking place in FBC furnaces and boilers, the book provides an extensive overview.

The first part of the book (Chapter 1) presents an overview of the characteristics and reasons for the development of FBC technology, the state of the art and prospects for the use of fluidized bed technology in a range of fields.

The second part is directed to readers engaged in research on fluidized beds combustion in particular, but this material will also be useful to those dealing with the research in the field of fluidized beds in general, as well as to those interested in coal combustion itself. This part of the book also presents a critical review of the current knowledge and investigation techniques employed in the field of fluidized bed combustion, as well as the trends in investigation of such processes.

The information provided about fluidized beds in this part of the book should also serve as a good basis for developing an understanding of the advantages and disadvantages of bubbling fluidized bed combustion, and suggests methods for optimizing and controlling FBC boilers. Without such knowledge it is not possible to understand the details and characteristics of the boiler design, and behavior in real full-scale FBC boilers. I also believe that engineers faced with problems in defining FBC boiler parameters in calculation and boiler design will benefit from reading Part Three, which presents the practical application of the processes presented in Part Two.

Chapter 2 is devoted to fluidized bed hydrodynamics. The main point presented here is that hydrodynamics lies at the basis of all other processes. Therefore, particular attention has been paid to bubble motion, and to particle and gas mixing processes.

Chapter 3 is devoted to a consideration of the heat transfer processes in fluidized beds. Given the challenges of this field it is understandable that much effort has been devoted to understanding heat transfer to immersed surfaces. Many empirical and experimental correlations for calculation of heat transfer coefficients for heat exchangers in fluidized bed boilers are presented here, and compared with available experimental results. However, in contrast to other books dealing with heat transfer in fluidized beds, this chapter pays special attention to heat transfer between large moving particle (fuel particle) and the fluidized bed media, and of course to the processes that are key to the performance and problems associated with heat transfer in full-scale FBC boilers.

Chapter 4 looks at the processes and changes experienced by a fuel particle, from its introduction into a hot fluidized bed up to complete burnout. Special attention has been given to the effect of the complex and heterogeneous nature of coal on the boiler design and also on the performance of such fuels in a fluidized bed. As such, this chapter will be of interest to anyone studying coal combustion processes.

Chapter 5, the first chapter of Part Three, provides a description and historical review of the concepts and designs of FBC hot-gas generators and boilers for different applications. It also presents methods for the choice of boiler concept and sizing, and describes auxiliary boiler systems and their characteristics. Particular attention is given to the issue of combustion efficiency and its dependence on fuel characteristics and boiler design. Methods for achievement of a wide range of load following are presented, and some important practical FBC problems are also discussed such as erosion of immersed surfaces and bed agglomeration issues. At the end of this chapter niche markets for bubbling fluidized bed boilers are discussed, paying special attention to the distributive heat and power production in small units, for using local fuels.

Chapter 6 is devoted to the detailed analysis of the influence of coal type and coal characteristics on the choice of FBC boiler parameters and design. This analysis is mainly based on extensive investigations on a wide range of different Yugoslav coals for fluidized bed combustion.

Chapter 7, the last chapter, looks at the critical issues surrounding emission control for SO_2, NO_x, CO, and particulates in FBC boilers.

At the end of each chapter devoted to the basic processes in fluidized beds (Chapter 2, Chapter 3, and Chapter 4) there are sections devoted to mathematical modeling of those problems. Great importance has been placed on mathematical modeling because in my opinion modern, differential mathematical models are now an essential engineering tool for calculation of and optimization of the many parameters relevant to both the operation and design of FBC boilers.

Writing a technical book is a difficult, lengthy and painful task. When the book is finished, the author must confess that the book's completion is at least in part due to contributions from many co-workers, friends and institutions, as well as specific researchers in the field of study that the book seeks to represent. I must mention that my first contact with fluidized bed combustion was at the department of Professor J. R. Howard at Aston University in 1976. I also acknowledge that I drew much of my inspiration from the creative, hard-working and friendly atmosphere at the VINČA Institute of Nuclear Sciences and, more particularly, in the Laboratory of Thermal Engineering and Energy. I thank my co-workers Dr. Borislav Grubor, MSc. Branislava Arsić, Dr. Dragoljub Dakič and Dr. Mladen Ilić for many years of successful cooperation in the field of fluidized bed combustion. Their scientific contributions to R&D efforts in FBC technology have an important place in this book, and highlight the considerable contribution that these workers have made to this field.

Recognition must be given to the benefits gained from taking part in the joint work of the International Energy Agency's "Implementing Agreement for Cooperation in the Development of Fluidized Bed Boilers for Industry and District Heating." Not only did this important forum allow me to be kept abreast of the many new scientific findings in the field of fluidized bed combustion, but it also gave me an opportunity to meet regularly with many colleagues and distinguished researchers in the field of fluidized bed combustion. During these meetings I was able to discuss the numerous unresolved problems in fluidized bed combustion, and attempt to define the essence of the complex physical and chemical processes that occur in these systems, and explore the best methods to investigate them.

Particular thanks are also due to Professor Bo Leckner from Chalmers University in Göteborg (Sweden), Professor Corr van den Bleek from Delft University (The Netherlands), Dr. Max van Gasselt from TNO (The Netherlands), Dr. Sven Andersson from Chalmers University, and Dr. E. J. "Ben" Anthony from CANMET (Canada).

When I started to write this book, there was no way of knowing that the economic and political situation in Yugoslavia would be so difficult over the past 12 years, or that the number of possible Yugoslavian readers would be so reduced by the split-up of the former Yugoslavia. However, it is hoped that this English edition will reach many readers in the new Balkan Peninsula states, since their economic situation and energy problems are similar.

The significant financial support provided by the Ministry for Science, Technologies and Development of Republic Serbia, both for the R&D incorporated in this volume and for the actual preparation of this book, was essential to its production. The Laboratory of Thermal Engineering and Energy also supported me during my many years of effort in writing this volume, and I would like to recognize my old friend, the former director of the Laboratory, Dr. Ljubomir Jovanović. I am also grateful to Dr. Milija Urošević and Miloš Urošević, dipl. eng., who are both old friends and colleagues, and the successful leaders of the Development Section of the CER, Čačak, factory, for their excellent cooperation and support of Yugoslavian research and development on FBC hot-gas generators and boilers.

Finally, I should acknowledge Mr. Vladimir Oka, dipl. eng., who translated Chapter 2–4 and 6 and all figure captions and tables, Mrs. Vesna Kostič, who translated Chapters 1 and 7, and Mrs. Rajka Marinkovič, who translated Chapter 5, for their excellent work on a difficult text. Special thanks are also due to Mr. Vladimir Živkovič, for layout and technical assistance in preparing the English edition.

Last and not least, I acknowledge the patience and support of my family, wife Jasmina, and sons Vladimir and Nikola, during the long and difficult period in which I wrote and prepared this volume.

Contents

P a r t   t w o

FUNDAMENTAL PROCESSES IN FLUIDIZED BED COMBUSTION BOILER FURNACES

Chapter 2

HYDRODYNAMICS OF GAS-SOLID FLUIDIZATION 37

Chapter 3

HEAT AND MASS TRANSFER IN FLUIDIZED BEDS 147

Chapter 4

FUNDAMENTAL PROCESSES DURING COAL COMBUSTION
IN FLUIDIZED BEDS. 211

Part three

FLUIDIZED BED COMBUSTION APPLICATIONS

Chapter 5

FLUIDIZED BED COMBUSTION IN PRACTICE 367

Chapter 6

INVESTIGATION OF COAL SUITABILITY FOR FLUIDIZED BED COMBUSTION 463

Chapter 7

HARMFUL MATTER EMISSION FROM FBC BOILERS 505

1.

DEVELOPMENT OF FLUIDIZED BED COMBUSTION BOILERS

1.1. **Problems of modern energy production
and the requirements posed for
coal combustion technologies**

Long periods of availability of cheap liquid and gaseous fuels have favorably affected industrial and technological development worldwide. At the same time, it has also resulted in an almost complete interruption of research and development of new technologies for coal and other solid fuels combustion. Research and development supported by coal producers and their associations have been insufficient to provide prompt development of new coal combustion technologies and to maintain the previously dominant position of coal in energy production.

Coal has been increasingly neglected for energy production, especially in heat production for industry and district heating systems. In many countries, coal was also suppressed for use in electric power production by large boiler units. Only countries with extensive coal reserves, traditionally oriented to coal as an energy source (for example, U.S.S.R., Great Britain, Germany, U.S.A.) continued to rely on coal, at least in large utility electric power systems. A similar orientation was also characteristic of some undeveloped countries rich in coal, which could not afford the use of oil even when it was relatively cheap. Figure 1.1 illustrates the loss of coal position showing the share of some fossil fuels for energy production in the U.S.A. in 1980 [1].

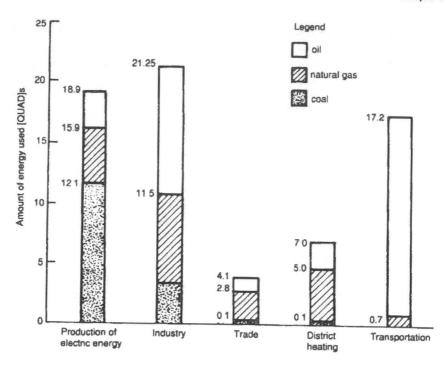

Figure 1.1.   *Share of different fossil fuels in energy production in U.S.A. in the year 1980. (1 QUAD – American unit for energy = 180·10⁶ barrels of oil – = 293·10⁹ kWh) (Reproduced by kind permission of the American Society of Mechanical Engineers from [1])*

   The share of certain fossil fuels in energy production varies among countries, according to available fuel reserves, local conditions, the type and level of technological and economic development and history. However, in general, it is quite clear that in the period before the first energy crisis, coal had lost market share in industry and for district heating of buildings and urban areas.

   The energy crisis at the beginning of the seventies, caused by an abrupt rise of liquid and gaseous fuel prices, has forced all of the leading countries in the world to reconsider their energy policy irrespective of their economic power and energy sources. The following principles have been generally accepted (at least until recently as concerns over greenhouse gases have now started to influence energy policy): (a) use domestic energy resources as much as possible, (b) reintroduce coal in all areas of energy production, (c) diversify the energy market by relying uniformly on several different energy sources and fuel suppliers, and (d) stimulate development and manufacturing of domestic energy-related equipment as a priority.

The high technological level of equipment for combustion of liquid and gaseous fuels, as well as the necessity for rational and efficient use of non-renewable energy resources, has resulted in very demanding requirements that must be fulfilled by equipment for energy production via coal combustion.

These requirements can be summarized as follows [1-6]:

(1) combust low-grade coals, with high content of moisture (up to 60%), ash (up to 70%) and sulphur (6-10%), effectively and inexpensively,

(2) effectively combust miscellaneous waste fuels, biomass and industrial and domestic wastes,

(3) achieve high combustion efficiency (>99%),

(4) achieve boiler flexibility to type and quality of coal, and assure alternate utilization of different fuels in the same boiler,

(5) Provide effective environmental protection from SO_2, NO_x and solid particles ($SO_2 < 400$ mg/m^3, $NO_x < 200$ mg/m^3, solid particles < 50 mg/m^3),

(6) achieve a wide range of load turndown ratio (20-100%), and

(7) enable automatic start-up and control of operational parameters of the plant.

Power plants, integrated into utility electric power systems, have to fulfill even more strict requirements [7]:

– high steam parameters, pressure up to 175 bar, temperature up to 540 °C,
– high combustion efficiency >99%,
– high overall boiler thermal efficiency >90%,
– desulphurization efficiency >90% SO_2 for coals with high sulphur content,
– desulphurization efficiency >70% SO_2 for lignites and coals with low sulphur content,
– NO_x emission <200 mg/m^3,
– high availability and reliability of the plant, and
– load turndown ratio of 1:3, with 5%/min. load change rate.

The price of produced energy, with the above requirements satisfied, must also be competitive with energy produced by plants burning oil or gas.

Prior to the energy crisis, independent of the low oil and gas costs, a fall in coal utilization for energy production and a narrowing of the field of application occurred because conventional coal combustion technologies were not able to fulfill the requirements mentioned above. Conventional technologies for coal combustion, by contrast, appeared to have effectively reached their commercial and technological maturity long ago. In spite of this, both grate combustion boilers and pulverized coal combustion boilers, did not meet modern requirements sufficiently well to maintain their market share for energy production. Before the advent of fluidized bed combustion (FBC) no significant new concept for coal utilization and combustion had appeared. Instead, conventional technologies were only improved and made more sophisticated by step changes, without the introduction of any truly new ideas [1, 8].

Pulverized coal combustion not only approached an effectively technical perfection, but its development has probably reached the limits for this technology in terms of size. Modern boilers of this kind are probably the largest chemical reactors in industry in general. The unit power of these boilers approaches 2000 MW_{th}. The furnace height and cross section reach 200 m and 200 m^2, respectively. Further increase of these dimensions is not probable. Boilers of this kind have a very high overall thermal efficiency (>90%) and high combustion efficiency (>99%), but they fail to comply with environmental protection requirements for SO_2 and NO_x emission without usage of very expensive equipment for flue gas cleaning. Only recently have acceptable cost-effective technical solutions for reduction of NO_x emission been developed. Flexibility of furnaces for different types of fuel do not fully meet contemporary requirements, while a turndown ratio, especially when burning low rank coals, can be achieved only with substantial consumption of liquid fuels. This highly effective means of combustion of different coals, from high rank coals to lignites, is problematic due to the requirement for the extremely expensive and energy-consuming preparation of the fuel and cannot be economically justified for units below 40 MW_{th}.

In the mid-power range (40-100 MW_{th}), before the introduction of FBC boilers, grate combustion boilers were used. The oldest coal combustion technology was not a match for liquid fuels in either technical, economic or ecological aspects. Grate combustion has many more disadvantages than combustion of pulverized coal: lower combustion efficiency, application limited only to high rank, coarse particle coals, without fine particles. Bulky and heavy movable parts are exposed to high temperatures. Ash sintering in the furnace is common. The price of the equipment for flue gas cleaning from SO_2, NO_x and ash particles is high compared to the price of the boiler itself and makes the energy production uncompetitive in the market.

Since the energy crisis has made usage of coal and other poor quality solid fuels indispensable, and since conventional technologies were unable to fulfill the requirements of contemporary energy production, investigation of new coal combustion technologies has become a prerequisite for further progress of energy production in many countries worldwide. Substantial governmental support, participation of boiler manufacturers, coal mines and large electric power production systems, as well as redirection of research in numerous scientific organizations and universities have enabled this "tidal wave" of research and development of new technologies for energy production − new coal and renewable energy source combustion and utilization technologies [3, 9].

Intensive studies of fluidized bed combustion were initiated, along with investigations of liquefaction and coal gasification, combustion of coal-water and coal-oil mixtures, MHD power generation, fuel cells, etc. Numerous international conferences on coal combustion and fluidized bed combustion [1, 5, 6, 10-13] have demonstrated that out of all technologies intensively studied since the beginning of the energy crisis in 1972, only the FBC has become commercially available, been able to technically and economically match conventional energy technologies, and to offer many superior features especially in terms of emissions and fuel flexibility.

1.2.   Development of FBC technology – background

The basic aim of FBC technology development in the U.S.A. was to enable utilization of coals with high sulphur content, while simultaneously fulfilling its strict environmental protection regulations. From the very beginning, work focused on the development of large boilers, mainly for utility electric energy production. In Great Britain the process was initiated by coal producers even before the onset of the energy crisis, with the explicit aim of enabling use of coal in industry, mainly for heat production in smaller power units. Another objective was the utilization of large amounts of waste coal, left after the separation, washing and enrichment of high-rank coals. Utilization of wood waste in the timber industry, peat and other waste fuels was favored in Scandinavian countries. In the Western European countries (Holland, Germany, France, Belgium, Austria) utilization of industrial and city waste was very important, in addition to interest in using fuels such as biomass and waste coals.

In undeveloped countries lacking other sources of energy, the basic impetus for development or use of FBC technology was the substitution of imported liquid and gaseous fuels, i. e., alleviating foreign trade balance problems and enhancing the utilization of domestic fuels (coal, mainly lignite and biomass) in small power plants [4, 14].

Technical, economic and ecological conditions for coal utilization, as well as reasons for FBC technology development, differ for small and medium power plants (industrial boilers and furnaces for heat production) and large boilers (for production of electric energy).

Liquid and gaseous fuels are highly competitive for boilers of low and medium capacity, whereas conventional coal boilers are not. Furnaces for burning liquid and gaseous fuels are smaller, simple in design and operation, possess high overall thermal efficiency, are fully automated and have large load turndown ratios. Environmental pollution is negligible except for nitrogen oxides. Conventional technologies for coal combustion cannot fulfill contemporary requirements and cannot compete with liquid and gaseous fuels in this power range. Therefore, a new technology for coal combustion, such as the FBC, should provide high combustion efficiency, satisfactory environmental protection, combustion of low quality fuels and flexibility for different fuels and loads. As we shall see, FBC boilers and hot-gas generators are by far superior to conventional coal combustion boilers in these aspects, and are a good match for plants burning liquid fuels.

In the high power range, the new technology should be competitive not only with conventional boilers burning liquid and gaseous fuels (which is no longer such a difficult requirement in light of surges in the price of these fuels, supply-related problems and hard currency requirements to pay for these boilers in the third world), but also with pulverized coal combustion boilers. The new combustion technology in this power range should deal with the following problems: reduction of the enormous size of the furnace, cost-effective environment protection and flexibility in utilization of different types of fuel.

Developments in FBC technology in the last twenty years, and the fact that FBC boilers and hot-gas generators became commercially available in the mid-eighties, helped confirm that this technology has successfully solved numerous problems related to coal combustion and energy production in general.

1.3.   A short review of FBC history

Long before the onset of the energy crisis in the seventies, when intensive research and development on FBC technology was initiated, the fluidized bed had been used as a suitable technology for different physical and chemical processes. In chemical engineering, the fluidization process as well as chemical and physical reactions in fluidized beds, had been extensively investigated and used immediately after the Second World War. A few plants using the fluidized bed in the chemical and oil industries were even built before the war [8, 15, 16].

Coal gasification, roasting of pyrite and zinc sulphite, catalytic cracking of hydrocarbons, catalyzed and non-catalyzed gas-particle reactions, drying, and mixing processes are only a few examples of reactions and technologies in which the fluidization process was used [17, 18]. In the course of development of these technologies, a great deal of information was accumulated, and experience gained in industrial exploitation, and various technical solutions were optimized or improved. This helped to serve as a solid basis for development of plants for fluidized bed combustion. Interestingly some companies (for example LURGI) entered the market for FBC using only their previous experience with fluidization in chemical engineering [16].

At the end of the fifties and the beginning of the sixties, the National Coal Board in Great Britain initiated studies on coal combustion in fluidized beds in order to increase coal consumption and regain the markets lost in competition with liquid fuels. Only at the beginning of the crisis in the seventies were these investigations to receive their maximum impetus, when researchers in many other countries joined the wave to develop this new technology (most notably U.S.A., Finland and Sweden).

In 1970 in Houston, Texas (U.S.A.) the Second International Conference on Fluidized Bed Combustion was held [13]. In the introductory lecture, one of the pioneers of this technology, Douglas Elliott described his expectations as follows:

(a) *Industry will be increasingly interested in FBC:*
- numerous firms and institutes will join the trend of investigating FBC, and
- this will result in faster and more diverse solutions of problems and substantial extension of the field of application will occur;

(b) *Numerous engineering and designing problems will be successfully managed:*
- design of the air distribution plate will be improved, but the air pressure drop across the distributor will remain the same,
- start-up systems will be developed which will not require auxiliary high-power oil burners,
- systems for control and operation of the process will be highly sensitive to a very narrow range of temperature changes of the bed,
- FBC boilers will have much smaller heat transfer surfaces than in conventional boilers, and
- coal has to be uniformly distributed over the bed surface, which is a problem necessitating serious considerations;

(c) *Numerous problems of organization of the combustion process will be solved:*
- reduced slugging and corrosion of heat transfer surfaces is an important advantage of FBC, but it is necessary to have at least 1000 hours of exploitation to verify this advantage in industrial conditions. FBC flexibility to utilization of different fuels has not been associated with any other combustion technology,
- high combustion efficiency will be achieved at low temperatures (750-850 °C); low-reactive fuels with high ash content will be burnt; and the problem of combustion of fine particles and incomplete volatile matter combustion will be solved,
- high heat capacity of FBC boilers will enable rapid power change without drastic alterations of combustion conditions,
- sulphur oxide retention is one of the most important advantages of FBC, but further development has to enable reduction of required amounts of limestone (Ca/S ratio); the cost of limestone usage and subsequent disposal has to be reduced and regeneration of used limestone may become cost effective,
- although very low, the NO_x content in the combustion products will not reach the equilibrium value for the bed temperature, since the fuel particle temperature is higher than bed temperature, and
- electrostatic precipitators will not be able to remove the fly ash of FBC with the same efficiency as seen in boilers with pulverized coal combustion;

(d) *FBC boilers will be used in a very wide spectrum of fields:*

- reconstruction of the existing boilers burning liquid fuels or conventional coal combustion boilers will be cost effective,
- replacement of the existing conventional coal combustion boilers will become cost effective with further improvement of FBC boilers, and
- small power boilers will be developed for industry and district heating; metallurgical furnaces and even households will be potential users of equipment for combustion of coal, liquid fuels and gas in fluidized bed.

At that time Douglas Elliott stated that FBC is fundamentally new technology, based on advantages resulting from the favorable conditions in which processes took place. He also stated that FBC had to become widely applied in other fields:

- heat exchangers with fluidized beds for cooling of flue gases and production of steam and other hot fluids,
- circulating fluidized beds – systems in which combustion is taking place in one bed, and heat transfer in another fluidized bed,
- production of steam with parameters needed for electric energy production,
- combustion of waste fuels, domestic waste and plastics,
- locomotives and ships are also potential users of FBC boilers, and
- high-temperature gas production.

During the same Conference, in another plenary lecture John Bishop predicted the following [13]:

- it will become possible to reach high specific heat generation; FBC boilers will achieve the specific heat production 3 MW_{th}/m^2 of the furnace cross section,
- industrial boiler design will be developed without horizontal immersed heat transfer tubes in the bed, which will reduce the cost and enable design of boilers with natural water circulation, and
- steam boilers will be developed with parameters corresponding to those needed for electric energy production: it will be possible to design a 1000 MW_{th} boiler which will require only a small excess air and will be able to achieve overall thermal efficiency higher than 90%.

At the same time John Bishop warned that:

- it will be very difficult to achieve high combustion efficiency; recirculation of fly ash will be insufficient for achieving high combustion efficiency. Therefore, special designs will be needed except for lignites and other high-reactive fuels,
- boiler manufacturers will initally tend to build in more heat transfer surfaces than needed,
- convective heat-transfer surfaces in the furnaces of FBC boilers should start with vertical sections in order to reduce elutriation of unburned particles and tube erosion, and

– pneumatic coal feeding under the bed surface, through upward-oriented nozzles, will not be justified for industrial boilers, and dried coal with narrow particle size range will be required.

Development of FBC boilers since the Second Conference has confirmed
almost all these predictions of the pioneers of the new technology, and in some aspects, even surpassed their expectations.

Introductory presentations at the 8th International Conference of Fluidized
Bed Combustion in Houston (1985) and at the 9th Conference in Boston (1987)
have substantiated that the development of FBC boilers has reached the commercial phase for both energy production in industry and utility electric energy production [1, 3, 5, 13, 19]. Numerous participants and their presentations at these
conferences (520 participants with 150 presentations in Boston) from a range of
scientific institutes, universities, R&D departments from leading boiler manufacturers, designing and engineering firms, as well as users of these boilers from both
industry and utility electric power systems, have shown enormous interest in development and application of FBC.

In the first designs of FBC boilers the inert bed material was in bubbling
fluidization regime. The inert bed material particles are in intensive chaotic motion, but the bed as a whole remains immobile and stationary. This type of boiler is
called the stationary bubbling FBC boiler (BFBC), or increasingly more common
first generation FBC boilers. At the end of the seventies a new type of FBC boiler
was introuced – the circulating fluidized bed combustion boilers (CFBC). In these
boilers the inert bed material is in the fast fluidization regime, solids move vertically upwards, and are then separated by cyclones and returned to the bottom of the
furnace. They are also called second generation FBC boilers. Fluidized bed combustion at elevated pressure in the furnace has been investigated concomitantly
with the development of the first generation FBC boilers. However, pressurized
furnaces (PFBC) have not fully entered the commercial phase and their future is un-
certain, although some manufacturers do offer this type of boiler. Plants of this kind
(about 10 of them worldwide) can still arguably be considered to be experimental
or demonstration units.

The development level of FBC technology can be judged by considering
data on the number of experimental, demonstration and commercial plants and
number of FBC boiler and furnace manufacturers.

According to the literature [15, 20] in 1980 there were a total of 33 FBC
boilers working at atmospheric pressure, with unit power of 1-100 MW$_{th}$ about 20
of them were experimental or demonstration units) and 6 pressurized boilers (all pilot or experimental) were working worldwide. Fifteen companies in Great Britain
alone were engaged in development and manufacturing of this type of boiler. All
boilers in this period were intended for heat production in industry or for district
heating. At the time, the largest boiler offered in the market was one of 40 MW$_{th}$.

In 1982 there were 120 FBC boilers worldwide, actually in operation or
in construction (18 demonstration and 102 commercial units). The steam ca-

pacity of these boilers ranged from 1.8 to 160 t/h, pressure up to 175 bar and temperature up to 540 °C [21]. At the time 36 manufacturers commercially offered these boilers (11 of these had bought licenses). As early as 1982, eleven manufacturers offered the boilers on the world market – circulating FBC boilers. Boilers of 1 to 500 t/h steam capacity, steam temperatures up to 540 °C and steam pressure of up to 180 bar were marketed. These boilers were recommended for combustion of the following fuels: coal, wood waste, biomass, liquid waste fuels, mud, coal slurry, coal washing residue, coke, petroleum coke, lignite.

In 1985 FBC boilers were already manufactured by 54 companies [22]. Twenty-one of them had bought licenses, 12 offered the second generation boilers (the circulating FBC boilers), while two of them (ASEA PFBC AB, Sweden, and Babcock Power Ltd., Germany) also offered pressurized FBC boilers. The first generation FBC boilers, with stationary bubbling FBC, were offered by practically all 54 manufacturers, and had capacities ranging from 1 MW_{th} to 150 MW_{th} (only exceptionally some offered boilers of 200 MW_{th} and even 600 MW_{th}). At the time CFBC boilers were offered with capacities of 30-400 MW_{th}. The steam parameters reached 175 bar and 540 °C.

By 1987 as many as 65 CFBC boilers were actually in operation and additionally 45 were under construction. Of these, 94 were steam units, with a total steam production of 12,800 t/h. The largest individual unit had a steam capacity of 420 t/h, that is 110 MW_e [9]. More recent data suggested that 112 CFBC boilers were in operation in 1990, of which the largest was 397 MW_{th} with steam parameters of 135 bar and 540 °C. The number of first generation FBC boilers was much higher.

The data clearly illustrate that first and second generation FBC boilers have entered the commercial phase in developed countries. The FBC boilers in operation have already accumulated several hundred thousand working hours in routine industrial exploitation. Industrial FBC boilers for production of hot water, steam and electricity have proved their features and advantages during years of operation. Experience has been gained with both bubbling and circulating FBC boilers. FBC boilers for electric energy production in utility systems have only recently been introduced into routine usage.

1.4.   Development of FBC technology in Yugoslavia

In Yugoslavia, research and development of FBC technology began in 1975, at the time when this technology had not yet entered the commercial phase, that is, when it was in the demonstration phase.

The first aim of investigations was to develop furnaces for production of hot gases and warm air. The reasons for such a direction is explained below:

- large amounts of liquid fuels were being used in Yugoslavia in agriculture for processing (drying, etc.) of agricultural products; in other branches of the economy, thermal processes commonly require hot gases that were usually produced by combustion of liquid fuels,
- furnaces for production of hot flue gases or hot clean air (especially in agriculture) are associated with favorable exploitation conditions; due to seasonal activities and frequent interruptions, interventions for correction of noted imperfections in design are feasible without additional expense caused by discontinuation of the process, and
- the furnace is the most important, and clearly "new part" of the FBC boiler; development of the FBC furnace solves most of the problems in the development of FBC boilers.

The development program was based on the following assumptions [4, 9, 14]:

- boiler design and construction in Yugoslavia is at a high level, so that boiler manufacturers were capable of producing FBC furnaces and boilers not necessarily relying on foreign licences,
- characteristics of Yugoslav coals, lignite above all, as well as characteristics of biomass and other waste fuels, necessitate original experimental studies, and
- the studies should be organized in such a way to provide the data needed for designing the furnace that will be appropriate for the available local coals.

By 1980 at the VINČA Institute of Nuclear Sciences, Belgrade, two experimental furnaces were constructed for investigation of solid fuel combustion in bubbling fluidized beds (2 kW_{th} and 200 kW_{th}), to be followed by two prototypes of FBC furnaces in the CER Čačak factory (0.5 MW_{th} and 1 MW_{th}). Using experimental data obtained on experimental furnaces built in the VINČA Institute, as well as operating experience obtained on the prototypes, the CER Čačak factory designed and commissioned the first two industrial furnaces for FBC burning coal, both of 4.5 MW_{th}. The furnaces were constructed to provide clean hot air (150 °C) for drying maize. Since then the CER Čačak factory has manufactured and commissioned more than 40 FBC furnaces of 1-4.5 MW_{th} capacity, burning coal and biomass.

At the end of 1982 the VINČA Institute in collaboration with CER Čačak and MINEL Boiler Manufacturers, Belgrade, initiated development of FBC boilers. In the VINČA Institute, in 1986 the first industrial FBC boiler of about 10 MW_{th} was built. The boiler was a reconstruction of the existing liquid fuel burning boiler. In 1988 MINEL started two FBC steam boilers of 2.5 MW_{th} each. MINEL also designed a FBC boiler with a steam capacity of 20 t/h burning wood waste from the cellulose industry. Most results of these studies, data and knowledge acquired during the realization of this research and development program in the period 1975 through 1992 are presented in this book.

Two more factories have initiated FBC boiler manufacturing based on foreign licences. As early as 1980, EMO Celje factory (now Slovenia), started a boiler prototype of 1.5 MW_{th}, but failed to go any further. In 1983 the Djuro Djaković factory Slavonski Brod (now Croatia), initiated their program of FBC boiler designs and started their experimental FBC boilers of 0.8 MW_{th} and 1.5 MW_{th}, while they started a FBC boiler of 20 t/h of steam in 1989. In 1988 MIN Niš factory manufactured a FBC boiler of 6 MW_{th} based on a foreign licence.

1.5.   Bubbling fluidized bed boilers – the state-of-the-art

The development of first generation FBC boilers was a gradual process that passed through the following phases

(a) investigation of hydrodynamics, heat transfer and combustion in large particle fluidized beds, in experimental apparatus and experimental furnaces,

(b) construction and investigation of pilot plants,

(c) construction of industrial-scale demonstration plants and their investigation in real industrial operation, and

(d) marketing of boilers of different types and parameters, as well as for different purposes.

In most countries, FBC technology development was the result of joint efforts of boiler manufacturers, electric power systems, state funds, scientific institutions and universities. Here, a modern approach to development of this technology was undertaken – realization of an investigation chain, starting with fundamental and applied research and ending with construction of demonstration and real industrial plants. Development of conventional equipment for energy production (industrial and utility boilers with grate firing and pulverized coal combustion boilers) has taken a different path. Industrial-scale units were built immediately, and their improvement was based on experience from practical operation and was accompanied by gradual increase of the unit capacity. Development was mainly financed and conducted only by the boiler manufacturers themselves.

The present state-of-the-art of first generation FBC boilers can be described as follows:

−   at the beginning of the eighties, first generation FBC boilers entered the commercial phase in the field of industrial application for heat and electrical energy production, as well as for district heating,

−   FBC technology has not yet reached full technical and commercial maturity and it is developing in accordance with market requirements and operating experience. This combustion technology has not yet exhausted all prospects of development and sophistication. We currently believe that it will be able to fulfill increasingly stricter market requirements in regard to combustion efficiency, emission control and cost effectiveness, and

− utilization of first generation FBC boilers for electric energy production in large utility electric power systems can still be regarded as in the demonstration program phase, in order to prove their reliability, availability and cost effectiveness [9].

It is generally believed that second generation FBC boilers (circulating) are more appropriate for utility applications. Further development of first generation boilers will be restricted to industrial applications for heat and electric energy production, as well as for district heating.

However, furnaces for clean air heating and hot gas production for agriculture and the process industries should not be overlooked with respect to the application of bubbling fluidized bed combustion [12, 14, 15, 20].

The parameters of commercial first generation FBC plants built so far, fulfill even the most strict requirements:

(a) *type and parameters of the working fluid:*

− air up to 400 °C,
− combustion products up to 900 °C,
− water up to 120 °C,
− saturated steam, and
− superheated steam up to 170 bar and 540 °C;

(b) *unit capacity:*

− 1-50 MW_{th} but some units have as much as 200 MW_{th} the largest boiler installed is 160 MW_e [3, 23, 24, 25];

(c) *steam capacity:*

− 2-160 t/h, but some units have even greater capacity.

The main reason why first generation FBC boilers will, most probably, be limited to production of heat and electric energy in industry is the fact that in the usual power range of industrial boilers (up to 100 MW_{th}) these bubbling FBC boilers are technically and economically superior to both conventional and the second generation FBC boilers.

Experience in construction and exploitation of bubbling FBC boilers in Germany by 1986 [26] suggested the superiority of first generation FBC boilers in the 1-20 MW_{th} power range (up to 50 MW_{th} if recirculation of unburnt particles is introduced). In the range over 50 MW_{th}, second generation FBC boilers are superior both technically and economically. During this period, 43 boilers with total capacity of 3227 MW_{th}, were either already working or commissioned in Germany, while German firms were constructing an additional 23 boilers abroad, their total capacity being 2048 MW$_{th}$.

1.6.   The features of first generation FBC boilers

The features of the fluidized bed combustion (advantages and disadvantages) result from the fact that fuel burns in a red hot bed of inert material (sand, ash, limestone) which is fluidized by upward air flow. The inert material does not participate in combustion, but provides highly favorable conditions for combustion. The fluidized bed is a special state of the mixture of particulate, loose solids and fluids in which the drag force of the particles is sufficient to support the weight of the particles. Solid particles are floating in chaotic movement, and the fluid/particle system in general undertakes some fluid-like properties.

Several modes of fluidized state are recognized with respect to gas velocity (fluidization velocity): stationary or bubbling bed; turbulent bed; and fast fluidization (or circulating fluidized bed). First generation FBC boilers are in the bubbling fluidization mode and are, therefore, called stationary bubbling FBC boilers. Second generation FBC boilers employ the fast fluidization regime, and are consequently called CFBC boilers.

Figure 1.2 illustrates a bubbling FBC boiler. In the lower part of the furnace, on the distribution plate, there is a fluidized bed of inert particulate material. Air

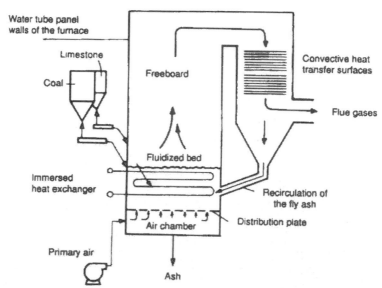

Figure 1.2.   *Schematic of the bubbling fluidized bed combustion boiler*

needed for combustion enters the furnace through the distribution plate and fluidizes the particles of inert bed material. Air velocity is lower than transport velocity of the particles, and the bed has a clearly defined, horizontal, although irregular free surface. Fuel burning (that is heat generation) mostly takes place in this fluidized bed of inert material.

When the surface of the furnace walls surrounding the fluidized bed is not sufficient to transfer the amount of heat required to maintain the FB temperature, typically at about 800 to 900 °C, heat must also be removed by the exchanger surfaces immersed in the fluidized bed.

Two ways of feeding the fuel are possible: over-bed or below the bed surface. For coarse, reactive coals, with or without only a small amount of fine particles (separated and washed coals), over-bed feeding and spreading on the bed surface are used. Thus, distribution of fuel over a larger area of the furnace cross-section is possible. For coal particles of 3-6 mm or less, fuel feeding below the bed surface is commonly used. Limestone for desulphurization is introduced in the same manner as the coal, and sometimes even with the coal.

Above the bed there is a freeboard with very low concentration of solid particles, where combustion of fine coal particles and volatiles is continued. Energy losses with unburned particles entrained with the combustion products can be reduced by their recirculation and reintroduction into the furnace for reburning.

A first generation FBC boiler is comprised of:

- a system for preparation, transport, mass flow rate control and feeding of coal,
- a system for transport, mass flow rate control and feeding of limestone,
- a start-up system,
- a system for air distribution,
- a fluidized bed furnace,
- a system for recirculation of unburned particles,
- a water circulation system (irradiated water-tube furnace walls, immersed heat exchangers and convective heat-transfer surfaces),
- a system for flue gas cleaning, and
- a system for removal of surplus or oversized inert material from the fluidized bed.

Advantages of fluidized bed combustion result primarily from the presence of fluidized inert material in the furnace. The main feature of the fluidized state (intensive mixing of the particles) ensures that in the entire space occupied by the fluidized bed, combustion takes place under the same favorable conditions – the same temperature and sufficient amount of oxygen. The large thermal capacity of the bed material and intensive heat transfer to the fuel particles, enable prompt and safe ignition of different and even low-grade and low-reactive fuels. In consequence, FBC boilers can effectively burn different low-grade coals and other poor quality fuels [4, 13, 14, 21, 26].

The possibility of utilization of different fuels, alternatively and/or simultaneously in the same boiler, is one of the most important features and advantages of

FBC boilers. The characteristic is shared by both first and second generation FBC boilers, the latter being superior in this respect. Bubbling FBC boilers can burn fuels with 60% moisture and up to 70% ash, with low heat capacity (lignites), coal waste from cleaning high-quality coals and coke, coal dust of low reactive coals, biomass of different origin, waste fuels, domestic waste, industrial waste, etc. The burning temperature is low, 800-900 °C, and below the ash sintering temperature, so that heat transfer surface slagging and fouling are avoided. Coal can be burned without prior "expensive preparation" (grinding or drying), in bulky pieces (crushed to 50 mm size), crushed to 3-5 mm when it is pneumatically injected into the bed, or pulverized if it is available in that state.

Bubbling FBC boilers without recirculation of unburned particles achieve combustion efficiency of 90%. Recirculation of unburned particles and their reintroduction into the furnace helps achieve combustion efficiency as high as 98%, depending on the coal type [13, 26, 27]. Highly-reactive coals (lignites) are characterized by high combustion efficiency, but high-rank coals and coke may not achieve even 85% efficiency without fly ash recirculation. With recycling ratios up to 5:1 (the ratio of the fly ash mass flow rate to coal mass flow rate) this type of coal can achieve even 99% combustion efficiency. A high percentage of fine coal particles (< 0.5 mm) is the primary cause of low combustion efficiency, especially when coal is fed on the bed, necessitating a recirculation system.

The heat transfer coefficient for the heat exchanger immersed in the fluidized bed is very high (~300 W/m^2K). Therefore, relatively small immersed heat-transfer surfaces may help remove from the bed as much as 50% of the total heat generated in the boiler. Heat transfer coefficients in the freeboard and in the convective pass of the boiler are similar to those of conventional boilers. The total amount of heat exchanged per unit area in these parts of the FBC boiler is lower due to lower gas temperatures, especially in the furnace itself. Generally, the size of heat transfer surfaces and amount of internals in the first generation FBC boilers are close to or somewhat below that of conventional boilers [6, 13, 26, 28-30].

One of the most important features of fluidized bed combustion is reduced emission of noxious combustion products, primarily SO_2, NO_x, chlorine compounds and other harmful compounds. By addition of limestone ($CaCO_3$) into the fluidized bed, in quantities leading to molar ratios of Ca/S up to 5, it is possible to achieve SO_2 retention of over 95% in the bed. According to regulations of numerous countries [25] the amount of $SO_2 + SO_3$ in the combustion products for FBC should not exceed 400 mg/m^3. Conventional boilers are allowed to have as much as 2000 mg/m^3, which confirms the substantial superiority of FBC boilers. First generation FBC boilers can even go below 400 mg/m^3 of SO_2 in flue gases [6, 13, 26, 28, 31-33]. It should also be kept in mind that first generation FBC boilers are likely to be much cheaper than conventional boilers if the latter must have equipment for flue gas desulphurization [26].

According to similar regulations, the NO_x emission of FBC boilers with unit power of 1-50 MW_{th}, must be below 300 mg/m^3, or even less than 200 mg/m^3 for boilers with larger unit power [26]. First generation FBC boilers can meet these requirements, albeit with some difficulty. They nevertheless do so more easily than the conventional boilers due to low burning temperature and possibility of arranging staged combustion by dividing combustion air into the primary and secondary [13, 26, 31-33]. The usual values of NO_x emission range betweeen about 300 and 700 mg/m^3 without fly ash recirculation and two-stage combustion.

First generation FBC boilers may follow load change in the range 70-100%, by changing bed temperature and fluidization velocity. With special designs, reduction of the bed height (that is reduction of size of immersed heat transfer surfaces) or bed division into several independent sections, the turndown ratio may even achieve 4:1.

The emission of CO is typically always below the upper allowed limit (the usual limiting value < 50 mg/m^3) due to high combustion efficiency. Appropriate design (cyclones and bag filters) can help reduce particle emission below the usually required values [26], that is below 50 mg/m^3 for boilers above 5 MW_{th}, or below 150 mg/m^3 for boilers less than 5 MW_{th}.

High thermal capacity of the red hot inert bed material in the furnace, enables prompt restart of the boilers after short interruption without usage of the start-up system. Thus, "warm" start-up is possible even after 24-48 hours of inactivity [6, 14, 31].

Conventional boilers burning solid or liquid fuels can be redesigned and retrofitted to bubbling fluidized bed coal combustion. Therefore, the FBC technology is important for switching from liquid fuels, increased efficiency of solid fuel combustion and reconstruction and revitalization of old conventional boilers.

All the above features of the first generation FBC boilers are summarized by the two general statements [28]:

— compliance with current strictest regulations of environmental protection and possibilities of further adjustment to even stricter future regulations, and
— high fuel flexibility – capability to burn high-rank and very low-grade fuels, including waste fuels.

If the above features are considered in combination with the fact that retrofit of conventional boilers is feasible, and that all the above properties also apply to units of low capacity (1-50 MW_{th}), it does not come as a surprise that numerous authors have pointed out that: "fluidized bed combustion (bubbling fluidized bed) is the only coal combustion technology which can both effectively and cost effectively replace liquid fuels complying concurrently within the strict requirements of current environmental protection regulations" [4, 6, 14, 15, 21, 26]. It is, at the same time, the only new coal technology that has entered the commercial phase.

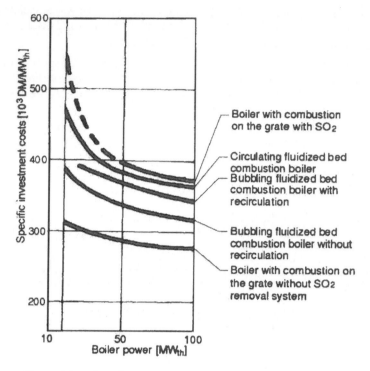

Figure 1.3. *Specific investment costs for different types of coal burning boilers*

To illustrate cost effectiveness of first generation FBC boilers from the results of an analysis [26] performed for the German market, see Fig. 1.3.

Before moving to a detailed elaboration of the disadvantages of first generation FBC boilers, we shall briefly review the advantages:

- combustion of low grade fuels (moisture content up to 60%, ash up to 70%),
- high fuel flexibility, both size, rank and type,
- low combustion temperature – possible combustion of coals with low ash,
- sintering temperature without fouling of the heat transfer surfaces,
- high heat transfer coefficients, and
- expensive pretreatment and preparation of fuels is not needed (drying, grinding),
- effective emission control of SO_2 in the furnace through removal by reaction with CaO,
- low NO_x emissions,
- burning of biomass and waste fuels,
- high combustion efficiency, and
- feasible retrofit of conventional boilers.

Every technology or an engineering design is, however, associated with some shortcomings. This is also the case for first generation FBC boilers. The following disadvantages have been generally recognized:

- relatively small amount of heat generated per unit area of furnace cross section when burning high volatile fuels,
- difficulties associated with design of high capacity units because of a large number of fuel feeding points that are required,
- small turndown ratio,
- relatively high SO_2 and NO_x emission levels,
- relatively insufficient fuel flexibility,
- relatively low combustion efficiency, and
- erosion of immersed heat transfer surfaces.

It is perhaps not surprising that since advantages of FBC result from the features of the fluidized bed, most of the shortcomings also arise from these characteristics. Bubbling fluidized beds are characterized by a markedly high ability for vertical particle mixing, both upwards and downwards, but lateral mixing is less intensive by an order of magnitude [17, 18]. Therefore, first generation FBC boilers require a large number of fuel feeding points. If the fuel is pneumatically injected into the bed, it is necessary to have a feeding point for each 1-2.5 m^2 of the furnace cross section [13]. When fuel is spread on the bed surface fewer feed points are needed – a feeder per 5-15 m^2. Since the specific heat generation per unit area of first generation FBC boiler furnace cross section is 0.5-2 MW_{th}, it is obvious that boilers with high unit capacity need a large number of feeding points. For example, the demonstration boiler in the Black Dog power plant (U.S.A.), 130 MW_e of power, has 12 coal spreaders, that is one for every 14 m^2 [34].

This large number of feeding points is the main reason why first generation FBC boilers are not being built for higher unit capacity (higher than 50-100 MW_{th}). The only exception is a first generation FBC boiler built with the capacity customary for utility electric power plants (100 MW_e).

An insufficiently wide unit capacity range and insufficient flexibility for fuels with different calorific value, arise from the practically impossible demand for the amount of heat removed from the fluidized bed via the immersed heat transfer surfaces. Requirements for current energy production in many countries are increasing in this respect. Therefore, FBC boilers should be more flexible in the quality of fuel used and in load following.

Relatively low combustion efficiency (approx. 90%) and desulphurization degree (approx. 90%) result from elutriation of fine particles from the bed. If intensive recirculation of the fly ash (approx. 5:1) and high molar ratios of Ca/S (approx. 4), respectively, are not achieved, then these values for the combustion efficiency and desulphurization degree may also be included in the list of first generation FBC boiler shortcomings, given the increasingly more strict regulations of environmental protection and economical fuel consumption [3, 10, 13, 26, 28, 29, 36-38].

Erosion of heat exchangers immersed in the fluidized bed results from intensive motion of the inert particles. Erosion is probably the greatest problem preventing further exploitation with first generation FBC boilers for which an effective solution has not yet been obtained [35].

The disadvantages of first generation FBC boilers should be noted with some reservations. Except for the first and the last problems noted above, the other listed features of first generation FBC boilers are described as shortcomings only because second generation FBC boilers are superior in every respect for large-scale applications. As compared to conventional coal-fired boilers, however, first generation FBC boilers remain superior even for these features. The large number of installed first generation FBC boilers and their performance in practice confirm that this type of boiler in the low and medium capacity range can effectively compete in the marketplace, dominating conventional coal-firing boiler technology, as well as those for liquid and gas fuel combustion.

1.7.   Reasons for circulating FBC boiler development

The shortcomings of first generation FBC boilers have been the subject of much research and development. Years of effort have resulted in increased sophistication, new designs and improved systems:

–   in order to improve combustion efficiency the system for fly ash recirculation has been introduced to allow reinjection of the unburned fuel particles into the furnace for reburning. However, it should be noted that first generation FBC boilers with fly ash recirculation, burning either low-reactivity fuels (bituminous coals, anthracite or coke) or coals with high percentage of fine particles (<1 mm), can achieve combustion efficiency of up to 98%. Introduction of fly ash recirculation has enabled higher SO_2 retention and better limestone utilization. Thus, first generation FBC boilers with fly ash recirculation can achieve sulphur retentions of as much as 95%. The molar Ca/S ratio of these boilers is also much lower than in boilers without the recirculation system [13, 26, 32, 33],

–   the problems of combustion efficiency and desulphurization efficiency have also been dealt with by the introduction of two or more fluidized beds placed one above the other [39]. Burn-up of unburned particles elutriated from the first (lower) bed is completed in the second (upper) bed together with desulphurization. Although manufacturers tend to call this type of construction second generation FBC, it is really only a variation of first generation FBC boilers,

–   a series of technical improvements has been tested in practice and on pilot plants in order to extend the range of turndown ratio for first generation FBC boilers. Change of the bed temperature, change of the bed height (by changing fluidization velocity or by removing the inert material from the bed), division of the furnace into several sections and bed slumping in some sections are examples of these attempts [33, 40]. Most commonly the division of the furnace into several separate compartments is used, enabling an increase in the turndown ratio to as much as 4:1. Due to exploitation-related problems (more complex design and therefore higher capital cost), load following remains a weak point of first generation FBC boilers,

– a series of technical innovations has been proposed to promote lateral mixing of particles and to reduce the number of feeding points to a technically acceptable level. Unequal air distribution through the distribution plate, slanting of the grate, differently shaped and distributed bed internals, have all been used to promote the intensity of lateral mixing [41], and

– the poorest results have been achieved in managing erosion of in-bed heat transfer surfaces. Protection of the tube surfaces with different coatings, refractory lining of the water-tube furnace walls and welding of protective vertical ribs or thickly placed cylindrical studs have all been used, but with varying degrees of success.

In spite of substantial improvement of first generation FBC boilers, two basic problems remain to be solved satisfactorily:

– improvement of lateral mixing of fuel, and
– separation of the combustion and heat transfer processes in time and space. In first generation FBC boilers these two processes both take place in the stationary bubbling fluidized bed.

In principle, new FBC boiler concepts could overcome the shortcomings of the first generation units resulting from these unsolved problems:

– major technical and design difficulties encountered in construction of high-capacity boilers,
– insufficient load turndown ratio, and
– insufficient fuel flexibility.

Introduction of the circulation of a mixture of inert material and fuel throughout the whole boiler unit, that is combustion based on fast fluidization, has provided conditions for successful management of the problems listed above.

1.8. Basic principles and description of circulating FBC boilers

In CFBC boilers combustion takes place in the vertical chamber (furnace) with a relatively small cross section and substantial height. The walls of the chamber are water-tube screened, and the lower part of the chamber is usually protected from erosion by fire bricks. A typical schematic of a CFBC

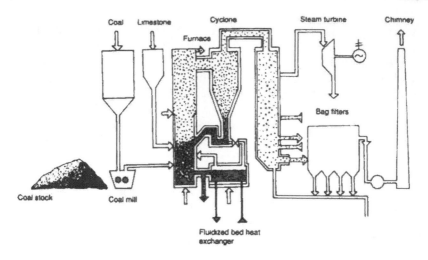

Figure 1.4. *Schematic of the circulating fluidized bed combustion boiler with auxiliary systems*

boiler is given in Fig. 1.4, together with the auxiliary systems for the boiler. Combustion takes place at 800-900 °C, similar to temperatures in first generation FBC boilers. Inert material in the furnace is a mixture of sand, ash, limestone and anhydrite. The inert bed particles are smaller than in first generation FBC boilers, while the velocity of the combustion products, that is fluidization velocity, is higher than the transport velocity of the inert particles. The inert particles are in fast fluidization regime and are, together with the fine unburned fuel particles, removed from the combustion chamber.

In one or more cyclones the solid material is separated from the gaseous combustion products and reintroduced into the furnace. Thus, recirculation of the solid particles (inert material and fuel particles) is realized in a closed circuit. Regulation of solids recirculation rate is achieved at the cyclone outlet by controlling the solid particle mass flow rate in the *stand-pipe* or through a special device called a pneumatic valve (loop seal) or, according to the design of the *L-valve* or *J-valve*.

Gaseous combustion products leave the cyclone to enter the convective part of the boiler (second pass). They exchange heat with convective heat transfer surfaces (preheater, superheater, economizer, air heater), passing subsequently through bag filters or electrostatic precipitators to the chimney.

Special features of the FBC boiler can be better explained by comparison with other modes of combustion. Figure 1.5 illustrates four modes of coal combustion explored so far:

— in grate-firing boilers, coal particles of 30-50 mm (no fine particles) remain immobile in a fixed bed. No inert material is present in the furnace, except for

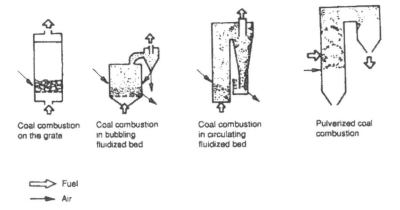

Figure 1.5.   *Schematic of the different coal combustion technologies*

ashes from the already burned coal. The air velocity is 4.5-6 m/s. Fuel is fed on the grate mechanically. Coal particles remain on the grate until combustion is completed,

- in boilers with stationary bubbling fluidized beds, combustion takes place in a fluidized bed of inert material with typical particle size 0.5-2 mm and bed height of 0.5-1 m. Fuel can be bulky particles as large as 50 mm or ground to 0-6 mm. Air velocity is 1.2-3.5 m/s. Fuel is fed into the furnace mechanically over the bed surface or pneumatically into the bed. Coal particles are in chaotic movement in the fluidized bed of inert material until they are burned up,

- boilers with circulating fluidized beds employ inert particle sizes of 50-300 μm and fluidization velocity in the range 3-9 m/s. Fuel and inert material permanently circulate. A typical fuel particle size is 6-10 mm. Coal particles circulate in a closed circuit together with inert material until they burn up, and

- pulverized coal combustion boilers employ very high gas velocities (10 m/s) and finely pulverized coal particles (<90 μm). No inert material is present in the furnace (except for the ash of the burning coal), and coal particles burn up in a single pass through the furnace.

Circulating FBC boilers have the following basic components and systems:

- a system for handling, preparation and feeding of coal and limestone into the furnace,
- combustion chamber (i. e., furnace) with water-tube or partially refratcory-lined walls,
- a system for distribution of the primary and secondary air,
- a start-up system,
- cyclones for separation and recirculation of solid material,

- pneumatic valves, L-valves or J-valves,
- external heat exchanger with fluidized bed (in one type of this boiler),
- convective heat-transfer surfaces (second pass),
- a system for flue gas cleaning, and
- a water-steam circulation system.

1.9.   Characteristics of second generation
FBC boilers

Gas velocity in second generation FBC boiler furnaces is much higher than the particle velocity; the particles move upwards, randomly and chaotically, individually and in groups (clusters) and many of them return downwards. The surface of the bed is not defined. At the bottom of the furnace, near the distribution plate, particle concentration and fluidization regime are similar to that of a bubbling stationary fluidized bed. The particle concentration gradually reduces towards the exit from the furnace. Relative gas-particle velocities are high, causing intensive transfer of heat from the gas to the particles and vice versa, and high combustion rate. Mixing of solid particles is also intensive, laterally and vertically.

The intensive mixing results in constant temperature in the furnace, from the distribution plate to the furnace exit. The same temperature is shared by the cyclones.

Conditions for combustion are very favorable – constant temperature, high relative velocities, presence of solid particles of inert material, and heat transfer by particle contact, intensive mixing of particles and air. Because of this, and because of solid particle recirculation, the fuel particles remain in the furnace until they burn up, or until they reach the size at which the cyclones cannot separate them from the combustion products. Circulating fluidized bed boilers typicaly achieve combustion efficiencies as high as 99.5%.

Intensive mixing of fuel and inert material in the fast fluidization regime and smaller furnace cross section make it possible for circulating fluidized bed boilers to have a small number of fuel feeding points. One feeding point is sufficient per 10-35 m^2 [7, 29], or per 150 t/h of steam production [37]. Some authors [13] claim that a single feeding point is sufficient for 50 MW_e, that is approximately 150 MW_{th}. These data suggest that the problems of design and construction of large boilers have been largely solved in second generation FBC boilers. At this time, initial designs for boilers in the 400-500 MW_e capacity range had already been proposed [29, 42, 43].

In CFBC boilers, combustion has been separated from heat transfer by adding heat exchangers in a back pass. In the furnace there are no in-bed heat transfer surfaces. A change of solid particle concentration in the furnace, which is simply realized by changing recirculation rate of solids in the primary circuit, significantly alters the heat transfer coefficient to the water-tube walls, as well as the amount of heat removed from the furnace. Thus, a wide range of load turndown ratios has been provided, 1:4 to 1:5, together with the high rate of load following of up to 10%/min. At the same time, the load change does not result in a change of combustion temperature, thus maintaining favorable conditions for combustion even with low load conditions [3, 6, 13, 28, 36].

The stated features of second generation FBC boilers have enabled successful combustion of different types of fuel, from those with high moisture content and low specific calorific value to high quality and low reactivity [21, 22]. Easy changes to the ratio of heat removed from the furnace (in the primary circle where solid particles circulate) and transfer to the convective pass, with constant temperature and high combustion efficiency, enable successful combustion of fuels of different qualities [7, 13, 28, 36].

Division of combustion air into the primary and secondary enables substantial reduction of NO_x formation in comparison to combustion in bubbling fluidized beds. Less than 200 mg/m^3 can be achieved [31-33].

Significantly smaller limestone particle size and recirculation of these particles make the process of SO_2 retention more intensive and the limestone utilization more complete. With a Ca/S molar ratio of 2-2.5 desulphurization efficiency can reach 99.5% [31-33]. In CFBC boilers no heat transfer surfaces transverse to the direction of particle motion are present in the primary circle of the solids circulation. Thus, erosion of the heat transfer surfaces is not so marked as it can be in the case of bubbling fluidized bed boilers.

To summarize, a short review of the advantages of second generation FBC boilers compared to first generation (bubbling bed) boilers is given below:

- higher combustion efficiency,
- higher sulphur retention degree,
- better limestone utilization,
- lower emission levels of NO_x and SO_2,
- wider range of load turndown ratio,
- design and construction of large units are feasible,
- a small number of fuel feeding points is needed, and
- lesser erosion of heat transfer surfaces.

It is, therefore, obvious that combustion in a circulating fluidized bed has additionally improved the favorable features of first generation FBC boilers while removing the basic disadvantages.

Unfortunately, the literature still lacks much of the data on disadvantages of the CFBC boilers as manufacturers have not in general been very forthcoming with such data.

CFBC boilers are much more complex plants than first generation boilers. Manufacturing is far more complex necessitating a higher technological level. This particularly applies to calculations, design and construction of large cyclones. Some authors suggest that the efficiency of these enormous cyclones, used in the primary circulation loop, is not as high as the manufacturers tend to claim. Use of pneumatic valves (L- or J-type) is also a relative novelty and far from mature technology. Calculation methods and optimum parameters of these valves have not been sufficiently investigated [44, 45]. In practice, there appear to be no reports on the commercial performance of these valves in the open literature.

Preparation of fuel and limestone is more complicated and more expensive for these boilers. Fuel and limestone crushing to particle sizes below 10 mm and 0.5 mm, respectively, is required.

The problem of CFBC boiler start-up has also been insufficiently studied. Some reports suggest that the start-up procedure lasts longer, and that the installed power of the start-up system (equipped with liquid firing burners) is an important part of the total boiler power. Therefore, the start-up of the CFBC boilers can also be considered as one of the insufficiently explored issues, if not an actual shortcoming of this type of boiler. Electric energy self-consumption, that is the so called parasitic power requirements for this type of boiler is also higher than for first generation boilers.

A brief review of the shortcomings includes:

–   an overly long start-up procedure,
–   more expensive fuel and limestone preparation,
–   higher self-consumption of electric energy, and
–   the efficiency of large cyclones is insufficiently researched and has not been verified.

1.10.   Circulating fluidized bed combustion boilers – the state-of-the-art

Development of boilers with circulating fluidized beds has two particularly interesting features:

–   the development was initiated by companies that had not manufactured boilers before, while major boiler manufacturers joined the development of this technology only at a later date, and
–   the development was marked by construction of large pilot plants (several MW_{th}). No investigations on laboratory or experimental plants were carried out at the beginning of the development of this technology.

The first steps in the development of this type of boiler are associated with Lurgi, a German company, when two pilot plants were mounted in Frankfurt in their research center. These plants were built in 1979. The larger plant had a capacity of 1.5 MW$_{th}$ [16, 37]. Lurgi based the development on ten years of experience in construction and operation of plants for calcination of aluminum hydroxide in circulating fluidized bed. As early as 1982 Lurgi built the first industrial CFBC boiler of 84 MW$_{th}$ in Luenen, Germany [37, 46].

The Ahlstrom Company (Finland) is also one of the pioneers of CFBC boiler construction. Based on their own experience [3, 47-51] they built a pilot plant in their laboratories in the town of Karhula. In 1979 they built another one in Pihlava with the following parameters of steam: pressure 8.5 MPa, production 20 t/h, and temperature 521 °C. Interestingly enough, the largest industrial CFBC boiler built by 1982 was constructed by Ahlstrom in the town of Kauttna, with steam production of 90 t/h, steam pressure 8.5 MPa and temperature 499 °C [51]. A utility company in the U.S.A., Colorado-Ute, also chose Ahlstrom over C-E/Lurgi to construct the largest CFBC boiler at the time, as a demonstration plant for electric energy production [3]. The boiler was built in the town of Nucla, and its power was 110 MW$_e$, steam production 420 t/h, steam pressure 10.4 MPa and steam temperature 540 °C [49].

In 1978 in Sweden, CFBC boiler development began at the Studsvik Institute of Energy. A prototype, 2.5 MW$_{th}$ boiler was started in 1981 [36]. In 1984 the Studsvik Institute signed an agreement with Babcock and Wilcox on license purchase [52], and the first boiler that resulted from this collaboration was started in 1986 [38, 52].

In 1980, Combustion Engineering (U.S.A.), a large international boiler manufacturer, made a license agreement with Lurgi and since that date the CE/Lurgi corporation has offered CFBC boilers.

These examples only serve to illustrate that major boiler manufacturers began second generation FBC boiler development later. The underlying reasons are hard to grasp. It is possible that the basic process applied in these boilers (circulating fluidized bed) was unfamiliar to conventional boiler makers. Also, these companies were primarily interested in manufacturing large boilers suitable for utility electric power systems, and began to build FBC boilers only when it became probable that the FBC boilers might attain these sizes.

By 1987 sixty-two CFBC boilers were operating, and an additional 45 were started by 1990. Of these 107 boilers 94 produced steam (12800 t/h in all). The largest working boiler produced 450 t/h [53]. The largest plant had four CFBC boilers with a total steam production of 770 t/h and electric power production of 94 MW$_e$ [54]. A boiler in the town of Nucla was used exclusively for electricity production for the electric power system of Colorado-Ute. It produced 420 t/h of steam and had unit power of 110 MW$_e$ [48].

Maximum steam parameters offered by the boiler manufacturers were: pressure 190 bar, temperature 580 °C, load turndown ratio up to 5:1. The manufacturers did not prescribe any restrictions as to the type and quality of fuel used. The constructed boilers used quite a range of different fuels: wood waste, peat, lignite, coals with high sulphur content, anthracite culm, waste obtained during coal separation, coke dust, waste fuels and garbage [22].

Designs for 400-500 MW_e boilers had already been reported by this time [43, 55] although it was more probable that most of the next generation boilers would cover the 150-300 MW_e range [53]. All boilers mentioned above were built based on data obtained with pilot plants that had much lower capacities, 2-10 MW_{th} [2, 16, 36-38, 46-52, 56] and on exploitation experience gained from industrial plants in operation.

The basic problem facing the CFBC boiler manufacturers at this time was not to achieve high power and high steam parameters, but verification of reliability and availability of the boilers and auxiliary systems for long-term exploitation needed within large utility electric power systems. Therefore, large demonstration units were being built in the range of 100-200 MW_e [3, 23, 48, 57, 58] to also be used for verification of technical features and advantages of FBC, the availability and reliability of the plants in the long run. The boiler in the town of Nucla was an example of this approach [48]. The electric power system of the Northern States Power Company [58] prepared the construction of a referential electric power plant with two CFBC boilers, 200 MW_e each, to be completed by 1990. Long-term programs for studying of these units were to cover: checking of the system performance, validity of scale-up the data obtained on small pilot plants onto industrial size plants, load following, and fuel flexibility. Given the above, the following conclusions were drawn:

- industrial CFBC boilers for heat production in industry (hot water, saturated steam, superheated steam and combined production of steam and hot gases, or combined production of heat and electric energy), have reached the commercial phase. Plants of this kind have been constructed since the eighties,
- CFBC boilers for electricity production only, which should be integrated into major utility electric power systems, are still in the demonstration phase, especially at larger scales. Several demonstration utility size boilers have already been built (200 MW_e). Manufacturers of these boilers already offer this kind of boiler, but electric power utilities appear to believe they should use them only after extensive investigation of the demonstration plants, and
- technical and economic comparative analyses of the designs, and exploitation data, suggested that the cost of energy produced in high power CFBC boilers and pulverized coal combustion boilers were quite similar. They match each other in several other parameters, as well. Which of these types of technology should be used depends on the actual exploitation conditions and investor requirements.

1.11. Application of the FBC boilers for energy production

FBC technology is still developing. In addition to the mentioned types of boilers – the first and second generation FBC boilers, PFBC boilers are also under investigation. The first designs of these boilers employed the bubbling fluidized bed, but more recently circulating fluidized bed boilers under elevated pressure have been proposed. Since only first and second generation boilers are widely used and commercially available, confirming their value during years of exploitation, their domains of application will be discussed briefly.

First and second generation FBC boilers are not a match for each other, as each can be used effectively only in a narrow capacity range. The reviewed advantages and disadvantages of the two types of FBC boilers clearly show the area in which each of them is markedly superior.

First generation FBC boilers are superior for the low and medium capacity range, up to 50 MW_{th}, for burning more reactive fuels and wastes, and for less strict emission regulation. Plant operation is also simpler.

Second generation FBC boilers are superior for higher capacities, for burning low-reactivity fuels and for situations in which there are strict environmental regulations.

In the range of 50-100 MW_{th} the two technologies are practically equally interesting. Which type of boiler will be selected depends on the actual exploitation conditions, as well as results of technical and economic analysis.

Application can be discussed in the light of several issues:

– what kind of fuel will be fired in FBC boilers?
– what is the purpose of the plant, that is, what kind of energy is it supposed to produce?
– in what capacity range are the FBC boilers superior? and
– is reconstruction of the existing boilers feasible or does a new boiler have to be built?

FBC boilers are markedly superior to all other combustion technologies in burning low quality coals, biomass and other waste fuels, as some of these fuels can be burned only in FBC boilers.

A wide variety of fuels can be fired in FBC boilers. It has already been noted that FBC boilers can burn wood waste, as well as anthracite and coke. Bubbling FBC boilers can employ a somewhat narrower range of fuels. Burning of diverse fuels in the same boiler can also be achieved. However, it should not be overlooked that combustion efficiency and overall thermal efficiency will be lower for low-reactivity fuels, if adequate measures are not undertaken to prevent unburned particle elutriation. Low-reactivity fuel combustion necessitates introduction of a fly ash recirculation system. Burning of high-volatile fuels necessitates introduction of secondary air in the freeboard. For combustion of fuels with different calorific value, a wide range of change of fuel feeding rate and heat transfer surfaces are necessary.

CFBC boilers burn equally effectively a wide range of fuels, from biomass to anthracite, without major design alterations. Effective combustion of low-reactivity fuels, such as anthracite and coke, is made possible by a high degree of recirculation, and it is possible for them to achieve combustion efficiency of up to 99.5%. It may seem somewhat surprising that most CFBC boilers (particularly those manufactured by Ahlstrom) burn high-volatile fuels (wood waste and peat). In first generation FBC boilers, volatile matter burns mostly above the bed and this may elevate the freeboard temperature to above 900 °C. At the same time, a smaller amount of heat is generated in the fluidized bed. CFBC boilers are characterized by the same combustion conditions in the whole furnace (constant temperature, high concentration of inert particles, high heat transfer coefficients). No problems have been encountered with excessive freeboard temperatures during combustion of highly volatile coals.

Combustion of coal with high sulphur content, when strict requirements for environmental protection have to be satisfied (<400 mg/m^3 SO$_2$), is feasible only in FBC boilers. A large number of analyses have confirmed that the investment for a FBC boilers is less expensive by 10-15% than investment for a conventional boilers of similar capacity equipped with the wet scrubbers as flue gas desulphurization units (see Fig. 1.3 [26, 58, 60]).

First generation FBC boilers are mainly used in industry and for district heating. Second generation FBC boilers are used for the same purpose, but are markedly superior in the high capacity range, and for utility electric power production.

FBC boilers can be used for heat and electric energy production. The existing first generation FBC boilers are both hot water and steam boilers while the second generation FBC boilers are mostly steam boilers. Practically achievable steam parameters in FBC boilers comply with the current requirements for production of heat in industry, as well as for electric energy in utility power systems. It can be stated that first and second generation FBC boilers can be successfully used as industrial boilers, for production of heat or for combined heat and power schemes. It was anticipated in the eighties [1, 5, 19] that in the following decade, most new utility boilers for electric energy production in the U.S.A. would be the CFBC boilers.

It has generally been accepted that second generation FBC boilers are superior to first generation FBC boilers for the high capacity range. According to available experience, in the range of 50 MW$_{th}$ and over, second generation FBC boilers are associated with lower unit energy production costs than the stationary, bubbling FBC boilers. They are cheaper by 10-15% than pulverized coal combustion boilers equipped with flue gas desulphurization [26, 28, 59-61].

Selection of the most appropriate type of the boiler should be based not only on its technical features and price of energy, but on the verified reliability and availability of the plant. This is very important in practical application, for which every interruption of energy supply results in major direct and indirect economic losses. High reliability of a plant is particularly important in large utility electric power systems.

The first and second generation FBC boilers have successfully demonstrated their advantages and reliability in energy production in industry and for district heating, that is in the power range up to 100 MW_{th}. Very large CFBC boilers, have not yet been fully accepted in large utility electric power systems as a match for conventional pulverized coal combustion boilers. However, it is expected that completion of several long term studies on demonstration plants will result in eventual recognition of CFBC boilers for this very important field of energy production. Results published so far indicate that FBC is a safe technology that will, undoubtedly, be applied on a wide scale in the next phase of power plant construction [49, 62-65].

An important area of FBC boiler application is the retrofit of the existing boilers on liquid or solid fuels, which are nonoperational due to either shortage of liquid fuels or their failure to comply with the increasingly strict regulations on environmental protection, or which are uneconomical to operate. Small and medium power boilers (up to 50 MW_{th}), grate firing boilers and boilers burning liquid fuels can be altered more easily to bubbling FBC technology. Reconstruction of large boilers firing liquid fuels is not technically or economically justified if bubbling FBC is considered. Reconstruction for pulverized coal combustion or circulating fluidized bed combustion is more acceptable. If the strict environmental protection regulations must also be fulfilled, CFBC is markedly superior. Revitalization or reconstruction of old pulverized coal combustion boilers into circulating fluidized bed boilers is a more acceptable solution [30, 34, 61, 66]. This allows utilization of different fuels, if the previous mine has been exhausted, cheaper fuels are available on the market, or if compliance with strict SO_2 and NO_x emissions regulation are required.

Previous analysis has shown that the current energy situation and modern requirements for industrial and utility boilers offer a wide field for effective and economical application of FBC boilers with bubbling or circulating fluidized beds.

This book presents an update of what is currently known on processes taking place in bubbling FBC boilers. Basic principles of engineering calculation methods and design of these boilers, basic design solutions and features of the boilers in operation are also considered. Several important factors have influenced the author to limit the content of the book to first generation FBC boilers:

– the energy situation in undeveloped countries in general, suggests utilization of first generation FBC boilers. They can successfully be used for substitution of liquid fuels in the low and medium power range; their design is simple, they are easy to operate, and comply with strict emission regulations. Application of

CFBC boilers in the above-mentioned countries will not ensue either promptly or smoothly, but it will, nevertheless follow a wide scale application of first generation FBC boilers,

– development of bubbling FBC boilers in Yugoslavia and other undeveloped countries, has resulted in independent manufacturing of these plants. Results of investigations in Yugoslavia, focusing on the specific features of lignite combustion, have been covered by the book since they are an important basis for design, calculation and exploitation of this kind of boiler. The possibilities for manufacturing and construction of a larger number of boilers, is also a major incentive for researchers, designers and others interested in the processes taking place in the boilers and their relevant features,

– there are many long-term studies and data on the performance of first generation FBC boilers and these have resulted in substantial information on the various features, advantages and disadvantages of these boilers. The analysis of the exploitation data is feasible with no major uncertainties, and at the same time, there are very few books which offer a review of what is currently known in this field,

– the fluidization technique is also used, in addition to energy production, in other fields of industry. Therefore, the content of the book will be interesting to experts in other related fields as well, and

– any elaboration of processes in circulating fluidized bed boilers must be based on processes and knowledge presented in this book. For example, combustion processes described in this book are very similar to processes in circulating fluidized bed combustion. Naturally, therefore any volume like this can be considered to precede a detailed consideration of processes in the CFBC boiler furnaces.

References

[1]   RE Harrington. A bright new future for coal and fluidized-bed combustion. Proceedings of 8th International Conference on FBC, Houston, 1985, Vol. 1, pp. 17-24.

[2]   A Kullendorff. Gotaverken CFB – A general review and introduction to the circulating fluidized bed boiler. Seminar Swedish Pulp and Paper Mission, in several towns of U.S.A. and Canada, 1985.

[3]   AM Manaker. Status of utility fluidized bed commercial development in the United States. Proceedings of ASME/IEEE Power Generation Conference, Milwaukee, (U.S.A.), 1985, 85-JPGC-13.

[4]   S Oka. Fluidized bed combustion, a new technology for coal and other solid fuel combustion (in Serbian). In: Energy and Development, Belgrade: Society Nikola Tesla, 1986, pp. 147-156.

[5]   WA Vangham. Keynote address. Proceedings of 8th International Conference on FBC, Houston, 1985, Vol. 1, pp. 1-5.

[6]   NB Smith, CB Thenem. FBC: A proven alternative. Proceedings of ASME/IEEE Power Generation Conference, Milwaukee (U.S.A.), 85-JPGCAPC-11.

[7]   JT Tang, F Engstrom. Technical assessment on the Ahlstrom pyroflow circulating and conventional bubbling fluidized bed combustion systems. Proceedings of 9th International Conference on FBC, Boston, 1987, Vol. 2, pp. 38-54.

[8]   N Čatipović, G Jovanović. Progress in fluidization technology and its influence on the development of other technologies (in Serbian). Hemijska industrija 6:151-156, 1980.

[9]   S Oka, B Grubor. State-of-the-art of CFBC boilers (in Serbian). Report of the Institute of Nuclear Sciences Boris Kidrič, Vinča, Belgrade, IBK-ITE-645, 1987.

[10]   D Anson. Rapporteur's report: Session II: Operating experience: 2nd Generation combustors. Proceedings of 3rd International FBC Conference, London, 1984, RAPP/II/1-14.

[11]   PF Fennelly. Fluidized bed combustion. J American Scientist Vol. 72, 3:254-261, 1984.

[12]   MA Conway. Has fluidized combustion kept its promise? Modern Power Systems Dec/Jan:19-22, 1984/85.

[13]   Sh Ehrlich. Fluidized combustion: Is it achieving its promise? Keynote address. Proceedings of 3rd International FBC Conference, London, 1984, KA/1/1–29.

[14]   S Oka, B Arsić, D Dakić. Development of the FBC hot-gas generators and boilers (in Serbian). Primenjena nauka 1:25-35, 1985.

[15]   S Oka. Fluidized bed combustion of solid fuels (in Serbian). Termotekhnika, 2:98-126, 1981.

[16]   L Reh, H Schmidt, G Daradimos, V Petersen. Circulating fluidbed combustion, an efficient technology for energy supply and environmental protection. Proceedings of Conference on Fluidization, London, 1980, Vol. 1, VI-2-1-11.

[17]   JF Davidson, D Harrison, ed. Fluidization, 2nd ed. London: Academic Press, 1985.

[18]   D Kunii, O Levenspiel. Fluidization engineering. New York: R. E. Krieger Publ. Co., 1977.

[19]   KE Yeager. FBC technology – The electric utility commitment, Proceedings of 8th International Conference on FBC, Houston, 1985, Vol. 1, pp. 11-16.

[20]   Department of Industry – Fluidized Bed Combustion Boilers for Industrial Uses, CEGB, 1982.

[21]   J Makansi, B Schwieger. Fluidized bed boilers. Power Aug:1-16, 1982.

[22]   Fluidized bed devices. Part A: Equipment offered. Modern Power Systems, Dec/Jan:67-77, 1984/85.

[23]   MD High. Overview of TVA's current activity in FBC. Proceedings of 8th International Conference on FBC, Houston, 1985, Vol. 1, pp. 6-10.

[24]   AM Manaker, PB West. TVA orders 160 MW$_e$ demonstration AFBC power station. Modern Power Systems Dec/Jan:59-65, 1984/85.

[25]   JW Bass, JL Golden, BM Long, RL Lumpkin, AM Manaker. Overview of the utility development of AFBC technology TVA. Proceedings of 9th International Conference on FBC, Boston, 1987, Vol. 1, pp. 146-152.

[26]   D Wiegan. Technical and economical status of FBC in West-Germany. Presented at International Conference on Coal Combustion, Copenhagen, 1986.

[27]   Fluidized Bed Combustion of Coal. Report of the National Coal Board, London 1985.

[28]   BN Gaglia, A Hal. Comparison of bubbling and circulating fluidized bed industrial steam generation. Proceedings of 9th International Conference on FBC, Boston, 1987, Vol. 1, pp. 18-25.

[29]   EA Zielinski, F Bush. Conceptual design of a 500 MW/e/ circulating fluidized-bed plant. Proceedings of 8th International Conference on FBC, Houston, 1985, Vol. 1, pp. 385-394.

[30]   P Basu, PK Halder. A new concept of operation of a pulverized coal fired boiler as circulating fluidized bed firing. Proceedings of 9th International Conference on FBC, Boston, 1987, Vol. 2, pp. 1035-1043.

[31]   B Leckner, LE Amand. Emissions from a circulating and stationary fluidized bed boiler: A comparison. Proceedings of 9th International Conference on FBC, Boston, 1987, Vol. 2, pp. 891-897.

[32]   B Leckner. Sulphur capture and nitrogen emissions from fluidized bed boilers – A comparison (preliminary data). Presented at the 12th IEA AFBC Technical Meeting, Vienna, 1986.

[33]   B Leckner, LE Amand. Emissions from a circulating and a stationary bed boiler-comparison. Chalmers University, Götteborg (Sweden), 1986, Report 186-158.

[34]   R Tollet, EM Friedman, D Parham, WJ Larva. Start-up activities at the black dog AFBC conversion. Proceedings of 9th International Conference on FBC, Boston, 1987, Vol. 1, pp. 153-160.

[35]   J Stringer. Current information on metal wastage in fluidized bed combustors. Proceedings of 9th International Conference on FBC, Boston, 1987, Vol. 2, pp. 687-698.

[36]   H Kobro. A discussion of the operation and performance of a 2.5 MW fast fluidized bed combustor and a 16 MW bubbling bed combustor. Proceedings of 3rd International Fluidization Conference, London, 1984, Vol. 1, DISC/14/110–120.

[37]   L Plass, R Anders. Fluid-bed technology applied for the generation of steam and electrical power by burning cheap solid fuels in a CFB boiler plant. Proceedings of 3rd Internation Fluidization Conference, London, 1984, Vol. 1, KN/II/1-1–11.

[38]   L Stromberg, H Kobro, et al. The fast fluidized bed – A true multifuel boiler. Proceedings of 8th International Conference on FBC, Houston, 1985, Vol. 1, pp. 415-422.

[39]   L Chambert. Development and experience of fluidized bed firing in Sweden. VDI Berichte, Nr. 601, 1986, pp. 460-474.

[40]   O Jones, RD Litt, JS Davis. Performance of Conoco's prototype MS-FBC oil field steam generator. Proceedings of 8th International Conference on FBC, Houston, 1985, Vol. 2, pp. 555-563.

[41]   J Werther, D Bellgerd. Feststoff Transport und Verteilung in wirbelschicht Feuerungen, VDI Berichte, Nr. 601, 1986, pp. 475-490.

[42]   D Turek, S Sopko, K Janssen. A generic circulating fluidized-bed system for cogenerating steam, electricity and hot air. Proceedings of 8th International Conference on FBC, Houston, 1985, Vol. 1, pp. 395-405.

[43]   K Atabay, H Barner. Advanced cycle circulating fluid bed for utility applications. Proceedings of 9th International Conference on FBC, Boston, 1987, Vol. 2, pp. 1021-1029.

[44]   PJ Jones, LS Leung. Down flow of solids through pipes and valves. In: JF Davidson, D Harrison, eds. Fluidization, 2nd ed. London: Academic Press, 1985, pp. 293-329.

[45]   MK Hill, RG Mallary, RR McKinsey. Development of the seal leg char recycle system. Proceedings of 9th International Conference on FBC, Boston, 1987, Vol. 2, pp. 862-866.

[46]   L Plass, G Daradimos, H Beisswenger. Coal combustion in the circulating fluid bed: Transfer of research and development results into industrial practice. Proceedings of 3rd European Coal Utilization Conference, Amsterdam, 1983, Vol. 2, pp. 31-56.

[47]   F Engstrom. Pyroflow-multifuel CFBC boiler with minimum impact on the environment. Proceedings of 3rd International Fluidization Conference, London, 1984, Vol. Late papers, DISC/37/335-342.

[48]   AH Gregory. Electric utilities largest circulating fluidized bed boiler – Construction update. Proceedings of 9th International Conference on FBC, Boston, 1987, Vol. 1, pp. 140-145.

[49]   TJ Boyd, WC Howe, MA Friedman. Colorado-Ute electric association CFB test program. Proceedings of 9th International Conference on FBC, Boston, 1987, Vol. 1, pp. 168-176.

[50]   L Bengtsson, F Engstrom, S Lahtineu, J Oakes. Commercial experience with circulating fluidized bed systems for cogeneration. Presented at American Power Conference, Chicago, 1981.

[51]   AJ Kelly. Fuel flexibility in a CFB combustion systems. Presented at Industrial Power Conference, New Orleans, 1982.

[52]   H Kobro, M Morris. Development of Studsvik CFB from prototype to commercial status. Proceedings of 9th International Conference on FBC, Boston, 1987, Vol. 1, pp. 161-167.

[53]   T Hirama, H Takeuchi, M Horio. Nitric oxide emission from circulating fluidized-bed coal combustion. Proceedings of 9th International Conference on FBC, Boston, 1987, Vol. 2, pp. 898-905.

[54]   JD Acierno, D Garver, B Fisher. Design concepts for industrial coal-fired fluidized-bed steam generators. Proceedings of 8th International Conference on FBC, Houston, 1985, Vol. 1, pp. 406-414.

[55]   A Kullendorff. Design of a 165 MW CFB boiler in Orebro, Sweden. Presented at 15th IEA AFBC Technical Meeting, Boston, 1987.

[56]   MM Delong, KJ Heinschel, et al. Overview of the AFBC demonstration projects. Proceedings of 9th International Conference on FBC, Boston, 1987, Vol. 1, pp. 132–139.

[57]   RP Krishnan, CS Daw, JE Jones. A review of fluidized bed combustion technology in the United States. In: WPM van Swaaij, NH Afgan, eds. Heat and Mass Transfer in Fixed and Fluidized Beds. New York: Hemisphere Publ. Co., 1985, pp. 433-456.

[58]   RJ Jensen, AE Swanson. The NSP reference plant concept using fluidized bed combustors. Proceedings of 9th International Conference on FBC, Boston, 1987, Vol. 1, pp. 109-111.

[59]   D Blauw, JO Donnell. Technology assessment and design of circulating fluidized bed boiler project for Iowa State University physical plant. Proceedings of 9th International Conference on FBC, Boston, 1987, Vol. 2, pp. 1089-1095.

[60]   Elaboration on the three firing systems, the spreader stoker, the bubbling bed and the circulating bed for a 50 t/h steam industrial power plant. Sulzer information. Presented at 14th IEA-AFBC Technical Meeting, Boston, 1987.

[61]   U Renz. State of the art in fluidized bed coal combustion in Germany. Yugo-
        slav-German Colloquium Low-Pollution and Efficient Combustion of Low-Grade
        Coals, Sarajevo, 1986. Sarajevo (Yugoslavia): Academy of Science of Bosnia and
        Herzegovina, Department of Technical Science, Special publications, 1987, pp.
        185-214.

[62]   MA Friedman, RN Melrin, DL Dove. The first one and one-half years of operation at
        Colorado-Ute electric association's 110 MW$_e$ circulating fluidized bed boiler. Pro-
        ceedings of 10th International Conference on FBC, San Francisco, 1989, Vol. 2, pp.
        739-744.

[63]   MA Friedman, TJ Boyd, TM McKee, RH Malvin. Test program: Status at Colo-
        rado-Ute electric association's 110 MW$_e$ circulating fluidized bed boiler. Proceed-
        ings of 10th International Conference on FBC, San Francisco, 1989, Vol. 2, pp.
        529-536.

[64]   M Bashar, TS Czarnecki. Design and operation of a lignite-fired CFB test program.
        Proceedings of 10th International Conference on FBC, San Francisco, 1989, Vol. 2,
        pp. 633-638.

[65]   S Oka. Circulating fluidized bed boilers – State-of-the-art and operational experi-
        ence (in Serbian). Elektroprivreda. 3-4:124-129, 1992.

[66]   P Basu. Design considerations for circulating fluidized bed combustors. Journal of
        the Institute of Energy 441:175-179, 1986.

2.

HYDRODYNAMICS OF GAS-SOLID FLUIDIZATION

The crucial difference between conventional solid fuel combustion and combustion in FBC boilers and furnaces is the fact that the latter implies that the combustion process and most heat transfer take place in inert, loose, particulate solids in the fluidized state. All features, advantages and disadvantages of this mode of combustion are related to the presence of the fluidized inert material in the furnace. The differences between first and second generation FBC boilers also arise from the fact that the processes take place in substantially different fluidization regimes.

Numerous auxiliary or accessory systems in FBC boilers also rely on processes involving the flow of two-phase systems of solid particles and gas. In bins and hoppers for coal, limestone and sand, in the pneumatic devices for solid material discharge from the furnace, as well as in systems for mass flow rate control and feeding of different particulate solid materials into the furnace, co-current or countercurrent downflow of dense packed beds, or pneumatic transport in lean phase or dense phase flow regimes are used. Systems for removal of bed material surplus from the furnace utilize the outflow of particulate solids in fluidized state or dense packed bed co-current downflow. The return of particles caught in the cyclones is also achieved partially by using countercurrent downflow in a dense packed bed of particles and partially by lean phase or dense phase pneumatic conveying. Finally, outflow and discharge of ash below the cyclones or bag filters are also examples of handling and flow organization of particulate solids.

It is, therefore, essential to elucidate basic notions about particulate solids and their features in order to differentiate between possible states of gas-solid parti-

cle mixtures. Here we will concentrate on the processes in the furnace itself. Therefore, Chapter 2 mainly deals with aerodynamics of the fluidized bed, which is the
basis for understanding any other processes in the furnaces of FBC boilers.

2.1.   Basic definitions and properties of
the particulate solids

The materials participating in the processes in FBC boilers (sand, coal,
limestone, ash) belong to a class of materials called loose (particulate) solids. The
hydrodynamics of fluidized beds, heat transfer in fluidized beds, coal combustion,
motion of particles in the bunkers, feeders, cyclones and separators, stand pipes and
other pipelines for transport of sand, limestone, coal and ash, all crucially depend
on the physical properties of solid particles.

These particulate solids (or loose, disperse) are a mechanical mixture of numerous solid particles. Many solid materials occur in nature and technical practice
in the loose form. Inorganic particulate solids in nature result from long-term natural processes: heating, cooling, thermal expansion, crushing, attrition and crumbling under the influence of atmospheric phenomena, river flows and sea waves.
Numerous modern technological operations result in loose, particulate solids –
crumbling, crushing, grinding, crystallization, precipitation, spraying, and drying.

In most cases, particulate solids are composed of numerous solid particles
of different shapes and very variable sizes. Most inorganic particulate solids appear
in nature in a wide spectrum of particle sizes. These substances are called
polydisperse materials. Loose materials resulting from technological operations
are usually polydisperse. Some technological processes enable the production of
particles that are similar in size and shape. Organic loose materials occurring in nature (seeds of various plants) will have similar sizes and shapes. Particulate solids
with uniform particle size and shape are called monodisperse materials.

Loose or powdered materials, not involved in some technological process,
usually occur as a "so-called" stagnant or fixed bed. Beds of reposed (fixed) loose
material are an important, and characteristic state of particulate solids that are the
basis for many subsequent considerations.

Physical and chemical characteristics of loose material and hydrodynamic
properties of solid particles are incorporated into formulae for the calculation of
numerous processes in FBC boilers. It is, therefore, indispensable to know them in
great detail. In the following pages basic definitions of these properties, modes of
calculation and determination will be provided. Descriptions of the reposed, fixed
bed of loose material and its features will be used for a comparison with substantially different features of fluidized beds.

2.1.1. Physical properties of the particulate solids

Bulk density of particulate solids is the mass of particles per unit of bed volume. Bulk density is always smaller than the true density of a solid particle, since the bed volume includes the volume of voids between the particles. Bulk density depends on the size and shape of the particles, state of particle surface, density of the solid particle and mode of particle "packing." If mode of particle "packing" is overlooked, major errors may ensue in determination of bulk density. Bulking of loose material to great depths, vibrations of the container walls etc., may result in settling of the bed, better packing of the particles and an increase of bulk density. The highest and lowest bulk density of particulate solids may differ by as much as 1.5 times.

According to definition, bulk density of a particulate solid may be calculated from the following expression:

$$\rho_b = \frac{m_b}{V_b} = \rho_p(1-\varepsilon)\qquad(2.1)$$

Rough classification of particulate solids may be accomplished according to their bulk density:

- light materials $\rho_b < 600$ kg/m^3,
- medium heavy materials 600 kg/m$^3 < \rho_b < 2000$ kg/m^3, and
- heavy materials $\rho_b > 2000$ kg/m^3.

Sand, limestone, coal and ash belongs to the medium heavy materials. Bulk densities of those, and some other materials are given in Table 2.1.

Particles of many particulate solids are porous. It is also therefore necessary to differentiate between particle density (including the volume of the pores), ρ_p and true particle density, ρ_s (skeletal density). The ρ_s density is frequently referred to as true, and ρ_p as apparent density, or even more commonly, particle density. Coal particles are a typical example of porous particles. It should also not be overlooked that the combustion process for some types of coal takes place not only on the external surface of the particle, but inside on the pore surface as well. In that case the data on the skeletal density, ρ_s, are needed as well as knowledge of how the particle density ρ_p is changing during the combustion process. Limestone particles are also porous. The two densities are related in the following way:

$$\rho_p = \frac{1+\rho_f\,\xi_p}{\dfrac{1}{\rho_s}+\xi_p}\qquad(2.2)$$

The void fraction of a fixed or fluidized bed is expressed as the ratio between the total volume of void space between the particles and the volume of the bed:

Table 2.1.   *Bulk density of some particulate solids*

Material	ρ_b [kg/m³]
Sand	1200-1400
Limestone	1200-1400
Coal	600-800
Ash	1200-1500
Iron ore (pulverized)	2800-3000
Table salt	800-900
Cement	1300-1900
White grain	770

$$\varepsilon = \frac{V_b - \Sigma V_p}{V_p} = 1 - \frac{\rho_b}{\rho_p} \qquad (2.3)$$

Void fraction of a fixed, stagnant bed of loose material depends on the size and shape of the particles, and also their size distribution, state of particle surface and mode of packing. Loose materials used in FBC boilers and other apparatuses with fluidized beds have 0.4-0.45 void fraction of the fixed bed.

The definition of the void fraction (2.3) applies to all states and regimes of the mixtures of solid materials and fluids, as well as for all regimes of fluidized bed. Void fraction is smallest in fixed beds of particulate solid material.

Particulate solids are characterized also with the following physical properties: moisture content, abrasion, stickiness, tendency of particles to aggregate. Magnitudes of some of these can only be described qualitatively.

2.1.2. Geometrical characteristics of the particulate solids

Individual particles of particulate solids can take various shapes: regular spheres, approximate spheres, sharp-edged crystals, squamous, fibrous, etc. Size of particles is, therefore, quite a general and vague term that can hardly be defined and quantitatively determined.

Materials composed of uniform spherical particles have a single, easily recognizable geometrical feature – the sphere diameter. The geometrical characteristics of irregularly shaped particles are not so simple to define. Irregular particles may also have numerous characteristic dimensions.

It has generally been accepted that the particle size should be defined by a mean equivalent diameter, and that irregular particles should be considered spheres with the diameter equal to the mean equivalent particle diameter. Numerous definitions of the mean equivalent particle diameter exist and these are given in Table 2.2.

Table 2.2. *Different definitions of the mean equivalent particle diameter*

No.	Term	Definition
1.	Arithmetical mean	$d_p = \dfrac{(d_1 + d_2)}{2}$ or $\\ d_p = \dfrac{(d_1 + d_2 + d_3)}{3}$
2.	Geometrical mean	$d_p = \sqrt{d_1 \cdot d_2}$ $\\ d_p = \sqrt{d_1 \cdot d_2 \cdot d_3}$
3.	Logarithmic mean	$d_p = \dfrac{d_1 + d_2}{\log(d_1 / d_2)}$
4.	Mean surface diameter	$d_p = \sqrt{\dfrac{A_p}{\pi}}$
5.	Mean volume diameter	$d_p = (6V_p / \pi)^{1/3}$
6.	Mean mass diameter	$d_p = (6 m_p / \pi \rho_p)^{1/3}$

The different definitions for the geometric characteristic of a particle are appropriate for description of different processes. It is usually more convenient to define a particle size according to its surface area, volume or mass, than according to purely geometric dimensions – d_1, d_2, d_3, width, length and height of a particle. Irrespective of the definition used, it is also important to know what kind of mean equivalent diameter was used for processing of experimental results and derivation of formulae. One should be consistent in selection and application of geometric characteristics.

In practice, the sieve analysis is most commonly us for determination of the particle size of particulate solids used in FBC boilers. T mean equivalent particle diameter is calculated then as the geometrical mean the size of orifices on adjacent sieves:

$$d_p = \sqrt{d_i d_{i+1}} \tag{2.4}$$

where d_i is the smallest opening size of the sieve through v ich the particle has passed, while d_{i+1} is the largest opening size through which the article fails to pass in the course of the sieving process.

The assumption that irregular particles can be considered as spherical, with the diameter corresponding to the mean equivalent diameter, does not imply that irregularity of the particles can always be disregarded. The hydrodynamic properties of irregularly shaped particles differ from those of spherical particles. When processes involving the particle external surface are considered, the fact that the surface area of irregular particles is larger than for spherical ones with the same volume must not be overlooked. The particle shape factor, sphericity, has been introduced to take into account aberration from the ideal sphere. It is customary to define the particle shape factor as the ratio of surface area of a sphere and surface of the particle having identical volumes [1, 2]:

$$\phi_x = \frac{A_s}{A_p} = \frac{V_p^{2/3}}{0.205 A_p}, \quad \text{where} \quad 0 \le |\phi_s| \le 1 \tag{2.5}$$

Table 2.3 gives particle shape factors for some materials [3].

Table 2.3.   *Particle shape factor of some loose materials*

Particle shape, material		ϕ_s	Particle shape, material		ϕ_s
Spherical, rounded: clay, chamotte, river sand		0.83-0.86	Pebbles	d_p = 12-20 mm	0.68
Sharp, rough, "toothed": anthracite, nonspherical sand		0.65		d_p = 3 mm	0.725
Sand	spherical	0.83	Broken glass		0.65
	angled	0.73	Polyvinyl chloride		0.68
	sharp-angled	0.60	Coal, dust, pulverized		0.65
	mean for all kinds	0.75	Coal dust, natural		0.73
Tungsten powder		0.89	Silica gel		0.18-0.33
Iron catalyzer		0.58	Raschig rings		0.3
Active carbon	d_p = 1-2 mm	0.64	Al_2O_3		0.43
	d_p = 1.5 mm	0.92	Angled		0.66
	d_p = 1.5-4.5 mm	0.79	Elongated		0.58
Oil shale	d_p = 2.5-11.2 mm	0.426	Squamous		0.43
	d_p = 34-62.5 mm	0.758	Metal-lurgical coke		
Stone coal	d_p = 6-11.5 mm	0.536		d_p = 6-11.5 mm	0.403

Defining geometrical features of polydisperse materials is more complex. In principle, it is impossible to describe such a material with a single geometrical characteristic, even if it is composed of uniform, regular spheres. The mean equivalent diameter of polydisperse particulate solids has to take into account the particle size distribution, that is the contribution of each class of particle size.

Table 2.4 gives different definitions of the mean equivalent diameter of polydisperse materials.

The mean equivalent diameters for a given polydisperse material calculated according to definitions given in Table 2.4 can vary a great deal. Different listed definitions originate from an effort to find the most appropriate geometric characteristic of polydisperse material for descriptions of different processes taking place in fluid-solid mixtures in different states and regimes. To describe the heat and mass transfer processes, definitions 3 and 7 are the most appropriate. Definition 8, however, is more appropriate for studying processes involving volume forces. In [2], definition 7 is consistently applied for description of all processes. When the experimental formulae are used, it is necessary to know which of the definitions of the mean equivalent diameter has been incorporated.

The mean equivalent diameter cannot accurately describe the geometrical properties of a polydisperse material. Loose materials with quite different particle size distribution may have the same mean equivalent diameter. For example, analyses of numerous processes necessitate knowledge of the exact proportion of the smallest particles known to be present, although they do not contribute substantially to the magnitude of the mean equivalent diameter.

Analyses of combustion processes of solid fuels in a fluidized bed necessitate knowledge of the content of particles <1 mm in the total mass of burning coal. The odds that these particles will be elutriated from the bed unburned are high. Knowing the mean equivalent diameter of coal is, therefore, not sufficient for the analysis of its behavior in the furnace. For more detailed analysis of the processes in which polydisperse particles take place, knowledge of the mean equivalent diameter, and the so called granulometric composition is necessary.

Granulometric composition provides information on distribution of particle sizes, and it is usually described in one of the following ways:

– differentially, as a probability density distribution of the particle sizes,
– integrally, i. e., cumulatively, as a resting curve, showing the distribution of the total mass that rests on the sieve with openings of size d_{pi} – for each class of the particle size. The total share of particles larger than this class is entered in the graph, and
– integrally, i. e., as a passing through curve, showing the distribution of the total mass that passes through the sieve with openings of size d_{pi} – for each class of the particle size. The total share of particles smaller than this class is entered the graph.

The three ways for presentation of particle size distribution are illustrated on Fig. 2.1.

Table 2.4. *Common definitions of the mean equivalent diameters of polydisperse, loose, materials*

No.	Name	Definition
1.	Arithmetical mean diameter	$= \dfrac{1}{n}\displaystyle\sum_{i=1}^{n} n_i d_{pi} = \sum_{i=1}^{n}\dfrac{\dfrac{y_i}{d_{pi}^2}}{\displaystyle\sum_{i=1}\dfrac{y_i}{d_{pi}^3}}$
2.	Logarithmic mean diameter	$= \dfrac{\displaystyle\sum_{i=1}^{n} n_i \log d_{pi}}{n} = \dfrac{\displaystyle\sum_{i=1}^{n}\dfrac{y_i}{d_{pi}^3}\log d_{pi}}{\displaystyle\sum_{i=1}\dfrac{y_i}{d_{pi}^3}}$
3.	Harmonic mean diameter	$= \dfrac{n}{\displaystyle\sum_{i=1}^{n}\dfrac{n_i}{d_{pi}}} = \dfrac{\displaystyle\sum_{i=1}^{n}\dfrac{y_i}{d_{pi}^3}}{\displaystyle\sum_{i=1}^{n}\dfrac{y_i}{d_{pi}^4}}$
4.	Mean surface diameter	$= \left\{\dfrac{\displaystyle\sum_{i=1}^{n} n_i d_{pi}^2}{n}\right\}^{0.5} = \left\{\dfrac{\displaystyle\sum_{i=1}^{n}\dfrac{y_i}{d_{pi}}}{\displaystyle\sum_{i=1}^{n}\dfrac{y_i}{d_{pi}^3}}\right\}^{0.5}$
5.	Mean volume diameter	$= \left\{\dfrac{1}{n}\displaystyle\sum_{i=1}^{n} n_i d_{pi}^3\right\}^{0.333} = \left\{\dfrac{1}{\displaystyle\sum_{i=1}^{n}\dfrac{y_i}{d_{pi}^3}}\right\}^{0.333}$
6.	Linear mean diameter	$= \dfrac{\displaystyle\sum_{i=1}^{n} n_i d_{pi}^2}{\displaystyle\sum_{i=1}^{n} n_i d_{pi}} = \dfrac{\displaystyle\sum_{i=1}^{n}\dfrac{y_i}{d_{pi}}}{\displaystyle\sum_{i=1}^{n}\dfrac{y_i}{d_{pi}^2}}$
7.	Mean diameter as a ratio	$= \displaystyle\sum_{i=1}^{n}\dfrac{n_i d_{pi}^3}{\displaystyle\sum_{i=1}^{n} n_i d_{pi}^2} = \dfrac{1}{\displaystyle\sum_{i=1}^{n} y_i d_{pi}}$
8.	Mass mean diameter	$= \left\{\dfrac{\displaystyle\sum_{i=1}^{n} n_i d_{pi}^4}{n}\right\}^{0.25} = \displaystyle\sum_{i=1}^{n} y_i d_{pi}$

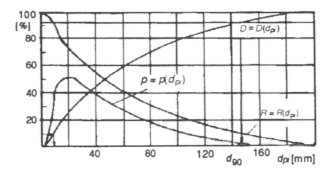

Figure 2.1.   *Three possible presentations of the particle size dis-*
tribution of the loose materials:
p – probability density distribution of the particle
size, R – distribution of the total mass that rests on the
sieve with openings of size d_{pi}, D – distribution of the
total mass that passes through the sieve with open-
ings of size d_{pi}

The terms "passing through curve" and "resting curve" originate from the most common mode of determination of granulometric composition – sieving through standard sieves. The "passing through" or "resting" curve, provide information on the share of the material that passes through or rests on the sieve with the d_i opening size in the total mass of the material.

The probability density size distribution, passing through and resting curves are interrelated as follows:

$$D(d_{pi}) = 1 - R(d_{pi}) = \int_0^{d_{pi}} p(d_{pi}) d(d_{pi}) \qquad (2.6)$$

In the course of granulometric analyses, particles are divided into classes according to their sizes. According to generally accepted recommendations, the number of classes should neither go below 5 nor exceed 20 [2]. A small number of classes cannot correctly represent granulometric composition of the material. A large number of classes may introduce into consideration particle sizes that are not representative of the material (e. g., a few extremely large particles which occur only exceptionally).

Materials obtained by grinding or crushing, usually have granulometric composition which can be represented by the well-known Rosin-Rammler distribution:

$$R(d_{pi}) = 100\exp(-bd_{pi})^n \qquad (2.7)$$

where R is expressed in [%]. The constants b and n depend on the type of material, its internal structure and coarseness of grinding. The $d = 1/b$ constant is also called the Rosin-Rammler mean diameter and represents particle size corresponding to $R = 36.8\%$.

If we know what has remained on a certain sieve after sieving (e. g., a 90 μm sieve, usually used for pulverized coal combustion analyses), the formula (2.7) can be represented as:

$$R(d_{pi}) = \left\{ \frac{R_{90}}{100} \right\}^{\left[\frac{d_p}{90} \right]^{-n}} \tag{2.8}$$

where R_{90} is the remaining material on a 90 μm sieve.

Granulometric composition of many materials obtained with different technological processes (e. g., spray driers) can be described by their Gauss probability density distribution.

The median diameter, d_{50}, is also often used for estimation of size of particulate solids. It marks the particle size at which 50% of the total number of particles, or the total mass or surface area, is due to particles smaller than d_{50} [2].

In addition to the already mentioned conditional representation of the size of particulate solids (d_{50}), the values of d_{30} or d_{40} are also used to describe the features of hydraulic transport. Measurements carried out on actual plants [4, 5] have shown that the actual values of the pressure drop, closest to the calculated ones, are those obtained when the particle sizes, below which lie 30-40% of material, are used in the calculation formulae. For evaluation of fineness of the coal grind in the thermal power plant mills, the R_{90} is used, i. e., what remains on a 90 μm sieve.

Estimation of polydisperse composition of particulate solids, i. e., the range of the particle size distribution, can be performed according to the ratio of the particle size corresponding to the share at 90% on the "passing through" curve, and the size of particles accounting for only 10% in the studied material:

$$j = \frac{d_{90}}{d_{10}} \tag{2.9}$$

The material is conditionally considered as monodisperse if $1 < j < 3$ [5]. According to [6], a material can be described as monodisperse if $d_{p\,max}/d_{p\,min} \leq$ 5-10. If granulometric composition is determined by sieving, the strictest condition of monodisperseness is, obviously, when all material remains between two adjoining standard sieves:

$$\frac{d_{p\,max}}{d_{p\,min}} \leq \delta \tag{2.10}$$

where δ is the ratio of the opening sizes of the two adjoining sieves in the standard sieve range. A wide range of particle sizes of different particulate solids, the differences in the shape and properties of the material, as well as a spectrum of acceptable definitions of the mean equivalent diameter, makes it impossible to adopt a single method for determination of particle size distribution by measurements.

The following methods have been used: sieving, air classification, centrifugation, elutriation, precipitation, impaction techniques, microscopic analysis, image analysis and light scattering [2, 4]. For materials with particles $\geq 40\,\mu m$, sieve analysis is most commonly used in engineering practice.

Sieve analysis is conducted in such a way to pour a measured quantity of solids collected on a series of standard sieves, each sieve having smaller openings that the one above. Selection of the size of sieves is made according to estimation of the range of particle sizes and the desired fineness of granulometric composition determination. A standard set of sieves is used for sieve analysis. The Soviet Standards GOST-3584-53 [4] require the ratio of the opening sizes of adjoining sieves to be $2^{0.5}$ for fine sieves and $2^{0.25}$ for coarser ones. Table 2.5 gives the values of opening sizes or sets of standard sieves according to some international standards [2].

Sieving analysis has several disadvantages that should be kept in mind in order to achieve the necessary reproducibility. Some materials tend to agglomerate, and it is also common for attrition and crumbling of particles to take place during sieving. Thus, the particle size distribution obtained may not accurately correspond to the actual situation. If particles $<40\,\mu m$ are found, some of the additional methods mentioned above have to be applied for determination of the detailed granulometric composition. In spite of these shortcomings, sieve analysis is used for investigations and calculations related to fluidized bed combustion as the most appropriate and convenient method.

Analysis of the experimental results obtained in measurements on FBC boilers in operation require knowledge of granulometric composition for the following materials: coal as received, coal at the furnace feeding point, fly ash at different locations along the fuel gas duct to the flue gas exhaust, inert material in the bed, inert material removed from the bed, limestone used for desulphurization. Determination of granulometric composition of these materials provides important information needed for designing the boiler, calculation of boiler parameters and analysis of the processes in the boiler. Therefore, the methods of granulometric analysis must be applied carefully and carried out in a consistent manner.

In the literature, different classifications of particulate materials are provided, according to the particle size [2,7]. One of the most commonly used classifications is the following: lumps ($d_{max} > 10$ mm), coarse grained ($d_{max} = 2$-10 mm), fine grained ($d_{max} = 0.5$-2 mm), powders ($d_{max} = 0.05$-0.5 mm) and pulverized material (dust) ($d_{max} < 0.05$ mm).

Table 2.5.   *Sets of standard sieves*

International standards (ISO)	German standards (DIN 4188)	U.S.A. standards ASTM-E-11-70	U.K. standards (BS-410-62)	
125 mm	–	5"	–	–
106 mm	–	4.24"	–	–
100 mm	–	4"	–	–
90 mm	–	3 ½"	–	–
75 mm	–	3"	–	–
63 mm	–	2 ½"	–	–
53 mm	–	2.12"	–	–
50 mm	–	2"	–	–
45 mm	–	2 3/4"	–	–
37.5 mm	–	1 ½"	–	–
31.5 mm	–	1 1/4"	–	–
26.5 mm	–	1.06"	–	–
25.0 mm	25.0 mm	1"	–	–
22.4 mm	–	7/8"	–	–
19.0 mm	20.0 mm	7/16"	–	–
–	18.0 mm	–	–	–
16.0 mm	16.0 mm	5/8"	–	–
13.2 mm	–	0.530"	–	–
12.5 mm	12.5 mm	½"	–	–
11.2 mm	–	7/16"	–	–
–	10.0 mm	–	–	–
9.5 mm	–	3/8"	–	–
8.0 mm	8.0 mm	5/16"	–	–
6.7 mm	–	0.265"	–	–
6.3 mm	6.3 mm	1/4"	–	–
5.6 mm	–	No. 3 ½	–	–
–	5.0 mm	–	–	–
4.75 mm	–	4	–	–
4.00 mm	4.0 mm	5	–	–
3.35 mm	–	6	3.35 mm	5
–	3.15 mm	–	–	–
2.80 mm	–	7	2.80 mm	6
2.36 mm	2.5 mm	8	2.40 mm	7
2.00 mm	2.0 mm	10	2.00 mm	8
1.70 mm	1.6 mm	12	1.68 mm	10

Table 2. 5. *Continued*

International standards (ISO)	German standards (DIN 4188)	U.S.A. standards ASTM-E-11-70	U.K. standards (BS-410-62)	
1.40 mm	–	14	1.40 mm	2
–	1.25 mm	–	–	–
1.18 mm	–	16	1.20 mm	14
1.00 mm	1.00 mm	18	1.00 mm	6
850 μm	–	20	850 μm	18
–	800 μm	–	–	–
710 μm	–	25	710 μm	22
–	630 μm	–	–	–
600 μm	–	30	600 μm	25
500 μm	500 μm	35	500 μm	30
425 μm	–	40	420 μm	36
–	400 μm	–	–	–
355 μm	–	45	355 μm	4
–	315 μm	–	–	–
300 μm	–	50	300 μm	2
250 μm	250 μm	60	250 μm	60
212 μm	–	70	210 μm	72
–	200 μm	–	–	–
180 μm	–	80	180 μm	85
–	160 μm	–	–	–
150 μm	–	100	150 μm	100
125 μm	125 μm	120	125 μm	20
106 μm	–	140	105 μm	150
–	100 μm	–	–	–
90 μm	90 μm	170	90 μm	170
–	80 μm	–	–	–
75 μm	–	200	75 μm	200
–	71 μm	–	–	–
63 μm	63 μm	230	63 μm	240
–	55 μm	–	–	–
53 μm	–	270	53 μm	300
–	50 μm	–	–	–
45 μm	45 μm	325	45 μm	350
–	40 μm	–	–	–
38 μm	–	400	–	–

2.1.3. Hydrodynamic properties of solid particles

The fluidized bed (fluidization state) is one of the possible states of a mixture of solid particles and fluid. According to its definition, it is a state, i. e., a process of interaction of numerous particles and fluid. In different modes of fluidization, particles move randomly and chaotically, either alone or in smaller or larger groups (clusters). The clusters disintegrate and reintegrate alternately, and/or randomly. The presence and motion of the surrounding particles significantly affect the interaction of particles and fluid.

For investigation of fluidization and for description of the phenomenon, it is important to know one of the basic hydrodynamic properties of a single particle – the free fall (or terminal) velocity. Knowing the free fall velocity and its physical implications is of utmost importance in understanding the fluidization process. The physical interpretation of the free fall velocity and fluidization is practically identical. In both cases it is a question of achieving a balance of the forces acting on a particle-gravity, buoyancy force and hydrodynamic resistance of a particle during motion. Free fall velocity of a particle and the minimum velocity for the fluidized state share the same physical essence, although the pertinent values for the same particles are quite different.

The free fall velocity, as a characteristic magnitude, is incorporated into many formulae which describe fluidized state and other possible states of a mixture of solid particles and fluids (for example, pneumatic transport). When the upward velocity of a fluid passing through the fluidized bed of particulate material (fluidization velocity) reaches the free fall velocity of a single, isolated particle, further increase of velocity will result in removal of that particle from the fluidized bed, followed by the larger ones, as well. Therefore, the free fall velocity determines the upper limit of the velocity range in which it is possible to maintain a fluidized state of a bed of particulate solids. Elaboration of the processes in FBC boilers and furnaces necessitates knowing the free fall (transport) velocity, especially for analysis of energy losses associated with unburned particles that are removed from the furnace. Analysis of ash behavior in the furnace, as well as analysis and calculations of other processes in accessory systems of the FBC boiler (pneumatic transport, cyclones, bins and hoppers, silos, bag filters, etc.) also necessitate knowledge of the particle terminal velocity.

The following forces act upon a single spherical particle within the gravity field during free fall in an infinite space of stagnant fluid:

– gravity force

$$F_R = \rho_p g V_p \tag{2.11}$$

– buoyancy force (Archimedes' force)

$$F_A = \rho_f g V_p \qquad (2.12)$$

and
  – resistance force

$$F_D = C_D \frac{d_p^2 \pi}{4} \frac{u_p}{2} \rho_f \qquad (2.13)$$

The gravity force and buoyancy force do not depend on the particle velocity, and they remain constant during the free fall if the fluid is incompressible ($\rho_f =$ const.). At zero time, if the free fall started from rest, the resistance force F_D equals zero and the motion of the particle has started due to an imbalance of forces:

$$F_g > F_A \quad \text{for} \quad \rho_p > \rho_f \qquad (2.14)$$

Free fall is a uniformly accelerated motion, and F_D increases during the fall, until the balance of forces is achieved:

$$F_g = F_A + F_D \qquad (2.15)$$

From this moment, the particles continue to move only due to inertia. The resultant force has become and remains zero. The particle continues to fall with a uniform velocity that is called the free fall (terminal) velocity.

The free fall velocity can be explained with a reversed sequence of events, also. If a particle is initially at rest on a porous barrier, and the fluid is moving vertically upwards, the particle will start floating when the fluid velocity reaches the free fall velocity. Thus, all the forces acting on the particle (F_g, F_A and F_D) will be balanced.

When expressions for the appropriate forces are introduced into the eq. (2.15), and the equation is reduced to a dimensionless form [7], the following expression is obtained:

$$C_D \, Re_t^2 = \frac{4}{3} \, Ar \qquad (2.16)$$

In the eq. (2.16) the right hand side depends only on the properties of particle and fluid. Therefore, it is possible to express the free fall velocity only by using particle and fluid properties ($d_p, \rho_p, \rho_f, \mu_f$). The problems arise from the fact that the drag coefficient C_D is a complex function of Reynolds number (that is, velocity), which cannot be expressed in a simple formula for a wide range of Reynolds numbers. The drag coefficient will also depend on the particle shape [7-9].

Figure 2.2 shows drag coefficient as a function of Reynolds number:

$$C_D = f(\text{Re}) \tag{2.17}$$

for a spherical particle and some other regular shapes of particles [4].

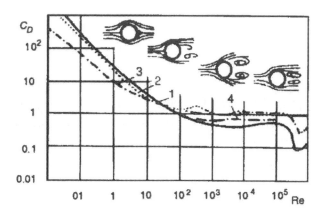

Figure 2.2.   *Drag coefficient of the single (isolated) spherical particle compared with the drag coefficients of non-spherical particles:*
1 – sphere, 2 – horizontal disk, 3 – infinite cylinder, 4 – cylinder of finite length

Equation (2.16) cannot be explicitly solved over the entire range of Re numbers and cannot provide a unique expression for a free fall velocity, u_t. Therefore, several approaches for free fall velocity determinations have been proposed: division of the Re number range into several subranges in which the curve $C_D = f(\text{Re})$, represented in Fig. 2.2, can be expressed in simple formulae, utilization of tables and nomograms [1, 2, 4, 7].

Different authors have proposed the division of the Re number range $0-10^7$ in different ways. According to the accepted division, formulae like (2.17) in these subranges may differ to a greater or lesser degree. A detailed review of the different approaches is given in [4, 7]. We shall only illustrate here the most commonly used division, which was originally proposed by M. Leva [10] and which, with few exceptions, is recommended in practically all books on fluidization [1-3, 6, 7, 11].

Table 2.6 gives the divisions offered by M. Leva and the pertinent formulae (2.17) for drag coefficient, together with formulae for calculations of free fall velocity for spherical particles resulting from them.

Table 2.6.   *Formulae for calculations of free fall velocity for spherical particle*

No.	Range of Reynolds number	$C_D = f(\text{Re})$	Formulae for calculation of u_t
1.	Laminar regime $0 < \text{Re} < 2$ $0 < \text{Ar} < 36$	$C_D = \dfrac{24}{\text{Re}}$	$\text{Re}_t = \dfrac{\text{Ar}}{18}$ or $u_t = g\dfrac{d_p^2}{18 v_f}(s-1)$
2.	Transition regime $2 < \text{Re} < 500$ $36 < \text{Ar} < 83000$	$C_D = \dfrac{18.5}{\text{Re}^{0.6}}$	$\text{Re}_t = 0.153\,\text{Ar}^{0.714}$ or $u_t = 0.153\dfrac{v_f}{d_p}\left[\dfrac{gd_p^3}{v_f^2}(s-1)\right]^{0.714}$
3.	Turbulent regime $500 < \text{Re} < 2\cdot10^5$ $83000 < \text{Ar} < 1.32\cdot10^9$	$C_D = 0.44$	$\text{Re}_t = 1.74\,\text{Ar}^{0.5}$ or $u_t = 1.74[gd_p(s-1)]^{0.5}$

The general interpolating formula is commonly used and recommended for calculations of free fall velocity in the whole range of Re numbers [6]:

$$\text{Re}_t = \frac{\text{Ar}}{18 + 0.61\text{Ar}^{0.5}} \tag{2.18}$$

Some authors [2] recommend 0.575 constant instead of 0.61. This reduces the formula for small Re numbers (2.18) to the No.1 formula from Table 2.6, and to formula No.3 from the same table, for large Re numbers.

According to eq. (2.16) and using C_D values for appropriate Re numbers from the graph in Fig. 2.2, a table can be made with Re_t and $C_D\text{Re}_t^2$ columns. Equation (2.16) and the calculations of Archimedes number allow one to obtain the $C_D\text{Re}_t^2$ value, and then, using the table, Re_t, u_t can be determined. These tables [7], as well as other types of nomogram, can be found in the literature [1, 2, 4].

Determination of the free fall velocity for particles of known diameter, or more commonly determination of diameter according to their known free fall velocity, can be achieved using the eq. (2.16) in a somewhat adjusted form:

$$\frac{\text{Re}_t^3}{\text{Ar}} = \frac{4\,\text{Re}_t}{3\mathcal{C}_D}$$ (2.19)

where the left hand side is independent of the particle diameter, and is:

$$\frac{\text{Re}_t^3}{\text{Ar}} = \frac{u_t^3}{g\nu_f} \cdot \frac{\rho_f}{\rho_p - \rho_f}$$ (2.20)

Figure 2.3 provides the dimensionless complex (2.20) in terms of Archimedes number, together with the dependence of Re_t on Archimedes number for spherical particles [2, 4].

Free fall velocity is also influenced by presence of solid walls and other particles. These effects are especially important during consideration of the problems of pneumatic transport [4, 7]; however, they will not be discussed here. We shall

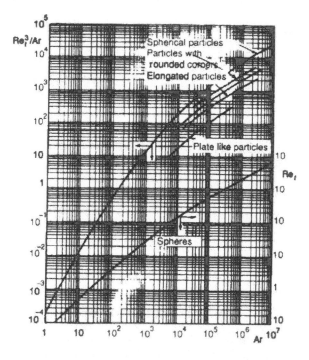

Figure 2.3. *Dependence of the dimensionless factor Re_t^3/Ar and Re_t number on Archimedes number. Plot is useful for quick estimation of the free-fall velocity (Reproduced by kind permission of Butterworth – Heinemann from [2])*

only mention that these two effects decrease the free fall velocity of a particle. Neglecting these two effects, and calculating the free fall velocity according to a formula for a single particle (Table 2.6) gives values that exceed the true value. Engineering calculations of pneumatic transport ensure the "safety" of the calculation in this way.

The influence of other particles can be taken into account with formulae recommended in Soviet literature [4, 6]:

$$u_{t,\varepsilon} = u_t \varepsilon^{4.75} \quad \text{for} \quad Ar < 10 \tag{2.21}$$

and

$$u_{t,\varepsilon} = u_t \varepsilon^{2.37} \quad \text{for} \quad Ar < 10^6 \tag{2.22}$$

For a wide range of Archimedes numbers the following formula is recommended:

$$C_D \, Re_{t,\varepsilon}^2 = \frac{4}{3} Ar[1 - 1.2(1-\varepsilon)^{2/3}]^2 \tag{2.23}$$

where index ε relates to presence of other particles.

A special problem, which has not been sufficiently studied, is a free fall velocity of a group of particles aggregating into a cluster in the fast fluidization regime. Conditions of aggregating, size and magnitude of clusters, or the free fall velocities of these clusters have not been fully elucidated [11, 12].

2.2. Onset and different regimes of gas-solid fluidization

2.2.1. Different possible states of the gas-solid mixtures

A mixture of solid material and fluid can exists only in motion. The two-phase system is characterized by the interaction of the solid particles and fluid. This interaction is mechanical and hydrodynamic, although exchange of heat and mass, as well as chemical reactions, can also take place simultaneously.

Interaction of solid particles and fluid in a mixture is manifested in a resistance force due to the presence of relative particle to fluid velocity. The resistance force acts in line with the relative velocity, but in the opposite direction. A mixture of solid particles and fluid necessitates the presence of this relative velocity, for which it is irrelevant whether only fluid is in motion or only solid particles, or both phases simultaneously. The direction of the motion is also of no consequence.

Between the two limiting states of the mixture – filtration of fluid through a fixed layer of solid particles and motion of the particles due to gravity force through an immobile fluid, there is a series of different forms (states) of mixtures of particulate solids and fluid. Many of these states, and not only fluidized bed as one of the possible states of the mixture, may be encountered in different industrial equipment with fluidized beds and, of course also in FBC boilers. Therefore, it is useful to outline all possible states of gas-solid mixtures (considering only processes in FBC boilers, in this instance we are dealing only with mixtures of particulate solids and gas). Further, knowing all of the possible states of the gas-solid mixture enables better understanding of fluidized bed behavior and its transition into other possible states of the mixture.

States of gas-solid mixtures can be differentiated by providing a list of their characteristics:

– according to the state of solid particles; particles can be thickly packed in a fixed bed, relatively immobile with respect to one another, or permanently in contact, or floating, chaotically moving, or occasionally in collision with each other,
– according to the direction of motion of the solid material; it is possible that solid material as a whole will have no directed, organized motion or that there is a permanent flow of material from the entry to the exit of the apparatus (two-phase flow),
– whether both phases (solid particles and gas) move simultaneously, or only one of the phases is in motion,
– do they move vertically or horizontally,
– do they move in the same or in the opposite directions,
– in a vertical system, the solid material can be free or, alternatively, its motion can be limited by some mechanical device (grate, rotary valve, etc.) placed either above or below, and
– according to the concentration or density of the mixture.

In order to define different states of the gas-solid mixture, it is not enough to classify them according to only one of the above features, since the remaining features can allow widely different regimes of solid-gas mixtures to be obtained. Table 2.7 gives a list of the range of states of solid-gas mixtures.

Diversity of possible states during vertical motion is obvious. The last column contains the number of the illustration in Fig. 2.4, representing the systems in which different states (Table 2.7) of the mixtures are realized.

Change of the state of a two-phase solid-gas mixture, and transition from one state to another, in the course of increased (decreased) gas velocity, can be followed by the well-known Zenz diagram [13, 14], Fig. 2.5. The diagram shows the change of pressure drop per unit length in dependence on gas velocity, for different, constant, particle mass flow rates. To better follow the pressure drop change, in the part of the diagram near the origin a linear scale was used (to the dotted line), followed by log scale.

Fixed and fluidized bed achieved by an upward gas flow through material lying on the grate (air distributor) is represented by line OADB on the diagram (Fig. 2.5), for zero net solid mass flow rate ($\dot{m}_p = 0$). At point B a transition takes place

Table 2.7. *Different possible regimes of gas-solid mixture*

Orientation of reactor		Direction of gas flow	Relative direction of gas and particle flow	Particle motion	Gas-solid mixture flow regime	Bed	Volume particle concentration (1 − ε) [%]	Number of scheme on Fig. 2.4
Vertical	Chaotic particle motion	Upward gas flow	Co-current upflow	Particle motion is restricted by mechanical device	Filtration of fixed bed	Dense packed	55-60	2.4a
	Directed particle motion				Stationary fluidized bed	Particles hover in the air	30-40	2.4b
					Fluidized bed with particle inflow at the bottom and outflow at the top of the bed	Particles hover in the air	25-35	2.4c
				Free motion of particles	Vertical dense phase upflow	Dense packed	40-50	2.4d
					Vertical lean phase pneumatic conveying		0-5	2.4e
			Countercurrent downflow		Dilute phase countercurrent particle downflow	Particles hover or are in motion	0-5	2.4f
				Particle motion is restricted by mechanical device	Fluidized bed with particle inflow at the top and outflow at the bottom of the bed		25-35	2.4g
		Downward gas flow	co-current downflow		Dense-phase co-current particle downflow	Dense packed	40-50	2.4h
					Dense-phase co-current particle downflow	Dense packed	40-50	2.4i
				Free motion of particles	Vertical downflow	Particles hover or are in motion	0-5	2.4j
Horizontal	Directed particle motion	Horizontal gas flow	Horizontal co-current		Horizontal pneumatic conveying		0-5	2.4k

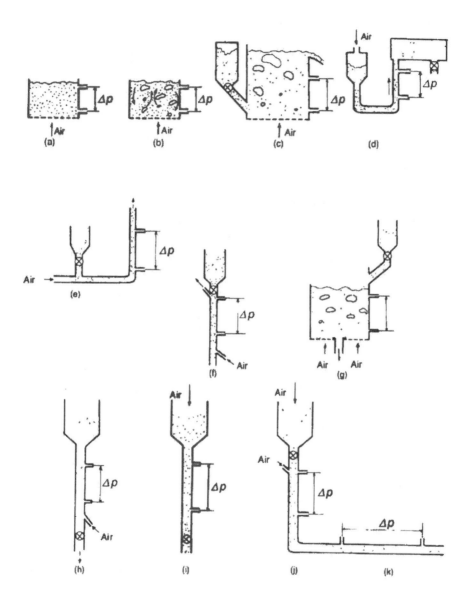

Figure 2.4. *Schematic of the possible practical realization of different two-phase gas-particle flow regimes*

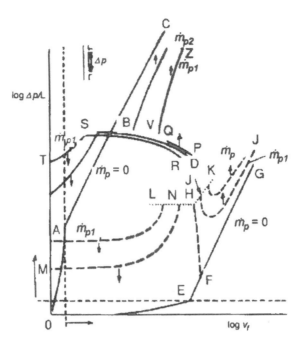

Figure 2.5.   *Zenz diagram for different two-phase gas-particle re-*
gimes in vertical vessels (Reproduced by kind permission
of the author prof. J. F. Davidson from [31])

from fixed bed to fluidized bed (Fig. 2.4a and b). In the fluidized bed region (BD)
the pressure drop is practically constant and equals the weight of all particles per
unit of tube cross section.

In furnaces of FBC boilers, during the start-up period, the inert material remains in fixed bed (OAB), changing to the fluidized state (BD) during the nominal
operation regime.

If fluidization takes place in tubes of relatively small diameter, at point D
the gas bubbles occurring in the fluidized bed tend to become so large that they occupy the entire cross section. This will induce the *unstable state* (DH) which cannot be represented by the diagram due to pressure drop oscillations.At increased
gas velocity, and with no supply of particles ($\dot{m}_p = 0$) the last solid particle will be
removed from the bed resulting in flow of pure gas (point F on the OEFG line).

A flow of solid particles maintaining a constant solids mass flow rate
($\dot{m}_p = const.$) results in *vertical pneumatic transport* (line JI in Fig. 2.5, that is
2.4e). Change of mass flow rate of solids results in pneumatic transport of higher
or lower concentration. This type of two-phase flow (combined with mixture
states represented in Fig. 2.4j and k) is utilized for feeding finely granulated fuels

(d_p < 10 mm) below the bed surface and for recycling of unburned particles from the flue gas filters for returning to the furnace.

The fluidization state with *material overflow from the bed surface* (2.4c) is represented with line (PQ), while removal of material from the bottom (Fig. 2.4g) is illustrated with line (SR). These states of mixture are commonly encountered in FBC boilers when coal ash remains in the bed and must be discharged in order to maintain a constant bed height.

Extending from line (SR), the line (ST) shows a pressure drop during out-flow of solid material in a counter-current dense phase packed bed with downflow (Fig. 2.4h). Discharge of coal, limestone or ash from the hoppers, and outflow of material from the bed bottom are examples of this kind of motion, with coun-ter-current gas flow to promote particle removal or cooling.

If the rotary valve is placed next to the hopper bottom, a mixture state as in Fig. 2.4f is produced, with a low concentration of solid particles. Line (NM) in Fig. 2.5 illustrates the pressure drop in this case.

Upward dense phase co-current gas-solid flow (Fig. 2.4d) is illustrated by line (VZ) in Fig. 2.5. This type of motion is rarely employed.

Downward co-current gas-solid flow (Fig. 2.4i and j) is more commonly ap-plied in FBC boilers for the same purposes as in the cases illustrated in Fig. 2.4h and f, where co-current gas flow usually prevents coal heating and ignition in the feed-ing system.

The diagram in Fig. 2.5 illustrates only mixtures with upward gas motion, so that states represented in Fig. 2.4h and f are not given. Appropriate diagrams can be found elsewhere [13, 14].

Horizontal pneumatic transport (Fig. 2.4j) corresponds to a curve similar to (JI), Fig. 2.5.

2.2.2. Fluidization regimes

The previous sections in this chapter, and Table 2.7 and Fig. 2.4, clearly show that fluidized bed is only one of the possible states of fluid-solid mixtures. Modern technologies use a mixture of solid particles and fluids in different states for a number of physical and chemical processes. In the last twenty years or so, one of these states – the fluidized state – has been used successfully for solid fuel com-bustion. However, FBC boilers have widely different features depending on whether they use the bubbling regime or fast fluidization regime. In order to un-derstand why these differences occur, and to determine the role of fluidized bed combustion among conventional combustion modes, we shall discuss below the basic features and nature of fluidized states and different regimes recognized in fluidized states, and correlate them to organization of the combustion process.

The fluidized state of loose material is effectively the opposite of the fixed bed state of loose material. Loose material reposed on a horizontal surface is characterized by immobile particles that lie on one another, touching each other at numerous points of contact, at which they exert frictional and adhesion forces. Gravity force, the weight of the particles and the overall weight of the bed, are transmitted in all directions via the contact points. In the fluidized state, however, the solid particles float, moving chaotically, clashing with each other, but their interactions are brief, and the interparticle forces are weak.

Freely poured loose material on a horizontal surface produces a cone. The angle of the conic surface and a horizontal plane is called the angle of repose of bulky material, β. The angle of repose depends on the friction and adhesion forces among the particles, and it is related to the fluidity, mobility or free discharge of loose material. The loose materials can be classified according to the angle of repose as follows:

- very free-flowing granules $25 < \beta < 30,$
- free-flowing granules $30 < \beta < 38,$
- fair to passable flow of powders $38 < \beta < 45,$
- cohesive powders $45 < \beta < 55,$ and
- very cohesive powders $55 < \beta < 70.$

Free-flowing material can flow out easily from bunkers and reservoirs through orifices of appropriately chosen sizes and shapes at the bottom of the reservoir or bunker. Cohesive and very cohesive powders necessitate special measures and technical arrangements to provide free and continuous discharge [15]. In the course of solid discharge from a bunker through the opening at the bottom, most of these materials make an arch that prevents further outflow.

In addition to the angle of repose, the angle of slide is also important for understanding the properties of a loose material during its outflow from a bunker. The angle of slide is defined as the angle from the horizontal of an inclined surface on which the material will slide due to the influence of gravity.

The angle of internal friction is defined as the equilibrium angle between flowing particles and bulk or stationary particles in a bunker. This angle is always greater than the angle of repose.

During filtration through a fixed bed of loose particulate solids lying on a porous bottom of a vessel, gas will find numerous pathways through void interspaces between the particles. If the bottom has homogenous porosity and significant resistance for gas flow, uniform distribution of gas will be provided over the whole surface of the bottom. Uniform distribution of gas will be provided over all cross sections of the bed due to the great number of particles and uniform porosity in a given cross section and along the bed height. The particles in the bed remain immobile. Gas flow resistance rises with increasing gas velocity, while the bed height and the shape of its free surfaces remain unaffected.

The pressure drop during gas filtration through a fixed bed of loose material, as a function of gas velocity, is schematically illustrated in Fig. 2.6. The same figure also shows the pressure drop for other fluidization regimes. It also can be seen that

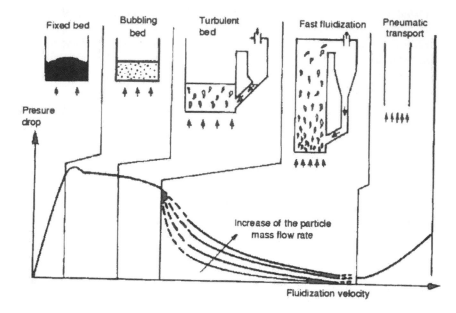

Figure 2.6. *Pressure drop in dependence of fluidization velocity for different fluidization regimes*

different states of solid-gas mixtures are used in conventional and contemporary modes of solid fuel combustion.

Figure 2.6 shows the pressure drop change for the following states of solid-gas mixtures:

− fixed bed,
− bubbling bed,
− turbulent bed,
− fast fluidization, and
− pneumatic transport.

The first and the last of the states listed are not included in the concept of fluidization that, however, covers the three other regimes from the above list.

When the filtration velocity reaches the critical value (minimum fluidization velocity) the particles begin to move. They separate from one another and chaotic motion ensues. The bed height rises. When adhesive forces among the particles are present, the pressure drop across the bed reaches its peak. With further increase of the fluidization velocity pressure drop across the bed will only maintain a constant value. Pressure drop equals the weight of bed material per unit cross section area. Bubbling fluidization ensues. Each particle of the material is floating, chaotically and moving intensively; the distance between individual particles is increasing and the bed height is expanding.

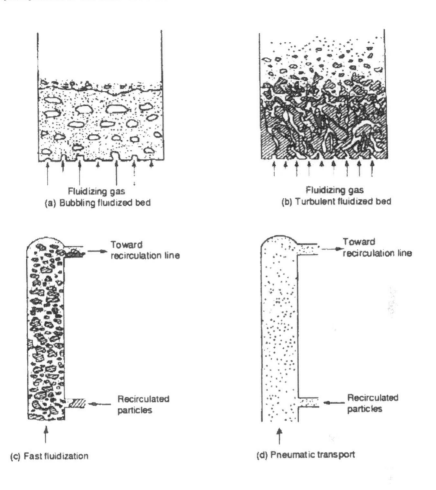

Figure 2.7. *Different fluidization regimes*

In an ideal case of fluidization for uniform size particles, the pressure drop in the region of bubbling fluidization does not change with the fluidization velocity. Reduction of the pressure drop, as illustrated in Fig. 2.6. results from elutriation of fine particles from the bed of polydisperse material, and is due to the reduced bed weight.

With velocities only slightly exceeding the minimum fluidization velocity, bubbles are created (minimum bubbling velocity) and the fluidized bed can be considered as if it were composed of two phases: bubbles with few or no particles at all (Fig. 2.7a) and particulate (emulsion) phase which is in a state of incipient fluidization. Bed expansion can be attributed to the presence of bubbles. The bubbles, originating in the immediate vicinity of the distribution plate, rise through the bed and, due to bubble coalescence, larger bubbles are formed. Alternatively, large bubbles break up and split into smaller ones. On the bed surface bubbles burst eruptively, ejecting the particles far above the bed surface. The motion of the bubbles produces very intensive particle circulation in fluidized bed. Behind the bubbles, in their "wake," particles move upwards, and around the bubbles, among them and especially near walls, the particles move downwards. The motion of bubbles contributes to axial mixing of particles and gas in fluidized bed. Transversal mixing is, however, much less intensive in the bubbling fluidized bed.

To properly consider bubble motion, we must differentiate between fast and slow bubbles, that is between fluidization of fine particles (powders) and of coarse particles [2, 11, 12, 16, 17]. The basic difference arises from bubble motion.

In a fluidized bed of fine particles the bubbles move faster than the gas in the emulsion phase (interstitial gas velocity in emulsion phase is considered to be equal to the minimum fluidization velocity). The gas that enters a bubble effectively does not mix with the gas outside the bubble and goes through the bed taking no part in bed processes. Thus, for example, during fluidized bed combustion oxygen in fast bubbles does not participate in combustion. In a fluidized bed of coarse particles, however, interstitial gas velocity in the emulsion phase exceeds bubble rise velocity. Therefore, in such bubbles continuous gas through flow is present [11].

In conventional boilers with solid fuel combustion on a fixed or moving grate, combustion takes place in fixed beds of large particles (d_p = 50-100 mm). Combustion air is introduced through the grate (partly above the bed), goes through the fixed layer of fuel and, together with slag and ash comprises the combustion environment.

In bubbling FBC boilers the combustion environment implies a fluidized bed of inert material (sand or ash). The quantity of fuel in the furnace accounts for 2-10% of the total mass of the inert material. The presence of the inert material, intensive mixing of particles and heat transfer by particle collisions, make this mode of combustion unique.

It has been generally believed that in the fluidization range, there are no major changes in hydrodynamic behavior of the bed between the minimum fluidization velocity and the transport velocity. Nevertheless, in the mid seventies it was realized that two essentially different regimes are present in this range: bubbling fluidized bed and turbulent fluidized bed regime [18-20].

The transition from bubbling to turbulent fluidization starts when, due to the increasing fluidization velocity in the emulsion phase, break-up of large bubbles and splitting into small ones and into differently shaped (elongated, irregular) voids take place. In the bubbling fluidization regime, bubble break-up is balanced by the process of bubble coalescence. When the break-up process dominates, transition to the turbulent regime starts. The bed state tends to homogenize. The turbulent regime is devoid of large bubbles [18-20]. Occasionally, relatively large bubbles may develop, which move upwards and then disappear. The same happens to minor voids in the bed: they can occasionally appear, merge, and then disappear.

Transition from bubbling fluidization to turbulent fluidization regime takes place over a relatively wide range of velocities [18]. It has not been determined exactly what causes and mechanisms lead to the regime change. Correlations that can be used for some loose materials to predict the velocities when the transition begins or when the transition into turbulent regime has been completed, are not yet sufficiently reliable [18-20]. Criteria for determination of the cutoff values of the transition areas have also not been established. According to studies presented in [19, 20] transition from the bubbling fluidization regime into the turbulent regime is characterized by extremely intense pressure oscillations (Fig. 2.8).

According to available data for fine particles, the ratios of characteristic velocities, U_k and U_c, to the free fall velocity approach and even exceed 10. For coarse particles, however, transition into the turbulent regime takes place when these ratios are lower (2-5); moreover, in some cases the transition into turbulent regime starts at velocities lower than the free fall velocity [20].

The turbulent fluidized bed regime is maintained over the velocity range from U_k to the transport velocity U_{tr} when the number of particles removed from the bed rises abruptly. Elutriation of particles from the bed takes place over the en-

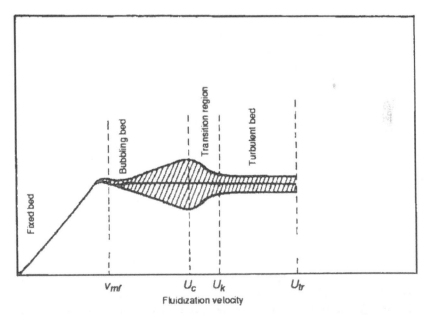

Figure 2.8.   *Pressure drop oscillations in transition from bubbling to fast fluidization*

tire fluidization range. In the turbulent fluidization regime, although the elutriation is more intense, the fluidized bed density (solid phase concentration) remains high and the free surface can be recognized. Free surface is not so sharply defined as in the case of bubbling bed. If the particles elutriated from the turbulent bed return with the same mass flow rate back to the bed (Fig. 2.6) the bed height will remain constant. The solid particle concentration above the bed will diminish with height, until a constant value is reached. If the number of particles returning to the bed is higher than the number of elutriated particles, the bed surface level will rise until the particle elutriation has increased sufficiently.

Figure 2.7b shows a characteristic picture of the turbulent fluidized bed regime. Large bubbles and plugs break up into smaller voids in the emulsion phase. Under the influence of gas velocity, the homogeneity of the emulsion phase also decays into "streaks and clusters." As a result of these processes, the turbulent bed becomes more homogenous, although it is composed of two phases. However, it can neither be described as continuous nor discontinuous [18-20]. Solid particle mixing and interaction of the phases and gas-solid contact are more intense than in the bubbling regime.

Large clusters of particles formed above the bed surface fall back into the bed, while smaller are elutriated with the gas. The mean effective size and density of these clusters is such that their free fall velocity is higher than fluidization velocity, and they generally return to the bed. Particle elutriation results from erosion of these clusters.

When the gas velocity reaches the free fall velocity of most of these clusters, particle elutriation from the bed rapidly increases. The transport velocity is achieved, and the turbulent regime changes into the fast fluidization regime (Figs. 2.6, 2.7c). The transport velocity is the margin between two substantially different categories of regimes in a gas-solid particle system: bubbling and turbulent fluidization on the one hand, and fast fluidization and pneumatic transport, on the other.

At velocities higher than transport velocity, if permanent inflow of solid material in the bed is absent, the whole bed inventory will be carried out from the bed. Depending on the recirculation mass flow rate of solid material (Figs. 2.6, 2.7c) the fast fluidization regime will be established with a higher or lower particle concentration. The appearance of the fast fluidization regime is not unlike the turbulent regime. Solid material is moving in clusters, but now most of the clusters are carried out of the furnace (apparatus). The bed surface is no longer clearly defined. The solid material concentration decreases gradually towards the exit. Particle mixing is very intensive, both axially and transversely. The particle clusters that are too large to be taken out at the established gas velocity, break up into smaller clusters or fall back making new clusters. This intensive backward motion of solid particles is one of the important features of the fast fluidization regime, contributing to the uniformity, both along the bed height and laterally, and to the intense gas-particle interaction. Size of clusters depends on fluidization velocity and mass flow of recirculating material.

The regimes present at velocities above the transport velocity differ substantially from the stationary fluidized bed regimes (bubbling and turbulent) in that the pressure drop and particle concentration depend on mass flow of the solid material $\dot{m}_p$ (Fig. 2.6) [18-20]. In the fast fluidization regime pressure drop increases with increase of recirculation mass flow rate, due to the increase of solid material concentration.

With increased gas flow rate and decrease of recirculation solids mass flow rate, the mixture concentration is lower, the clusters are smaller and increasingly greater numbers of particles exist as independent entities (i. e., the number of clusters decreases) (Fig. 2.6, 2.7d). The motion of the gas-solid mixture increasingly changes into dilute-phase pneumatic conveying, the particles move individually and rectilinearly, and back motion ceases.

When the voidage fraction is $\varepsilon = 0.95$, an abrupt change of the mixture concentration takes place with a gas velocity rise. It can, therefore, be proposed that this is a conditional margin at which fast fluidization becomes pneumatic transport [1, 22].

The existing formulae for determination of transport velocity (i. e., transition into the fast fluidization regime) are not sufficiently reliable. Intensive experimental investigations of the fast fluidization regime are still in progress. For each new material it is necessary to carry out a unique set of experiments [20], since the effect of particle size and shape has not been fully elucidated. The same applies to the influence of bed dimensions, system pressure or other parameters [18, 20].

According to results provided in [20] the transport velocity rises with the particle size and density, but its ratio to the free fall velocity becomes smaller.

The transport velocity depends on the size of the clusters formed in transition from the turbulent to the fast fluidization regime. An appropriate model is still lacking, and we do not as yet quite understand the mechanism of cluster formation. Therefore, an adequate method to predict the transport velocity is not yet available.

In addition to the bubbling fluidized bed regime applied to first generation FBC boilers, the turbulent regime, fast fluidization regime and pneumatic transport have been used in the combustion technologies developed so far.

First generation FBC boilers are designed to operate in the turbulent regime when significant increase of heat generation per square metre of furnace cross section area is required. In these plants, due to the high elutriation rate of unburned particles, utilization of a recirculation system is mandatory (illustrated in Fig. 2.6) [23, 28].

The fast fluidization regime is utilized in second generation FBC boilers – circulating FBC boilers. Here, substantially more intense transverse mixing of the particles, utilization of the whole furnace volume for intense combustion and change of the particle concentration with change of solids mass flow rate of recirculating material, which is used for load following, are the basic advantages brought about by the fast fluidization regime in the solid fuels combustion.

Pulverized coal combustion, as the basic technology of combustion, is currently the most commonly used, primarily in large utility electric power generation plants. Combustion takes place in the regime of pneumatic transport of pulverized fuel. The largest boilers (capacities up to 600 MW$_e$) employ this combustion mode.

2.2.3. Relative gas-particle velocity

Comparison of the different modes of combustion and their analysis from the point of view of the state of gas-particle mixture and fluidization regimes, offers an insight into their similarities and dissimilarities, as well as the causes of their advantages and disadvantages. It is also possible to identify parameters that may serve for the comparison of different combustion technologies. Here we will refrain from a detailed elaboration of their technical parameters, which will be discussed in greater detail in the following chapters. The most important hydrodynamic parameters are: the presence and motion of the inert material in the furnace and gas velocity. In all FBC technologies the presence of large quantities of inert material is an essential feature. The gas velocity in grate firing boilers is 2-5 m/s. In first generation FBC boilers it is 1.5-2 m/s, in second generation FBC boilers the gas velocity is typically 4.5-9 m/s, and in pulverized coal combustion boilers, the gas velocity is about 7-15 m/s. However, due to different particle sizes of both inert material and fuel associated with different combustion technologies, a correct evaluation of the in-furnace processes requires a knowledge of the relative gas-particle velocity. Therefore, we shall discuss the change of gas-particle relative velocity in a fixed bed of particulate solids, in different fluidization regimes and in the pneumatic transport regime.

The relative gas-particle velocities influence the intensity of gas-particle mixing, the processes of heat and mass transfer between gas and particles themselves, as well as combustion intensity. Thus, the relative velocity influences both intensity and efficiency of the processes in these systems as well as the pertinent conditions for effective performance of such combustion technologies.

Figure 2.9 illustrates the change in relative velocity for different states of the mixture, according to the solid particle concentration in the mixture and solids mass flow rate. The same figure shows which states of a gas-particle mixture are represented in particular types of chemical engineering and energy producing systems (chemical reactors, boilers).

In a fixed layer the gas-particle relative velocity equals the gas velocity – filtration rate. In a bubbling fluidized bed the relative velocity approximates the true fluidization velocity (net particle velocity equals zero, since the particles move chaotically), but it can be smaller than the free fall velocity of a single particle. In a turbulent fluidization regime, the relative velocity is higher (due to occurrence of a large number of clusters and back-mixing of the particles).

The highest values of relative velocities (10-20 times higher than the free fall velocity based on average mean equivalent particle diameter) occur in the fast fluidization regime at high values of the particle mass flow rate. The relative veloc-

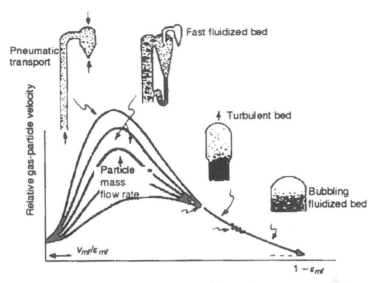

Figure 2.9. *Relative gas-particle velocity for different gas-particle flow regimes*

ity falls with lowering particle concentrations to approximate the free fall velocity of an isolated particle in the dilute-phase pneumatic transport. Although direct evidence is not available, the relative velocity in the fast fluidization regime can be correlated with the free fall velocity of a cluster. The change of the gas velocity or solids mass flow rate leads to the change of cluster size spectrum, that is their free fall velocities. The backward particle motion also influences the particle relative velocity since average particle velocity is lower [17, 24-27]. Experimental investigations [17] suggest that relative gas-particle velocity is 8-40 times greater than the free fall velocity for particles of 55 μm, and as much as 40-300 times greater for particles of $d_p = 20$ mm. For larger particles ($d_p = 290$ μm) relative velocity is less, 1-4 times greater than the free fall velocity. These relations suggest the sizes of clusters formed in the fast fluidization regime.

2.3. The bubbling fluidized bed

2.3.1. *General characteristics and macroscopic behavior of the bubbling fluidized bed*

Let us consider a bed of loose material lying on a porous plate with high flow resistance, enabling uniform distribution of gas flowing through complex irregular channels among the particles (Fig. 2.10).

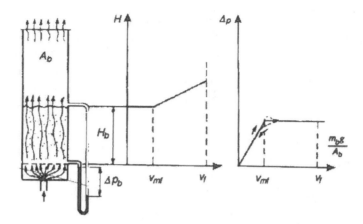

Figure 2.10. *Bed height and pressure drop across the bed in transition from fixed to fluidized bed*

With a velocity increase, the resistance to gas filtration through interspaces among the particles increases, linearly at the beginning while the gas flow regime is laminar, then as the square of the velocity in the turbulent filtration regime. The particles will remain undisturbed. Only the positions of some particles may change to rearrange their position to present the least resistance to gas flow. The bed height also remains unaltered. Gas flow through the fixed bed can be considered like every other flow through tubes and channels, except that these channels are complex, irregularly shaped and of changing diameter, a system of channels with numerous interconnections.

When gas velocity reaches the critical value at which the pressure drop equals the bed weight per unit of cross section area, fluidization takes place. Each particle itself and the bed as a whole "float." For each individual particle, flow resistance is balanced by its weight (taking into account the Archimedes buoyancy force). The particles which have been lying on top of each other start to move chaotically. The collisions among them are numerous and frequent, but brief. With further increase of gas velocity, the mean distance between the particles increases. Voids in the bed increase in comparison to the voids in the fixed bed, and the bed expands. The free surface of the bed, which was originally irregular (or conical in case of free pouring) becomes horizontal. Gas-solid fluidization of most loose materials is also accompanied by bubble generation (that is larger volumes of void space free of particles), simultaneously with the attainment of the minimum fluidization velocity. This state is known as non-homogenous, aggregate or bubbling fluidization. Occurrence of bubbles, however, does not accompany liquid-solid fluidization. There are some loose materials in which bubbles do not occur when the minimum fluidization velocity has been achieved. Over a narrow velocity range, before minimum velocity at which bubbles do occur, v_{mb} (incipient, minimum bubbling velocity), homogenous fluidization takes place accompanied by substantial bed expansion. With the occurrence of the first bubbles the bed height drops abruptly, but, with further velocity increase it continues to rise. Experience in fluidization of different materials has suggested that properties of fluidized beds are significantly influenced by the type of material, size, shape and features of particles. Thus, the behavior of one material in bubbling fluidization cannot be used as a parameter for behavior of any other, even apparently similar material.

Geldart [29, 30] was the first to comprehensively study the behavior of different materials in the course of fluidization, and suggest a classification of particulate solids according to density and size of particles. Geldart's classification (Fig. 2.11) divides particulate solids into four groups [2, 11, 12, 31]:

- group A comprises materials with particles of small mean size and low density (ρ_p < 1400 kg/m³). During fluidization of these materials homogenous fluidization can be attained with substantial bed expansion before the occurrence of bubbles. Bubble rise velocity exceeds the interstitial gas velocity in the emulsion phase. A maximum bubble size, however, does appear to exist,
- group B includes numerous materials with particles of medium size and medium density. Ordinary river or sea sand is a typical representative of the group. Bubbles occur immediately after the minimum fluidization velocity has been reached. The bubble rising velocity is greater than the interstitial gas velocity in the emulsion phase. There is no evidence of a maximum bubble size,
- group C includes highly cohesive, fine powders which do not fluidize easily. They are prone to bed channeling, and

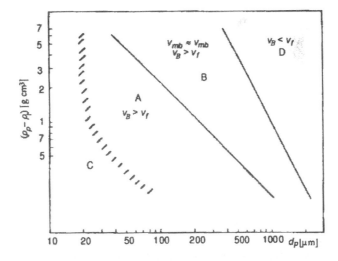

Figure 2.11. *Geldart's particle classification diagram (Reprinted from the Journal of Powder Technology [29]. Copyright 1978, with permission from Elsevier Science)*

– group D comprises loose materials with extremely dense and coarse particles. Their important feature is that the bubbles rise slowly, much more slowly than the interstitial gas velocity in the emulsion phase.

FBC boilers utilize loose particulate solids included in groups A, B or D according to Geldart's classification. Materials used in boilers with bubbling fluidization mainly belong to group B, but in the range bordering with group D, and some even actually belonging to group D. Boilers with circulating fluidized beds also utilize materials from group B, but approaching the group A features.

Bubbling fluidized beds have a series of properties which has encouraged development of different technologies in which physical and chemical processes take place in a fluidized bed. A fluidized bed provides an extremely appropriate environment for numerous reactions and for handling of loose material. The reactions are usually heterogeneous gas-solid particle surface reactions. The following conditions are considered most favorable for these reactions: large total surface of particles and chaotic motion of particles associated with frequent collisions, enabling intense mixing of gas and particles and as a consequence high heat transfer.

The bubbling fluidized bed has found its place in many chemical technologies and processes: cracking and reforming of hydrocarbons, coal carbonization and gasification, ore roasting, Fischer-Tropsch synthesis, aniline production, polyethylene production, calcination, coking, aluminum anhydride production, powder granulation, vinyl chloride and melamine production, incineration, nuclear fuel processing, combustion of solid, liquid and gas fuels. The fluidized bed is also used for the following physical processes: drying, adsorption, cooling, freezing, transport, and thermal treatment.

The bubbling fluidized bed (hereinafter simply referred to as fluidized bed), due to mobility of the particles, has numerous features typical of liquids. The presence of bubbles and their motion, in addition to chaotic particle motion, result also in directed, organized particle circulation. The free surface of the fluidized bed is roughly horizontal (Fig. 2.12a), roughly, since it is irregular due to the presence of bursting bubbles and does not present a very sharp transition. At the transition between the fluidized bed and the space above the bed (freeboard), depending on fluidization velocity, there is a wider or narrower zone in which concentration of particles typical for a fluidized bed falls towards zero far from the bed surface. This transition zone, the splash zone, is the area of abrupt, but not drastic, discontinuous decrease of particle concentration.

Two bodies, one with lower, and the other with higher density than the bed density can lie on the free surface of the fixed bed. If the bed is fluidized, the higher density body will sink to the bottom of the vessel. The less dense body will float on the surface (Fig. 2.12c). If some other loose material is poured on the surface of the particulate solids in a fixed bed, it will remain on the surface. In the fluidized state, however, the two materials will immediately mix and a fluidized bed of the mixture will result. The mixture will be homogenous over the entire bed volume (Fig. 2.12b). If the lateral walls of the vessel, for an "immobile" fixed bed of particulate

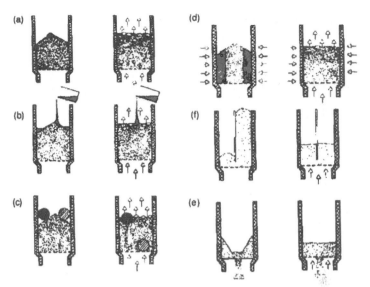

Figure 2.12. *Some overall characteristics of the fluidized bed compared with the behavior of the fixed bed*

solids are heated, it will take a relatively long time for the material located round the vessel axis to become heated. Only after a substantial period of time will a uniform temperature distribution in the bed be established. When a fluidized bed is heated, however, a uniform temperature distribution in the entire bed volume is obtained very quickly (Fig. 2.12d).

Loose materials will pour out of the vessel until an arch is formed, which will eventually prevent further outflow. In the case of free flowing materials, the vessel will effectively empty, but some material will still remain in the corners (Fig. 2.12e). Fluidized material, under the same conditions, will flow out like a liquid. If the opening is located on a lateral wall at the level of the free surface, the surplus will flow out of the vessel. In two linked vessels, a fluidized bed of particulate solids will behave like liquid in linked vessels, that is, the free surface in both vessels will achieve the same level (Fig.2.12f).

2.3.2. Minimum fluidization velocity

As has already been noted, the minimum fluidization velocity, v_{mf}, of a particulate solid is the velocity at which all particles begin to float[*]. When fluidization

---

[*] It has become customary in engineering practice, although physically not justified, when studying the fluidization process to refer all velocities on the whole cross section, as if no solids were present. These velocities are lower than the true gas velocities in the interspace between the particles. These apparent velocities will be marked as "v," while the true gas velocities will be marked as "u" in this book

is established, the pressure drop will remain constant if gas velocity continues to increase (Fig. 2.10). Thus, the minimum fluidization velocity can be simply determined by using a diagram of the measured pressure drop as a function of' fluidization velocity.

For a bed of ideal monodisperse particulate material with insignificantly small interparticle forces, the line presenting pressure drop across the fixed bed breaks abruptly when the minimum fluidization velocity is achieved (Fig. 2.13a). Determination of the minimum fluidization velocity of a polydisperse material, with irregularly shaped particles and rough particle surface, or with strong cohesive forces, and for the materials of the Geldart's group C, is somewhat more complex. Intense cohesive forces among the particles will result in significantly higher pressure drop before the minimum fluidization velocity has been attained. When fluidization has been established, the pressure drop will assume a normal value ("b" curve). During velocity decrease hysteresis will occur "c" curve).

When polydisperse materials are fluidized, the transition is gradual. Smaller particles begin to float at lower velocities. The pressure drop curve is similar to "c" curve ("d" curve). Minimum fluidization velocity is generally determined in these cases at the crossing point of the extrapolated left and right branches of the pressure drop curve. During fluidization of polydisperse materials at velocities substantially above the minimum fluidization velocity, the pressure drop diminishes ("e" curve), due to elutriation of fine particles. In case of nonuniform fluidization, or if bed channeling takes place, the pressure drop plot will be similar to the "c" and "d" curves.

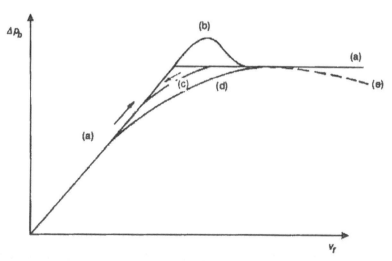

Figure 2.13. *Some characteristic curves of the bed pressure drop dependence on the fluidization velocity*

The shape of the fluidization diagram (pressure drop versus fluidization velocity) provides a great deal of information on the nature and character of the fluidized bed and features of loose material in the fluidized state. Therefore, it is quite useful if FBC boilers are equipped with pressure drop measurement across the bed of inert material. Monitoring of the pressure drop is particularly important during the boiler start-up period.

The minimum fluidization velocity of a wide range of diverse materials has been determined since the inception of fluidization studies, and the literature offers a wide range of correlations for determination of the minimum fluidization velocity (e. g., references [2, 11, 31, 33]). Originally, all such correlations were empirical, but the functional form proposed by Wen and Yu [34] is now widely accepted. These workers obtained this form by equating the correlations for the pressure drop of fixed and fluidized bed.

If we assume that the pressure drop through complex, irregular interspaces among the particles of a fixed bed can be expressed using the commonly employed Darcy formula:

$$\Delta p = \lambda \frac{L}{D} \frac{v^2}{2} \rho_f \qquad (2.24)$$

which applies for a streamline flow through a pipe with circular (or almost circular) cross section, the Carman-Kozeny equation is obtained for the pressure drop across a fixed bed (see references [2, 7, 35]):

$$\Delta p_b = H_b \frac{180(1-\varepsilon)^2}{\varepsilon^3} \frac{\mu_f v_f}{(\phi_s d_p)^2} \qquad (2.25)$$

where the following relations have been used:

- the total volume of interspaces between the particles

$$V_\varepsilon = V_b - \Sigma V_p = \varepsilon V_b \qquad (2.26)$$

- the mean hydraulic diameter

$$D_h = \frac{6V_\varepsilon}{\Sigma A_p} = \frac{\varepsilon d_p}{\phi_s (1-\varepsilon)} \qquad (2.27)$$

and
- the total surface of particles in the bed

$$\Sigma A_p = \Sigma \phi_s d_p^2 \pi \qquad (2.28)$$

The numeric coefficient in the Carman-Kozeny equation (2.25) was obtained mainly based on experiments carried on with fine powders, Geldart's group

A, for laminar flow. Therefore, it does not provide a satisfactory result when used for pressure drop measurements across a bed of particles with size >150 μm.

Here, Ergun's equation is used (see references [2, 35]) to cover practically all materials, since the second term accounts for inertia forces that become important in the turbulent flow regime:

$$\Delta p_b = H_b \frac{150(1-\varepsilon)^2}{\varepsilon^3} \frac{\mu_f v_f}{(\phi_s d_p)^2} + H_b \frac{175(1-\varepsilon)}{\varepsilon^3} \frac{\rho_f v_f^2}{\phi_s d_p} \qquad (2.29)$$

In these equations, the mean equivalent diameter of polydisperse particulate material is calculated according to definition 7, Table 2.4. Gas velocity v_f is based on the total cross section of the vessel (or a tube), as if there were no particles in it, as it has become customary in fluidization practice.

The pressure drop across a fluidized bed of particulate solids equals the weight of the bed material, reduced by the buoyancy forces, per unit of bed surface:

$$\Delta p_b = (1-\varepsilon)(\rho_p - \rho_f)g H_b \qquad (2.30)$$

Following the assumptions of Wen and Yu [34], at incipient fluidization, that is at the minimum fluidization velocity v_{mf}, the values of the pressure drop calculated according to (2.29) and (2.30) must be equal:

$$150\frac{(1-\varepsilon_{mf})^2}{\varepsilon_{mf}^3} \frac{\mu_f v_{mf}}{(\phi_s d_p)^2} + 175\frac{1-\varepsilon_{mf}}{\varepsilon_{mf}^3} \frac{\rho_f v_{mf}^2}{\phi_s d_p} = (1-\varepsilon_{mf})(\rho_p - \rho_f)g \quad (2.31)$$

Wen and Yu [34] have shown that for widely different loose materials the following holds:

$$\frac{1}{\phi_s \varepsilon_{mf}^3} \approx 14 \quad \text{and} \quad \frac{1-\varepsilon_{mf}}{\phi_s^2 \varepsilon_{mf}^3} \approx 11 \qquad (2.32)$$

according to which, with slight rearrangements, eq. (2.31) can be written as:

$$\text{Re}_{mf} = \frac{d_p v_{mf} \rho_f}{\mu_f} = (33.7^2 + 0.0408 \text{Ar})^{0.5} - 33.7 \qquad (2.33)$$

The expression (2.33) can be used for calculation of the minimum fluidization velocities when the characteristics of particles and gas are known. The correlation (2.33) is obtained on the basis of 284 experimental points over the range of Re = 0.001–4000, and the minimum fluidization velocity can be calculated with a SD = ±34%. In recent times, this expression is frequently used in practice. The general form of this expression is also used for presentation of results of experimentally determined minimum fluidization velocity of different materials, where the constants specific for each given material are determined.

If it is necessary to know more accurately the correct minimum fluidization velocity, measurements are still inevitable. Numerous formulae for the calculation of the minimum fluidization velocity proposed in literature so far, illustrate the attempts to obtain more precise formulae for actual materials and actual working conditions for pertinent technologies in which fluidized beds are used. At the same time, it has also become obvious that generalized formulae do not give satisfactory accuracy. Table 2.8 gives some of the best known correlations listed in literature. Most of these correlations are obtained for fine particles and powders, mostly Geldart's group A, since most processes in industry use this kind of material, that is they employ fluidized beds of these materials. Only recently with development of fluidized bed combustion technology has it become necessary to know the minimum fluidization velocity for coarse particle beds. Figure 2.14 [32] gives a comparison of some of the formulae given in Table 2.8. The significant disagreement in the results is obvious, especially for particles greater than 0.5 mm.

Investigations of minimum fluidization velocity at elevated temperatures pertinent to fluidized bed combustion are rare. Figure 2.14 shows the results of experimental determination of the minimum fluidization velocity of silica sand with particles 0.3-1.2 mm in size, and in the 20-500 °C temperature range [32]. The re-

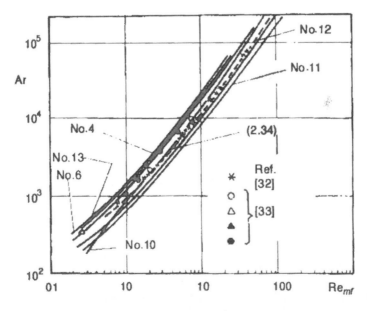

Figure 2.14. *Experimental data for minimum fluidization velocity for silica sand in the temperature range 20-500 °C [32], compared with the formula (2.34) and some correlations given in Table 2.8*

Table 2.8   *Various experimental correlations for minimum fluidization velocity*

No.	Author	Ref.	v_{mf}	Gas	Particles Material	d_p [μm]
1.	Baerg	33	$0.361 \dfrac{[d_p\,\rho_f(1-\epsilon_{mf})]^{1.23}}{\rho_f}$, $Re_{mf} < 20$	Air	Catalyzer, aluminum and iron powder glass beeds	6-880
2.	Miller, Logwinuk	33	$0.00125 \dfrac{d_p^2\,(\rho_p - \rho_f)^{0.9}\rho_f^{0.1}}{\mu_f} g$	Air, He, CO₂, ethane	Al_2O_3, SiO_2	97-249
3.	Leva	33	$0.0079 \dfrac{d_p^{1.88}\,(\rho_p - \rho_f)^{0.94}}{\mu_f^{0.88}}$, $Re_{mf} < 10$	Air, He, CO₂	Silicagel, sand, anthracite, catalyzer	51-970
4.	Goroschko	33	$(v_f/d_p)\dfrac{Ar}{150(1-\epsilon_{mf})/\epsilon_{mf}^3 + (1.75 Ar \epsilon_{mf}^3)^{0.5}}$			
5.	Frantz	33	$0.001065 \dfrac{d_p^2\,(\rho_p - \rho_f)}{\mu_f} g$, $Re_{mf} < 32$	H₂, N₂, Ar, ethane, mixture of gases	Sand, catalyzer	46-305
6.	Davis, Richardson	33	$0.00078 \dfrac{d_p^2\,(\rho_p - \rho_f)g}{\mu_f}$	Air	Catalyzer	
7.	Pillai, Raya Rao	33	$0.000701 \dfrac{d_p^2\,(\rho_p - \rho_f)g}{\mu_f}$, $Re_{mf} < 20$	Air	Aluminium and iron powder	58-1100
8.	Baeyens	33	$0.0009 \dfrac{d_p^{1.88}\,(\rho_p - \rho_f)^{0.934}g^{0.934}}{\mu_f^{0.87}\rho_f^{0.066}}$, $Re_{mf} < 10$	Air		
9.	Kumar, Sen Gupta	33	$0.005 \dfrac{\mu_f}{d_p\,\rho_f} Ar^{0.78}$	Air	Sand, sugar	
10.	Saxena, Vogel	35	$(\mu_f/d_p\rho_f)[(25.28^2 + 0.0571 Ar)^{0.5} - 25.28]$	Air	Dolomit	88-1480
11.	Babu	36	$(\mu_f/d_p\rho_f)[(25.25^2 + 0.0651 Ar)^{0.5} - 25.25]$			
12.	Todes	6	$(\mu_f/d_p\rho_f)\dfrac{Ar}{1400 + 5.25\sqrt{Ar}}$			

sults from these experiments are best presented by correlation in the form (2.33), but with different coefficients:

$$\mathrm{Re}_{mf} = (24^2 + 0.049\mathrm{Ar})^{0.5} - 24 \tag{2.34}$$

Specific values of the coefficients are influenced by particle shape.

Minimum fluidization velocity of large particles (Geldart's group D) necessitates further investigations especially at high temperatures (500-1000 °C).

Numerous processes in the chemical and process industries that take place in fluidized beds and, to a somewhat lesser degree fluidized bed combustion technology itself, require knowledge of maximum fluidization velocity at which fluidization can be maintained. Conditionally, this upper limit for monodisperse material can be approximated by the free fall velocity of a single isolated particle, or, for polydisperse materials, by free fall velocity of the tiniest particle. Comparison of the physical essence of the minimum fluidization velocity to the free fall velocity of a single isolated particle reveals that these physical properties share the same nature. In both cases, these are velocities at which, though in different conditions, a balance of forces acting on a particle exposed to the vertical gas flow is realized. Therefore, quite logically very early investigators tried to find out the relation between the minimum fluidization velocity and the free fall velocity. Thus, P. H. Pinchbeck and F. Popper [37] carried out series of experiments and established the following ratios of these velocities:

— for fine particles, i. e., $\mathrm{Re}_t < 4$

$$\frac{u_t}{v_{mf}} \approx 91.6 \tag{2.35}$$

and

— for coarse particles, i. e., $\mathrm{Re}_t > 1000$

$$\frac{u_t}{v_{mf}} \approx 8.72 \tag{2.36}$$

Further studies of Wen and Yu [34] and Godard and Richardson [38] confirmed these relationships suggesting their dependence on the bed porosity at incipient fluidization (that is porosity of a fixed bed). The above relations can be generalized, starting from equation (2.31) if the free fall velocity is appropriately introduced (see [2, 31]):

$$1.75 \frac{1}{\phi_s \varepsilon_{mf}^3} \left\{ \frac{v_{mf}}{u_t} \right\}^2 \mathrm{Re}_t^2 + 150 \frac{1-\varepsilon_{mf}}{\phi_s^2 \varepsilon_{mf}^3} \frac{v_{mf}}{u_t} \mathrm{Re}_t = \mathrm{Ar} \tag{2.37}$$

Figure 2.15 [31] illustrates this correlation for spherical particles compared with experimental results of several authors.

If the analysis of Wen and Yu is used, given by expression (2.32) and appropriate expressions for the free fall velocity given in Table 2.6, the correlation (2.37) can be reduced to a general form applicable to materials composed of nonspherical particles:

$$a \left\{ \frac{v_{mf}}{u_t} \right\}^2 + 1650 \left\{ \frac{v_{mf}}{u_t} \right\} + c = 0 \qquad (2.38)$$

where constants a and c have values given in the following table:

Flow regime	a	c	Re_t-number range	Ar-number range
Laminar	$1.37 \cdot Ar$	-17.86	<2	<36
Transition	$3.72 \cdot Ar^{0.714}$	$-6.58 \cdot Ar^{0.286}$	2-500	36-83000
Turbulent	$42.63 \cdot Ar^{0.5}$	$-0.575 \cdot Ar^{0.5}$	>500	>83000

The results given in Fig. 2.15 show that for fluidized beds of coarse particles, most commonly used in bubbling FBC boilers, the range of practically feasible velocities is relatively narrow, about 10 v_{mf}, which should be kept in mind during selection of working parameters of the furnace, especially in the light of the polydisperse nature of actual inert materials used in practice.

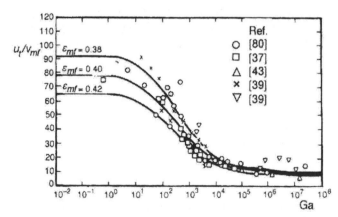

Figure 2.15. *Ratio of terminal velocity to minimum fluidization velocity versus Galileo number (Reproduced by kind permission of the author prof. J. F. Davidson from [31])*

Minimum bubbling velocity is present only in fluidized beds of fine powders (Geldart's group A) and for fluidized bed combustion is of no practical relevance, because bed materials belong to groups B or D, and working conditions are far from the v_{mb}-v_{mf} velocity range. Determination of the minimum bubbling velocity is connected with large experimental errors and is usually performed visually, so that the few studies do not yield reliable data [2, 11, 31]. Only one of the recently recommended formulae will be given here (according to Abrahamsen and Geldart, [11]):

$$\frac{v_{mb}}{v_{mf}} = \frac{2300 \, \rho_f^{0.126} \, \mu_\phi^{0.523}}{d_p^{0.8} \, g^{0.934} \, (\rho_p - \rho_f)} \exp{(0.716\,\Phi)} \tag{2.39}$$

where Φ is the mass content of particles smaller than 45 μm.

2.3.3. Bed expansion

It has become customary to point out that a bubbling fluidized bed has a sharply marked horizontal free surface. In practice, this surface cannot be described as either horizontal or clearly marked. Rising through the bed, bubbles grow up to substantial sizes and burst in an intense manner profoundly perturbing the bed surface, and ejecting all particles from their upper surface out into the freeboard. Larger particles fall back into the bed, and numerous small particles are elutriated and removed from the furnace. Therefore, the free surface of the bed is very turbulent and irregularly shaped, similar to the surface of boiling liquid. At the same time, the assumption of an abrupt, discontinuous change of particle concentration cannot be justified. At the free surface zone there is an area called the splash zone. In the splash zone the particle concentration gradually changes from the values characteristic for a bubbling fluidized bed to values characteristic for the area far from the bed surface. Further, it is difficult to define a unique position for a free surface, namely, the bed height, due to substantial oscillations at the surface. Surface position can be determined experimentally as the cross section of two linear changes of pressure with height, in the bed and above it. Due to the substantial pressure oscillations in the bed, determination of the position of the bed free surface is therefore associated with major difficulties and errors. Nevertheless, it remains a fact that a fluidized bed expands as the fluidization velocity increases. Expansion of a fluidized bed and the change of bed height with the increase in fluidization velocity are important for designing of equipment employing fluidized beds. In particular, this parameter determines the correct placement of the heat exchanger tube bundle (if any) immersed in the fluidized bed, as well as the correct calculations of heat transferred to the furnace walls.

Further, elaboration of bed expansion should discriminate between two situations:

(a) fluidized beds of powders belonging to Geldart's group A expand with the gas velocity increase above the minimum fluidization velocity so that, before the minimum bubbling velocity has been achieved, porosity (that is bed height) increases linearly with the velocity, Fig. 2.16 [11, 39]. Bed expansion results from the increase of the mean distance between particles, and

(b) for other types of particulate solids (Geldart's groups B and D), as well as for group A powders after bubble formation, bed expansion mainly results from the presence of bubbles.

Increase of bed height can be expressed by the change of bed porosity or bed density:

$$\frac{H}{H_{mf}} = \frac{1-\varepsilon_{mf}}{1-\varepsilon} = \frac{\rho_{mf}}{\rho_\varepsilon}$$ (2.40)

In cases where bubbles are present, this relation can further be elucidated by introducing the mean volume fraction occupied by bubbles in the total bed volume, δ_B:

$$\frac{H}{H_{mf}} = \frac{1-\varepsilon_{mf}}{(1-\delta_B)(1-\varepsilon_p)}$$ (2.41)

where ε_p, is void fraction of the particulate (emulsion) phase of the bed.

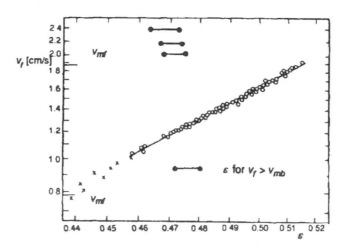

Figure 2.16. *Expansion of the fluidized bed of particles (group A according to Geldart classification), for velocities less than minimum bubbling velocity (Reproduced by kind permission of the author prof. J. F. Davidson from [31])*

If we assume, according to the standard form of Davidson's two-phase theory of fluidized beds [1, 11, 12, 31, 41] that the particulate (emulsion) phase is in the state of incipient fluidization, with porosity ε_{mf} (this applies for materials in groups B and D [11]), then relationship (2.41) is reduced to:

$$\frac{H}{H_{mf}} = \frac{1}{1-\delta_B} \quad \text{or} \quad \frac{H-H_{mf}}{H} = \delta_B \tag{2.42}$$

The spread of experimental data on bed expansion, that is change of void fraction in the fluidized bed, is very marked. In addition to the already mentioned reasons, the fluidized bed diameter (that is cross section magnitude), air distributor shape and bed temperature may also substantially influence bed expansion [6]. Data on bed expansion also differ for materials in groups B and D due to different bubble rise velocities in fluidized beds composed of these materials. Complexity and insufficient knowledge of the nature of the phenomenon and a large number of potentially influential parameters, as well as the difficulties associated with experimental studies, are the reasons for the large number of correlations of different forms used to describe the existing experimental results [1, 11, 31, 32, 40, 41].

According to the fundamental, simplified two-phase Davidson's theory, the volume fraction occupied by bubbles may be expressed using the fluidization velocity and the bubbles rise velocity, from the balance of gas flow through emulsion and the bubbling phase.

If v_B is rise velocity of an isolated bubble in a bed of immobile particles, then, according to Nicklin [42] and Davidson [31, 40] the absolute velocity of a group of bubbles in a fluidized bed is:

$$v_B^* = v_f - v_{mf} + v_B \tag{2.43}$$

As will later be explained in the discussion on bubble motion, gas will flow upwards through "slow" bubbles ($v_B < u_f$) with the velocity of $3v_{mf}$ [40]. In case of "fast" bubbles, however, a bubble carries upward "its own" gas with v_B^*.

Supposing that bubbles are free of particles ($\varepsilon_B = 1$), and that the emulsion (particulate) phase is in the state of incipient fluidization ($\varepsilon_p = \varepsilon_{mf}$), the following relationship applies:

$$\varepsilon = \delta_B + (1-\delta_B)\varepsilon_{mf} \tag{2.44}$$

The volume gas flow rate through the bed can be expressed as a sum of flow rates through the emulsion phase and bubbles. In cases of "slow" bubbles, in addition to the upward movement of the bubbles, gas flow through the bubble with $3v_{mf}$ must also to be taken into account. According to the mass flow rate balance it appears that:

$$v_f = (1 - \delta_B) v_{mf} + \delta_B (v_B^{\bullet} + 3 v_{mf}) \qquad (2.45)$$

while the analogous expression for "fast" bubbles is:

$$v_f = (1 - \delta_B) v_{mf} + \delta_B v_B^{\bullet} \qquad (2.46)$$

Expressions (2.43), (2.44) and (2.45) render bed porosity as functions of gas velocity and bubble rise velocity, which is, for "slow" bubbles:

$$\varepsilon = \varepsilon_{mf} + (1 - \varepsilon_{mf}) \frac{v_f - v_{mf}}{v_f + v_{mf} + v_B} \qquad (2.47)$$

and according to (2.43), (2.44) and (2.46) for a "fast" bubble bed:

$$\varepsilon = \varepsilon_{mf} + (1 - \varepsilon_{mf}) \frac{v_f - v_{mf}}{v_f \, 2 v_{mf} + v_B} \qquad (2.48)$$

The relationships derived above show that porosity of a fluidized bed depends on the nature, size and velocity of bubbles, which is probably one of the major causes of the great differences seen among experimental results.

In deriving correlations (2.47) and (2.48), the assumptions pertinent to two-phase Davidson's model were used in order to illustrate the complexity of the problem of describing changes in fluidized bed porosity, and Davidson's model and bubble motion will be discussed more comprehensively later. For now it can be said that these two correlations are increasingly used [41] for presentation of experimental results and prediction of bed expansion, since previously used correlations failed to yield satisfactory results.

Changes in fluidized bed porosity are classically represented by the Richardson and Zaki correlation [31, 43]:

$$\frac{v_f}{u_t} = \varepsilon^n \qquad (2.49)$$

where "n" depends on the type of loose material and bed porosity at incipient fluidization. Using Ergun's expression (2.31) or (2.37), "n" can be calculated in function of Reynolds (or Archimedes) number:

$$n = \frac{- \log \dfrac{Re_t}{Re_{mf}}}{\log \varepsilon_{mf}} \qquad (2.50)$$

Figure 2.17 shows the change of exponent "n" as a function of the Galileo number, according to (2.50) for different values of ε_{mf} and compares it to experimental results reported by several authors [31, 43-46].

Similar empirical approaches use the following correlations:

$$\varepsilon - \varepsilon_{mf} = C(v_f - v_{mf})^n \tag{2.51}$$

or

$$Re = C\,(Ar\varepsilon^n)^p \tag{2.52}$$

where the unknown constants depend on the type of material and experimental conditions. A better physical sense is achieved in attempts to apply expressions for pressure drop (2.29) and (2.30) [6, 11], for an arbitrary bed porosity, deriving the following expression:

$$Ar = \frac{150(1-\varepsilon)}{\varepsilon^3}\,Re + \frac{175}{\varepsilon^3}\,Re^2 \tag{2.53}$$

which can be expressed explicitly in terms of Reynolds number:

$$Re = \frac{Ar\,\varepsilon^3}{150(1-\varepsilon) + \sqrt{175Ar\,\varepsilon^3}} \tag{2.54}$$

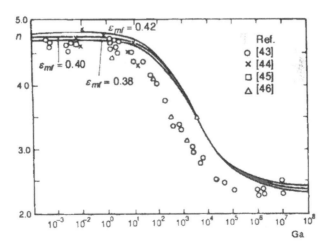

Figure 2.17. *Exponent "n" in Richardson and Zaki correlation as a function of Galileo number (Reproduced by kind permission of the author prof. J. F. Davidson from [31])*

Table 2.9. *Various correlations for fluidized bed porosity and bed expansion*

No.	Author	Ref.	Recommended correlation	Fluid	Experimental conditions
1	M. Leva	49	$\dfrac{v}{u_t} = 0.09\dfrac{\varepsilon^6}{1-\varepsilon}$, Re < 1	Air, He, CO_2	Sand, catalizer, coal
2.	J Happel	49	$\dfrac{v}{u_t} = \dfrac{3-4.5(1-\varepsilon)^{1/3}+4.5(1-\varepsilon)^{5/3}-3(1-\varepsilon)^2}{3+2(1-\varepsilon)^{5/3}}$, Re < 1		$D = 63$ and 102 mm, theory
3.	L. I Erkova	49	$Re = C\,(Ar\varepsilon^3)^n$, 1 < Re < 290	Air, CO_2, water, Kerosene	
4	T. Blickle	49	$\varepsilon^2 - \varepsilon_{mf}^2 = \dfrac{v-v_{mf}}{k\sqrt{g}\,d_p}$, Re > 290		$D = 231$ mm, theory
5	M. E. Aerov	49	$\varepsilon = \varepsilon_{mf}\left[\dfrac{18Re+0.36Re^2}{Ar}\right]^{0.21}$		Theoretical, homogeneous fluidization
6	O M Todes	9	$\varepsilon = \varepsilon_{mf}\left[\dfrac{Re+0.02Re^2}{Re_{mf}+0.02Re_{mf}^2}\right]^{0.1}$		Heterogeneous fluidization
7	W. K. Lewis	49	$\dfrac{H-H_{mf}}{H_{mf}} = \dfrac{1.07}{d_p^{0.5}}(v-v_{mf})$	Air	Glass beds $D = 61$ and 114 mm
8	S. S. Zabrodski	33	$\varepsilon - \varepsilon_{mf} = k(v-v_{mf})^n$, $n = 0.85\text{-}1.0$	Air	Sand, $D = 100$ mm Influence of temperature
9	S. S. Zabrodski	50	$\dfrac{\varepsilon_2}{\varepsilon_1} = \left(\dfrac{18Re_2+0.36Re_2^2}{18Re_1+0.36Re_1^2}\right)^{0.21}$, for T_2		
10	A. M. Xavier	51	$\dfrac{H-H_{mf}}{H_{mf}} = \dfrac{v-v_{mf}}{v_B}$, d_B = const		Theoretical
11	A. M. Xavier	51	$H = H_{mf} + \dfrac{5b}{3}[(H+B)^{0.6} - B^{0.6}] - 5b^2[(H+B)^{0.2} - B^{0.2}] + 5b^{2.5}\,tg^{-1}[(H+B)^{0.2}/b^{0.5}] - tg^{-1}[(B^{0.2}/b^{0.5})]$ $b = 1917(v-v_{mf})^{0.8}/g^{0.1}$, $B = 4\sqrt{A_O}\cdot d_B \neq const$		Theoretical

Todes [6] suggests two simplified, but less exact formulae:

$$Re = \frac{Ar\varepsilon^{4.75}}{18 + 0.6\sqrt{Ar\varepsilon^{4.75}}} \tag{2.55}$$

and

$$\varepsilon = \left[\frac{18\,Re + 0.36\,Re^2}{Ar}\right]^{0.21} \tag{2.56}$$

It is important to recognize that formulae (2.51) through (2.56) hold only for bubble-free fluidized beds. According to experimental studies of heterogeneous fluidization (with bubbles), Todes [6] recommends a formula like (2.56), but with the exponent of 0.1 that can be expressed as follows:

$$\varepsilon = \varepsilon_{mf}\left[\frac{Re + 0.02\,Re^2}{Re_{mf} + 0.02\,Re_{mf}^2}\right]^{0.1} \tag{2.57}$$

Modern approaches, which take into account existence, nature and rise velocity of bubbles, are based on formulae like (2.47) and (2.48), while differences between authors originate from different correlations for bubble diameter and bubble rise velocity. The following expression is commonly used for bubble rise velocity:

$$v_B = 0.71\sqrt{gD_B} \tag{2.58}$$

while the bubble diameter is calculated according to Darton [47, 48]:

$$D_B = 0.54(v_f - v_{mf})^{0.4}\,(H + 4\sqrt{A_0})^{0.8}\,g^{-0.2} \tag{2.59}$$

The problem of prediction of fluidized bed expansion remains unsolved. Table 2.9 [32] offers formulae usually given in literature, which show that a new, more accurate physical model is needed for the description of the complex phenomenon of bed expansion. Figure 2.18 shows results of measurements [32] of void fraction in fluidized beds of silica sand ($d_p = 0.43$ mm) at bed temperatures of 20 and 500 °C. The results are compared to values obtained using formulae given in Table 2.9. Great differences among the formulae are quite obvious, but it still remains a fact that none of these agrees well with experimental results.

More recent investigations [2, 11, 32] have shown that substantial influence on the bed expansion is exerted by the number, size, velocity and nature of bubbles. Therefore, the shape and nature of the distribution plate and/or immersed heat transfer surfaces in the bed, associated with other features in the fabrication that af-

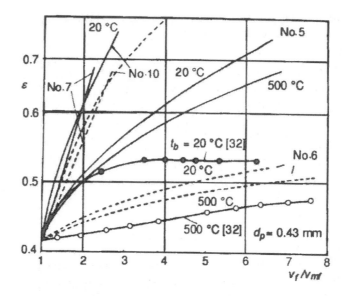

Figure 2.18. *Comparison of the experimental data and some correlations for bed expansion from Table 2.9, in temperature range 20-500 °C [32]*

fects the bubble size and nature, substantially influence bed expansion during
fluidization. Since this is a feature that is in common with fluidized bed boilers and
furnaces, designers should bear it in mind. Also, poor or nonuniform fluidization,
or bed channeling, influences change of void fraction, that is, bed expansion. Data
on character and intensity of bed expansion at elevated temperatures for particles in
Geldart's group D are unfortunately very scarce.

2.3.4. Particle elutriation from fluidized bed

In order to explain the process of particle elutriation from a fluidized bed,
and out of the system (furnace, chemical reactor) comprising the bubbling fluidized
bed, it is necessary to consider the behavior and motion of particles and gas near the
bed surface and in the whole freeboard volume. Elutriation of particles from the
fluidized bed surface is a complex phenomenon, although it results from only two
causes:

-- on the bed surface due to intense bursting and erupting of bubbles for which
   "real local" gas velocity can be significantly greater than superficial
   fluidization velocity based on the total bed cross section, which leads to conditions in which larger particles can also be ejected into the freeboard, and

– all particles with free fall velocity smaller than the average gas velocity in the freeboard will be elutriated away from the surface.

These two simple causes lead to a very complex picture of particle movement in the freeboard (Fig. 2.19). From the bed surface eruption of bubbles arises, not simply the elutriation of single particles, but rather clusters. Moving upwards, these clusters dissipate and disintegrate, and slow down. Some clusters move upward continuously, gradually dissipating, and some dissipate only after they reach the maximum ejection height, and are beginning to fall towards the bed surface. Particles separated from the clusters and single particles entrained from the bed surface move upwards among clusters until the largest begin to fall towards the surface. Only the smallest particles, with terminal velocities lower than the gas velocity above the bed, are "carried over" or transported out of the reactor (furnace), even if the freeboard is sufficiently high.

In every freeboard cross section there are clusters moving upwards and clusters falling, particles moving upwards and downwards. With increasing distance from the bed surface the amount of falling clusters and rising particles is greater, while the total number of clusters is lower. At a certain height, there are no more clusters, and still higher above the surface there are no more falling particles either. Near the wall, however, due to lower local velocities the share of falling particles and clusters will be higher. This picture of motion in the freeboard is also good for fluidized beds of monodisperse materials. But in this case, after a certain height particles will be no longer present if the gas velocity is lower than the particle terminal velocity. In real conditions, for a fluidized bed of polydisperse materials, or if abrasion of particles takes place, the finest particles will be removed from the reactor irrespective of the freeboard height.

The described character of particle motion above the bed, chaotic motion of particles in the bed and bubble motion in the bed, lead to major spacial and temporal changes of particle concentration, i. e., the gas-solid mixture density of the bubbling fluidized bed de-

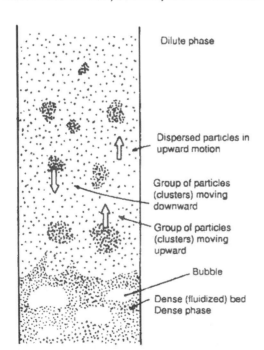

Dilute phase

Dispersed particles in upward motion

Group of particles (clusters) moving downward

Group of particles (clusters) moving upward

Bubble

Dense (fluidized) bed Dense phase

Figure 2.19. *Particle motion in vicinity and above the bubbling fluidized bed surface*

creases gradually (practically exponentially) above the bed surface to a constant value at the distance from the bed surface where pneumatic transport of small particles takes place. Near the distribution plate, the fluidized bed density is somewhat lower than the constant characteristic value for the fluidized bed.

Major oscillations occur near the free surface, primarily due to large bubbles, and irregular shape and unstable position of the free bed surface. Figure 2.20a shows experimentally determined change of density in the bubbling fluidized bed and near the bed surface for different fluidization velocities [6, 51]. Measurements were performed using a capacitance probe. The small scale of the diagram prevents more precise presentation of low concentrations in the region distant from the bed surface. Figure 2.20b illustrates the measured fluctuations of density.

Figure 2.21 shows the density change in the bed and above it; change of mean, net, mass flow rate of particles above the bed are also shown. Some of the terms related to the elutriation process are also explained in the figure.

The net mass flow rate of particles in a bubbling bed equals zero, and the bed density is the highest. In the splash zone upward and downward particle motion is intense, ensuring that the bed surface is irregular and unstable. With increasing distance from the bed surface, the net mass flow rate of particles through the freeboard cross section is decreasing. An increasingly higher number of small particles is moving upwards and also higher number of larger particles is going downwards.

In every single cross section two types of particles can be distinguished, according to the mechanism causing their motion and their chance of elutriation. Particles erupted out of the bed under the influence of bubbles, having free fall velocity higher than the mean gas velocity (and these will return sooner or later to the bed)

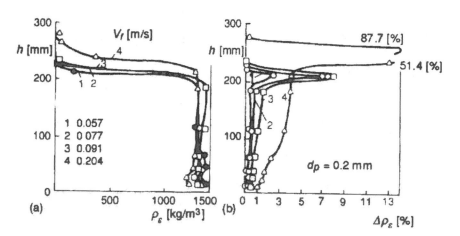

Figure 2.20. *Local bulk density of the fluidized bed (a) and fluctuations of the local bed density (b) in depth and near free bed surface [6,51]*

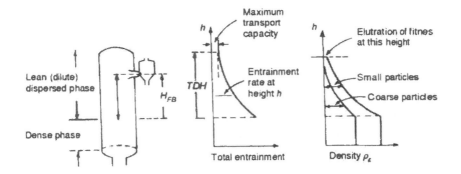

Figure 2.21. *Particle concentration (mean density) in a bubbling fluidized bed boiler furnace and total entrainment flux of solids (particles) in the freeboard*

are called entrained particles, and the process is called entrainment. Each fluidized bed, each particle size and each particular fluidization condition are characterized by the maximum height which can be reached by particles of determined size – transport disengaging height (*TDH*). In a fluidized bed of monodisperse material (or material with a narrow fraction of particles, in which even the smallest particle achieves a free fall velocity higher than the mean gas velocity) above the *TDH* both mass flow rate and particle density equal zero.

In every cross section, in addition to the entrained particles (which will, if moving upwards still fall back into the bed again) there are particles which will be taken out of the system as their free fall velocity is lower than the mean gas velocity. This process is called elutriation. Above the *TDH*, particle flow is composed only of elutriated particles, i. e., those with free fall velocities lower than the mean gas velocities – v_f. Also, particle elutriation is limited by the maximum saturation carrying capacity of gas flow.

Mass flow rate and size distribution of particles which will be elutriated from the system depend on the height at which the furnace gas outflow is situated. If the exit is below the *TDH*, particles with free fall velocities higher than the mean gas velocity will be elutriated, i. e., the lower the freeboard, the higher the particle loss.

Particle entrainment from the surface of the bubbling fluidized bed has not been completely elucidated so far. It has not been precisely determined whether particles ejected into the freeboard come from the "nose" or the "wake" of the bubble. It is, nevertheless, generally believed [6, 31, 52, 53] that most particles ejected with bubbles out of the bed come from the "wake." Some studies show [52] that the conditions of bubble entrainment occur only at sites at which two or more bubbles merge into one near the bed surface when merging abrupt local gas acceleration takes place.

An extensive review of particle entrainment and elutriation studies is provided in [2]. These studies were conducted over a wide range of parameters: bed diameter 0.035-0.90 m, bed height 0.03-0.91 m, freeboard height 0.3-6 m, particle size 19-3170 μm, fluidization velocity 0.040-4.76 m/s. Only empirical equations, however, are still in use for the calculation of *TDH* and the mass flow rate of elutriated or entrained particles. The only acceptable model of the phenomenon has been proposed by Kunii and Levenspiel [1].

Numerous data on the transport disengaging height have been reported. The proposed correlations reported in literature should, however, not be used beyond the range of parameters for which the experiments were carried out. The expression recommended by Leva and Wen [31], based on the assumption that particles move according to the law of vertical upward shot, with the initial velocity u_o^* at the bed surface:

$$TDH = \frac{\rho_p \, d_p^2}{18 \, \mu_f} \left[u_o^* - (u_f - v_f) \ln\left(1 + \frac{u_o^*}{u_t - v_f} \right) \right] \tag{2.60}$$

can be truly useful, since the value of the local initial ejection velocity of a particle from the bed surface is not known.

D. Geldart [11] derived an empirical correlation for particles in group D:

$$TDH = 1200 \, H_{mf} \, \text{Re}_p^{1.55} \, \text{Ar}^{-1.1} \tag{2.61}$$

applicable in the range $15 < \text{Re}_p < 300$, $19.5 < \text{Ar} < 6.5 \cdot 10^5$, $H_{mf} < 0.5$m, and obtained for particle size in the range 75 μm $< d_p <$ 2000 μm.

A. P. Baskakov [52] proposed the expression in the form:

$$TDH = H_{mf}^{0.58} \, D_b^{0.42} \left(m \frac{v_f - v_{mb}}{v_{mb}} + n \right) \tag{2.62}$$

obtained for 50-400 mm bed height and 76-450 mm bed diameter. The *m* and *n* constants are given in the following table:

d_p [μm]	134	229	396	515	976	1330
v_{mf} [m/s]	0.04	0.10	0.27	0.45	0.96	1.21
m	1.20	1.64	3.59	5.34	33.10	29.70
n	1.78	2.18	3.25	2.59	2.15	2.40

Naturally, random values of gas velocity at the fluidized bed surface leading to particle ejection into the freeboard comply with the Gaussian probability density distribution. As a result of such entrainment velocity distribution, the particle concentration (mixture density) in the freeboard will diminish exponentially:

$$\rho_\varepsilon = \rho_0 e^{-a_i h} \qquad (2.63)$$

With increase of the freeboard height, at a given distance from the bed surface, particle concentration increases if other parameters remain unchanged. If the freeboard height exceeds *TDH*, the density at any distance from the bed surface reaches the peak value. In this case, the following relationship holds:

$$\rho_{\varepsilon max} = \rho_{0\,max} e^{-a_i h} \qquad (2.64)$$

Lewis et al. [1, 54] have shown that:

$$\rho_{\varepsilon\,max} - \rho_\varepsilon = const$$

along the whole height above the bed, and that this constant difference increases with a decrease of freeboard height.

In accordance with the particle density above the bed, the total entrainment flux of the particles will fall exponentially with increasing distance from the bed surface:

$$F_i = F_{i\infty} = (F_{i0} - F_{i\infty})\exp(-a_f h) \qquad (2.65)$$

Dependence of the same type can be proposed for the entrainment flux of different fractions of polydisperse loose materials. For particles with d_{pi} diameter, it follows that:

$$F_i = F_{i\infty} = (F_{i0} - F_{i\infty})\exp(-a_f h) \qquad (2.66)$$

where

$$F_i = \sum F_i y_i \qquad (2.67)$$

with the assumption that particles of certain fractions do not influence each other, i. e., that constant a_f remains independent of the particle size.

Entrainment flux of solids for $h > TDH$ is expressed as:

$$F_{i\infty} = E_{i\infty} y_i \qquad (2.68)$$

Entrainment has been studied in numerous experiments in order to determine the elutriation rate constant E_i. Excellent detailed reviews of these experiments are given in [2, 55]. Table 2.10 gives some of the most popular correlations recommended for determination of elutriation rate constant, $E_{i\infty}$ [2, 11, 12, 31, 55]

$$F_{i0} = E_0 y_i \qquad (2.69)$$

Table 2.10. *Recommended empirical correlations for elutriation rate constant $E_{i\infty}$.*

Ref.	Correlation
56	$$\frac{E_{i\infty}\, g d_p^2}{\mu_f (v_f - u_t)^2} = 0.0015\, \mathrm{Re}_t^{0.6} + 0.01\, \mathrm{Re}_t^{1.2}$$
57	$$\frac{E_{i\infty}}{\rho_f v_f} = 1.26 \cdot 10^7 \left(\frac{v_f^2}{g d_p \rho_p}\right)^{1.88} \quad \text{for} \quad \frac{v_f^2}{g d_p \rho_p} \leq 581.8 \cdot 10^{-3}$$ $$\frac{E_{i\infty}}{\rho_f v_f} = 4.31 \cdot 10^4 \left(\frac{v_f^2}{g d_p \rho_p}\right)^{1.15} \quad \text{for} \quad \frac{v_f^2}{g d_p \rho_p} \geq 581.8 \cdot 10^{-3}$$
58	$$\frac{E_{i\infty}}{\rho_f (v_f - u_t)} = 1.52 \cdot 10^{-5} \left(\frac{(v_f - u_t)^2}{g d_p}\right)^{0.5} \mathrm{Re}^{0.725} \left(\frac{\rho_p - \rho_f}{\rho_f}\right)^{1.15}$$
59	$$\frac{E_{i\infty}}{\rho_f (v_f - u_t)} = 4.6 \cdot 10^{-2} \left(\frac{(v_f - u_f)^2}{g d_p}\right)^{0.5} \mathrm{Re}^{0.3} \left(\frac{\rho_p - \rho_f}{\rho_f}\right)^{0.15}$$
60	$$\frac{E_{i\infty}}{\rho_f v_f} = 23.7 \exp\left(-5.4 \frac{u_t}{v_f}\right)$$
61	$$E_{i\infty} = 33 \left(1 - \frac{u_t}{v_f}\right)^2$$
62	$$\frac{E_{i\infty}}{\rho_f v_f} = 9.43 \cdot 10^{-4} \left(\frac{v_f}{g d_p}\right)^{1.65}$$
55	$$E_{i\infty} = \rho_p (1 - \varepsilon_i) u_{pi}, \quad \text{where is}$$ $$\varepsilon_i = \left(1 + \frac{\lambda (v_f - u_t)^2}{2 g D_h}\right)^{-1/4.7}$$ $$\frac{\lambda \rho_p}{d_p^2}\left(\frac{\mu_f}{\rho_f}\right)^{2.5} = 5.17 \cdot \left(\frac{\rho_f (v_f - u_t) d_p}{\mu_f}\right)^{-1.5} D_p^2$$ $$\text{for} \quad \frac{\rho_f (v_f - u_t) d_p}{\mu_f} \leq \frac{2.38}{D_b} \quad \text{and}$$ $$\frac{\lambda \rho_p}{d_p^2}\left(\frac{\mu_f}{\rho_f}\right)^{2.5} = 12.3 \left(\frac{\rho_f (v_f - u_t) d_p}{\mu_f}\right)^{-2.5} D_h$$ $$\text{for} \quad \frac{\rho_f (v_f - u_t) d_p}{\mu_f} \geq \frac{2.38}{D_h}$$

and the entrainment rate constant of solids at the bed surface:

$$E_0 = 0.3 \, D_B \, d_p^{-1.17} \, (v_f - v_{mf})^{0.66}$$
(2.70)

for the bubbling fluidized bed, and

$$E_0 = 36.8 \, D_b \, d_p^{-2.47} \, (v_f - v_{mf})^{1.69}$$
(2.71)

for cases when bubble size reaches the diameter of the fluidized bed (occurrence of plugs).

In [55] there is also the following recommended correlation:

$$\frac{E_0}{A_b D_{Bo}} = 3.07 \cdot 10^{-9} \frac{\rho_f^{3.5} \cdot g^{0.5}}{\mu_f^{2.5}} (v_f - v_{mf})^{2.5}$$
(2.72)

The constant a_f ranges from 3.5-6.4 m^{-1} [2], while in [55] it is recommended that $a_f = 4$ m^{-1}.

Transport disengaging height can be defined as the height at which the specific mass flow rate of particles F_t reaches 99% of $F_{t\infty}$:

$$TDH_t = \frac{1}{a_f} \ln \left\{ \frac{E_0 - E_{t\infty}}{0.01 E_{t\infty}} \right\}$$
(2.73)

There are large disagreements between experimental data and correlations for specific elutriation flux [2, 12, 31, 62], and these empirical correlations should be used very carefully within the experimental conditions for which they have been derived.

The process and intensity of particle elutriation and entrainment are influenced by geometry of the reactor, bed diameter and depth, internals of different shape in the bed and above it (grids, baffles, packing), particle size distribution of the bed material and position (height above the bed surface) of openings for gas outflow.

The occurrence of plugs, related to decrease of bed diameter and increase of its depth, influences elutriation rate. Increase of bubble size at the bed surface also promotes particle entrainment. Therefore, the internals (immersed heat exchanger tubes) causing bubble splitting, also influence decrease of particle elutriation [2]. According to other investigations [11], immersed tubes in the bed failed to influence elutriation, but if placed above the bed they resulted in reduction of elutriation. Uneven gas distribution and poor fluidization may also promote particle elutriation.

2.3.5. Bubbles in a fluidized bed

The chaotic structure of bubbling fluidized beds, an alternative occurrence of gas phase and particles in the same point, sharp boundaries between the space filled with particles and freeboard without particles are the basic features of the fluidized state which we call a bubbling fluidized bed. In addition to being present in all interspace among particles, the gas phase occurs in isolated larger volumes, like bubbles in boiling liquid. Apparently, bubbles are the basic feature of a fluidized bed. At the same time, bubbles are the basic cause of numerous specific features of a fluidized bed and the most intriguing phenomenon which has not been elucidated as yet.

In the preceding pages, external, macroscopic features of the bubbling fluidized bed have been described: minimum fluidization velocity, pressure drop, bed expansion, particle elutriation. Most of these phenomena are only manifestations of presence and motion of bubbles in a fluidized bed, but in analyzing those phenomena it was not necessary to study the local structure of the fluidized bed and specific properties of bubbles.

The study and understanding of heat and mass transfer processes in a fluidized bed, the mixing of particles and chemical reactions, either in gas phase or on particle surfaces, are not feasible if bubble motion and local structure of fluidized beds are not thoroughly understood. These processes depend on the interaction of gas and particles, directed and chaotic particle motion and the heat and mass transfer between the gas phase (bubbles) and emulsion (particulate, dense) phase of the fluidized bed.

The local structure of a fluidized bed, gas-particle interaction, heat and mass exchange between gas (lean) phase and dense (particulate, emulsion) phase of a fluidized bed depend essentially on the properties of bubbles, their shape, growth, rise velocity, merging (coalescence), splitting and bursting. In light of the importance of fluidized beds for organization of effective heat and mass transfer processes and chemical reactions, the study of behavior and properties of bubbles has been a focus of interest of investigators and an important link for development of numerous technologies.

Bubbles occur either immediately, or just after the minimum fluidization velocity has been reached, or for some type of particles (Geldart's group A) at somewhat higher velocities. Occurrence of bubbles is purely randomized. There are no predetermined places on the air distribution plate where bubbles will occur, if the air openings are uniformly distributed. Irregularity in manufacturing of air distributors, or somewhat lower number of the openings per unit area may be places at which the occurrence of bubbles is more probable. When they occur, bubbles are approximately spherical, but later on, as they grow, they take on a shape similar to a gas bubble in liquid. Irregular and randomly distributed in the fluidized bed mass, they move in rows mainly vertically upwards. The presence of obstacles, other bubbles or directed flow of particles in the emulsion phase induces a local, brief aberration of bubbles from vertical upward motion and their motion along irregular trajectories with frequent aberrations from the vertical line. It is believed that while moving upwards through a fluidized bed, the bubbles grow mainly due to merging (coalescence). Therefore, the number of bubbles duly decreases towards the bed surface. Large bubbles become unstable, and minor disturbances, always present in the fluidized bed, lead to their break-up. In fluidized beds with small lateral dimensions, bubble dimensions may reach the size of bed cross section, when the bubbling regime of fluidized bed becomes the slugging regime. Motion of particles in the slugging regime has completely different features than the free bubbling regime, and also influences other processes – heat and mass transfer, mixing and chemical reaction. Differences in bubble size, shape and mode of motion, i. e., in the structure of the fluidized bed, are the basic reasons why, with other hydrodynamic and thermal conditions unchanged, it is simply not possible to extrapolate the results obtained on laboratory-scale fluidized beds to large plants used in practice.

Study of bubbles in fluidized bed can be divided into studies of a series of different processes: the origin and appearance of bubbles, the rise velocity, growth and coalescence, the break-up (splitting) and bursting of bubbles at bed surface.

The mechanism of bubble appearance in the fluidized bed is not clear. Experiments show that the visible bubbles occur in fluidized systems in which the particle fluid density ratio is high. Larger particles, larger bed porosity and lower fluid viscosity also promote bubble appearance. The influence of interparticle forces is, however, not well understood. Theoretical attempts to explain the appearance of bubbles, establish criteria for determination of the conditions which favor their occurrence and discover factors influencing further behavior and features of bubbles have widely different starting points. Developed theories rely on quite different assumptions on different manifestations of interactions in the two-phase gas-particle system: (a) the occurrence of visible bubbles is related to the nature and magnitude of interparticle forces [63, 64], (b) the stability of two-phase system subjected to small disturbances is studied [65-68], or (c) it is assumed that these systems always contain bubbles which, however, become visible only when the bubble diameter/particle diameter ratio exceeds some cutoff value [69, 70]. Some attempts have been made to explain the bubbles as shock waves occurring during propagation of dynamic waves in two-phase systems [71, 72].

There is an interesting group of theories which interprets the appearance of bubbles with processes taking place near the distribution plate and the design characteristics of the plate itself. Here, the occurrence of bubbles is related to the process of penetration of gas jets discharged from the distribution plate openings. Detailed reviews of these theoretical attempts and respective experimental investigations are given in [2, 73]. This approach is interesting since: (a) it relates the appearance and features of bubbles to the shape and features of the distribution plate;

and (b) it provides the possibility of evaluating the smallest distance from the plate at which the immersed heat transfer surfaces in the bed should be placed. At smaller distances, the immersed surfaces will be exposed to major erosion. Thus, for practical fluidized bed combustion applications, a determination of the gas jet penetration depth is a much more important issue than the bubble appearance itself.

A gas jet discharged from the distribution plate openings penetrates the two-phase gas-particle system taking along a lot of solid particles. At a certain distance from the opening the jet becomes unstable, breaking up into a series of bubbles. The jet penetration depth depends on physical properties of gas and particles, geometry and size of the opening and system pressure. Numerous experimental investigations have been aimed at determining the maximum jet penetration depth before break-up into bubbles. In [74] it is recommended that the following experimental formula for the maximum jet penetration depth be employed:

$$\frac{L_{max}}{d_0} = 814.2 \left(\frac{\rho_p d_p}{\rho_f d_0} \right)^{-0.585} \left(\frac{\rho_f v_0 d_0}{\mu_f} \right)^{-0.654} \left(\frac{v_0^2}{g d_0} \right)^{0.47} \tag{2.74}$$

while [75] recommends:

$$L_{max} = 6.5 \left(\frac{\rho_f}{\rho_p - \rho_f} \frac{v_0^2}{g d_0} \right)^{0.5} \tag{2.75}$$

These characteristic distances were obtained in experiments with idealized conditions: a single opening in the distribution plate forming bubbles, while the whole bed is maintained in the state of incipient fluidization. In spite of that, these data are useful for calculations in real fluidized beds and evaluation of the minimum distance of the immersed heat transfer surfaces from the distribution plate.

The basic parameters of a bubble in a fluidized bed are: shape, rising velocity and size. For a more detailed elaboration of fluidized bed processes the share of bed volume taken by bubbles should also be known. Obtaining the magnitude of these four parameters has been the object of numerous experimental studies and there have also been attempts to predict them theoretically using the physical properties of the gas and particles and fluidization parameters. Special measurement methods have also been developed: X-ray photography, capacitance probes for detection of shape, size and velocity of bubbles [76-78]. Due to major difficulties experienced in these investigations, it is more common to study a simplified situation corresponding to a single bubble formed and rising under conditions of incipient fluidization. Experiments have shown that the shape of these bubbles is similar to the shape of bubbles in an actual fluidized bed. Small bubbles are close to spherical, larger are vertically flattened and distorted, with inverted bottom, while the largest are spherical cap-shaped. Although the experiments are usually performed in two-dimensional fluidized beds (for easier visual monitoring) the results obtained can be used for the analysis of real, three-dimensional beds. Theoretical analyses also usually apply to isolated single bubbles. They are based on the analogy with gas bubbles rising in liquid. Similarity of features and bubble behavior in liquids and fluidized bed justify this approach, and it is observed that the shape is similar. In both cases small bubbles rise slowly, and larger ones rise more rapidly. A cloud of bubbles or a vertical chain of bubbles in both liquid and fluidized bed may also merge into larger bubbles. The influence of walls on bubble motion is also the same. Numerous studies have confirmed [79, 80] that the rise velocity of a single bubble in fluidized bed can be expressed by a formula derived for a gas bubble in liquid [81]:

$$v_{B\infty} = 0.711\sqrt{gD_B} \qquad (2.76)$$

Strictly speaking, D_B, bubble diameter is the diameter of an equivalent sphere with the same volume as the real bubble. In fluidized beds, the constant factor in the formula (2.76) ranges from 0.57-0.85 [1, 79, 80]. Nevertheless, it has been generally accepted that the rise velocity of bubbles can be calculated according to this expression.

Differences between bubbles in these two cases are primarily reflected in the actions of surface tension in liquid and gas exchange between bubbles and particulate (emulsion) phase in fluidized bed. In spite of these differences, theoretical studies based on the analogy with gas bubbles in liquid successfully describe bubbles in fluidized bed. The first theoretical model describing dynamics of an isolated bubble in a fluidized bed was proposed by Davidson and Harrison [79]. This model has subsequently been subjected to several improvements, but even in its simple form it manages to predict the basic features of bubbles. Due to its simplicity it is very useful for understanding of phenomena and features of bubbles, and it will, therefore, be used here as a basis for further presentation.

Davidson's model is based on very simple assumptions:

– a bubble is spherical (or circular if two-dimensional), and it contains no particles,
– the particulate (emulsion) phase in which the bubble is rising is in the state of incipient fluidization behaving as an incompressible non-viscous fluid with density of $\rho_p(1 - \varepsilon_{mf})$,
– gas flow between particles in the emulsion phase is the incompressible, viscous flow and the respective flow resistance can be calculated by the classical Darcy's formula,
– further away from the bubble, fluidized bed pressure changes linearly like hydrostatic pressure, according to (2.30), and
– gas pressure in the bubble is constant.

Motion of particles around the spherical bubble is described by equations for the motion of nonviscous fluid, i. e., as potential flow. The velocity potential in spherical coordinates is:

$$\Phi_p = -v_{B\infty} \cos\theta \left(r + \frac{R_B^3}{2r^2} \right) \qquad (2.77)$$

for a three-dimensional bubble, and in cylindrical coordinates:

$$\Phi_p = -v_{B\infty} \cos\theta \left(r + \frac{R_B^2}{r} \right) \qquad (2.78)$$

for a two-dimensional bubble with circular cross section.

The motion of particles around the spherical bubble as seen by an observer moving with the bubble (a) and fixed observer (b) is illustrated in Fig. 2.22 [1].

The pressure distribution around the bubble is described by Laplace's equation. The pressure in the lower half of the bubble is lower than the pressure in the surrounding emulsion phase, while in the upper half it is higher. It is assumed that the pressure in the bubble is constant, approaching a linear distribution with increasing distance from the bubble. Figure 2.23 illustrates pressure distribution around bubbles according to the measurements of Reuter [82]. The analytical expression for pressure distribution is the following:

$$p = -\rho_p (1-\varepsilon_{mf})g \left(r - \frac{R_B^3}{r^2} \right)\cos\theta \qquad (2.79)$$

The difference in pressures inside the bubble and around it cause gas exchange at the bubble-emulsion interface. Gas from the emulsion enters the lower

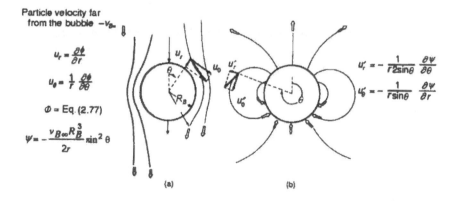

Figure 2. 22. *Particle flow around a spherical bubble according to Davidson and Harrison model (Reproduced by kind permission of the author Prof. Dr. Octave Levenspiel from [1])*

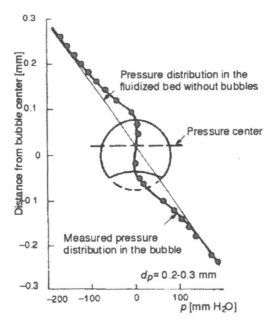

Figure 2.23. *Pressure distribution around bubble according to Reuter's measurements (Reproduced by kind permission of the author Prof. Dr. Octave Levenspiel from [1])*

half of the bubble, while it leaves the bubble in the upper half. The gas velocity distribution around the bubble, obtained by Davidson's model, illustrates this fact. Figure 2.24 shows streamlines of gas flow around an isolated bubble [1].

Characteristics of interstitial gas flow in the emulsion phase around an isolated spherical bubble, i. e., streamlines represented in Fig. 2.24, depend on the v_B/u_f, i. e. ratio of bubble rising velocity and true interstitial gas velocity between particles in the emulsion phase which is, according to Davidson's model, equal to the minimum fluidization velocity.

In the extreme case, when the bubble is fixed, the total quantity of gas entering the lower half of the bubble will leave the bubble. At the extreme, when the bubble rise velocity through the emulsion phase is very high, gas from the emulsion will not enter the bubble at all. Inside the bubble, closed gas circulation will be present.

Between these two extremes, when slow bubble motion is present, $v_B < u_{mf}$, an annular gas shell moves together with the bubble. Gas from this ring permanently flows through the bubble, and then returns to the emulsion. The remaining part of the gas which flows through the bubble is actually fresh gas from the emul-

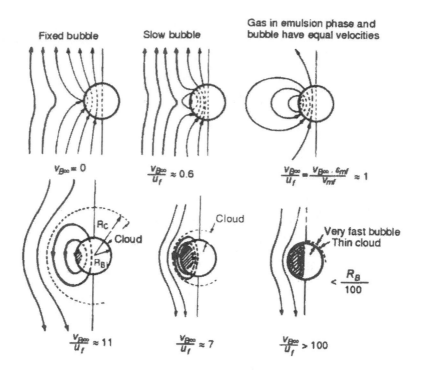

Fixed bubble   Slow bubble   Gas in emulsion phase and
                                           bubble have equal velocities

$v_{B\infty} = 0$ $\dfrac{v_{B\infty}}{u_f} \approx 0.6$ $\dfrac{v_{B\infty}}{u_f} = \dfrac{v_{B\infty} \cdot \varepsilon_{mf}}{v_{mf}} \approx 1$

$\dfrac{v_{B\infty}}{u_f} \approx 11$ $\dfrac{v_{B\infty}}{u_f} \approx 7$ $\dfrac{v_{B\infty}}{u_f} > 100$

Figure 2.24. *Gas flow around an isolated spherical bubble according to the Davidson and Harrison model for different ratios of bubble velocity to real (interstitial) gas velocity (Reproduced by kind permission of the author Prof. Dr. Octave Levenspiel from [1])*

sion which "rinses" (washes up through) the bubble. When bubble velocity becomes greater than the interstitial gas velocity in the emulsion, $v_B > u_{mf}$, a so called "cloud" is formed around the bubble. It is a spherical area around the bubble inside which gas circulates, in and out of the bubble. Only gas from the cloud circulates through the bubble. Gas from the remaining part of the emulsion avoids the cloud and does not circulate through the bubble. With increasing bubble velocity, the cloud diameter decreases, and for very fast bubbles gas circulates effectively only inside the bubble.

The cloud versus bubble diameter ratio is given by the following expressions:

$$\left(\frac{R_c}{R_B}\right)^2 = \frac{v_{B\infty} + u_{mf}}{v_{B\infty} - u_{mf}} \tag{2.80}$$

for a two-dimensional bubble, and

$$\left(\frac{R_c}{R_B}\right)^3 = \frac{v_{B\infty} + 2u_{mf}}{v_{B\infty} - u_{mf}} \tag{2.81}$$

for a three-dimensional bubble.

Mass exchange between bubbles and emulsion phase, i. e., circulation of emulsion gas through the bubble, is very important for effective organization of chemical reactions in a fluidized bed. In this respect, there is an essential difference between fluidized beds for small and large particles. In a fluidized bed of large particles circulation of gas from the emulsion through the bubbles is significant; the bubbles are intensively rinsed (washed), and a smaller amount of gas will pass the fluidized bed taking no part in the bed reactions. In a fluidized bed of small particles, bubbled gas by-passes the bed taking no part in reactions occurring in the bed. The participation of gas in the cloud is determined only by the intensity of gas diffusion between bubbles (i. e., clouds) and emulsion.

Davidson's theoretical model [79] predicts gas flow rate through a bubble:

$$q_B = 4v_{mf}R_B = 4u_{mf}\varepsilon_{mf}R_B \tag{2.82}$$

for a two-dimensional bubble, and

$$q_B = 3v_{mf}\pi R_B^2 = 3u_{mf}\varepsilon_{mf}\pi R_B^2 \tag{2.83}$$

for a three-dimensional bubble.

The mean gas flow velocity through the maximum bubble cross section is $2v_{mf}$ (for a two-dimensional bubble), i. e., $3v_{mf}$ (for a three dimensional bubble), which means that the bubble allows the passage of two or three times more gas than through the same cross section in the emulsion phase.

Although very simple, Davidson's model correctly explains the basic features and behavior of bubbles in a fluidized bed. Visual observations and quantitative measurements also confirm the main results of this theory. Numerous experiments [83, 84] have established the presence of a cloud and even determined the size of the cloud surrounding a bubble. Experimental results agree with expressions (2.80) and (2.81) [11, 83]. Further, measurements of pressure around the bubble (Fig. 2.23) [82] correspond closely to the predictions of Davidson's model.

An upward gas flow through the bubble explains why the particles comprising the "bubble roof" do not "dive" through and "bury" the bubble. It is also clear why gas in bubbles can pass through the fluidized bed while entering into no reaction with the rest of the bed.

The Davidson model's greatest idealization is its assumption that a bubble is spherical. Figure 2.25 shows the shape of an actual bubble [84]. The lower part is dented, forming a kidney shape. Due to gas inflow, particles behind the bubble are in turbulent motion, and pressure is lower in that area influencing the upward drift of particles into the "wake" behind the bubble. "The bubble wake" usually implies

Figure 2.25.
*X-ray photograph of a
real three-dimensional
bubble in a fluidized bed
[84] (Reproduced by
kind permission of the
author Prof. Dr. Octave
Levenspiel from [1])*

the flat-bottomed volume behind the bubble which complements it in the sense that it allows the bubble to approximate the shape of a complete sphere. The motion of particles in the "wake" contributes to particle mixing in the fluidized bed. Particles in the wake may account for as much as 1/3 of bubble volume [80]. The presence of a "cloud" and "wake" during bubble motion must be taken into account in the analysis of bubble processes in a fluidized bed and their mathematical modelling.

Very large bubbles cannot exist very long and they will disintegrate into smaller bubbles. It has also been suggested that bubbles may be destroyed by particles from the wake carried up to the interior from the bottom by gas circulation, when the bubble rise velocity, v_B, exceeds the particle free fall velocity [70].

The simple Davidson's model has been improved by several investigators, but major improvement in the prediction of the behavior of a single bubble has not ensued. Collins [85, 86] and Stewart [83] assumed that a bubble is shaped like a kidney, which is closer to the actual shape of a bubble (Fig. 2.25), with the remaining assumptions taken from Davidson. Murray maintained the bubble spherical shape, but instead of the given boundary condition for the pressure change characteristic of Davidson's model, he also solved the momentum equation for particle motion in the emulsion phase [87]. Finally, Jackson [65] kept the spherical shape of a bubble, assuming, however, a changeable porosity in the emulsion phase, and rejecting the assumption that the particulate phase behaves like an incompressible fluid.

Changes of gas pressure above the bubble in this model are assumed to be equal to the change of momentum of particle motion.

The agreement of these models with measurements [83, 88] is shown in Fig. 2.26, confirming that these, essentially simple models, predict the cloud size fairly well.

It has become customary to express the cloud volume as a part of bubble volume:

$$V_c = V_B \frac{\alpha}{\alpha - 1} \qquad (2.84)$$

However, models of single bubble behavior proposed by Davidson and others who have developed the original model, do not allow for calculation of the size of "wake," i. e., the quantity of particles in the bubble "wake." According to analogy with the behavior of gas bubbles in liquid, the shape and size of the "wake" is a function of Reynolds number, and the size of bubble "wake" in a fluidized bed has also been estimated.

If a flat bottom for a bubble is assumed, the bubble volume can be calculated from the following expression:

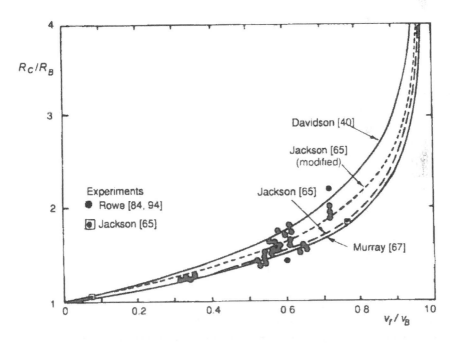

Figure 2.26. *Ratio of cloud radius to bubble radius calculated for an isolated bubble according to different bubble models, compared with experimental results (Reprinted from Chem. Eng. Sci. [88]. Copyright 1962, with permission from Elsevier Science)*

$$V_B = \frac{\pi R_B^2}{3} (2 - 3\cos\theta_w + \cos^3\theta_w) \qquad (2.85)$$

while the ratio of "wake" and bubble volumes is:

$$\alpha_w = \frac{2 + 3\cos\theta_w - \cos^3\theta_w}{2 - 3\cos\theta_w + \cos^3\theta_w} \qquad (2.86)$$

According to [11, 89] the angle θ_w can be calculated from the empirical formula applicable for bubbles in liquid:

$$\theta_w = 50 + 190\exp(-0.62\,\mathrm{Re}_B^{0.4}) \qquad (2.87)$$

in the region $1.2 < \mathrm{Re}_B < 100$.

Measurements [11, 89] also show that the bubble "wake" volume can be related to the bubble volume as:

$$\alpha_w = 0.037\,\mathrm{Re}_B^{1.4} \qquad (2.88)$$

For most bubbles in a fluidized bed $\mathrm{Re}_B < 20$, and it is assumed that the bubble "wake" in a fluidized bed can be determined from expression (2.86), or (2.88). However, it is difficult to determine the effective viscosity of the emulsion phase. Measurement of θ_w angle allows the determination of effective viscosity from (2.87). According to [80] the effective viscosity of the emulsion phase ranges from 0.7-1.3 Ns/m^2.

Obviously, the issue of "wake" volume determination in a fluidized bed has not yet been fully resolved, especially in the light of the fact that differences do occur between a fluidized bed of fine particles (group B) and large particles (group D) [11]. Group D bubbles are almost spherical ($\theta_w \to 180°$) [90], but in group B fluidized beds, θ_w ranges from 90-130° [80].

The behavior of a single bubble which is formed and rises in a fluidized bed at incipient fluidization, has been studied experimentally in great detail, and the mathematical models proposed can be considered to describe the actual physical events fairly well.

The behavior of bubbles in a real fluidized bed is, nevertheless, different and bubbles with different rise velocities, shapes and sizes are present. A bubble caught in the "wake" of a preceding larger bubble will be accelerated, so that the mean bubble rising velocity is higher than the velocity calculated according to expression (2.76). Further, due to coalescence, bubbles grow with increasing distance from the distribution plate, and their rise velocity also increases. Close to the distribution plate, bubbles rise vertically upward, and experience lateral velocity components, but maintain their main motion in the vertical direction. In spite of the differences mentioned here, experimental and theoretical knowledge resulting from the investigation of isolated bubble behavior are still commonly used to describe free bubbling in a fluidized bed.

Davidson and Harrison [79] were the first to assume that the actual bubble rise velocity in a real fluidized bed has to be higher than predicted by expression (2.76):

$$v_B = v_{B\infty} + (v_f - v_{mf})$$ (2.89)

Calculation of bubble rise velocity according to expressions (2.76) and (2.89) requires information on size, i. e., bubble diameter. Numerous investigations have been conducted to study the size and growth of bubbles for different designs of distribution plates as a function of distance from the plate. Table 2.11 gives the best known formulae.

Table 2.11. *Empirical correlations for calculation of bubble diameter*

Authors [Ref]	Correlation
Mori, Wen, [91]	$D_B = D_{B\infty} - (D_{B\infty} - D_{Bo})e^{-3\frac{x}{D_s}}$ $D_{Bo} = 1.38 g^{-0.2}[A_o(v_f - v_{mf})]^{0.4}$ for distribution plate with orifices $D_{Bo} = 0.376 (v_f - v_{mf})^2$ for porous distribution plate
Rowe [78]	$D_B = (v_f - v_{mf})^{0.5}(h + h_o)^{3/4} g^{-0.2}$
Darton [48]	$D_B = 0.54 (v_f - v_{mf})^{0.4}(h + 4\sqrt{A_o})^{0.8} g^{-0.2}$
Werther [92]	$D_B = 8.53 \cdot 10^{-3}[1 + 27.2(v_f - v_{mf})]^{1/3} \cdot$ $[1 + 6.84(h + h_o - h_B)]^{1.21}$

Formulae for calculation of the size (diameter) and growth of bubbles with increasing distance from the distribution plate were obtained for fluidized beds of particles from Geldart's group B. All are empirical and demand knowledge of the influence of the shape of the distribution plate [93]. For fluidized beds of large particles (Geldart's group D) used in furnaces with fluidized bed combustion, the following formula is recommended for the size and growth of slow bubbles:

$$D_B = 0.0326 (v_f - v_{mf})^{1.11} h^{0.81}$$ (2.90)

The bubble "merging" process (coalescence) which results in a decrease of bubble count and increase of bubble size with increasing distance from the distribution plate has not been sufficiently well studied at this time. Nor have the physical causes of this process been completely understood. The following observations can be made. Two bubbles of similar size, if sufficiently close, will merge. The larger, faster bubble will attract, to its "wake," a smaller, slower bubble which it is passing by. A bubble in the "wake," leeward from the other bubble, will accelerate and will be captured by the preceding bubble. It has also been noted, but not explained, that the resulting bubble has 10-20% larger volume than the sum of the two predecessors [93]. Numerous experimental studies [12, 48, 76, 78, 90-92, 94] are available which provide abundant results and the empirical formulae given in Table 2.11, but all have failed to explain the essence of the bubble merging and growth process. Unfortunately, only a few theoretical models of this phenomenon have been proposed [12, 93], and they are essentially based on Davidson's model of an isolated bubble or have some other limitation. Thus, Darton's mechanistic model [48] which gives the formula given in Table 2.11 has a constant which must be determined experimentally. In spite of such recognized limitations, the various formulae are commonly used for calculations in theoretical models describing processes (chemical reactions, combustion, heat transfer) taking place in reactors and furnaces using fluidized beds.

As it has already been noted, bubbles in freely bubbling fluidized beds do not grow indefinitely large. Large bubbles become unstable and will disintegrate (break up) [93, 95].

According to [91], the maximum bubble size in a freely bubbling fluidized bed can be calculated according to the following formula:

$$D_{B\infty} = 1.49 [D_b^2 (v_f - v_{mf})]^{0.4} \tag{2.91}$$

In fluidized beds with baffles or immersed heat transfer surfaces, the mean size of bubbles and maximum bubble size are smaller than in freely bubbling fluidized beds [95]. The presence of immersed surfaces promote bubble disintegration. It has been generally accepted that a bundle of horizontal tubes immersed in a bed prevents bubble growth providing uniform distribution of bubbles in the cross section [96]. Larger bubbles break up in clashes with the tubes, and their size can be considered constant and equal to the horizontal distance between the tubes [97].

2.3.6. Gas and particle mixing in fluidized bed

Intense mixing is the most important feature of fluidized beds. Although the important role of solid particle mixing is well understood, the importance of gas mixing should not be overlooked. The processes of gas and particle mixing are closely related and interactive. Numerous important features of the fluidized state result from this intense mixing, i. e., gas and particle motion. Uniform temperature field, high heat transfer coefficients to immersed surfaces, fast heating of cold particles injected into the heated bed, and uniform distribution of fuel particles during fluidized bed combustion, all result from the intense and chaotic motion of particles. The enhanced rate of chemical reactions on the particle surface and gas-gas homogeneous chemical reactions, depends significantly on the process of gas mixing in a fluidized bed. Mixing in a fluidized bed is decisive in drying processes and thermal processing of metals in fluidized bed reactors.

Mixing processes are particularly important for the organization of the combustion process of solid fuels in furnaces with fluidized beds. The intensity and properties of gas and particle mixing (e. g., the motion of gas in and associated with bubbles) have been discussed in the previous section, while here we discuss the behavior of gas in the emulsion-particulate phase of the bed that determines the following processes in the fluidized bed furnaces: mixing of fuel and inert material particles, uniform distribution of fuel in the bed, selection of mode and location for the fuel feeding, selection of the number of locations for fuel feeding, ash behavior in fluidized bed, heat transfer to immersed surfaces, combustion of volatiles, erosion of heat transfer surfaces, etc.

It has been generally accepted and experimental evidence supports the belief that bubble motion is the basic cause of particle mixing in the fluidized bed. In the discussion on solid particle motion the fact that two types of particle motion are present should be remembered: chaotic motion of individual particles and directed motion (circulation) of groups of particles. The presence and motion of bubbles is the basic cause of the latter type of motion. As they move upwards, bubbles drag particles from the emulsion phase in their "wake." Due to the general stationary condition of the bubbling fluidized bed, the upward motion of particles has to be compensated by a downward motion in some other part of the bed. Thus, quasistationary particle circulation patterns occur in the emulsion phase of the fluidized bed. The intensity, shape, character and number of these solid circulation patterns depend on numerous parameters, but the major impact of the size and geometry of fluidized beds should be pointed out. This is one of the basic reasons why it is not possible to realize hydrodynamic similarity in "small" beds in laboratory size equipment with beds in industrial size plants. H_b/D_b ratio and absolute size of bed influence the mixing processes.

Studies of the gas motion and mixing in the emulsion phase should involve three processes: molecular motion in the gas, turbulent mixing in the gas and directed motion of gas resulting from organized particle circulation. Molecular motion and turbulent mixing of gas are important for the processes on the bed particle surface or fuel particles in the bed. These processes determine whether fuel particle combustion will take place in the diffusion or kinetic regime. Macroscopic gas motion in the emulsion phase is determined by particle motion. In their wakes, bubbles drag along both particles and gas flowing through the emulsion phase. In downward motion, particles may significantly interrupt gas motion, which is directed upwards in the emulsion phase. Very intense particle backmixing patterns may even cause, in some areas of the bed, downward gas motion. The processes of macroscopic gas mixing in the emulsion phase significantly influence distribution and combustion of volatiles released from solid fuel particles.

Mixing in the fluidized bed has been studied extensively experimentally. Most experimental studies are directed to particle mixing. In spite of this, the mixing process is still the primary object of study, since the basic mechanisms, physical causes and particle motion dynamics remain elusive. Detailed reviews of experimental results are given in [18, 99], and a review of studies performed in the U.S.S.R. is given in [3, 6]. Experimental investigation of mixing in fluidized beds has been carried out primarily for particle A beds (Geldart's classification) due to their wide application in chemical reactors. The study of mixing in large particle beds (groups B and D) has only recently drawn increasing interest, with the advent of fluidized bed combustion technology, and studies on the mixing properties of large particle beds remain quite rare [100].

Initial appreciation of the fluidized bed as a homogenous medium (due to the uniform temperature distribution throughout the bed and intense mixing and heat transfer processes) was the reason it was first accepted that the mixing process can be described in the manner previously used to study mixing in gases and liquids. Thus, experimental methods and theoretical approaches were based on analogy with molecular, chaotic motion in gases and liquids.

Numerous experiments in which motion of labelled isolated particles or groups of particles (colored, heated or radioactive) was monitored, first established the fact that particles in a bubbling fluidized bed move in two ways: chaotic and directed [1, 6]. Bondareva [101], Kondukova [102], and Massimilla [103] performed the key experiments in this area. These and subsequent experiments have shown that the chaotic motion of particles can be represented in the following way:

$$(\Delta l)^2 = (\Delta l_d)^2 + (\Delta l_c)^2 = D_e \Delta t + v_{cr}^2 \Delta t^2 \qquad (2.92)$$

After dividing by Δt and averaging, the mean square distance a particle crosses per unit of time is obtained

$$\frac{\overline{\Delta l^2}}{\Delta t} = D_e + v_{cr}^2 \Delta t \qquad (2.93)$$

which is a constant value in cases of pure molecular diffusion. In fluidized beds, obviously due to the existence of directed circulatory motion, the mean square distance is not constant, but linearly dependent on time and characteristic velocity of circulatory motion.

Maximum measured velocities of particles range between 30-50 cm/s, and the circulation velocities 5-20 cm/s [6, 101, 102]. The effective coefficient of particle diffusion was measured in several experiments [6] and was of the order of magnitude of $D_e = 1$ cm^2/s which is a few orders of magnitude below the actual intensity of particle mixing in fluidized beds.

In addition to monitoring the motion of isolated particles, mixing in fluidized beds was also studied in the following ways:

(a) monitoring of concentration changes of labelled particles in two different areas of fluidized beds, i. e., measuring of labelled particle mass flow rate through a horizontal plane [1, 104],

(b) measuring of time (probability density distribution) a particle stays in a fluidized bed (residence time) [1, 98, 105], and

(c) measuring of spreading (dispersion) of labelled particles. Coloring, heating, isotopes, mixing of electroconductive and nonconductive particles were the method used [1, 98, 99].

In spite of obvious differences between the processes of molecular diffusion and mixing in fluidized beds, the most common way of representing the results of investigation was by the introduction of an effective coefficient for the macroscopic gas mixing in fluidized beds (effective solids dispersion coefficient), D_s [1, 6, 98].

These studies are based on representation of the mixing process using the classical one-dimensional diffusion equation:

$$\frac{\partial C_s}{\partial t} = D_s \frac{\partial^2 C_s}{\partial x^2} \tag{2.94}$$

Experiments clearly show that:

– mixing in fluidized beds is a markedly anisotropic process. Mixing is an order of magnitude more intensive in the axial (vertical) direction than horizontally, and

– the effective diffusion coefficient D_s is not a constant, but depends on numerous parameters, mostly on bed dimensions and fluidization velocity.

An idea of the magnitude of the effective coefficient of macroscopic particle mixing in fluidized beds can be obtained from the following few figures illustrating the results of experiments of several authors. Figure 2.27 [1] presents results of the measurement of the effective axial particle dispersion coefficient by Lewis et al. [106].

The effective radial particle dispersion coefficient is, according to the measurements of Mori and Nakamura [1, 107] (presented in Fig. 2.28), lower by an order of magnitude. The comparison of experimental results for different bed materials [64], related to the product of fluidization velocity and bed diameter is shown in Fig. 2.29.

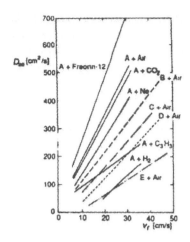

A — catalyst, d_p = 0.11 mm
B — glass spheres, d_p = 0.04 mm
C — glass particles, mixture, d_p = 0.092 mm
D — glass spheres, 0.076 mm
E — glass spheres, 0.155 mm

Figure 2.27.
Effective axial dispersion coefficient of particles according to Lewis' different fluidized bed materials and different fluidizing gases (Reproduced with permission of the AIChE from [106]. Copyright 1962 AIChE. All rights reserved)

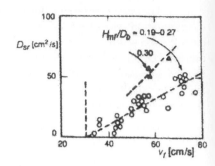

Figure 2.28.
Effective radial dispersion coefficients of particles as a function of fluidization velocity according to (Reproduced with kind permission of the Society of Chemical Engineers, Japan, from journal Kakagu Kogaku [107]. All rights reserved)

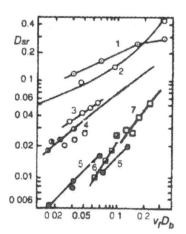

1 — Silica gel, wide fraction, d_p = 30 μm

2 — Catalyst, d_p = 40 μm

3 — Catalyst, d_p = 60 μm

4 — Silica gel, narrow fraction, d_p = 80 μm

5 — Silica sand, d_p = 450 μm

6 — Flint, d_p = 350 μm

Figure 2.29. *Effective radial dispersion coefficients of particles for different bed materials as a function of the fluidization velocity to bed diameter product [3]*

All reported experiments show a marked influence of fluidized bed size on the process of solid particle mixing [6, 98, 99]. Although it has not yet been established which dimensions in a fluidized bed influence mixing processes most, it has been generally accepted that the mixing coefficients obtained in experiments with smaller fluidized beds cannot be extrapolated to large, industrial-scale plants. Figure 2.30 [6] compares the values of coefficients for axial dispersion (at fluidization velocity $v_f = 2v_{mf}$) reported by different authors, obtained in experiments with 3 cm to 3 m diameter fluidized beds.

According to available experimental and theoretical studies it is impossible to predict the intensity of mixing of solid particles as a function of bed size, particle size and characteristics and fluidization velocity. Based on experimental results from different authors and his own theoretical examination of the process of solid particle mixing, M. O. Todes [6] suggested that the order of magnitude of effective coefficients of solid particle axial dispersion can be expressed as follows:

$$D_{sa} \approx \frac{1}{60} \sqrt{L^3 g \frac{(\rho_p - \rho_f)(1-\varepsilon)}{n\rho_p \varepsilon}} \tag{2.95}$$

where L is the characteristic (less) dimension of bed, bed diameter or height, and $n = 0.1$.

The influence of fluidization velocity is usually expressed as follows [6]:

$$D_{sa} \approx \frac{1}{60} \sqrt{L^3 g \frac{(\rho_p - \rho_f)(1-\varepsilon)}{n\rho_p \varepsilon}} \left(\frac{v_f}{v_{mf}} - 1 \right)^m \tag{2.96}$$

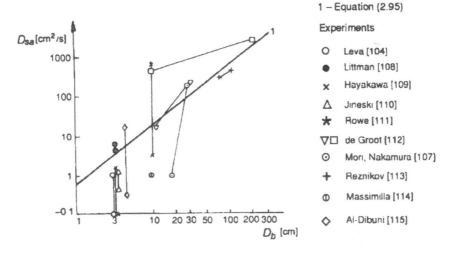

1 – Equation (2.95)

Experiments

O   Leva [104]
●   Littman [108]
x   Hayakawa [109]
△   Jineski [110]
★   Rowe [111]
▽□   de Groot [112]
⊙   Mori, Nakamura [107]
+   Reznikov [113]
⊕   Massimilla [114]
◇   Al-Dibuni [115]

Figure 2.30. *Effective axial dispersion coefficient of particle according to different authors in fluidized beds of different sizes, 3 cm to 3 m [6]*

where experimentally determined exponent m, usually equals 1, although some authors reported that $m = 2$.

The influence of the size and shape of particles here is not properly understood, though it should be kept in mind that most experiments were carried out with particles classified as Geldart's group A [98-100].

Various types of models have been used in this situation. Myasnikov [116] and Cibarov [117] tried to use the methods of statistical physics. While, based on the classical methods of fluid mechanics, Jackson [118] and Tamarin [119] as well as Buevitch [120] studied the interaction of continuous and discrete (solid) phase. The diffusion model was most commonly used to accumulate a large number of data points on the effective solid dispersion coefficient, D_s. These models are based on the classical concept of diffusion and applied to fixed and moving bubbling fluidized beds [3, 6, 52]. Detailed reviews of these models are given in [1, 3, 10, 52, 98].

Physically the most appropriate and currently the most commonly used models are based on experimental evidence as the solid particles circulate in fluidized beds under the influence of bubbles. The presence of bubbles in fluidized beds makes the system not only locally non-homogenous, but macroscopically markedly non-homogenous. Numerous studies of bubble motion in recent years [122-125, 127] have shown extremely irregular distribution of bubbles in fluidized beds.

In the vicinity of the air distribution plate (Fig. 2.31 [125]) the majority of bubbles move next to the walls, while at greater distances, due to coalescence (merging) and transverse motion of the bubbles, the main stream of bubbles is cen-

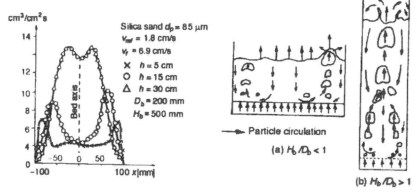

Figure 2.31.
Bubble concentration in cross section of the bubbling fluidized bed (Copyright 1986. Reproduced from [125] by permission of Routledge, inc., part of the Taylor & Francis Group)

Figure 2.32.
Particle circulation in bubbling fluidized bed (a) shallow bed, (b) deep bed (Reprinted from Chem. Eng. Sci. [128] Copyright 1978, with permission from Elsevier Science)

tralized round the deep bed axis ($H_b/D_b > 1$). Similar processes in shallow beds with a large cross section area ($H_b/D_b < 1$) lead to formation of several intensive columns (chains) of large bubbles (Fig. 2.32). The same figure also illustrates possible circulation patterns in deep beds. The complexity of solid particle circulation and the patterns induced by bubble motion is represented in Fig. 2.33 from the measurements of Masson [125, 128].

The first model taking into account the non-homogenous nature of fluidized bed and the presence of circulating motion of particles was proposed by Davidson and Harrison [79], and Kunii and Levenspiel [1]. These, so called two-phase fluidized bed models were subsequently improved in various ways using new experimental data, but their physical essence was not altered. Therefore, we shall stick to the description of the basic model according to Kunii-Levenspiel [1]. Detailed review of these improved models is given in [2, 121].

Kunii-Levenspiel's model [1] is based on the experimentally-established fact that a rising bubble drags particles in its "wake," and in the released void emulsion particles also move upward. This particle flow upward has to be compensated by corresponding particle flow downward.

Rowe and Partridge [80] measured that the bubble "wake" contained rising particles with the volume of roughly 25-30% of the bubble volume (Fig. 2.34).

Later on, Rowe [129] found that in addition to the particles "in the wake" ($\approx 0.25 V_B$), the emulsion particles behind the bubble also move upward, their volume being approximately $0.35 V_B$. Based on these facts and on Davidson's model of bubble motion, Kunii and Levenspiel proposed a model of particle motion in the emulsion phase. The basic features of the proposed model are illustrated in Fig. 2.35:

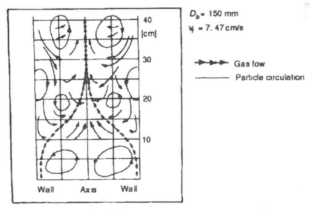

Figure 2.33. *Particle circulation in a bubbling fluidized bed (Reprinted from Chem. Eng. Sci. [128]. Copyright 1978, with permission from Elsevier Science)*

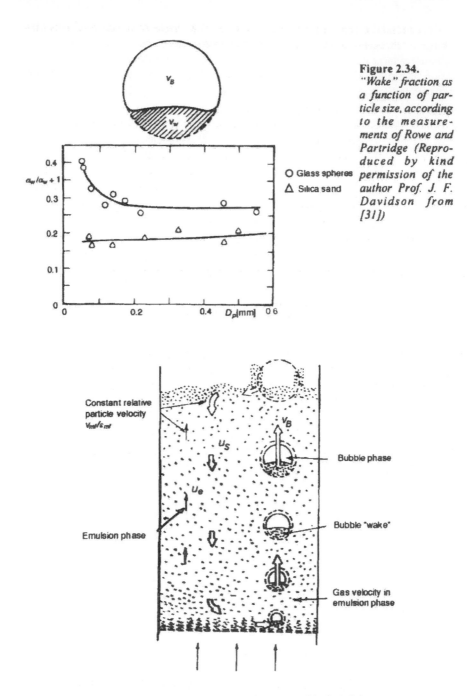

Figure 2.34.
*"Wake" fraction as
a function of par-
ticle size, according
to the measure-
ments of Rowe and
Partridge (Repro-
duced by kind
permission of the
author Prof. J. F.
Davidson from
[31])*

○ Glass spheres
△ Silica sand

Constant relative
particle velocity
v_{mf}/ε_{mf}

u_S

u_e

Emulsion phase

v_B

Bubble phase

Bubble "wake"

Gas velocity in
emulsion phase

Figure 2.35. *Kunii and Levenspiel's fluidized bed model*

(a) Each bubble drags particles along. The ratio of volume of particles "dragged in the wake" and the bubble volume is:

$$\alpha_w = \frac{V_w}{V_B} \tag{2.97}$$

Advanced models [2, 121] take into account particulate emulsion dragged behind the "wake." It is assumed that the porosity in the "wake" is equal the porosity of the emulsion phase: $\varepsilon_w = \varepsilon_{mf}$.

(b) During formation, a bubble from the emulsion phase drags along particles in the "wake" behind it. These particles move upwards through the bed with velocity that equals the rise velocity of bubbles, with permanent exchange of particles with the emulsion phase. On the bed surface particles from the "wake" join and mix with the emulsion phase particles and start moving downwards with the gas velocity.

(c) Due to the downward particle movement, the relative gas velocity in the emulsion phase is calculated from the following expression:

$$u_e = u_{mf} - u_s = \frac{v_{mf}}{\varepsilon_{mf}} - u_s \tag{2.98}$$

If we assume that the emulsion particles are immobile ($u_s = 0$), then $u_e = v_{mf}/\varepsilon_{mf}$, as has been assumed in Davidson's bubble motion model. The (2.98) assumption implies that downward gas motion is also possible, with intense mixing, i. e., downward particle motion ($u_s > v_{mf}/\varepsilon_{mf}$).

(d) the original model describes the fluidized bed with fast bubbles and thin cloud when $v_B/v_{mf} > 5$, i. e., $v_f/v_{mf} > 2$. The mass balance of particles passing the cross section of the bubble is:

$$(1 - \delta_B - \alpha_w \delta_B) u_s = \alpha_w \delta_B v_B \tag{2.99}$$

implying that:

$$u_s = \frac{\alpha_w \delta_B v_B}{1 - \delta_B - \alpha_w \delta_B} \tag{2.100}$$

The total gas flow rate can be represented as a sum of gas flow rate in emulsion and gas flow rate in bubbles and in the "wake":

$$v_f = (1 - \delta_B - \alpha_w \delta_B)\varepsilon_{mf} u_e + (\delta_B + \alpha_w \delta_B \varepsilon_{mf}) v_B \tag{2.101}$$

Using relations (2.98), (2.99) and (2.101), the bubble rise velocity can be expressed in terms of fluidization velocity and the minimum fluidization velocity:

$$v_B = \frac{1}{\delta_B}[v_f - (1-\delta_B - \alpha_B)v_{mf}]$$ (2.102)

Since with a higher fluidization velocity, v_f is significantly higher than the second term in the brackets in (2.102), and with low fluidization velocities the volume fraction occupied by bubbles is $\delta_B = 0$, as a first approximation for both cases the result is:

$$v_B \approx \frac{v_f - v_{mf}}{\delta_B}$$ (2.103)

which is identical to the relation obtained when it was assumed that particle movement in the emulsion phase is zero.

Expressions (2.98), (2.99), and (2.102) permit one to express the emulsion gas velocity as:

$$u_e = \frac{v_{mf}}{\varepsilon_{mf}} - \left\{ \frac{\alpha_w v_f}{1-\delta_B - \alpha_w \delta_B} - \alpha_w v_{mf} \right\}$$ (2.104)

If we assume that $\alpha_w = 0.2\text{-}0.4$, $\varepsilon_{mf} = 0.5$ and $\delta_B = 0$, it appears that downward gas motion ($u_e < 0$) is possible if:

$$\frac{v_f}{v_{mf}} > 6\text{-}11$$ (2.105)

The key assumption of the two-phase model of particle motion in a fluidized bed is the model of particle exchange between "wake" and emulsion.

It is assumed that particles present in the region limited by the bubble D_B diameter and the cloud D_c diameter, enter the bubble "wake" and mix thoroughly with the "wake" particles leaving it with the same flow rate with which they enter it.

According to this assumption [130] the coefficient of interchange of solid particles from emulsion and particles from bubble "wake" can be expressed as a volume of the exchanged particles per unit of time and bubble volume:

$$K_{ew} = 3\frac{(1-\varepsilon_{mf})}{(1-\delta_B)\varepsilon_{mf}}\frac{v_{mf}}{D_B}$$ (2.106)

If the effective axial mixing coefficient of particles D_{sa}, is defined as:

$$D_{sa} = \left| \frac{\text{volume of particles moving upwards}}{\text{coeficient of interchange per unit bed volume}} \right|^2 \cdot \left| \frac{\text{velocity of particles moving upwards}}{\text{total particle volume in bed}} \right|^2$$

or:

$$D_{sa} = \frac{\delta_B \alpha_w (1-\varepsilon_{mf}) v_B}{\delta_B K_{ew} (1-\delta_B)(1-\varepsilon_{mf})} \qquad (2.107)$$

and when the expression for K_{ew} (2.106) is replaced, and (2.103) used, it appears that:

$$D_{sa} = \frac{\alpha_w^2 \varepsilon_{mf} \delta_B}{3 v_{mf}} D_B v_B^2 = \frac{\alpha_w^2 \varepsilon_{mf}}{3\delta_B v_{mf}} D_B (v_f - v_{mf})^2 \qquad (2.108)$$

Expression (2.108), with the appropriate values for bubble diameter, D_B, is in agreement with the experimental results for particle diffusion investigations [1].

In accordance with this model, the radial particle dispersion coefficient, resulting from bubble passage is given in the following expression [1]:

$$D_{sr} = \frac{3}{16} \frac{\delta_B}{1-\delta_B} \frac{v_{mf} D_B}{\varepsilon_{mf}} \qquad (2.109)$$

Values obtained with this expression agree with the results in Fig. 2.28 [18].

Dispersion and mixing of gases in fluidized beds, especially for that part of gas which flows through the emulsion phase, are very important for chemical processes, both gas-gas processes and gas-solid particle processes, e. g., combustion. If we assume, for the sake of this discussion, that gas which crosses the fluidized bed in bubbles does not participate in chemical reactions, then gas mixing in the emulsion phase is decisive for such chemical processes. Combustion of volatiles (which are rapidly released during combustion of geologically young coals in fluidized bed), i. e., their distribution in the bed volume and opportunity to contact oxygen and mix with it, depends primarily on the process of gas mixing in the emulsion phase. It should, however, always be kept in mind that close relationship exists between gas and particle mixing.

Experimental studies and models used to describe gas mixing processes are similar to those for particle mixing. Non-stationary and stationary experiments for labelled gas dispersion, measurement of its concentration or residence time in the bed, provided approximate data on the order of magnitude of the effective gas diffusion coefficient in both radial and axial directions [1, 98, 99].

The measured apparent diffusion coefficient for macroscopic gas mixing in the axial direction in a fluidized bed, D_g, ranges from 0.1-1 m^2/s for beds of 0.1 m to 10 m. Radial diffusion coefficients are by contrast a few orders of magnitude lower, 10^{-3}-10^{-4} m^2/s [99].

Information on gas mixing in fluidized beds is scarce and varies a great deal. As in the case of particle mixing, the nature of the influence of the basic parameters of fluidized bed – fluidization velocity, particle size and bed dimensions – have not been reliably determined. However, it has been determined, that mixing is more intense in larger plants and for higher fluidization velocities. It should also be kept in mind that most experiments were carried out with particles of group A according to Geldart's classification.

Although gas mixing is subjected partially to particle motion, owing to the effect of bubbles and obvious gas exchange between bubbles and the emulsion phase, gas mixing coefficients in fluidized beds are somewhat higher than effective particle mixing coefficients. Ratios of these coefficients are given in Fig. 2.36 [99].

All available knowledge and facts on processes in fluidized bed suggest that classical modes of description pertinent to monophasic, homogenous systems are not appropriate. During gas motion through fluidized beds it is clear that two separate flows are present. Motion of gas in bubbles and motion of gas through the emulsion phase; nevertheless interaction of the two flows also occurs. Therefore, if the nature of gas flow through the bubbles and emulsion phase are known, the problem of gas mixing in fluidized beds is reduced to determination of the exchange coefficient of gas interchange between bubble gas and emulsion phase gas. Hence, it is natural that the first models of Orcutt and Davidson [131] and Kunii and Levenspiel [132], describing gas exchange between bubbles and emulsion gas,

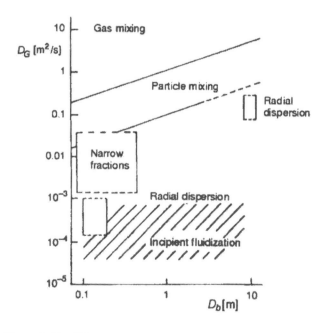

Figure 2.36. *Effective gas and particle diffusion coefficients in bubbling fluidized beds of different sizes (Reproduced by kind permission of the author Prof. J. F. Davidson from [11])*

were based on the original Davidson model. Future elaborations of the gas interchange model between bubbles and the emulsion phase, as well as definition of the
gas (mass) transfer coefficient, should take into account the nature of gas motion
which has been adopted for both bubble and emulsion phases. The magnitude and
character of the mass transfer coefficient depends on these assumptions. Models
proposed by Davidson, and Kunii and Levenspiel were proposed for fluidized beds
with fast bubbles and intensive mixing in the emulsion phase ($v_f > 2v_{mf}$, $v_B > 5v_{mf}$),
where one can assume for bubble phase that there is "plug-flow" (i. e., no circulation within the bubble, and gas velocity in the bubble is constant), while the emulsion phase is characterized by a state of "ideal" mixing (i. e., all parameters in the
emulsion phase are distributed evenly). The gas exchange coefficient between the
bubble and emulsion phase can be defined differently, and can be calculated per
unit volume or per unit area.

Gas interchange between bubbles and the emulsion phase, according to
Davidson's, and Kunii and Levenspiel's suggestions about flow regimes in both
phases, can be expressed by the following equations, depending on the type of volume used in the definition:

(a) defined per unit volume of the bubble, V_B,

$$-\frac{1}{V_B}\frac{dN_{AB}}{dt} = -v_B\frac{dC_{AB}}{dl} = (K_{Be})_B\,(C_{AB} - C_{Ae}) \qquad (2.110)$$

(b) defined per unit volume of the bed, V_b,

$$-\frac{1}{V_b}\frac{dN_{AB}}{dt} = -(v_f - v_{mf})\frac{dC_{AB}}{dl} = (K_{Be})_b\,(C_{AB} - C_{Ae}) \qquad (2.111)$$

and

(c) defined per unit volume of the emulsion phase, $V_e = (1 - \delta_B)V_b$,

$$-\frac{1}{V_e}\frac{dN_{AB}}{dt} = -(v_f - v_{mf})\frac{H_b}{H_{mf}} = (K_{Be})_e\,(C_{AB} - C_{Ae}) \qquad (2.112)$$

There are some obvious relationships:

$$(K_{Be})_b = \delta_B\,(K_{Be})_B = (1-\delta_B)(K_B)_e \qquad (2.113)$$

and

$$1-\delta_B = \frac{H_{mf}}{H_b}, \quad \frac{\delta_B}{1-\delta_B} = \frac{v_f - v_{mf}}{v_B} \qquad (2.114)$$

If the mass transfer coefficient is defined per unit of exchange area (surface
of bubble or external surface of cloud), the mass transfer equation is:

$$-\frac{dN_{AB}}{dt} = -\frac{d(V_B C_{AB})}{dt} = S_B (k_{Be})_B (C_{AB} - C_{Ae}) \qquad (2.115)$$

where the obvious link is present:

$$V_B (K_{Be})_B = S_B (k_{Be})_B \qquad (2.116)$$

For description of mass transfer between bubble and emulsion phase, some nondimensional parameters are also used:

X_B – number showing how many times the quantity of gas in a bubble is exchanged during passage of the bubble through the bed, per unit volume of the bubble:

$$X_B = \frac{(K_{Be})_B}{v_B / H_b} \qquad (2.117)$$

or:

X_b – number showing how many times the quantity of gas in a bubble has been exchanged during passage of the bubble through the bed, per unit volume of the bed:

$$X_b = \frac{(K_{Be})_b}{v_f / H_b} = \frac{(K_{Be})}{v_{mf} / H_{mf}} \qquad (2.118)$$

An obvious relation exists, implying (2.103):

$$X_b = \left[1 - (1 - \delta_B) \frac{v_{mf}}{v_f} \right] X_B \approx \left[1 - \frac{v_{mf}}{v_f} \right] X_B \approx X_B \qquad (2.119)$$

In his papers Davidson [79, 131] considered gas exchange between gas in a bubble and gas in the emulsion phase assuming that no difference between these gases is present (classical two-phase model). It was also assumed that mass transfer is realized by convective gas flow through the bubble and by molecular diffusion. The total amount of exchanged gas is the result of the sum of these processes:

$$Q_b = (q_b + k_{Be} S_{Be}) \qquad (2.120)$$

Mass flow rate q_B is defined by expression (2.83) according to Davidson's model, while the mass transfer coefficient by molecular diffusion is calculated according to analogy with gas diffusion for a bubble, rising in a fluid:

$$k_{Be} = 0.975 D_G^{0.5} \left[\frac{g}{D_B} \right]^{-0.25} \qquad (2.121)$$

The mass transfer coefficient between bubble and emulsion phase, defined with eq. (2.110), can then be expressed as:

$$(K_{Be})_B = \frac{Q_B}{V_B} = 4.5\,(\frac{v_{mf}}{D_B}) + 5.85\left(\frac{D_G\,g^{0.5}}{D_B^{2.5}}\right)^{0.5} \qquad (2.122)$$

The second classical model of Partridge and Rowe [133] assumes that gas in the bubble and cloud are thoroughly mixed (have the same properties) and considers mass transfer at the interface between cloud and the emulsion phase: the mass transfer coefficient is calculated according to analogy with mass transfer on the surface of a liquid droplet rising in another liquid. The empirical correlation used for the calculations is:

$$\mathrm{Sh}_c = \frac{k_{Sc}D_c}{D_G} = 2 + 0.69\mathrm{Sc}^{1/3}\,\mathrm{Re}_c^{1/2} \qquad (2.123)$$

Then, the mass transfer coefficient per unit of cloud volume is

$$K_{ce} = \frac{k_{ce}\pi D_c^2 \varepsilon}{V_c} = \frac{3.9\,\varepsilon\,D_G\,\mathrm{Sh}_c}{V_c^{2/3}} \qquad (2.124)$$

The model of Kunii and Levenspiel [1, 132] is a combination of the previous two models, introducing three phases: bubble, bubble cloud and emulsion. The model incorporates two processes: (1) gas exchange between bubble and cloud, strictly complying with Davidson's model; and (2) gas exchange between cloud and the emulsion phase, assuming that the process is analogous to mass transfer from a vertical cylinder of the same height and diameter as the spherical cloud [1].

In the case when two resistances to mass transfer between bubble and emulsion are lined up in sequence, the mass transfer coefficient in equation (2.110) is equal to:

$$\frac{1}{K_{Be}} = \frac{1}{K_{Bc}} + \frac{1}{K_{ce}} \qquad (2.125)$$

If the mass transfer coefficients in relation (2.125) are defined per unit volume of bubble, V_B, then the exchange coefficient between bubble and cloud $(K_{Bc})_B$ has to be calculated according to Davidson's formula (2.122), and the exchange coefficient between cloud and emulsion according to:

$$(K_{ce})_B \approx 6.78\left(\frac{\varepsilon_{mf}D_{Ge}\,v_B}{D_B^3}\right)^{1/2} \qquad (2.126)$$

where the effective gas diffusion coefficient in emulsion, D_{Ge}, is taken to be between $\varepsilon_{mf}D_G$ and D_G.

Experimental verification [1, 12] failed to clarify the issue completely, i. e., which of the classical models, or subsequently developed models [2, 12] are better suited to real processes. Models preferring the convective mass transfer processes

(Davidson's) yield excessively large values for apparent overall coefficients of gas interchange, $(K_{Be})_B$. Models implying that the mass transfer process is controlled by molecular diffusion (Partridge and Rowe, Kunii and Levenspiel) yield overall coefficients of gas interchange which are far below the experimentally obtained ones [12].

In [12], Sit and Grace [134] used the latest experimental results to suggest the following combination of convective and diffusion processes of gas exchange between bubble and emulsion:

$$(k_{Be})_B = \frac{v_{mf}}{3} + \left(\frac{4\,\varepsilon_{mf} D_G v_B}{\pi D_B} \right)^{1/2} \tag{2.127}$$

i. e.,

$$(K_{Be})_B = 2\frac{v_{mf}}{D_B} + \frac{12}{D_B^{3/2}} \left(\frac{\varepsilon_{mf} D_G v_B}{\pi} \right)^{1/2} \tag{2.128}$$

These formulae are recommended for situations where there are numerous bubbles in a freely bubbling fluidized bed, where v_B and D_B are the mean values of these parameters along the bed height.

In all the classical models mentioned here, the basic assumption of Davidson's two-phase model of bubbling fluidized bed has been integrated a priori, i. e., emulsion is in the state of incipient fluidization. In this case, distribution of gas flow is such that all excess of gas above that necessary for incipient fluidization goes through bubbles. However, experimental evidence is gathering which suggests that gas flow through emulsion is greater than in incipient fluidization [12, 94] which may cause the discrepancy between models describing the gas exchange between bubbles and emulsion and those advocating the expressions (2.127) and (2.128) [12].

When considering fluidized bed combustion processes, it should always be kept in mind that most experimental results reported here, as well as theoretical models, have been obtained and developed for the conditions pertinent to fluidized beds of group A (Geldart's classification) solids, and that bubbling fluidized bed combustion usually takes place with group B or even group D. Besides, the influence of immersed heat transfer surfaces on hydrodynamics of fluidized bed, bed expansion, size and rise velocity of bubbles, gas and particle mixing processes can be very important [52, 135]. In industrial equipment the influence of air distribution plate design also cannot be overlooked [3, 136]. The literature provides extensive analysis of these influences and they shall not be presented here in great detail. A few short remarks, however, will be given in order to indicate the difference between fluidized beds of small (<0.5 mm) and large particles (>1 mm) [100]:

– in a large-particle fluidized bed, bubbles move more slowly than gas in the emulsion phase; their growth is defined by expression (2.90), and bubble rise velocity can be calculated according to Davidson's model,

– bed height expansion, i. e., mean void fraction, ε_b of large particle beds is relatively greater with the same v_f/v_{mf} ratio than for small particle beds,

– in gas exchange between emulsion and bubbles in the large particle beds, an important role is played by the convective mass transfer process, i. e., the "washing" of bubbles with gas from emulsion; it is, therefore, logical that gas mixing models integrating this process are more appropriate,

– according to experiments [137] the gas diffusion coefficient D_{Cie} in fluidized beds of large particles ranges between 0.005 and 0.03 m^2/s, where the major contribution is that of chaotic motion (meandering), while turbulent mixing is of a lesser intensity by an order of magnitude, and

– in fluidized beds of large particles, mixing of solid particles has the same nature as that in small-particle beds, but there are insufficient data to substitute for the information for group A bed materials which have been accumulated for years.

These few comments highlight a general lack of experimental studies of large-particle fluidized bed hydrodynamics which could be used to designing models adjusted to a fluidized bed of large particles.

Uniform temperature in the whole volume occupied by a fluidized bed in which solid fuel combustion takes place and uniform distribution of oxygen in a bed are frequently highlighted as the basic reasons for high combustion efficiency, since in the whole bed volume the same, favorable conditions for the combustion process are present. In practice, however, in large furnaces, these favorable conditions may be difficult to achieve. Significant differences in temperatures between certain parts of the bed, and overheated zones occur which predispose the bed to ash sintering and defluidization (bed blockage), as well as interruption of the operation of the system as a whole; further a large proportion of volatiles fails to realize contact with oxygen in the bed, leaves the bed unburned and burns in the freeboard. In addition to the risk of sintering, these phenomena reduce the possibility of introduction of heat transfer surfaces in bed. The most common cause of malfunction of industrial-scale bubbling FBC plants is: poor and nonuniform fluidization due to improper selection and design of the distribution plate and insufficiently uniform distribution of fuel resulting from inadequate particle mixing of both inert material and fuel on the one hand, and inadequate mixing of gases and volatiles on the other.

Hot spots, resulting from poor mixing, occur usually near the fuel feeding points, irrespective of whether fuel is fed onto or into the bed. Near these points high concentrations of char and volatile matter occur.

High concentrations of fuel particles lead to localized high heat generation, and temperatures higher by 100-200 °C in the bed mass can occur. Insufficient air in these regions results in incomplete combustion and the production of a high percentage of CO, which presents both environmental problems, and the problem of combustion in the freeboard, to allow complete burnout of CO to CO_2. As well, a high concentration of volatiles always occurs when geologically young coals or biomass and wood wastes are burned. Volatiles cannot be easily mixed with oxygen, and tend to combust only partially in the bed, and such volatile matter rises like a plume through the bed and leaves it unconverted, as combustion usually takes place only at the boundaries of the "plume" (Fig. 2.37).

The correct and intense particle circulatory motion in fluidized bed combustion furnaces is of utmost importance for the selection of the location, mode and number of feeding points. Experience in design and exploitation of bubbling FBC boilers and furnaces suggests that feeding of small-sized fuel ($d_p < 2$-3 mm) below bed surface, close to the air distribution plate, requires one feeding point per m^2 of bed surface. In the case of coarse fuel feeding onto the bed surface, a single feeding point may suffice for 5-6 m^2 of bed surface.

The relatively small area which can be "covered" with a single feeding point results from poor lateral (radial) particle mixing in bubbling fluidized beds. An effective "covering" of a larger area can be achieved by better spreading of fuel on the bed surface or intense "dragging" of fuel particles into the bed mass by complex circulatory flows.

In order to reduce negative consequences of poor lateral mixing of fuel in fluidized beds, a series of design solutions has been realized in practice to enable more intense large-scale particle circulation. Figure 2.38 shows an oblique distribution plate, uneven air distribution, and different kinds of internals (baffles, tube bundles, etc.).

According to investigation of circulating flows of solid particles in fluidized beds, Masson [125, 128] recommends the following guidelines for fuel feeding point disposition:

– pneumatic feeding of pulverized fuel should be performed in the vicinity of the distribution plate, a few centimetres away, in the regions where circulating particle flow rises towards the bed surface; in smaller furnaces this coincides with the furnace axis,

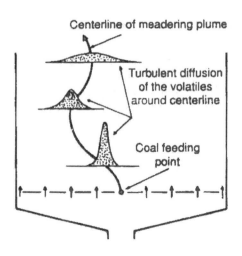

Figure 2.37.
Meandering and spreading of the volatiles plume in bubbling fluidized bed above the underbed coal feeding point

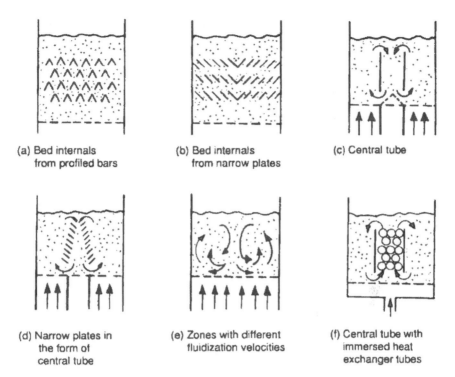

(a) Bed internals
from profiled bars

(b) Bed internals
from narrow plates

(c) Central tube

(d) Narrow plates in
the form of
central tube

(e) Zones with different
fluidization velocities

(f) Central tube with
immersed heat
exchanger tubes

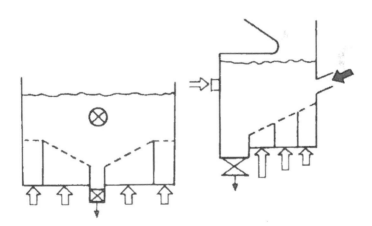

Figure 2. 38. *Different distribution plate designs and different bed internals for intensify-*
ing radial mixing and particle circulation in bubbling fluidized beds (Re-
printed from Chem. Eng. Sci. [128]. Copyright 1978, with permission from
Elsevier Science)

- fuel feeding on the bed surface (without fuel particle spreading) should be performed next to the furnace walls, as close to the bed surface as possible, so that the fuel is "dragged" by the shortest possible route into the bed depth,
- fuel feeding below the bed surface using screw feeders, through the furnace walls, should be performed either immediately below the bed surface or immediately above the distribution plate,
- simultaneously with feeding problems, organization of drainage of excess inert material (ash) should be managed, taking care to prevent fuel reaching the opening for bed material removal via the shortest route, and
- openings for the removal of excess inert bed material at the bottom of the bed should be located in the regions of upward particle motion, to prevent fresh fuel from the bed surface entering directly the removal opening.

Circulating flows of solid particles also influence distribution and erosion of immersed heat transfer surfaces [128, 135, 138-140]. Tube bundles should not be located in the regions of the main solid particle circulating flows.

2.4.   Mathematical modelling of the fluidized bed

The preceding sections have presented basic experimental facts on macroscopic features of hydrodynamics of bubbling fluidized beds, as well as basic knowledge on gas and particulate solid motion processes. Attempts to describe some of the processes with mathematical models have also been illustrated. Logically, the question that follows is: is it possible to describe fluidized bed as a complex whole in a single, self-contained, consistent mathematical model? In light of the fact that fluidized beds in industrial-scale plants are always used for chemical processes (combustion of solid fuels, among others, which is the main focus of this book), and that hydrodynamics are decisive for realization of these processes, the importance of a good, physically substantiated model of a fluidized bed becomes even more important. Further, it should be expected that according to processes taking place in fluidized beds, different aspects of hydrodynamics will have a major impact on different processes. Also, mathematical models of solid fuel combustion, which will be discussed subsequently (in Chapter 4), must be based on a good model of fluidized bed hydrodynamics.

Study of physical and chemical processes has always been associated with proposal of theoretical, mathematical models, more so with the development of numerical mathematics and powerful computers. The aim of modelling may differ. Mathematical modelling has a high educational importance; it develops an approach to problems through analysis of physical processes and their subsequent synthesis, and develops comprehension of interactions of different processes and parameters. Hence, mathematical models are indispensable for explanation and generalization of experimental results, design of experimental programs and planning of future experiments. From the practical engineering point of view, mathematical models should be used for calculations, and design and control of operation of real plants. Mathematical models also have a more practical purpose – the development of engineering and design tools which will help calculation and design of real plants. All individual tasks and problems solved in the course of mathematical model development for the calculation and design of real plants should also ideally make it possible to extrapolate experimental results obtained in laboratory-scale equipment on large-scale industrial plants, i. e., to enable scale-up.

A more complete review of engineering calculation methods based on experimental correlations for the macroscopic parameters of fluidized bed and the usage of data, constants and correlations obtained during investigation of operational plants can be found elsewhere, in the books [3, 6, 52]. In the English language literature extensive reviews and comparative analyses of mathematical models are also provided in the books [1, 2, 11, 12, 31, 121] and various review articles [141-145]. This book, however, does not provide detailed elaboration of proposed or currently well-developed mathematical models. Instead, it cover some of the basic problems of modeling, modes of managing them and the basic assumptions of the models, to the extent needed to better understand the problems of fluidized bed combustion modelling.

A physically appropriate model of hydrodynamic phenomena in fluidized beds should involve a mathematical description of the following processes: origin, growth and rise of bubbles, their actual shape, motion of solid particles, their mixing and organized directed circulating flows, gas flow in the bubbles, around the bubbles and in dense, particulate (emulsion) phase of bed, gas exchange between bubbles and the emulsion phase, merging (coalescence) and break-up of bubbles, their random distribution in the bed, different sizes of bubbles in bed cross sections and along the bed height, gas and particle motion in the bed and bursting of bubbles on irregular, turbulent bed surface, gas back-flow, etc. Obviously, without introducing some simplifying assumptions and neglecting some of the less-important processes and influencing factors (i. e., without modelling of the overall processes) such problems cannot be solved, at least not yet. This has been illustrated by difficulties encountered in attempts to describe hydrodynamic states of bubbling fluidized beds only according to basic equations of fluid mechanics and interaction of individual particles and fluid [116-120].

It was understood very early on [153] that fluidized beds, at least in an hydrodynamic sense, with a view to mathematical modelling, do not behave as simple two-phase gas-solid particle system, i. e., neither as a system through which gas flows like a "piston," nor like an ideally mixed system with uniform distribution of parameters throughout the whole volume. It was then suggested that a fluidized bed should be assumed to be a two-phase system where one phase will be composed of bubbles (i. e., gas which crosses the bed in them), and the other of emulsion (i. e., gas and particles together). The emulsion phase is in a state of incipient fluidization, and all excess of gas is incorporated in bubbles. The basic problem of fluidized bed modelling is the gas exchange between two phases. This was the focus of a large number of models, trying to solve the problem of gas exchange between bubbles and emulsion phase [147-151]. Although analyses performed so far [121, 152] indicate that these models were sufficient for calculations of slow chemical reaction in fluidized beds, they were unable to describe many important practical processes. Major shortcomings of these models imply that they do not account for the actual shape of interface between two phases (bubbles and emulsion) and instead must depend on empirical correlations for the mass transfer between phases which then prevent effective scale-up of such models.

A substantial contribution to the better (physically correct) description of hydrodynamic processes in a bubbling fluidized beds has been achieved by incorporation of Davidson's and Harrison's model of bubble motion in the general two-phase model of fluidized beds. On this basis, different models have been proposed assuming that the gas phase in bubbling fluidized beds is enclosed in a number of spherical bubbles or in bubbles of similar shape [40, 131, 132, 153].

The following assumptions are the basis of these and subsequent models that with slight alteration of details have been able to provide better descriptions of physical processes in fluidized beds:

(1) all excess gas, above that necessary for incipient fluidization, goes through the bed in bubbles; and the bed is divided into two phases: bubbles and emulsion,

(2) all bubbles in the bed are of the same size, uniformly distributed in the bed volume,

(3) the number of bubbles (i. e., volume fraction occupied by bubbles) in the bed can be calculated from the mass balance:

$$N_B V_b v_B = \delta_B v_B = v_f - v_{mf} \qquad (2.129)$$

(4) the absolute velocity of bubbles rising though the bed can be calculated by expressions (2.76) and (2.89),

(5) bed expansion beyond height at the incipient fluidization results only from the presence of bubbles and can be calculated according to expression (2.42),

(6) bubbles are spheres,

(7) mass transfer between the bubble phase and emulsion phase is achieved though the bubble surface in two ways: by gas convection (according to expression (2.82)) and by diffusion (in model [131]) according to expression (2.121),

(8) it is assumed that gas in the bubble phase is ideally mixed, and in the emulsion phase either piston flow or ideal mixing is assumed,

(9) gas in the emulsion phase flows at incipient (minimum) fluidization velocity, and

(10) the mass transfer coefficient between bubble and emulsion phase can be calculated from the expression (2.122). The above assumptions have been adopted in the model of Orcutt and Davidson [40, 131]. Subsequently, numerous variations of the so-called two-phase fluidization model have been proposed, among which the most important are [133, 153-160].

Although it has been assumed in most models that the fluidized bed is composed of two phases (bubbles and emulsion), there is a difference in the way that the bubble cloud and bubble "wake" are considered. In some models cloud and "wake" are added to the bubble [133, 155, 157], while others add cloud to the emulsion [131, 153-155], although they handle the mass transfer between bubble, cloud and "wake" and within both phases in different ways. Only Kunii and Levenspiel [132] and Fryer and Potter [153] treat the fluidized bed as a three-phase model: bubble-cloud-emulsion. Almost all models accept as a fact that bubbles are either completely particle-free or contain so few particles that their influence can be neglected.

One crucial assumption is the division of the gas flow into two parts: bubble phase and emulsion phase. Although recent experimental data suggest that gas flow rate in the emulsion phase is larger than necessary for incipient fluidization [12, 94], almost all models (used for modelling of chemical reactors) assume that $v_e = v_{mf}$; some even assume that gas flow through the emulsion equals zero, $v_e = 0$ [132, 157, 159]. This assumption is obviously critical for modelling of fluidized bed combustion, since it is believed that burning of coal char takes place only in the emulsion phase, and thus in a bed of large particles, the gas flow rate through the emulsion phase is intense. In models [132, 157] downward gas motion in the emulsion phase is also predicted, according to expression (2.98) and the model of gas mixing in emulsion phase suggested by Kunii and Levenspiel. Most other models assume piston-like gas flow in emulsion [40, 131, 133, 155, 156] and/or ideal mixing [40, 131, 153, 155]. In some models axial diffusion is also allowed [132, 153, 155]. The size and shape of bubbles and clouds in most models are considered according to the theory of Davidson and Harrison, while only [133, 156] deviate from the spherical shape of bubbles incorporating Murray's theoretical approach [87]. All models imply fully mixed gas in the bubble and cloud, while the "wake" of bubbles is added to the cloud in most models.

Substantial differences among the models in treatment of mass transfer between bubbling and emulsion phases have already been described in the previous sections. Davidson's model which does not assume any resistance to gas exchange between cloud and emulsion, is employed in models [40, 135, 155], the model of Kunii and Levenspiel in models [132, 153], and the diffusion model of Partridge and Rowe in [133]. Other models apply empirical correlations for mass transfer between bubble and emulsion, incorporating cloud into the bubble.

All two-phase models of fluidized beds based on the bubble motion model are characterized by the fact that bubble dimension is a crucial parameter. It is also obvious that the real situation cannot be represented by a single characteristic dimension, since the bed contains bubbles of various sizes, from the tiniest near the distribution plate to largest near the bed surface. This assumption of constant bubble size in the bed is the major flaw of most models [40, 131, 132, 153]. This assumption is critical for combustion models in shallow beds. Recent models try to overcome this flaw by predicting bubble growth in the course of bubble rise through bed to a certain characteristic height, according to one of the formulae given in Table 2.11. Darton's expression [46] for bubble growth is commonly used. Above this characteristic height bubbles of a constant maximal size according to formulae like (2.91) are assumed. Depending on this feature Horio and Wen [161] classified all models in three groups:

(a) models of the first order in which parameters do not depend on the bubble size [146-148, 150, 151],

(b) models of the second order in which the bubble diameter is a characteristic dimension, but mean bubble size for the whole bed is adopted [131, 132, 153], and

(c) models of the third order in which bubble growth along the bed height is taken into account [154-160, 162].

Introduction of bubble diameter into models enabled transfer of results from one system to another, scaling up to industrial-size plants, assuming that in these conditions bubble size is similar, irrespective of the overall bed size.

Although some models [162] introduce the influence of bed diameter on bubble size, the scale-up problem has not yet been solved. In order to understand this situation, it should only be noted that all models are one-dimensional, with averaging of all parameters over the bed cross section, where nonuniform distribution of bubbles in the bed is not taken into account and neither are the nature and intensity of circulating organized particle motion which depend on the bed diameter and height. These phenomena are also important for fluidized bed combustion of solid fuels, since particle mixing and fuel particle residence time in the bed depend on circulating motion of bed particles.

From the various published reviews and analyses [1, 2, 11, 12, 121, 131] and [141, 145], it can be concluded that in spite of satisfactory agreement with experimental results, more knowledge on the physics of fluidized bed processes is needed in order to achieve desired certainty in prediction of parameters for industrial-scale fluidized bed plants. This is primarily related to processes of origin and growth of bubbles and mixing of particles and gas in bed.

One possible route of model improvement is division of the fluidized bed into three areas: near the distribution plate, the main bed mass and near the free surface of the bed. Modelling of the region near the plate [12], in which gas flows in jets and where bubbles are not yet formed [2, 73, 74], is particularly important for calculation appropriate for industrial plants and for their distribution plates with bubble-caps. There are only a few [12] models proposed for the region near the distribution plate and for the region next to the bed surface. Modelling of fluidized bed combustion has, so far, paid much greater attention to regions near the bed surface and in the freeboard.

Nomenclature

a_f	entrainment constant showing the change of entrainment flux of solids F_t with the distance from the bed surface, [1/m]
A_o	specific cross section area of the distribution plate per orifice, [m²]
A_b	cross section area of the bed, [m²]
A_p	particle surface area, [m²]
ΣA_p	total surface area of all particles in the bed, [m²]
A_s	surface area of spherical particle of the volume V_p, [m²]
C_A	molar concentration of the gas component A per unit volume, [mol/m³]
C_D	particle drag coefficient
C_s	particle mass concentration, [kg/m³]
$\bar{d}$	Rosin-Rammler mean particle diameter, [m] or [mm]
d_t	opening size of standard sieves, [mm]
d_p	mean equivalent particle diameter (diameter of spherical particle), [m] or [mm]
d_{pi}	particle diameter of the fraction (class) i, [m] or [mm]
d_o	diameter of the orifice at distribution plate, [m]
d_1, d_2, d_3	dimensions of the particle of irregular shape, in three perpendicular directions, passing through the centre of gravity, [m]
D	tube diameter, [m]
$D(d_{pi})$	total mass of particles that pass through the sieve with openings of size d_{pi}, [%] or [kg/kg]
D_b	bed diameter, [m]
D_B	bubble size (diameter of sphere having the bubble volume) in fluidized bed, [m]
D_{Bo}	bubble size (diameter) at the bed surface, [m]
D_{Bx}	maximum bubble diameter in the fluidized bed, [m]
D_c, R_c	diameter of the gas cloud around bubble, [m]
D_e	effective solids mixing coefficient, [m²/s]
D_g	effective coefficient of the macroscopic gas mixing in fluidized bed, [m²/s]
D_G	molecular gas diffusion coefficient, [m²/s]
D_{Ge}	effective gas diffusion coefficient in emulsion phase, [m²/s]
D_h	hydraulic diameter, [m]
D_s	effective solids dispersion coefficient, [m²/s]
D_{sa}	effective axial dispersion coefficient of solids, [m²/s]
D_{sr}	effective radial dispersion coefficient of solids, [m²/s]
E_{rx}	elutriation rate constant for particle size fraction d_i, for $h > TDH$, [kg/m²s]
E_o	entrainment rate constant of solids at bed surface, [kg/m²s]
F_A	buoyancy force, Archimedes force, [N]
F_D	resistance force, [N]
F_g	gravity force, [N]

F_i entrainment flux of solids of size fraction d_{pi}, at distance h from the bed surface, [kg/m²s]

F_{io} entrainment flux of solids F_i, at bed surface, [kg/m²s]

$F_{i\infty}$ entrainment flux of solids F_i, for $h > TDH$, [kg/m²s]

F_t total entrainment flux of solids at distance h from the bed surface, [kg/m²s]

$F_{t\infty}$ total entrainment flux of solids F_t, for $h > TDH$, [kg/m²s]

F_{to} total entrainment flux of solids F_t, at bed surface, [kg/m²s]

g acceleration of gravity, [m/s²]

h height, [m]

h_o parameter characterizing distribution plate, [m]

h_B distance above distributor at which bubbles form, [m]

H_b(or H) bed height (fixed or fluidized), [m]

H_{FB} freeboard height, [m]

H_{mf} bed height at minimum fluidization, [m]

k_{Be} mass transfer coefficient between bubble and emulsion, based on bubble surface, [m/s]

k_{ce} mass transfer coefficient between cloud and emulsion, based on cloud surface, [m/s]

K_{Be} overall coefficient of gas interchange between bubble and emulsion, based on bubble volume, [m³/m³s]

K_{Bc} overall coefficient of gas interchange between bubble and cloud, based on bubble volume, [m³/m³s]

K_{ce} overall coefficient of gas interchange between cloud and emulsion, based on bubble volume, [m³/m³s]

K_{ew} coefficient of interchange of solids between cloud-wake region and emulsion, [m³/m³s]

l distance, [m]

Δl average distance of particle motion in fluidized bed in time period Δt, [m]

Δl_d average distance of particle chaotic motion in fluidized bed in time period Δt, [m]

Δl_c average distance of bubble induced particle motion in fluidized bed in time period Δt, [m]

L distance, [m]

L_{max} maximum jet penetration depth from the orifice in distribution plate, [m]

m_b mass of the bed (fixed or fluidized), [kg]

m_p mass of the particle, [kg]

$\dot{m}_p$ mass flux of particles per 1 m² of the cross sectional area, [kg/m²s]

n_i number of particles with size d_{pi}, $\Sigma n_i = n$

N fluidization number (= v_f / v_{mf})

N_A number of moles of the component A, [mol]

N_B number of bubbles in fluidized bed per unit volume of the bed, [1/m³]

p pressure, [N/m²]

$p(d_{pi})$ probability for particle to be in size range around d_{pi}, $\int p(d_{pi})d(d_{pi}) = 1$

Δp_b pressure drop across the bed (fixed or fluidized), [N/m²]

q_B volume gas flow rate through the bubble, [m³/s]

Q_B volume of gas interchange between bubble and emulsion phase per unit time, [m³/s]

r, θ, ψ spherical and cylindric coordinates

$R(d_{pi})$ total mass of particles that rest on the sieve with openings of size d_{pi}, [%] or [kg/kg]

R_B radius of idealized spherical bubble, [m]

R_c radius of gas cloud around idealized spherical bubble, [m]

s particle to fluid density ratio

S_B bubble surface, [m²]

S_{Be} surface of gas interchange between bubble and emulsion phase, [m²]

TDH distance above bed surface beyond which entrainment rate becomes relatively unchanging (transport disengaging height), [m]

t time, [s]

Δt time period, [s]

u_e real (interstitial) gas velocity in emulsion phase, [m/s]

u_f real (interstitial) gas velocity in fluidized bed, [m/s]

u_o^* velocity of particles ejected from the bed surface, [m/s]

u_{mf} real (interstitial) gas velocity at incipient fluidization, ($= v_{mf}/\varepsilon_{mf}$), [m/s]

u_p particle velocity, [m/s]

u_s particle velocity in emulsion phase, [m/s]

u_t free fall (terminal) velocity of an isolated particle, [m/s]

u_{te} free fall (terminal) velocity of group of particles, [m/s]

U_c velocity at which starts transition from bubbling in turbulent fluidization, [m/s]

U_k velocity at which turbulent fluidization is already established, [m/s]

U_{tr} transport velocity, at which fast fluidization is already established, [m/s]

v_B rise velocity of group of bubbles in fluidized bed, [m/s]

$v_{B\infty}$ rise velocity of isolated bubble in fluidized bed, [m/s]

v_B^* absolute rise velocity of group of bubbles in fluidized bed, [m/s]

v_{cr} mean velocity of particle circulating motion, [m/s]

v_f superficial velocity of fluidizing gas (fluidization velocity), [m/s]

v_{mb} minimum bubbling velocity, [m/s]

v_{mf} superficial gas velocity at incipient fluidization (minimum fluidization velocity), calculated for solids free bed cross section, [m/s]

v_o velocity of gas issuing from the opening of distribution plate, [m/s]

V_b bed volume (fixed, fluidized), [m³]

V_B bubble volume, [m³]

V_c volume of the gas cloud around the bubble, [m³]

V_e volume of the emulsion phase (= the bed volume at incipient fluidization), [m³]

V_g total volume of voids between particles in the bed, [m³]

V_p volume of the particle, [m³]

ΣV_p total volume of all particles in the bed, [m³]

V_w volume of "wake" associated with the bubble, [m³]

X_B, X_b nondimensional coefficients of gas interchange between bubble and emulsion phase, per bubble volume and bed volume, respectively

y_i mass fraction of the particles of the size d_{pi}, in unit mass of the loose material, $\Sigma y_i = 1$, [kg/kg]

Greek symbols

α ratio of cloud volume to bubble volume

α_w ratio of bubble "wake" volume to bubble volume

β angle of repose of the loose material

δ_B volume fraction occupied by bubbles

ε void fraction

ε_B void fraction of bubbles (=1)

ε_{mf} void fraction at incipient fluidization

ε_p void fraction of the emulsion (particulate) phase

ε_w	void fraction of the emulsion in bubble "wake"
Θ_w	angle measured from the nose of the bubble to edge of the base
λ	friction factor
μ_f	dynamic viscosity of fluid, (gas), [kg/ms]
ξ_p	pore volume per unit mass of particle
ρ_b	bulk density of fixed bed, [kg/m³]
ρ_ε	bulk density of fluidized bed, [kg/m³]
ρ_f	fluid (gas) density, [kg/m³]
ρ_{mf}	density of bed at incipient fluidization, [kg/m³]
ρ_o	density of fluidized bed just above the bed surface in the splash zone, [kg/m³]
ρ_p	particle density (including volume of the pores), [kg/m³]
ρ_s	true particle density (skeletal density), [kg/m³]
ϕ_s	particle shape factor
Φ_p, Ψ_p	velocity potential and stream function for particle flow around bubble, [m²/s²]

Dimensionless criterial numbers

$$Ar = \frac{g d_p^3 \rho_f (\rho_p - \rho_f)}{\mu_f^2}$$ Archimedes number

$$Ga = \frac{d_p^3 \rho_f^2 g}{\mu_f^2}$$ Galileo number

$$Re = \frac{u d_p}{v_p}$$ Reynolds number for particle, or Re_p

$$Re_t = \frac{u_t d_p}{v_f}$$ Reynolds number for particle based on the terminal velocity

$$Re_{mf} = \frac{d_p v_{mf}}{v_f}$$ Reynolds number for particle based on minimum fluidization velocity

$$Re_B = \frac{D_B v_B}{v_f}$$ bubble Reynolds number

$Re_{t,\varepsilon}$ Reynolds number based on the terminal velocity for group of particles,

$$Sc = \frac{\mu_f}{\rho_f D_G}$$ Schmidt number

$Sh_c = k_{cc} D_c / D_G$ Sherwood number based on the diameter of bubble cloud

References

[1] D Kunii, O Levenspiel. Fluidization Engineering. New York: R. E. Krieger Publ. Co., 1977.

[2] AP Cheremisinoff, PN Cheremisinoff. Hydrodynamics of Gas-Solids Fluidization. London: Gulf. Publ. Co. 1984.

[3] NP Muhlenov, BS Sazhin, VF Frolov, eds. Calculation of Fluidized Bed Reactors (in Russian). Leningrad: Khimiya, 1986.

[4] GL Babuha, MI Rabinovich. Hydromechanics and Heat Transfer in Polydisperse Two-Phase Flow (in Russian). Kiev: Naukovaya dumka, 1969.

[5] PG Kiseljev, ed. Handbook for Hydraulic Calculations (in Russian). Moscow: Ehnergiya, 1974.

[6] OM Todes, OB Citovich. Reactors with Coarse Particle Fluidized Beds (in Russian). Leningrad: Khimiya, 1981.

[7] S Oka. Transport of Fluids and Solid Particulate Material Through Pipes (in Serbian). Belgrade: Faculty of Traffic Engineering, 1988.

[8] GK Batchelor. An Introduction to Fluid Dynamics. Cambridge: Cambridge University Press, 1967.

[9] SL Soo. Fluid Dynamics of Multiphase Systems. Blaisdell Publ. Co., Waltham, Mass.-Toronto-London (translation in Russian). Moscow: Mir, 1971.

[10] M Leva. Fluidization. New York: McGraw-Hill, 1959.

[11] JF Davidson, R Clift, D Harrison, eds. Fluidization, 2nd ed. London: Academic Press, 1985.

[12] JG Yates. Fundamentals of Fluidized-Bed Chemical Processes. London: Butterworths, 1983.

[13] FA Zenz. Regimes of fluidized behavior. In: JF Davidson, D Harrison, eds. Fluidization, 1st ed. London: Academic Press, 1971, pp. 1-23

[14] FA Zenz, DF Othmer. Fluidization and Fluid-Particle Systems. New York: Reinhold, 1960.

[15] W Reisner, M von Elsenhart Rothe. Bins and Bunkers for Handling Bulk Materials – Practical Design and Techniques. Claustal (Germany): Trans Tech Publications, 1971.

[16] N Čatipović G Jovanović. Progress in fluidization technology and its influence on the development of other technologies (in Serbian). Hemijska industrija 6:151-156, 1980.

[17] AM Squires, Applications of fluidized beds in coal technology. Proceedings of International Symposium Future Energy Production – Heat and Mass Transfer Problems, Dubrovnik (Yugoslavia), 1975.

[18] J Yerushalmi, A Avidan. High-velocity fluidization. In: JF Davidson, R Clift, D Harrison, eds. Fluidization, 2nd ed. London: Academic Press, 1985, pp. 226-292

[19] J Yerushalmi, DH Turner, AM Squires. The fast fluidized bed. Ind. Eng. Chem., Process Design Development 1:47-53, 1986.

[20] J Yerushalmi, NT Cancurt. Further studies of the regimes of fluidization. Powder Technology Vol. 24, 2:187-205, 1979.

[21] H Takeuchi, T Harama, T Chiba, J Biwas, LS Leung. A quantitative definition and flow regime diagram for fast fluidization. Powder Technology Vol. 47, 2:195-199, 1986.

[22] L Stromberg, H Kobro, et al. The fast fluidized bed – a true multifuel boiler. Proceed-
      ings of 8th International Conference on Fluidized-Bed Combustion, Houston, 1985,
      Vol. 2, pp. 415-422.

[23] D Wiegan. Technical and economical status of FBC in West-Germany. Presented at
      International Conference on Coal Combustion, Copenhagen, pp. 1-33, 1986.

[24] Y Yosufi, G Gau. Aérodynamique de l'écoulement vertical de suspensions
      concentrees gas-solids – I. Régimes d'écoulement et stabilité aérodynamique.
      Chem. Eng. Sci. 9:1939-1946, 1974.

[25] Y Yosufi, G Gau. Aérodynamique de l'écoulement vertical de suspensions
      concentrees gas-solids – II. Chute de pression et vitesse relative gas-solids. Chem.
      Eng. Sci. 9:1947-1956, 1974.

[26] SR Sanker, TN Smith. Slip velocities in pneumatic transport. Part I. Powder Tech-
      nology Vol. 47, 2:167-177, 1986.

[27] SR Sanker, TN Smith. Slip velocities in pneumatic transport. Part II. Powder Tech-
      nology Vol. 47, 2:179-194, 1986.

[28] AM Manaker, PB West. TVA orders 160 MW_e demonstration AFBC power station.
      Modern Power System Dec/Jan:59-65, 1984/85.

[29] D Geldart. Types of gas fluidization. Powder Technology Vol. 7, 2:285-292, 1973.

[30] J Baeyens, D Geldart. An investigation into slugging fluidized beds. Chem. Eng.
      Sci. 1:255-265, 1974.

[31] JF Davidson, D Harrison, eds. Fluidization, 1st ed. London: Academic Press, 1971.

[32] D Dakić, B Grubor, S Oka. Investigation of incipient fluidization and expansion of
      coarse particle bed (in Russian). In: SS Kutateladze, S Oka, eds. Transport Processes
      in High Temperature and Chemically Reacting Turbulent Flows. Novosibirsk
      (U.S.S.R.): Siberian Branch of the U.S.S.R. Academy of Sciences, 1982, pp. 76-90.

[33] WS Grewal, SC Saxena. Comparison of commonly used relations for minimum
      fluidization velocity of small solid particles. Powder Technology Vol. 26,
      2:229-238, 1980.

[34] CY Wen, YW Yu. Mechanics of fluidization. Chem. Eng. Progr. Symp. Series
      100-125, 1966.

[35] SC Saxena, GS Vogel. The measurements of incipient fluidization velocities in a bed
      of coarse dolomite at temperature and pressure. Trans. Inst. Chem. Eng. 3:184-195,
      1977.

[36] P Babu, B Shah, A Talwalker. Fluidization correlations for coal gasification materi-
      als – Minimum fluidization velocity and fluidized bed expansion ratio. AIChE
      Symp. Series, 1978, pp. 176-186.

[37] PH Pinchbeck, F Popper. Critical and terminal velocities in fluidization. Chem. Eng.
      Sci. Vol. 6, 1:57-67, 1956.

[38] KF Godard, JF Richardson. Use of slow speed stirring to initiate particulate
      fluidization. Chem. Eng. Sci. 1:194-195, 1969.

[39] KF Godard, JF Richardson. Correlation of data for minimum fluidizing velocity and bed expansion in particulate fluidized systems. Chem. Eng. Sci. 2:363-371, 1969.

[40] JF Davidson, D Harrison. Fluidized Particles. Cambridge: Cambridge University Press, 1963.

[41] AM Xavier, DA Lewis, JF Davidson. The expansion of bubbling fluidized beds. Trans. Inst. Chem. Eng. 4:274-280, 1978.

[42] DJ Nicklin. Two-phase bubble flow. Chem. Eng. Sci. 17:693-706, 1962.

[43] JF Richardson, WN Zaki. Sedimentation and fluidization. Trans. Inst. Chem. Eng. 1:35-47, 1954.

[44] JF Richardson, RA Meikle. Sedimentation and fluidization. Part III. The sedimentation of uniform fine particles and of two-component mixtures of solids. Trans. Inst. Chem. Eng. 2:348-359, 1961.

[45] H Steinour. Particle size analyses of the Fe powders in powder metallurgy. Ind. Eng. Chem. 3:618-625, 1944.

[46] EW Lewis, EW Bowermann. Fluidization of solid particles in liquids. Chem. Eng. Progr. 3:603-610, 1952.

[47] A Bar-Kohen, LR Glicksman, RW Hudges. Semiempirical prediction of bubble diameter in gas fluidized beds. Int. J. Multiphase Flow 1:101-114, 1981.

[48] RC Darton, RD La Nauze, JF Davidson, D Harrison. Bubble growth due to coalescence in fluidized beds. Trans. Inst. Chem. Eng. 4:274-280, 1977.

[49] ME Aerov, OM Todes. Hydrodynamic and thermodynamic basis of the bubbling fluidized bed reactor operation (in Russian). Leningrad: Khimiya, 1968.

[50] SS Zabrodski. High-temperature reactors with bubbling fluidized beds (in Russian). Moscow: Ehnergiya, 1971.

[51] VG Kobulov, OM Todes. Structure of a fluidized solids layer in relation to the type of gas distribution. Journal of Applied Chemistry 5:1075-1084, 1966.

[52] VA Baskakov, BV Berg, AF Rizkov, NF FilipovskiJ. Heat- and Mass-transfer processes in bubbling fluidized beds (in Russian). Moscow: Metalurgiya, 1978.

[53] AT Basov, BN Morhevka, GH Melik-Ahnazarov, et al. Investigation of the structure of nonuniform bubbling fluidized beds (in Russian). Himicheskaya promyshlennost 8:619-622, 1968.

[54] WK Lewis, EG Gillibrand, PM Lang. Entrainment from fluidized beds. Chem Eng. Progress Symp. Series 38:65-77, 1962.

[55] CY Wen, LH Chen. Fluidized bed freeboard phenomena and elutriation. AIChE Journal 1:117-128, 1982.

[56] S Yagi, T Aochi. Elutriation of particles from a batch fluidized bed. Presented at the Soc. of Chem. Eng. (Japan) Spring Meeting, 1955.

[57] FA Zenz, NA Weil. A theoretical-empirical approach to the mechanism of particle entrainment from fluidized beds. AIChE Journal 4:472-479, 1958.

[58] CY Wen, RF Hashinger. Elutriation of solid particle from a dense phase fluidized bed. AIChE Journal 2:220-232, 1960.

[59] I Tanaka, H Shinohara, H Hirosue, Y Tanaka. Elutriation of fines from fluidized bed. J. of Chem. Eng. of Japan 1:51-65, 1972.

[60] G Geldart, et. al. The effect of fines on entrainment from gas fluidized beds. Transactions of Inst. Chem. Eng. 4:269-275, 1979.

[61] M Colakyan, N Čatipović, G Jovanović, T Fitzgerald. Elutriation from a large particle fluidized bed with and without immersed heat transfer tubes. Presentad at the AIChE 72 Meeting, San Francisco, 1979.

[62] L Lin, JT Sears, CY Wen. Elutriation and attrition of char from a large fluidized bed. Powder Technology 1:105-116, 1980.

[63] PJ Buysman, GAL Peersman. In: AAH Drinkenburg, ed. Proceedings of International Symposium on Fluidization, Eindhoven (the Netherlands). Amsterdam: Neth. Univ. Press. 1967, pp. 38-43.

[64] K Rietema. Application of mechanical stress theory to fluidization. In: AAH Drinkenburg, ed. Proceedings of International Symposium on Fluidization, Eindhoven (the Netherlands). Amsterdam: Neth. Univ. Press. 1967, pp. 154-167.

[65] R Jackson. The mechanics of fluidized beds. Part 1. The stability of the state of uniform fluidization. Part 2. The motion of fully developed bubbles. Trans. Inst. Chem. Eng. 1:13-35, 1963.

[66] RH Pigford, T Baron. Hydrodynamic stability of a fluidized bed. Ind. Eng. Chem. Fundamentals 1:81-89, 1965.

[67] JA Murray. On the mathematics of fluidization. Part 1. Fundamental equations and wave propagation. J. Fluid Mech. Vol. 21, Part 3:465-493, 1965.

[68] O Molerus. The hydrodynamic stability of the fluidized bed. In: AAH Drinkenburg, ed. Proceedings of International Symposium on Fluidization, Eindhoven (the Netherlands). Amsterdam: Neth. Univ. Press. 1967, pp.134-143.

[69] HC Simpson, BW Rodger. The fluidization of light solids by gases under pressure and heavy solids by water. Chem. Eng. Sci. 1:179-184, 1961.

[70] D Harrison, JF Davidson, JW De Kock. On the nature of aggregative and particulate fluidization. Trans. Inst. Chem. Eng. 2:202-215, 1961.

[71] GB Wallis. Two-phase flow aspects of pool boiling from a horizontal surface. Proceedings of Symposium of Interaction of Fluids and Particles. London: Institute of Mechanical Engineering, 1962, pp. 9-15

[72] J Verloop, PM Heertjes. On the origin of bubbles in gas-fluidized beds. Chem. Eng. Sci. 5:1101-1107, 1974.

[73] L Massimilla. Gas jets in fluidized beds. In: JF Davidson, R Clift, D Harrison, eds. Fluidization, 2nd ed. London: Academic Pres, 1985, pp. 133-172

[74] CY Wen, et al. Jetting phenomena and dead zone formation on fluidized bed distributor. Proceedings of 2nd Pacific Chem. Eng. Conference, New York: AIChE, 1977, Vol. 2, pp. 1182-1197.

[75]   WC Yang, DL Keairns. Design and operating parameters for a fluidized bed agglom-
       erating combustor/gasifier. In: JR Grace, JM Matsen, eds. Fluidization. New York:
       Plenum Press, 1978, pp. 305-314.

[76]   PN Rowe, DJ Everett. Fluidized bed bubbles viewed by X-rays, Part I, Part II, Part
       III. Trans. Inst. Chem. Eng. 1:42-60, 1972.

[77]   D Kunii, K Yoshida, I Hiraki. The behavior of freely bubbling fluidized beds. Pro-
       ceedings of International Symposium on Fluidization, Eindhoven (the Nether-
       lands). Amsterdam: Neth. Univ. Press. 1967, pp. 243-256.

[78]   PN Rowe. Prediction of bubble size in a gas fluidized bed. Chem. Eng. Sci.
       4:285-288, 1976.

[79]   JF Davidson, D Harrison. Fluidized Particles. Cambridge: Cambridge University
       Press, 1963.

[80]   PN Rowe, BA Partridge. An X-ray study of bubbles in fluidized beds. Trans. Inst.
       Chem. Eng. 3:157-190, 1965.

[81]   RM Devis, GI Taylor. The mechanics of large bubbles rising through extended liq-
       uids in tubes. Proc. Roy. Soc., A 200:375-383, 1950.

[82]   H Reuter. Rate of bubbling rise in the gas-solid fluidized bed. Chem. Ingr. Tech.
       10:1062-1066, 1965.

[83]   PSB Stewart. Isolated bubbles in fluidized beds – Theory and experiment. Trans.
       Inst. Chem. Eng. 2:T60-T66, 1968.

[84]   PN Rowe. Gas-solids reaction in a fluidized bed. Chem. Eng. Progr. 3:75-83, 1964.

[85]   R Collins. The rise velocity of Davidson's fluidization bubbles. Chem. Eng. Sci.
       8:788-789, 1965.

[86]   R Collins. An extension of Davidson's theory of bubbles in fluidized beds. Chem.
       Eng. Sci. 8:747-755, 1965.

[87]   JD Murray. On the mathematics of fluidization. Part 2. Steady motion of fully devel-
       oped bubbles. J. Fluid Mech. Vol. 22, Part 1:57-80, 1965.

[88]   PN Rowe, BA Partridge, E Lyall. Cloud formation around bubbles in gas fluidized
       beds. Chem. Eng. Sci. 12:973-985, 1964.

[89]   R Clift, JR Grace, ME Weber. Bubbles, Drops and Particles, New York: Academic
       Press, 1978.

[90]   RR Cranfield, D Geldart. Large particle fluidization. Chem. Eng. Sci. 4:935-947,
       1974.

[91]   S Mori, CY Wen. Estimation of bubble diameter in gaseous fluidized beds. AIChE
       Journal 1:109-115, 1975.

[92]   J Werther. Influence of distributor design on bubble characteristics in large diameter
       gas fluidized beds. In: JF Davidson, DL Kerains, eds. Fluidization. Cambridge:
       Cambridge University Press, 1978, pp. 7-12.

[93]   R Clift, JR Grace. Continuous bubbling and slugging. In: JF Davidson, R Clift, D
       Harrison, eds. Fluidization, 2nd ed. London: Academic Press, 1985, pp. 73-132.

[94]   PN Rowe, CXR Yacono. The bubbling behavior of fine powders when fluidized. Chem. Eng. Sci. 12:1179-1192, 1976.

[95]   D Harrison, JR Grace. Fluidized beds with internal baffles. In: JF Davidson, D Harrison eds. Fluidization, 1ˢᵗ ed. London: Academic Press, 1971, pp. 599-626.

[96]   O Sitnai, AB Whitehead. Immersed tubes and other internals. In: JF Davidson, R Clift, D Harrison, eds. Fluidization, 2ⁿᵈ ed. London: Academic Press, 1985, pp. 473-493.

[97]   A Bar-Cohen, RW Hughes. Predicting asymptotic bubble size in tube/bubble interactions within fluidized beds. Israel J. of Technology 137-145, 1980.

[98]   OE Potter. Mixing. In: JF Davidson, D Harrison eds. Fluidization, 1ˢᵗ ed. London: Academic Press, 1971, pp. 293-381.

[99]   JJ van Deemter. Mixing. In: JF Davidson, R Clift, D Harrison, eds. Fluidization, 2ⁿᵈ ed. London: Academic Press, 1985, pp. 331-355.

[100] TJ Fitzgerald. Coarse particle systems. In: JF Davidson, R Clift, D Harrison, eds. Fluidization, 2ⁿᵈ ed. London: Academic Press, 1985, pp. 413-436.

[101] OM Todes, AK Bondareva. Conduction and heat transfer in bubbling fluidized beds (in Russian). Journal of Engineering Physics 2:105-110, 1960.

[102] NB Kondukov, AN Kornilaev, et al. Investigation of particle motion in bubbling fluidized beds. I. Experimental methods and particle trajectories (in Russian). Journal of Engineering Physics 7:13-18, 1963. II. Particle kinematic (in Russian). Journal of Engineering Physics 7:25-32, 1960.

[103] L Massimilla, JW Westwater. Photographic study of solid-gas fluidization. AIChE Journal 2:134-142, 1960.

[104] M Leva, M Grummer. Correlation of solids turnover in fluidized systems – Relation to heat transfer. Chem. Eng. Progr. 2:307-313, 1952.

[105] SR Tailbi, MAT Cocquerel. Solid mixing in fluidized beds. Trans. Inst. Chem. Eng. 2:195-201, 1961.

[106] WK Lewis, ER Gilliland, H Girouard. Entrainment from fluidized beds. Chem. Eng. Progr. Symp. Series 38:65-87, 1962.

[107] Y Mori, K Nakamura. Solid mixing in fluidized bed. Kakagu Kogaku 11:868-875, 1965.

[108] H Littman. Solid mixing in straight and tapered fluidized beds. AIChE Journal 6:924-929, 1964.

[109] T Hayakawa, W Graham, GL Osberg. Resistance-probe method for determining local solid particle mixing rates in a batch fluidized bed. Canad. J. Chem. Eng. 2:99-103, 1964.

[110] G Jineski, L Teoreanu, E Ruckenstein. Mixing of solid particles in a fluidized bed. Canad. J. Chem. Eng. 2:73-76, 1966.

[111] PN Rowe, BA Partridge, AG Cheney, et al. The mechanisms of solid mixing in fluidized beds. Trans. Inst. Chem. Eng. 9:271-283, 1965.

[112]   JH De Groot. In: AAH Drinkenburg, ed. Proceedings of International Symposium on Fluidization, Eindhoven (the Netherlands). Amsterdam: Neth. Univ. Press. 1967, pp. 348-356.

[113]   IL Reznikov, et al. Mixing coefficient of karnalit in industrial size furnace with bubbling bed (in Russian). Himicheskaya promyshlennost 4:300-302, 1972.

[114]   L Massimilla, S Bracale. Il mescolamento della fase solido nei sistemi solido-gas fluidizzati. Liberi e Frenati, Ric. Sci. 1509-1525, 1957.

[115]   MR Al-Dibuni, J Garside. Particle mixing and classification in liquid fluidized beds. Trans. Inst. Chem. Eng. 2:94-103, 1979.

[116]   VG Levich, VP Myasnikov. Kinetic theory of fluidized state (in Russian). Himicheskaya promyshlennost 6:404-408, 1966.

[117]   VA Cibarov. Kinetic model of fluidized bed (in Russian). Vestnik LGU, Ser. mat. meh. astron. 13:106-111, 1975.

[118]   R Jackson. Hydrodynamic stability of fluid-particle systems. In: JF Davidson, R Clift, D Harrison, eds. Fluidization, 2nd ed. London: Academic Press, 1985, pp. 65-119.

[119]   AI Tamarin, YS Teplickij. On hydrodynamic of fluidized bed (in Russian). Journal of Engineering Physics 6:1005-1111, 1971.

[120]   YA Buyevich. Statistical hydrodynamics of disperse systems. Part 3. Pseudo-turbulent structure of homogenous suspensions. J. Fluid Mech., Vol. 56, Part 2, 313-336, 1972.

[121]   WPM van Swaaij. Chemical reactors. In: JF Davidson, R Clift, D Harrison, eds. Fluidization, 2nd ed. London: Academic Press, 1985, pp. 595-629.

[122]   J Werther. Influence of the bed diameter on the hydrodynamics of gas fluidized beds. AIChE Symp. Series 3-62, 1974.

[123]   J Werther. Bubbles in gas fluidized beds. Part I, Part II. Trans. Inst. Chem. Eng. 2:149-169, 1974.

[124]   PN Rowe, HA Masson. Interaction of bubbles with probes in gas fluidized beds. Trans. Inst. Chem. Eng. 3:177-185, 1981.

[125]   HA Masson. Fuel circulation and segregation in FBC. In: M Radovanović ed. Fluidized Bed Combustion. New York: Hemisphere Publ. Co., 1986, pp. 185-250.

[126]   M Radovanović ed. Fluidized Bed Combustion. New York: Hemisphere Publ. Co., 1986.

[127]   D Geldart, MJ Rhodes. From minimum fluidization to pneumatic transport – A critical review of the hydrodynamics. Proceedings of 1st International Conference CFB, Halifax (Canada). 1985, pp. 21-31.

[128]   HA Masson. Solid circulation studies in a gas-solid-fluid bed. Chem. Eng. Sci. 5:621-623, 1978.

[129]   PN Rowe. Estimation of solids circulation rate in a bubbling fluidized beds. Chem. Eng. Sci. 3:979-980, 1973.

[130]   JC Yoshida, D Kunii. Stimulus and response of gas concentration in bubbling fluidized bed. J. of Chem. Eng. of Japan 11-23, 1968.

[131]   JC Orcutt, JF Davidson, RL Pigford. Reaction time distribution in fluidized catalytic reactors. Chem. Eng. Prog. Symp. Series 38:1-17, 1962.

[132]   D Kunii, O Levenspiel. Bubbling bed model. Model for flow of gas through a fluidized bed. Ind. Eng. Chem. Fundamentals 2:446-452, 1968.

[133]   BA Partridge, PN Rowe. Chemical reaction in a bubbling gas-fluidized beds. Trans. Inst. Chem. Eng 335-347, 1966.

[134]   SP Sit, JR Grace. Effect of bubble interaction on interphase mass transfer in gas fluidized beds. Chem. Eng. Sci. 2:327-335, 1981.

[135]   O Sitnai, AB Whitehead. Immersed tubes and other internals. In: JF Davidson, R Clift, D Harrison, eds. Fluidization, 2nd ed. London: Academic Press, 1985. pp. 473-493

[136]   AB Whitehead. Distributor characteristics and bed properties. In: JF Davidson, R Clift, D Harrison, eds. Fluidization, 2nd ed. London: Academic Press, 1985, pp. 173-199

[137]   G Jovanović, NM Čatipović, TJ Fitzgerald, O Levenspiel. In: JR Grace, JM Matsen, eds. Fluidization. New York: Plenum Press, 1980, pp. 225-237

[138]   M Trifunović. Corrosion and erosion problems in fluidized bed furnaces – Review of the operating experience (in Serbian). Report of the Institute of Nuclear Sciences Boris Kidrič, Vinča, Belgrade, IBK-ITE-665, 1987.

[139]   J Stringer. Current information on metal wastage in fluidized bed combustors. Proceedings of 9th International FBC Conference, Boston, 1987, pp. 685-695.

[140]   B Leckner, F Johnsson, S Andersson. Erosion in fluidized beds-influence of bubbles. Presented at EPRI Fluidized-Bed Materials Workshop, Port Hawkesbury (Canada), 1985.

[141]   CY Wen, LH Chen. Flow modeling concepts of fluidized beds. In: NP Cheremisinoff, R Gupta, eds. Handbook of fluids in motion. Ann Arbor: Sci/ Butterworths Publ., 1983.

[142]   JR Grace. Fluidized bed reactor modelling: An Overview. American Chem. Soc. Washington DC. AChS Symposium Ser. 168:3-37, 1981.

[143]   PH Rowe. Fluidization studies carried out at London university (in Russian). Izvestiya AN U.S.S.R., Mekhanika zhidkosti i gaza 6:50-60, 1972.

[144]   DL Pyle, BRE Jones. Fluidized bed reactors. A Review Advan. Chem. Ser. Proceedings of 1st International Conference Reaction Eng. Washington, D. C., 1970, Washington, 1972, Vol. 109, pp. 106-130

[145]   JG Yates. Fluidized bed reactors. The Chem. Engineer 303:671-677, 1975.

[146]   RD Toomey, HF Johnstone. Gaseous fluidization of solid particles. Chem. Eng. Progr. 2:220-231, 1952.

[147]   CY Shen, HF Johnstone. Gas-solids contact in fluidized beds. AIChE Journal 2:349-356, 1955.

[148]   JJ van Deemter. Mixing and contacting in gas-solid fluidized beds. Chem. Eng. Sci. 1:143-158, 1961.

[149]   HF Johnstone, JD Batchelor, WY Shen. Low-temperature oxidation of ammonia in fixed and fluidized beds. AIChE Journal 2:318-323, 1955.

[150]   WG May. Fluidized bed reactor studies. Chem Eng. Progr. 1:49-56, 1959.

[151]   H Kobayashi, F Arai. Effects of several factors on catalytic reaction in a fluidized bed reactors. Kakagu Kogaku 11:885-891, 1965.

[152]   PH Calderbank, FD Toor. Fluidized bed as catalytic reactor. In: JF Davidson, D Harrison eds. Fluidization, 1st ed. London: Academic Press, 1971.

[153]   C Fryer, OE Potter. Countercurrent backmixing model for fluidized bed catalytic reactors. Applicability of simplified solutions. Ind. Eng. Chem. Fundamentals 3:338-344, 1972.

[154]   T Mamuro, I Muchi. Mathematical model for fluidized-bed catalytic reactor. J. Ind. Chem. (Japan) 1:126-129, 1965.

[155]   FD Toor, PH Calderbank. Reaction kinetics in gas-fluidized catalyst beds. Part II: Mathematical models. In: AAH Drinkenburg, ed. Proceedings of International Symposium on Fluidization, Eindhoven (the Netherlands). Amsterdam: Neth. Univ. Press. 1967, pp. 373-386.

[156]   H Kobayashi, F Arai, T Chiba, Y Tanaka. Estimation of catalytic conversion in gas-fluidized beds by means of two-phase model. Effect of bed diameter. Kakagu Kogaku 3:274-280, 1969.

[157]   K Kato, CY Wen. Bubble assemblage model for fluidized bed catalytic reactors. Chem. Eng. Sci. 8:1351-1369, 1969.

[158]   S Mori, I Muchi. Theoretical analyses of catalytic reaction in fluidized bed. J. Chem. Eng. of Japan. 2:251-267, 1972.

[159]   M Horio, CY Wen. An assessment of fluidized-bed modeling. AIChE Symp. Series 61:9-21, 1977.

[160]   J Werther. Stromungsmechanische Grundlagen der Wirbelschichttechnik. Chem. Ind. Tech. 11:193-202, 1977.

[161]   M Horio, CY Wen. An assessment of fluidized-bed modeling. AIChE Symp. Series 161:9-21, 1977.

[162]   WPM van Swaaij, FJ Zuiderweg. The design of gas-solids fluidized beds – Prediction chemical conversion. Fluidization and its applications. Proceedings of International Symposium on Fluidization, Toulouse (France), 1973, pp. 454-467.

3.

HEAT AND MASS TRANSFER IN FLUIDIZED BEDS

In solid fuel fluidized bed combustion boilers 30-50% of the total generated heat is transferred to the exchanger surfaces which are in contact with the bed of inert material. Heat transfer surfaces may be tube bundles immersed into the fluidized bed, or water-tube furnace walls in contact with the bed. However, although most important, heat transfer to the heat exchangers is not the only process of heat transfer taking place in FBC boiler furnaces. Hence it is necessary to pay attention to all transfer processes that are important for fuel combustion and heat transfer with the exchanger surfaces.

3.1. Heat transfer processes in bubbling fluidized bed combustion boiler furnaces

In furnaces of FBC boilers with stationary, bubbling beds, specific conditions for heat transfer exist. Compared to the conventional boilers there are two major differences: combustion temperature is lower (800-900 °C), and solid particle concentration (not only in the fluidized bed) is much higher. So, heat transfer by radiation is a less important process in FBC boilers. In conventional boilers heat energy is transferred by two mechanisms – gas convection and radiation, while in FBC boilers three mechanisms are in effect: radiation, gas convection and heat transfer by contact of solid particles.

   Average furnace temperature in FBC boilers is not dramatically changed along the height, i. e., with the distance from the distribution plate. In the fluidized bed a temperature of 800-900 °C is maintained, depending on the ash sintering temperature and optimal conditions for SO_2 bonding with CaO. Temperature is practically uniform in the whole bed volume. Due to intensive heat transfer from the hot particles of inert bed material to the cold combustion air entering through the distribution plate, even in the immediate vicinity of the plate the local bed and gas temperatures are equal to the average bed temperature. Temperature drop in the vicinity of the furnace walls is also limited to very narrow regions. When burning fuels with high volatile content (lignites, biomass) freeboard temperature can be higher than bed temperature. In furnaces used for production of high-temperature gases, with walls covered with fire brick and other insulation materials, temperature in the freeboard can be up to 100-200 °C higher than the bed temperature. Temperature immediately above the bed in FBC boilers with water-tube walls is not much different from that in the bed. Further away from the bed's surface, temperature differences along the height and in the cross-section are much greater than in the fluidized bed. In this part of the furnace heat transfer conditions are close to those in conventional boilers.

   The greatest differences in heat transfer conditions in FBC boiler furnaces come from different particle concentrations. In the region occupied by the fluidized bed, solid particle concentration is around 10^3 kg/m^3, while at the furnace exit it is around 0.1 kg/m^3 [1]. Between these two values solid particle concentration changes exponentially, and especially large changes are in the splash zone, near the free bed surface. Because of the change in solid particle concentration, the role of any particular mechanism of heat transfer in FBC boiler furnaces changes sharply along the furnace height.

   In the fluidized bed region heat transfer by solid particles prevails, and in the freeboard, far from the bed surface the radiation mechanism prevails. One should have in mind that the glowing bed surface also radiates to the surrounding walls in the region above the bed and to the other irradiated heat transfer surfaces in the furnace. Intensity of the heat transfer process can also greatly differ along the furnace height. In fluidized beds, heat transfer coefficients to the immersed surfaces range from 250 to 700 W/m²K [1, 2]. In the freeboard heat transfer coefficients are less than 100 W/m²K [1].

   When considering heat transfer processes in fluidized bed furnaces (and that also relates to experimental data on heat transfer in the fluidized bed given in this chapter) it should be borne in mind that in FBC boilers particles of type B, according to Geldart, are usually used. Inert material particle size also ranges widely, with a significant percentage of particles smaller than 1 mm, but also with a high content of very large particles when fuels with large particles and a high content of tramp material (stones) are used.

In the FBC boiler furnace heat is transferred to tube bundles immersed in the fluidized bed and to the water-tube furnace walls. Heat transfer mechanisms in these two cases are similar, but due to different surface geometry, different character of the flow of particles in contact with these surfaces and the difference in bubble size and movement, heat transfer intensity may be different.

Heat transfer between gas and particles has a very important role in processes taking place in the fluidized bed. In the case of fluidized bed combustion, it is the heating of colder air entering through the distribution plate, coming into contact with the hot inert material particles. The very high intensity of this heat transfer process (among other factors, also due to a very large contact area up to 45000 m^2/m^3) is the cause of very rapid air heating. The temperature difference between particles and air in most parts of the bed is very small. Only very near the distribution plate (up to 5-10 cm distant), temperature differences are larger, and local bed (and air) temperature is lower than the average bed temperature.

Due to intensive heat transfer between bed particles, and particle mixing, bed temperature is practically constant in the whole bed volume. However, in cases of intensive local heat generation, which is often the case in reactive fuel combustion and combustion of coals with high content of volatiles, even these intensive heat transfer processes are not enough to make bed temperature constant over the bed volume. Local bed overheating may lead to problems in furnace and boiler operation. Therefore, the study of heat transfer (diffusion) between different parts of the fluidized bed is necessary in order to estimate the dangers of local overheating.

In fluidized bed furnaces and boilers, fuels with high ash and moisture content can also be burned, because cold fuel particles are rapidly heated in the fluidized bed of inert material particles. Heat transfer from bed particles to fuel particles needs to be analyzed and studied in order to correctly model the combustion process. This process is especially important during furnace or boiler start-up.

As this short review shows, in fluidized bed boiler furnaces several specific heat transfer processes take place: heat transfer between gas and particles, heat transfer between the emulsion and bubbling phases, heat transfer between different parts of the bed, heat transfer between the bed and fuel particles, heat transfer to immersed tube bundles and water-tube walls, in the bed as well as above it. The basic characteristic of these processes is the presence of solid particles and heat transfer by particle contact and mixing. In this chapter these processes will be studied in more detail.

Heat transfer in the fluidized bed has been the most studied process since the first ideas for using fluidized beds in different technological processes. In the course of several decades, an enormous number of mostly experimental works have been published. In the last several years interest in heat transfer processes in bubbling fluidized beds gradually decreased, although there are not enough reliable data for beds of larger particles, which are important for fluidized bed combustion boilers. There exist detailed reviews and analyses of published works, as well as experimental correlations for calculating heat transfer, of which the best known are [3-10]. Heat transfer in the fluidized bed has also been widely investigated in Yugoslavia, so exhaustive reviews of this field also exist [11-15].

3.2. Heat transfer between gas and solid particles in bubbling fluidized beds

The temperature field in the fluidized bed is practically homogenous due to intensive mixing of particles in both horizontal and vertical directions. At temperature levels of several hundred or even a thousand degrees Celsius, in conditions of even fluidization, temperature differences do not exceed 2-5 °C. Gas temperature at the exit from the fluidized bed is practically equal to the bed temperature, i. e., temperature of the solid particles. These facts speak of very high ability of the fluidized bed of solid particles to exchange heat with the fluidizing gas. This ability to exchange high quantities of heat between gas and particles is caused by a very large specific surface used for heat transfer (3000-45000 m²/m³), although average heat transfer coefficients for gas to particle (or vice versa), calculated per m² of particle surface, are relatively small (6-25 W/m²K). The large heat capacity of solid particles also contributes to small differences in gas and particle temperatures in the fluidized bed. Gas temperature follows particle temperature and not vice versa. Heat transfer from gas to particles is never the limiting factor in organization of processes in the fluidized bed.

Experimental data and correlations for calculating heat transfer in fluidized (as well as in immobile, fixed) beds of solid particles enable us to draw two conclusions: (a) data of different authors differ up to several orders of magnitude, and (b) for small Reynolds numbers heat transfer coefficients are significantly smaller than for the single particle, which is hard to explain physically [8, 11, 12].

The main causes for such results should be sought in problems in measuring gas temperature in the fluidized bed, data processing and interpretation, as well as in supposed physical models of the process in organization and processing of experimental data. Temperature measurement problems in the fluidized bed and data interpretation have been exhaustively analyzed in [7, 8, 11, 12] and will not be considered here. We will only note that the main reason for large differences between heat transfer coefficients for single particles and particles in the fluidized bed is considered to lie in the fact that, in data processing using energy balance equations, axial molecular heat transfer through the gas is not taken into account [8, 9, 16].

The range of Reynolds numbers covered by the results of numerous authors is shown in Fig. 3.1. Particular references and correlations may be found in [8, 11, 12]. Figure 3.1 shows the extent to which experimental results for heat transfer between gas and particles in the fluidized bed differ from the standard correlation for a single particle:

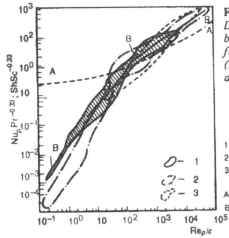

Figure 3. 1.
Dependence of heat transfer coefficients between gas and solid particles for different flow regimes, on Reynolds number (Reproduced by kind permission of the author Prof. J. F. Davidson from [17])

1 – data for bubbling fluidized bed
2 – data for fixed bed
3 – data for dense moving bed and pneumatic transport
A-A – heat transfer from single particle, eq (3 1)
B-B-B – heat transfer in fluidized bed, eqs (3 3) and (3 4)

$$\mathrm{Nu}_p = 2 + 0.74\,\mathrm{Re}_p^{0.5}\,\mathrm{Pr}^{0.33} \tag{3.1}$$

For the whole range of Re_p numbers from 10^{-1} to 10^2 in [7, 17] the following formula is suggested:

$$\mathrm{Nu}_p = 0.03\,\mathrm{Re}_p^{1.3} \tag{3.2}$$

which takes into account that for most gases $\mathrm{Pr} \approx 0.7$ holds true. Based on more detailed analysis of many authors' results, including numerous investigations in U.S.S.R., Gel'perin and Einstein [8] recommend two formulae:

$$\mathrm{Nu}_p = 16\cdot10^{-3}(\mathrm{Re}_p/\varepsilon)^{1.3}\,\mathrm{Pr}^{0.33} \tag{3.3}$$

for $\mathrm{Re}_p/\varepsilon < 200$, and:

$$\mathrm{Nu}_p = 0.4(\mathrm{Re}_p/\varepsilon)^{2/3}\,\mathrm{Pr}^{0.33} \tag{3.4}$$

for $\mathrm{Re}_p/\varepsilon > 200$.

Results of some authors differ from these correlations by as much as ±100-200%. These correlations describe heat transfer for both fixed and fluidized beds as well as two-phase gas-particles flow.

If axial molecular heat transfer in gas is taken into account the following formula for Nusselt number may be obtained [9, 16]:

$$\mathrm{Nu}_p = (7-10\varepsilon+5\varepsilon^2)(1+0.7\,\mathrm{Re}_p^{0.2}\,\mathrm{Pr}^{0.33})+ \\ +(1.33-2.4\varepsilon+1.2\varepsilon^2)\,\mathrm{Re}_p^{0.7}\,\mathrm{Pr}^{0.33} \tag{3.5}$$

which is true for $\varepsilon = 0.35\text{-}1$. This expression shows good agreement with the correlation for fixed beds for small Re_p numbers [18]:

$$Nu_p = 2 + 11 Re_p^{0.6} Pr^{0.33} \qquad (3.6)$$

obtained based on the experimental data taking into account axial conduction in gas for $Re_p > 15$. In the $Re_p > 100$ range correlation (3.5) agrees with most results found in the literature [9]. The results and analyses given in [8] show that taking into account the axial conduction removes the unnatural difference of heat transfer coefficient for single particles and particles in the fluidized bed.

By simple analysis it can be shown that, in spite of small heat transfer coefficients, near the inlet of cold gas through the distribution plate, gas and particle temperatures very soon become practically equal. Temperature difference between gas and particles decreases a hundred times after some ten millimetres, i. e., 5-10 particle diameters [4, 10]. The distance at which gas and particle temperatures are virtually equal is larger for larger particles because of smaller total particle surface, and increases with the increase in particle size. Since bed height in industrial installations is 0.5 to 1 m, it is safe to assume that temperature is equal across the entire bed height.

In bubbling fluidized beds, equaling of the temperature in the bubble and particle, and gas and particle temperatures in the emulsion phase, should be considered separately.

Due to intensive circulation of gas through the bubble, (i. e., gas exchange between bubble and emulsion, which has been considered in detail in the previous chapter), it is not difficult to conclude that gas temperatures in the bubble and emulsion phase very quickly reach the same value. Slow bubbles in a bed of large particles are completely "washed" by the gas from the emulsion phase, and hence in this case temperature equalization is faster than for fast bubbles, in which gas communicates only with the surrounding emulsion, or with the bubble's cloud. Besides, particles falling through the bubble as it moves upwards (although their volume concentration is smaller than 1%), due to high heat capacity contribute to faster temperature equalization [6, 19]. As may be seen from Fig. 3.2 [4], the distance at which temperature difference between gas in the bubble and particles in the emulsion phase is reduced has an order of two bubble diameters. Since the bubbles near the distribution plate have small diameter (~1 cm), gas temperature becomes equal to emulsion particle temperature at practically the same distance as in the emulsion phase (compare with Fig. 3.2a).

Above industrial-type distribution plates with equally spaced orifices with large diameter, and nozzles or bubble caps (usually used for fluidized bed boilers), temperature equalization is also very quick.

Heat transfer between jets of gas flowing out of the distribution plate and the bed particles is very intensive, and according to [6] ends at the distance:

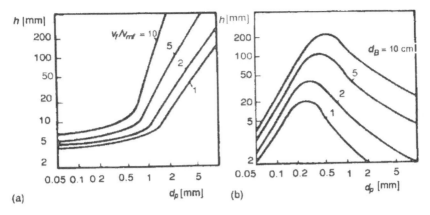

Figure 3.2. *Distance from distribution plate at which inlet gas to particle temperature difference in the fluidized bed becomes 100 times smaller. Calculated for air and glass beads [8]; (a) homogenous fluidization, (b) bubbling fluidized bed*

$$L_{t\max} = v_R r_R T_b / 0.168 u_t \mathrm{Ga}^{-0.1} T_R \qquad (3.7)$$

In solid fuel fluidized bed combustion, it is considered, based on sufficient experimental data, that char combustion after devolatilization takes place in the emulsion phase. Combustion of volatiles partially takes place in the bubbles as well.

Air used for combustion enters through the distribution plate at much lower temperature than the bed temperature and partly traverses the bed in the bubbles. Hence it is necessary to take into account heat and mass transfer between the bubbles and emulsion phase when dealing with the combustion process in the fluidized bed. As has already been remarked, heat transfer between bubbles and emulsion and also between the gas and particles in emulsion, is very intensive and at small distances from the distribution plate temperatures of gas in emulsion and bubbles and particles are practically equal. Because of that heat transfer between emulsion and bubbles is not crucial for fluidized bed combustion processes. Mass transfer processes between bubbles and emulsion are by far more important. Oxygen consumption is always higher in the emulsion, so combustion efficiency and necessary excess air greatly depend on oxygen transfer between bubbles and emulsion.*

---

* Remark: Mass transfer between fluidized bed particles and gas will not be considered in this book. In fluidized bed combustion, inert material particles do not take part in chemical reactions. Processes of mass transfer are interesting, for example, for drying of wet materials in the fluidized bed. We will also point out that it is not quite clear if there is an analogy between heat and mass transfer between gas and particles in the fluidized bed. However, often for calculation of mass transfer the formulae obtained for heat transfer are used, where Nu number is replaced by Sh number and Prandtl number by Schmidt number [8]

Heat transfer between bubbles and the emulsion phase may be treated taking as the basic assumption the two-phase fluidized bed model and gas exchange between bubbles and emulsion, described in detail in the previous chapter (Section 2.3.6).

If we assume, according to Kunii and Levenspiel [3, 15] that gas from the bubble exchanges heat with the particles in the "cloud" around the bubble, and that heat is later transferred from the "cloud" to the emulsion phase, then, by the rule of addition of thermal resistances, the total heat transfer coefficient between the bubble and the emulsion phase may be defined (calculated per unit of bubble volume):

$$\frac{1}{(H_{Be})_B} = \frac{1}{(H_{Bc})_B} + \frac{1}{(H_{ce})_B} \tag{3.8}$$

analogous to eq. (2.125) for gas transfer between bubble and emulsion. Also, analogous to the previous discussions of gas exchange between bubble and emulsion we can assume that heat is transferred between the bubble and the cloud due to gas flow through the bubbles and by molecular diffusion:

$$(H_{Bc})_B = \frac{q_B \rho_g c_g + h_{Bc} S_{Bc}}{V_B} \tag{3.9}$$

Analogous to expression (2.121), the heat transfer coefficient between bubble and the cloud, taking into account molecular diffusion, is defined as:

$$h_{Bc} = 0.975 \rho_g c_g \left(\frac{\lambda_g}{\rho_g c_g}\right)^{1/2} \left(\frac{g}{D_B}\right)^{1/4} \tag{3.10}$$

Another expression for heat transfer coefficient between bubble and the cloud can be obtained analogous to expression (2.122):

$$(H_{Bc})_B = \frac{v_{mf} \rho_g c_g}{D_B} + 5.85 \frac{(\lambda_g \rho_g c_g)^{1/2} g^{1/4}}{D_B^{5/4}} \tag{3.11}$$

Heat transfer between the cloud and the emulsion phase is carried out mostly by diffusion, and can be approximately calculated using the formula analogous to (2.126):

$$(H_{ce})_B \approx 6.78 (\lambda_g \rho_g c_g)^{1/2} \left(\frac{\varepsilon_{mf} v_B}{D_b^3}\right)^{1/2} \tag{3.12}$$

When considering temperature change of the gas in the bubble, except heat exchange between bubble and the emulsion phase described by (3.8) and (3.12), heat transfer with the particles falling through the bubble can be taken into account:

$$-\rho_g c_g v_B \frac{dT_{gB}}{dl} = \gamma_B \alpha_p a_p \eta_h (T_{gB} - T_p) +$$
$$+ (H_{Bc})_B (T_{gB} - T_p) \tag{3.13}$$

Described processes of gas and heat transfer between bubbles and the emulsion phase (Sections 2.3.6 and 3.2) as well as the expressions and models for calculating the amount of transferred mass and energy enable the modelling of combustion to take these processes into account in a physically correct manner.

3.3. Heat and mass transfer between fuel particles and a bubbling fluidized bed

For understanding solid fuel fluidized bed combustion processes and for their modelling, heat transfer processes between fuel particles and the emulsion phase of the fluidized bed of hot inert material are the most important. After the cold fuel particle gets to the hot fluidized bed, the process of particle heating takes place, and the particle is heated to its ignition temperature. During particle heating, drying and devolatilization processes also take place. These processes depend on the intensity of heat transfer from the fluidized bed to the fuel particle. After particle ignition heat emission from the fuel particle, which is at higher temperature, takes place. Besides the heat transfer process, during combustion the oxygen diffusion process towards the fuel particle is also important, as well as the diffusion of combustion products from the fuel particle to the bed volume.

Because of that, we will discuss in more detail the heat and mass transfer processes between fuel particles and the fluidized bed. There was little interest for research into heat and mass transfer processes of fluidized bed particles and particles of different kinds (mobile and immobile, larger or smaller than the bed particles) until intensive research of solid fuel fluidized bed combustion began. So experimental data on these process are scarce. In older literature these problems are not addressed at all or it is recommended that for particles much larger than bed particles, formulae for immersed surfaces (tubes, spheres) should be used. For particles of size comparable to bed particles formulae for heat transfer between gas and bed particles have been recommended [7, 8, 17].

In the last several years there are more and more experimental investigations of these processes [20-24]. In [22, 24] exhaustive reviews of work in this field have been given.

Experimental investigations have been carried out with moving and fixed particles, using naphthalene particles for mass transfer process research [24], and using spherical or cylindrical heaters for heat transfer investigations [25], monitoring cooling and heating of spherical particles with inserted thermocouples [14, 24, 26], and also by monitoring char or graphite particle combustion [20, 21, 23].

3.3.1. Mass transfer between fuel particles and bubbling fluidized beds

When discussing the mass transfer process of a fuel particle in the fluidized bed, it is assumed that the fuel particle exists solely in the emulsion phase. Under these conditions convective mass transfer is very low due to small gas velocities (v_{mf}), especially in beds of small particles of type B according to Geldart. Besides, inert material particles "screen" the fuel particle and slow down the access of oxygen and outflow of CO_2. The first attempts to encompass mass transfer in mathematical models of fluidized bed combustion have assumed that mass is transferred by molecular diffusion, i. e., that Sh = 2 [27]. The presence of inert material particles has been accounted for by multiplying with ε_{mf}, Sh = $2\varepsilon_{mf}$. Later, especially due to the fact that solid fuel combustion mostly takes place in a fluidized bed of large particles where the influence of convection should not be neglected, the correlation analogous to (3.1) for the heat transfer has been used [28]:

$$Sh = (2 + 0.69 \, Re^{0.5} \, Sc^{0.33}) \varepsilon \qquad (3.14)$$

Table 3.1 gives the correlations that different authors have used to describe mass transfer to the fuel particle in the fluidized bed.

Table 3.1. *Empirical correlations for calculating mass transfer from large particles to the fluidized bed*

	Reference	Correlation for Sh-number	Note
1.	Avedesian & Davidson [27] Basu, Broghton, Elliott [29]	$Sh = 2\,\varepsilon_{mf}$	Sh = const.
2.	Gordon & Amundson [30]	$Sh = 2$	
3.	Yagi & Kunii [31] Leng & Smith [32]	$Sh = 2 + 0.6 Re^{0.5} Sc^{0.33}$	Velocity in Re is not defined
4.	Congalidis & Georgakis [33]	$Sh = 2 + 0.6 (Re/\varepsilon_{mf})^{0.5} Sc^{0.33}$	
5.	Chakraborty & Howard [34]	$Sh = 2\varepsilon + 0.69 Re^{0.5} Sc^{0.33}$	ε – bed porosity
6.	Pillai [28], Basu [35]	$Sh = (2 + 0.69 Re^{0.5} Sc^{0.33}) \varepsilon$	
7.	La Nause & Jung [36]	$Sh = 2\varepsilon + 0.69 (Re/\varepsilon)^{0.5} Sc^{0.33}$	
8.	La Nause & Jung [37]	$Sh = 2\varepsilon_{mf} + 0.69 \, (Re/\varepsilon_{mf})^{0.5} Sc^{0.33}$	
9.	La Nause, Jung, Kastl [23]	$Sh = \varepsilon_{mf} + 4\varepsilon_{mf} [d(v_{mf}/\varepsilon_{mf} + v_B)/\pi D_G]^{0.5}$	
10.	Tamarin [20, 21]	$Sh = 0.248 (ScAr)^{1/3} (d/dp)^{0.5}$	

Remark: Characteristic dimension in Sh and Re numbers is the diameter of active particle d

The greatest shortcoming of the above correlations is that they do not take into account the size ratio of fuel and inert material particles. It is obvious that mass transfer conditions, gas flow around fuel particles and their contact with the inert bed material particles, will be different if $d < d_p$ or $d > d_p$. In the case where the fuel particle size is close to the size of fluidized bed particle size, formulae for mass transfer between bed particle and gas may be used (see Section 3.2).

The most detailed experimental research of mass transfer to particles of different diameter in the fluidized bed of inert material particles to date has been conducted by W. Prins [24]. Investigations have been carried out in fluidized beds of different density materials (glass beads, alumina, porous alumina, silica sand) with particle diameter ranging between 100 and 740 μm. The fluidized bed was 12.7 cm in diameter and 15 cm in height. The diameter of spherical naphthalene particles for which mass loss (mass transfer) has been measured, was $d = 2$-20 mm. The ratio of naphthalene and bed material particle diameters ranged from 2 to 300. Fluidization velocity was $v_f / v_{mf} = 1.5$-7.2.

Experiments have shown that fluidization velocity has no influence on mass transfer. This fact confirms the statements of many authors who claimed that mass transfer intensity is the same for fixed and fluidized beds [38] and that the same correlations may be used. Measurements of W. Prins [24] in fixed beds have confirmed such opinions.

The greatest influence on mass transfer comes from the size of active (fuel) particles and inert material particles. The influence of inert material particles is surely the consequence of minimum fluidization velocity – increase in particle size yields an increase in mass transfer. With the increase in active (fuel) particle size, mass transfer decreases, most probably because of the increase in boundary layer thickness at the particle surface.

Figure 3.3 [24] shows the influence of both diameters, that of the active (fuel) particle and fluidized bed particle. The influence of active particle diameter stops at:

$$\frac{\varepsilon_{mf} \, d}{v_{mf}} > 0.025 \, \text{s} \tag{3.15}$$

As a result of his experiments, W. Prins [24] has managed to correlate with the following non-dimensional correlation, with scatter of $\pm 15\%$:

$$\varepsilon_{mf} \, (j_D)_{mf} \, \text{Re}_{mf}^m = 0.105 + 1.505 \left(\frac{d}{d_p} \right)^{-1.05} \tag{3.16}$$

where:

$$m = 0.35 + 0.29 \left(\frac{d}{d_p} \right)^{-0.5} \tag{3.17}$$

Figure 3.4 [24] shows this correlation compared with the experimental data based on which it has been obtained.

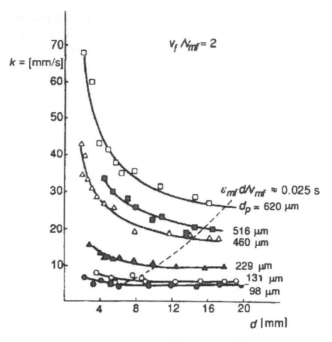

Figure 3.3. *Dependence of mass transfer coefficient for naphthalene spheres in fluidized bed of glass beads on sphere diameter, and for different bed particle sizes. According to the measurements of W. Prins (Reproduced by kind permission of the author Dr. W. Prins from [24])*

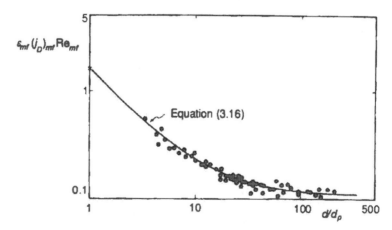

Figure 3.4. *Generalization of experimental results for mass transfer from naphthalene spheres in fluidized bed of glass beads. Comparison with correlation given by W. Prins (3.16) (Reproduced by kind permission of the author Dr. W. Prins from [24])*

Expressions (3.16) and (3.17) hold true in the $0.1 < \mathrm{Re}_{mf} < 20$ and $1 < d/d_p < 300$ range, but for $d/d_p \to \infty$ also shows good agreement with other authors' results for mass transfer from immersed large bodies. Good agreement has also been shown with the results of mass transfer measurements during char particles combustion [36], which have been obtained at much higher temperatures.

This exhaustive experimental research has also shown that the results of mass transfer measurements for fixed particles differ from those obtained with free-moving particles, and are consistently higher by 20-50%, probably for the influence of bubbles and higher relative gas velocity.

The conventional approach to physical description of the mass transfer process between fuel and fluidized bed particles (emulsion phase) is based on molecular diffusion and convective transfer, analogous to mass transfer of single particles in cross flow, taking into account that inert material particles are "in the way" of mass transfer. Formulae from Table 3.1 are based on such a model, except for the La Nause, Jung and Kastl formula (No. 9). La Nause, Jung and Kastl [23] have proposed a different model similar to the "packet" heat transfer model (which will be discussed in Section 3.5). According to this model mass transfer to the fuel particle is based on two mechanisms:

— particle "packets" carry fresh gas from the bed towards the fuel particle; movement of these "packets" of particles is caused by the moving of bubbles. This mechanism constitutes the "particle" component of mass transfer, and
— the second component is classical, convective mass transfer with the gas flowing through the emulsion phase.

Both mass transfer mechanisms are depicted in Fig. 3.5.

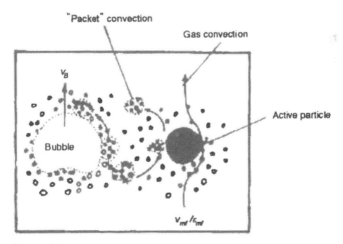

Figure 3.5.
Different mass transfer processes from a large "active" particle in a fluidized bed. According to the model of La Nause, Jung and Kastl (Reprinted from [23]. Copyright 1978, with permission from Elsevier Science)

3.3.2. Heat transfer between fuel particles and bubbling fluidized bed

During combustion of fuel particles in the fluidized bed heat transfer processes occur in both directions: (1) during particle heating, drying and devolatilization heat is transferred from the fluidized bed to the fuel particle, and (2) after ignition of volatiles, and especially during char combustion, fuel particle temperature is higher than the bed temperature and heat is transferred from the fuel particle to the fluidized bed.

In general, heat is transferred between fuel particle and fluidized bed by three mechanisms: gas convection, particle contact and radiation. Heat transfer by radiation has not been sufficiently researched and so is mostly neglected. Depending on the fuel particle size mentioned mechanisms may or may not significantly influence heat transfer. For fuel particles equal or smaller than inert material particles, convective heat transfer is most important, while for the large particles heat transfer through particle contact has a major influence.

Exhaustive experimental research of W. Prins [24] enables insight into the influence of the most important parameters on fuel particle to bed heat transfer. In fluidized beds with different particle density (glass beads, alumina), 130-1011 mm in diameter, heat transfer from large particles has been investigated in the temperature range of 300-900 °C and for fluidization velocities $v_f < 0.8$ m/s. Heat transfer has been measured for silver particles ($d = 4$-8 mm) and graphite particles ($d = 4$-20 mm), having inserted thermocouples. Particles have been virtually free since very thin thermocouples have been used (0.1, 0.5, and 1.5 mm).

Similar to the heat transfer to large immersed surfaces (tubes, tube bundles), the heat transfer coefficient for the fuel particle (in this experiment replaced with silver or graphite sphere) has a maximum at certain, optimal, fluidization velocities. This maximum increases with the decrease in bed particle size. Figure 3.6 [24] shows the dependence of heat transfer coefficient for free graphite particle on fluidization velocity and the influence of the inert material particle size.

Figure 3.7 shows the influence of the silver (stationary) and graphite (mobile) particle diameter on the maximum heat transfer coefficient in the fluidized bed of glass beads of variable size.

Comparing these results with the mass transfer measurements, the following differences can be noted:

- fluidization velocity does not influence mass transfer but has major influence on heat transfer,
- with the increase of particle size of inert material, mass transfer increases but the heat transfer decreases, and
- heat transfer coefficients of free particles are some 10% lower than for stationary particles (for mass transfer the difference has been 20-50%).

These differences are the consequence of different heat and mass transfer mechanisms. In the case of mass transfer the basic process is gas convection and in-

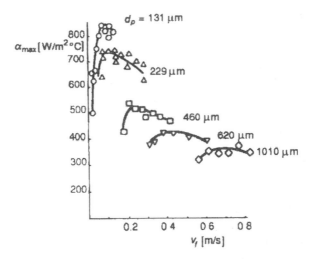

Figure 3.6.
Dependence of heat transfer coefficient for free silver sphere in fluidized bed of glass beads on fluidization velocity and bed particle size. According to the measurements of W. Prins (Reproduced by kind permission of the author Dr. W. Prins from [24])

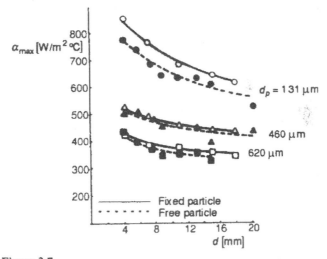

Figure 3.7.
Dependence of maximum heat transfer coefficients for free large particles in fluidized bed of glass beads on large particle diameter and bed particle size. According to the measurements of W. Prins (Reproduced by kind permission of the author Dr. W. Prins from [24])

ert material particles disturb the process. In the case of heat transfer inert material particles play a significant role as well as heat transfer by particle contact. Gas convection is less significant, so the fuel particle mobility is less important.

The different nature of heat and mass transfer processes shows that it is not justified to speak of the analogy between them with respect to immersed fuel particles (bodies) and the fluidized bed.

W. Prins has presented his experimental results by the following non-dimensional expressions, for a bed temperature of 300 °C:

(a) maximum heat transfer coefficients for fixed spherical particles (Fig. 3.8):

$$Nu_{max} Ar^{-m} = 4.175 \left(\frac{d}{d_p} \right)^{-0.278} \qquad (3.18)$$

where:

$$m = 0.087 \left(\frac{d}{d_p} \right)^{0.128} \qquad (3.19)$$

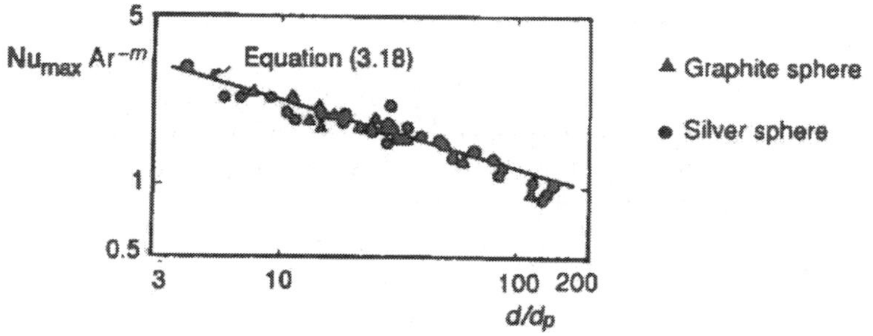

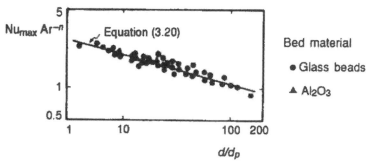

Figure 3.8. *Dependence of Nu_{max} for fixed spherical particle in fluidized bed on sphere to bed particle diameter. Experimental data of W. Prins compared with correlation (3.18)(Reproduced by kind permission of the author Dr. W. Prins from [24])*

(b) maximum heat transfer coefficient of a free spherical particle (Fig. 3.9):

Figure 3.9. *Dependence of Nu_{max} for free spherical particle in fluidized bed on sphere to bed particle diameter. Experimental data of W. Prins [24] compared with correlation (3.20) (Reproduced by kind permission of the author Dr. W. Prins from [24])*

$$\text{Nu}_{max}\, \text{Ar}^{-n} = 3.539 \left(\frac{d}{d_p} \right)^{-0.257} \tag{3.20}$$

where:

$$n = 0.105 \left(\frac{d}{d_p} \right)^{0.082} \tag{3.21}$$

Results of experiments carried out at fluidized bed temperatures 300, 600 and 900 °C have been correlated with the same expressions by dividing the left hand side of correlations (3.18) and (3.20) with the factor:

$$f_T = 0.844 + 0.0756 \left(\frac{T_b}{273} \right) \tag{3.22}$$

By comparing the results for silver and graphite particles, which have very different emissivities, W. Prins has concluded that radiation has no influence up to a fluidized bed temperature of 900 °C.

The suitability of using the obtained results – described by correlations (3.18) and (3.20) – together with (3.22) may be judged from Fig. 3.10 where these results are compared to the results of heat transfer coefficient measurements based on the experiments during graphite particle combustion [20, 24, 25, 41, 42].

Results obtained by W. Prins [24] obtained from the combustion experiments are in good agreement with the proposed formula obtained by measuring the heat transfer coefficient, while the results of other authors are some 30% above the line defined by eq. (3.20) and factor (3.22). According to the analysis by W. Prins, the main reason lies in the use of the wrong heat release for carbon combustion, because it is usually assumed that the combustion is complete, i. e., that only CO_2 is formed. In experiments of W. Prins the concentration of both CO and CO_2 has been measured and so good agreement has been obtained with the measured heat transfer coefficient.

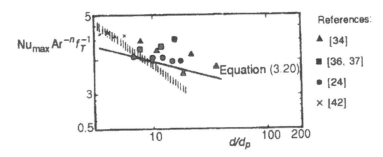

Figure 3.10. *Comparison of maximum heat transfer coefficients for large particles in fluidized bed determined in combustion experiments and correlation (3.20)(Reproduced by kind permission of the author Dr. W. Prins from [24])*

In literature, for small bodies and particles comparable to the bed particle size, it is recommended to calculate heat transfer by the formulae given in Section 3.2 for heat transfer between gas and bed particles [3, 4, 7, 8]. If the active particle diameter is close to the fluidized bed particle diameter, in Soviet literature [4] for fluidized beds of small particles (group B according to Geldart) it is recommended to calculate heat transfer using the Nusselt number values $Nu_p = 10$, and for beds of large particles according to the interpolation formula:

$$Nu = 10 + 0.23(ArPr)^{1/3} \tag{3.23}$$

In mathematical modelling of particle heating and devolatilization for large coal particles (>1 mm) Agarwall et al. [39] have used, for Re < 100, formula (3.2) and for Re >100 the following formula:

$$Nu = 2.0 + 1.8 Re^{0.5} Pr^{0.33} \tag{3.24}$$

In [40], for mathematical modelling of devolatilization processes in large coal particles (3-11 mm), Botterill's recommendation [7] has been used, that heat transfer coefficient may be calculated using formulae for bodies (tubes) immersed in the fluidized bed.

Based on the assumption of Tamarin and Galerstein [20, 21] about the analogy of heat and mass transfer between fuel particles and fluidized bed, La Nause and Jung [22] have, using expression No. 10 from Table 3.1, proposed the following formula for calculating heat transfer coefficient:

$$Nu = \frac{0.248}{3855} q_R \frac{D_G}{\lambda_g} (C_{O_2} - C_{O_{2p}}) Ar^{0.33} \frac{d_p^{0.27}}{d^{0.65}} \tag{3.25}$$

In deducing this expression they have used their own experimentally determined correlation for the temperature difference of coke particles and the fluidized bed during combustion:

$$\Delta T = T - T_b = 670 d_p^{0.23} d^{0.15} \tag{3.26}$$

where the diameters are in centimetres ($0.07 < d_p < 0.2$ cm, and for coke particles $0.4 < d < 1.2$ cm).

Tamarin and Galerstein [20, 21] recommend a simple formula (for $d = d_p$):

$$Nu = 0.27 Ar^{0.32} \tag{3.27}$$

These data and the exhaustive review of experimental research of heat transfer between small bodies and the fluidized bed (e. g., [25]) or measurements of burning particle temperature (e. g., [20, 21]) given in W. Prins' work [24], show that this problem is still not investigated thoroughly enough. It is not possible to, reliably, recommend the formula for heat transfer coefficient between fuel particles and the fluidized bed.

Heat and mass transfer to the fuel particles is now more often the subject of research due to intensive development of fluidized bed combustion mathematical modelling. Presented results of W. Prins are a major advance in research of these mechanisms and form a firm basis for developing mathematical models of combustion. However, new experimental research is required, especially combustion experiments, in order to enable more reliable calculation of heating rate, fuel particle temperature, devolatilization rate and fuel particle burning rate.

3.4. Apparent conductive heat transfer in bubbling fluidized beds

It is a well known property of fluidized beds that the temperature is uniform over the entire bed volume. Many advantages of solid fuel combustion in the fluidized bed come from uniformity of the temperature field. However, this property of fluidized beds must not be taken for granted: measurements show that in the fluidized bed temperature is highly uniform in the vertical direction, but in lateral direction there may exist significant temperature differences. Uniformity or non-uniformity of temperature in fluidized beds is caused by heat transmission between different bed regions. Numerous measurements have shown that effective heat conductivity in a fluidized bed is much higher than either gas heat conductivity or solid particle heat conductivity. Obviously, the basic mechanism is not the molecular heat conductivity, but heat transfer by solid particles and their contacts. That is why we speak of effective heat conductivity of the fluidized bed. Also, knowing that solid particle mixing in the axial direction is much more intensive than in radial direction (see Section 2.3.6, previous chapter), the experimentally established fact that effective heat conductivity in axial direction is an order of magnitude greater (1500-38000 W/mK) than in radial direction (50-5000 W/mK) [5, 6, 43], may be explained.

Heat transfer processes in the fluidized bed and problems of non-uniform temperature distribution are more important for organization of processes in metallurgical furnaces, furnaces for thermal processing of metals, equipment in process and chemical industries, in drying equipment with the fluidized bed, etc. In these devices heating or cooling of the bed with immersed exchanger surfaces or electrical heaters is used and large quantities of cooler, wet or warmer material are constantly introduced, which may lead to significant temperature differences. It is sometimes necessary, due to technological requirements, to create temperature differences between the material inlet and outlet. Discussions of heat transmission processes are, therefore, mostly dedicated to such processes in the literature [4-6, 14, 15, 43].

In fluidized bed boilers, in normal operating conditions, there is no reason for creating of large temperature differences either in the axial or radial direction. Heat transfer surfaces in the bed, if any, are uniformly spaced and water-tube furnace walls uniformly take away heat in all directions. Basic causes of high temperature differences in FBC boilers are: non-uniform fluidization, and non-uniform fuel distribution and heat generation, either because of particle segregation in the axial direction or because of small numbers and large spacing of fuel feeding points.

Non-uniform temperature fields in fluidized bed combustion are most often caused by insufficient fuel feeding points. Due to locally high concentrations of fuel at the feeding points, especially for fuels with high volatile content, local overheating occurs with possible ash and inert material sintering.

If the assumption, that heat transmission in the fluidized bed is due to solid particle mixing, holds true, i. e., that heat transfer occurs through transport of heated particles, then specific heat flux may be represented by the expression:

$$q = -D_s \frac{dI_p}{dx} \tag{3.28}$$

or, since:

$$I_p = i_p \rho_p (1-\varepsilon) \quad \text{and} \quad i_p = c_p t \tag{3.29}$$

it follows that:

$$q = -D_s \rho_p c_p (1-\varepsilon) \frac{dt}{dx} = -\lambda_{ef} \frac{dt}{dx} \tag{3.30}$$

where:

$$\lambda_{ef} = -D_s \rho_p c_p (1-\varepsilon) \tag{3.31}$$

is the effective heat conductivity of the fluidized bed. From expression (3.31) it may be seen that effective thermal diffusivity of the fluidized bed is equal to the effective coefficient of particle diffusion:

$$a_{ef} = \frac{\lambda_{ef}}{\rho_p c_p (1-\varepsilon)} = D_s \tag{3.32}$$

which makes possible prediction of the effective thermal diffusivity based on the effective coefficient of solid particle mixing, using experimental data given in Section 2.3.6, or based on expressions (2.95), (2.96), (2.108), and (2.109).

Literature lacks data on the measurements of effective heat conductivity. To date measurements have only been carried out in small-scale devices, so it is hard to transpose results obtained to industrial-scale devices. Although many experiments [6, 44] determined maximum effective heat conductivity, at particular specific fluidization velocity according to Baskakov [6], there is no firm consensus on the actual existence for this maximum.

The influence of fluidization velocity, bed height and particle size, according to measurements of A. P. Baskakov, in a fluidized bed of corundum particles, may be seen in Fig. 3.11. With increase in particle size from 60 μm to 320 μm heat conductivity greatly decreases. With a decrease in bed height, effective heat conductivity increases. The determined maximum is much less pronounced for large particles and smaller bed heights. The existence of a maximum may be explained by the influence of two opposite processes: (1) increase of particle mixing with the increase in fluidization velocity, and (2) simultaneous decrease of bed density.

All experiments, however, show uniform increase of effective thermal diffusivity with the increase of fluidization velocity. B. A. Borodulya [6] gives the following experimental formula:

$$\frac{\rho_g \, a_{ef}}{\mu_g} = 1.4 \left(\frac{v_f}{v_{mf}} - 1 \right)^{0.35} \mathrm{Ar}^{0.12} \left(\frac{H_b}{D_b} \right)^{0.05} \tag{3.33}$$

obtained for a bed of silica sand ($d_p = 200\text{-}1420$ μm).

Physical models of heat transmission in fluidized beds are based on models of particle mixing described in Section 2.3.6. The review of these models is given in [14, 15].

In the literature one may often see attempts to calculate temperature fields in the fluidized bed using Fourier's equation for heat conduction [5, 6]. Examples of calculating temperature change in the radial direction in metallurgical furnaces, using expression (2.96) are also shown, which have yielded results in good agreement with experiments. In [14, 15], Fourier's equation for fluidized bed has been solved

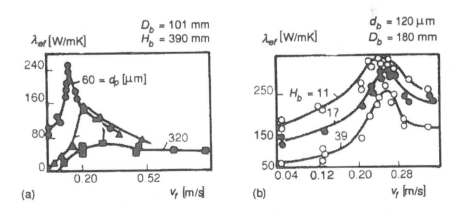

Figure 3.11. *Effect of fluidization velocity on effective radial heat conductivity in bubbling fluidized bed [6]; (a) for different bed particle sizes, (b) for different bed heights*

for the anisotropic domain, with different effective heat conductivity in radial and axial directions. Effective heat conductivity has been calculated by the Kunii-Levenspiel model of particle mixing (Section 2.3.6), using expressions (3.32), (2.108) and (2.109). Bed porosity, bubble content in bed volume and bubble size have been calculated using expressions (2.57), (2.59), and (2.44). Numerical calculation of the temperature field has shown good agreement with experimental data.

3.5.   Heat transfer between fluidized bed and surface

3.5.1. Mechanisms of bed-to-surface heat transfer

One of the most important properties of solid particle fluidized beds is very intensive heat transfer to immersed surfaces. Since fluidized beds are at the same time very suitable for numerous physical and chemical processes with heat consumption or generation, it is very often the case that heat is removed or added, using heat transfer surfaces immersed in the bed. These may be: tube bundles, spiral pipe exchangers or pipes making up the walls of a reactor or furnace (water-tube walls).

Intensive heat transfer to the immersed surfaces is the consequence of great heat capacity and mobility of particles. Particle heat capacity is around 1000 times greater than the heat capacity of gases.

Mechanisms by which heat is transferred from the fluidized bed to immersed surfaces (or vice versa) are numerous and very complex. Virtually all known heat transfer mechanisms are in effect:

- heat transfer by particle motion and contact (particle convection), by which the heat is transferred from the fluidized bed mass to the exchanger surfaces, α_{pc},
- heat transfer by gas convection, by which the heat is transferred from the gas to the exchanger surface, α_{gc}. This process may be divided into two components:
  - heat transfer by gas in bubbles, and
  - heat transfer by gas in emulsion,
  and
- heat transfer by radiation, α_{rad},

so that the total heat transfer coefficient may be represented by the sum:

$$\alpha_{gc} = \alpha_{gB} f_o + \alpha_{ge} (1 - f_o)$$

$$\alpha = \alpha_{pc} + \alpha_{gc} + \alpha_{rad} \tag{3.34}$$

Due to the different physical nature of heat transfer mechanisms between the fluidized bed and immersed surfaces, with the change in flow, geometrical and physical parameters of the bed, great differences in heat transfer intensity occur. The heat transfer process, except on the flow parameters, particle mixing characteristics and overall, organized circulating flows in the bed, also depends on numerous physical properties of particles and gas, and physical and geometrical characteristics of the heat transfer surface and the fluidized bed. With changes in those parameters, the role and influence of certain heat transfer mechanisms also change. It is thought that at fluidized bed temperatures below 600-700 °C (or at temperature differences smaller than 700 °C) the radiation component has no significant influence on heat transfer intensity [7, 10, 46].

We will first discuss the characteristics of heat transfer processes in the conditions where we may neglect heat transfer by radiation. The greatest influence on heat transfer intensity is that of the fluidization velocity and particle size. With the change of fluidization velocity (from filtration velocity of the fixed bed to free fall velocity), the following characteristics have been experimentally determined:

- at the minimum fluidization velocity, heat transfer increases sharply. Obviously, changes in heat transfer mechanisms occur. In the fixed bed heat is transferred to immersed surfaces by conduction through the particle material, by particle contact and by convective gas flow between bed particles. With the onset of fluidization, the key mechanism is particle convection (motion), and
- with further increase in fluidization velocity, the heat transfer coefficient continues to increase, but reaches a maximum at some optimal fluidization velocity. At fluidization velocities greater than optimal, heat transfer intensity to the immersed surfaces decreases. In the range of velocities lower than the optimal, the key mechanism is particle convection, the intensity of which increases with the increase in particle mixing. For fluidization velocities higher than optimal, the key mechanism of heat transfer is gas convection. Due to lower fluidized bed density in this range of velocities (smaller number of particles per unit volume) influence of heat transfer by particle convection decreases. Figure 3.12

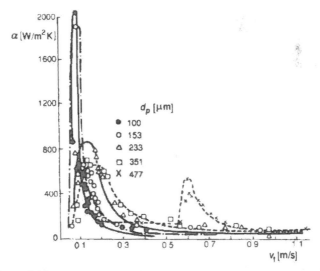

Figure 3.12.
Effect of fluidization velocity on heat transfer coefficient between fluidized bed and immersed surface, for different bed particle sizes. Experimental data of Bondareva [44]

shows the change in heat transfer coefficient to a surface immersed in a bed of silica sand, depending on the fluidization velocity. Shown are experimental results of Bondareva [44], for particles of different size (d_p =100-477 µm).

Results of these and many other experiments have determined the following key characteristics of heat transfer to immersed surfaces: (1) Significant change in heat transfer coefficient is noted with the change in particle size. With increase in particle diameter, optimal fluidization velocity increases, and the maximum heat transfer coefficient is significantly smaller. Particle size influences the change in the relative influence of different heat transfer mechanisms. Figure 3.13 shows the share of different mechanisms in total heat transfer according to the measurements of Baskakov [47]. In fluidized beds of small particles (<0.1 mm, group A according to Geldart) particle convection makes for more than 90% of heat transfer, while in beds with large particles greater than one millimeter (group D according to Geldart), it makes only 20% of particle heat transfer. (2) Physical properties of particles, above all their heat capacity, also influence the heat transfer intensity. (3) Measurement of heat transfer coefficient to surfaces with low heat inertia [48, 49] has shown large fluctuations over time (up to three orders of magnitude), which speaks of the important role of particles, their mixing and contact with exchanger surfaces to the total heat transfer intensity.

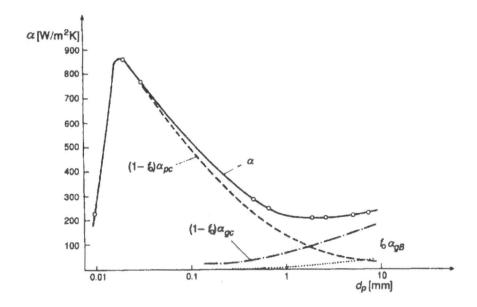

Figure 3.13. *Effect of different heat transfer mechanisms on total heat transfer between fluidized bed and immersed surface. According to the measurements of Baskakov [47]*

Particles in the bed volume (distant from exchanger surfaces) exchange heat with the gas and each other in collisions and by gas conduction. Particle contacts happen often enough and last long enough to make particle temperature uniform.

Particle groups, by bubble influence and by the mixing process, move near the heat transfer surface which is at a quite different temperature. Due to high temperature gradient intensive heat transfer occurs. If the particles stay longer in "contact" with the surface (due to the surface size or shape and mixing process intensity), surface and particle temperature become equal and heat transfer stops. Hence, the greatest temperature differences, and the most intensive heat transfer occur when particle motion is intensive and contact with the surface short-lasting, i. e., at higher fluidization velocities.

The point of contact between particle and immersed surface is too small to significantly influence the heat transfer process. For heat transfer between particles and an immersed body, conduction through the gas film between them is also deemed significant, in other words, effective thickness through which the heat is transferred by gas conduction. Because of that, the heat transfer coefficient sharply increases with the decrease in particle size, and also the overall influence of this mechanism to total heat transfer (Figs. 3.12 and 3.13).

For very small particles (group C according to Geldart) the heat transfer coefficient sharply decreases, due to the influence of molecular forces between particles and the decrease in intensity of particle motion.

The role of gas convection becomes significant when the gas flow between particles becomes turbulent. Transition from laminar to turbulent flow occurs at particle diameters greater than about 800 µm. For particles of this size and larger, the convective component of heat transfer by gas becomes significant (Fig. 3.13) and total heat transfer coefficient does not depend much on the particle size.

Many authors [5, 6, 46] recommend Todes' formula for optimal fluidization velocity, at which heat transfer is most intensive:

$$(\text{Re}_p)_{opt} = \frac{\text{Ar}}{18 + 5.22\sqrt{\text{Ar}}} \tag{3.35}$$

although, according to [46], this expression should not be used for particles smaller than 400 µm.

For particles of group B according to Geldart, the Zabrodsky formula [50] is recommended:

$$\alpha_{max} = 37.6 \left(\frac{\rho_p}{\rho_g} \right)^{0.2} \lambda_g^{0.6} d_p^{-0.36} \tag{3.36}$$

where the numerical constant has dimension $[(\text{W/K})^{0.4} \, \text{m}^{-1.04}]$, and which is obtained experimentally under the following conditions: immersed body diameter 20-60 mm, particle size 280-2440 µm, particle density 2000-4000 kg/m^3, and Ar = 80·10^6.

For particles of group A, Botterill [46] recommends the Khan formula [51]:

$$\mathrm{Nu}_{\max} = 0.157 \mathrm{Ar}^{0.475} \tag{3.37}$$

For group D particles, Botterill [7, 46, 52] recommends, for Archimedes numbers $10^3 < \mathrm{Ar} < 2 \cdot 10^6$, the following formulae:

$$\alpha_{g\mathrm{cmax}} = \frac{\lambda_g}{d_p^{0.5}} \, 0.86 \mathrm{Ar}^{0.39} \tag{3.38}$$

with dimension of the numerical constant $[\mathrm{m}^{-0.5}]$,

$$\alpha_{p\mathrm{cmax}} = \frac{\lambda_g}{d_p} \, 0.843 \mathrm{Ar}^{0.15} \tag{3.39}$$

and

$$\alpha_{\max} = \alpha_{g\mathrm{cmax}} + \alpha_{p\mathrm{cmax}} \tag{3.40}$$

For calculating gas convection Baskakov [53] recommends the following formulae:

$$\mathrm{Nu}_{gc} = 0.0175 \mathrm{Ar}^{0.46} \mathrm{Pr}^{0.33} \tag{3.41}$$

for $v_f > v_{opt}$ and

$$\mathrm{Nu}_{gc} = 0.0175 \mathrm{Ar}^{0.46} \mathrm{Pr}^{0.33} \left(\frac{v_f}{v_{opt}} \right)^{0.3} \tag{3.42}$$

for $v_{mf} < v_f < v_{opt}$, which should be used together with expression (3.40).

3.5.2. Heat transfer to immersed surfaces – experimental results

The discussion of heat transfer mechanisms and formulae presented in the previous section has shown that numerous physical properties of gas and particles influence the intensity of heat transfer processes.

Heat conductivity in gas also has great influence on heat transfer. Heat transfer to the immersed heat transfer surface greatly depends on heat conduction through the gas gap during particle-to-surface and particle-to-particle collision. Much smaller is the influence of heat conductivity of particles, which is much higher, so particle material is no obstacle for heat transfer. Influence of temperature on the heat transfer process is a consequence of the change in heat conductivity of the gas.

The ability of the fluidized bed to exchange heat with surfaces in contact also greatly depends on the particle heat capacity. The greater the particle heat capacity $(c_p \rho_p)$ the greater is the ability to, in a short time, receive the heat from the exchanger surface and transfer it to the bed volume (and vice versa). Specific heat of the gas has a smaller influence on heat transfer, especially since for different gases, $c_g \rho_g$ changes very little.

It has already been mentioned that with the increase in temperature, intensity of heat transfer also increases, mainly due to the increase in heat conductivity in gas [7, 9]. Increase in pressure above atmospheric has little influence on particle convection, but the change in gas density significantly changes heat transfer intensity by gas convection [9], which leads to an increase in heat transfer coefficient to immersed surfaces. Pressure influences heat transfer also because of the changed bubble size.

Section 3.5.1 has dealt with the influence of fluidization velocity and particle size on heat transfer.

The literature contains much data of numerous experimental investigations, for heat transfer surfaces of different shape and position and in different experimental conditions. Authors have correlated their results in different manners, depending on which factors of influence they have investigated. Most often, relations of the following form are used:

$$\frac{\alpha d_p}{\lambda_g} = f\left(\frac{c_g \mu_g}{\lambda_g}, \frac{\rho_g d_p v_f}{\mu_g}, \frac{\rho_p}{\rho_g}, \frac{c_p}{c_g}, \frac{\lambda_p}{\lambda_g}, \frac{D_b}{d_p}, \frac{D_T}{d_p}, \frac{H_{mf}}{D_T}, \frac{v_f}{v_{mf}}, \varepsilon, \varepsilon_{mf}\right) \quad (3.43)$$

which was obtained using the theory of similarity and dimensional analysis.

Numerous reviews of current experimental research [7, 8, 9, 12, 13] show significant differences in the value of the heat transfer coefficient obtained in different experiments (more than 10 times). Differences are caused mainly by the difficulties in defining and measuring bed temperature and by some researchers not paying attention to certain parameters and hence not taking them into account when organizing an experiment [12]. For many experimental formulae the exact experimental set-up and conditions are not known.

Heat transfer to vertical walls in contact with fluidized bed. Table 3.2 gives the formulae obtained in the research of heat transfer to the walls in contact with the fluidized bed. These are mainly older experiments, carried out in small-scale fluidized beds, so their results can hardly be used for calculating heat transfer to water-tube walls of industrial-scale furnaces.

According to [57] and [3] expression No. 5 in Table 3.2 correlates around 95% of the data used, with the accuracy of ±50%. Since it has included data of many experiments of different authors, expression No. 5, recommended by Wen and Leva, may be regarded as the most reliable for use. However, it must be borne in mind that experiments have been carried out in small devices and mainly small particles were used.

Heat transfer to a single immersed tube. For practical, engineering purposes much more important are data on heat transfer to single tubes or tube bundles. Investigations of heat transfer to surfaces of this geometry have been abundant, and

Table 3.2. *Empirical correlations for heat transfer to the walls of the fluidized bed*

No.	Author	Correlation	Conditions
1.	Van Heerden. Nobel. and van Krevelen [38]	$$Nu_p = 0.58\,Pr^{0.5}(B\,Re_p)^{0.45}\left(\frac{c_p}{c_g}\right)^{0.36}\left[\frac{\rho_p(1-\varepsilon_{mf})}{\rho_g}\right]^{0.18}$$ B-shape factor has the following values: 0.62-0.78 for carborundum, 0.39-0.58 for coke and 0.59 for Fe$_3$O$_4$	Material: Fe$_3$O$_4$, carborundum, coke $d_p = 0.05$-0.8 mm; Gas: air, Ar, CH$_4$, N$_2$ + H$_2$; $D_b = 5.8$ cm
2.	Dow and Jacob [54]	$$Nu_p = 0.55\,Re_p^{0.8}\left(\frac{\rho_p c_p}{\rho_g c_g}\right)^{0.25}\left(\frac{D_b}{d_p}\right)^{0.03}\left(\frac{H_b}{D_b}\right)^{-0.65}\left(\frac{1-\varepsilon}{\varepsilon}\right)^{0.25}$$	Material: Fe; $d_p = 0.07$-0.17 mm Gas: air; $D_b = 5.08$; 7.62 cm
3.	Toomey and Johnstone [55]	$$Nu_p = 3.75\left(Re_{mf}\log\frac{v_f}{v_{mf}}\right)^{0.47}$$	Material: glass; $d_p = 0.06$-0.85 mm Gas: air; $D_b = 12$ cm
4.	Levenspiel and Walton [56]	$$Nu_p = 0.6\,Pr\,Re_p^{0.3}$$	Material: glas, coal; $d_p = 0.15$-4.34 mm Gas: air; $D_b = 10.3$ cm
5.	Wen and Leva [57]	$$Nu_p = 0.16\,Pr^{0.4}\,Re_p^{0.76}\left(\frac{\rho_p c_p}{\rho_g c_g}\right)^{0.4}\left(\frac{v_f^2}{g d_p}\right)^{-0.2}\left(\frac{H_{mf}}{\eta\,H_b}\right)^{0.36}$$ where $\eta = \eta\,(v_f/v_{mf}\cdot d_p)$ and is determined from the diagram given in [57] and [3]	Correlated results of a number of authors: Dow, Jacob [54], van Heerden [38], Toomey, Johnstone [55] and their own results

the most important correlations for calculating heat transfer coefficients can be found in already mentioned monographs and reviews.

Table 3.3 shows correlations of many authors for calculating the heat transfer coefficient for a single horizontal tube immersed in a fluidized bed, which they have suggested, based on their own experimental results or the results of several authors. Correlations given are for horizontal tubes, vertical tubes, horizontal finned tubes and small immersed bodies.

The comparison of these correlations is very complex, due to different experimental conditions and different non-dimensional numbers used.

Correlations from No. 1 to No. 9 pertain to the range of fluidization velocities from minimal to optimal. In this range the heat transfer coefficient increases with an increase of fluidization velocity.

Differences between different authors' correlations may be seen in Fig. 3.14, taken from [8] where correlations, developed until 1967 by many authors, have been compared (among them are also correlations Nos. 1, 2 and 9 from Table 3.3.). From Fig. 3.14 it is obvious that correlations from these years did not take into account all significant factors. Later experimental research (correlations Nos. 3, 4, 5 and 7) take into account the influence of bed dimensions and particle physical properties.

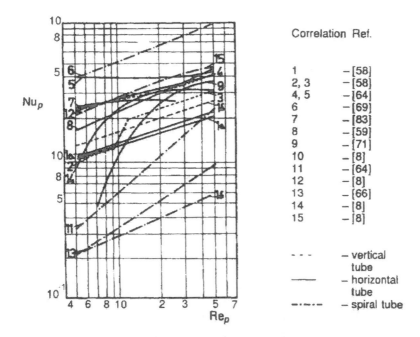

Figure 3.14. *Heat transfer between fluidized bed and immersed surface in velocity range from v_{mf} to v_{opt}. Comparison of different correlation Nu = f(Re), (Reproduced by kind permission of the author Prof. J. F. Davidson from [8])*

Table 3.3. *Empirical correlations for heat transfer to single immersed tube*

No.	Author	Correlation	Conditions
		HORIZONTAL SMOOTH TUBE	
1.	Vreedenberg [58, 59]	$\mathrm{Nu}_{D_T} = 0.66\mathrm{Pr}^{0.33}\left[\mathrm{Re}_{D_T}\dfrac{\rho_p}{\rho_g}\left(\dfrac{1-\varepsilon}{\varepsilon}\right)\right]^{0.44}$	$\mathrm{Re}_{D_T} < 2000$, fine and light particles of group A (Geldart), $D_T = 16.9,\ 33.6,\ 51$ mm, $d_p = 64\text{-}316$ μm
2.	–''–	$\mathrm{Nu}_{D_T} = 420\mathrm{Pr}^{0.3}\mathrm{Re}_{D_T}^{0.3}\left(\dfrac{1}{Ar^*}\right)^{0.3}$	$\rho_p = 1600\text{-}5150$ kg/m³, $v_f = 0.01\text{-}0.239$ m/s, $T_b = 40\text{-}340°C$, $\mathrm{Re}_{D_T} > 2500$, bigger and heavier particles, group C (Geldart)
3.	Kurochkin [60]	$\dfrac{a D_T}{\lambda_p} = 0.0214\left(\dfrac{v_f c_p \rho_p D_T}{\lambda_p}\right)^{0.21}\left(\dfrac{D_T}{d_p}\right)$	Sand: $d_p = 0\text{-}400$ μm, 40-1000 μm, 1000-3000 μm, Cylindrical tubes
4.	–''–	$\dfrac{a D_h^*}{\lambda_p} = 0.0412\left(\dfrac{v_f c_p \rho_p D_h^*}{\lambda_p}\right)^{0.21}\left(\dfrac{D_h^*}{d_p}\right)$ D_h^* = 4 times cross section over circumference = $4A_T/O_T$	Sand: $d_p = 0\text{-}400$ μm, 40-1000 μm, 1000-3000 μm, Elliptical pipes
5.	–''–	$\dfrac{a D_h^*}{\lambda_p} = 0.0412\left(\dfrac{v_f c_p \rho_p D_h^*}{\lambda_p}\right)^{0.21}\left(\dfrac{D_h^*}{d_p}\right)$ D_h^* = 4 times cross section over circumference = $4A_T/O_T$	Sand: $d_p = 0\text{-}400$ μm, 40-1000 μm, 1000-3000 μm, Flat tubes
6.	Ternovskaya and Korenberg [61]	$\mathrm{Nu}_p = 2.9\left(\dfrac{1-\varepsilon}{\varepsilon}\mathrm{Re}_p\right)^{0.4}\mathrm{Pr}^{0.33}$	Pyrite slag
7.	Ainstein [62]	$\mathrm{Nu}_p = 5.76(1-\varepsilon)\dfrac{1}{\varepsilon^{0.34}}\mathrm{Re}_p\mathrm{Pr}^{0.33}\left(\dfrac{H_b}{D_b}\right)^{0.16}$	

No.	Reference	Equation	Conditions
8.	Grewal and Saxena [65]	$\dfrac{\alpha D_T}{\lambda_g} = 47(1-\varepsilon)\left(\dfrac{v_f D_T \rho_p}{\mu_g}\dfrac{\mu_g^2}{d_p^3\rho_p^2 g}\right)^{0.325}\left(\dfrac{\rho_p c_p D_T^{3/2} g^{1/2}}{\lambda_g}\right)^{0.23}Pr^{0.3}$	$\rho_p = 2650$ kg/m³
9.	Tishchenko and Khvastukin	$Nu_p = 1.7 Ar^{0.84}$	
10.	Gel'perin [64]	$Nu_p = 4.38\left(\dfrac{1-\varepsilon}{\varepsilon}\right)Re_p^{0.32}\left[\dfrac{1}{6(1-\varepsilon)}\right]^{0.32}$	
11.	Traber et al. [66]	$Nu_{p,max} = 0.021 Ar_b^{0.4} Pr^{0.33}\left(\dfrac{D_b}{d_p}\right)^{0.13}\left(\dfrac{H_{mf}}{d_p}\right)^{0.16}$	
12.	Chechetkin [67]	$Nu_{p,max} = 0.2 Re_{opt}^{0.5} Pr_{opt}^{0.4} d_p^{-0.69}$; $Re_{opt} = 0.209 Ar^{0.52}$	
13.	Shlapkova [63]	$\alpha_{max} = 2 d_p^{0.2} D_T^{-0.11} d_p^{-0.4}$	$D_T = 6.8$ and 3.4 mm Sand, $d_p = 200$ and 126 μm
14.	Maskaev and Baskakov [68]	$Nu_{p,max} = 0.21 Ar^{0.32}$	$1.4\cdot10^5 \leq Ar \leq 3\cdot10^8$; $d_p = 2$-13 mm
15.	Grewal and Saxena [65]	$\alpha_{max} = 0.9\dfrac{\lambda_g}{d_p}\left(Ar\dfrac{D_{12.7}}{D_T}\right)^{0.21}\left(\dfrac{c_p}{c_g}\right)^{45.5 Ar^{-4.7}}$	$3\cdot10^2 < Ar < 1\cdot10^5$ Particles: glass, dolomite, sand, Al₂O₃, carbide, $d_p = <1$ mm; $D_{12.7}$ – tube diameter $D_T = 12.7$ cm
16.	Sarkits [69]	$Nu_{p,max} = 0.0087 Ar_b^{0.42} Pr^{0.33}\left(\dfrac{c_p}{c_g}\right)^{0.45}\left(\dfrac{D_b}{d_p}\right)^{0.16}\left(\dfrac{H_{mf}}{d_p}\right)^{0.45}$	$15 < Ar_b < 1000$
	[9]	$Nu_{p,max} = 0.019 Ar_b^{0.5} Pr^{0.33}\left(\dfrac{c_p}{c_g}\right)^{0.1}\left(\dfrac{D_b}{d_p}\right)^{0.13}\left(\dfrac{H_{mf}}{d_p}\right)^{0.16}$	$2.6 < Ar_b < 8.5\cdot10^5$

Table 3.3. *Continued*

No.	Author	Correlation	Conditions
17.	Botterill et al. [46]	$\alpha_{max} = \alpha_{pc,max} + \alpha_{gc}$ $\alpha_{p,max} = \frac{\lambda_g}{d_p} 0.843 Ar^{0.15}$ $\alpha_{gc} = \frac{\lambda_g}{d^{1/2}} 0.86 Ar^{0.39}$	$10^6 \leq Ar_b \leq 2 \cdot 10^6$ Gas: air, argon, CO_2, freon Particles: copper balls, sand, glass $T < 50$ °C d_p – group C (Geldart)
18.	Todes et al. [90]	$\alpha_{max} = 4.16\sqrt{\lambda_p c_p \rho_p}\, Ar^{-0.077}$, or $\alpha_{max} = 212\lambda_g^{0.645}\left(\frac{\rho_p}{\rho_g}\right)^{-0.077} d^{-0.123}$	$10 < Ar_b < 10^6$ where constant 212 has dimension [(W/K)$^{0.346}$m$^{-1.223}$], and constant 4.16 has dimension [s$^{-0.5}$]
19.	Baerg et al. [71]	$\alpha_{max} = 120.81\ln\left(7.05 \cdot 10^{-6}\, \frac{\rho_p}{d_p}\right)$	
20.	Zabrodsky, Antonishin, and Parnas [50]	$\alpha_{max} = 37.6\left(\frac{\rho_p}{\rho_g}\right)^{0.2} \lambda_g^{0.6} d_p^{-0.36}$	for $d_p \leq 800$ mm where constant 37.6 has dimension [(W/K)$^{0.4}$m$^{-1.04}$]
21.	Khan, Richardson, and Shakiri [51]	$Nu_{max} = 0.157 Ar^{0.475}$	for particles of group A according to Geldart

VERTICAL SMOOTH TUBE

No.	Reference	Formula	Conditions
22.	Mickley and Trilling [72]	$$\alpha = 0.0281 \left(\frac{\rho_\varepsilon^2}{d_p^3} \right)^{0.238}$$	Vertical tube, $D_T = 12.5$ mm; Glass beads, $d_p = 40$-450 μm
23.	Gel'perin [70]	$$Nu_p = 7.08 \left(\frac{1-\varepsilon}{\varepsilon} \right) Re_p^{0.285} \left(\frac{v_f}{v_{mf}} \right)^{-0.2} \left(1 - \frac{2r}{D_T} \right)^{0.36} \left[\frac{1}{6(1-\varepsilon)} \right]^{0.285}$$ r – distance from the tube to the center of bed	Vertical tube

HORIZONTAL FINNED TUBE

No.	Reference	Formula	Conditions
24.	Bartel et al. [73]	$$Nu_p = \frac{10(1-\varepsilon)^{0.5}}{(1-\lambda Re_p^2)^2}$$ $\lambda = 58300\,\delta^{2.91 l (293.8\,\delta - 8.37)}$, $\gamma = -0.0006758 - 1.872\,l(-107.28 + 3.22)$ δ – fin thickness, l – fin length	$d_p = 114$-470 mm; $l = 10$-25 mm; $d = 0.6$-0.8 mm; $D_T = 16$ mm, 315 fins/m

SMALL IMMERSED BODIES

No.	Reference	Formula	Conditions
25.	Varrigin and Martjushin [74]	$$Nu_{p,max} = \frac{\alpha_{max} d_p}{\lambda_g} = 0.86\,Ar^{0.2}$$ for other gases α_{max} should be multiplied with $(\lambda_g / \lambda_a)^{0.6}$	Spherical immersed body; Gas: air; $30 < Ar < 2 \cdot 10^5$
26.	Pillai [108]	$$Nu_{p,max} = 0.365(T/273)^{0.82}\,Ar^{0.22}$$ T – body temperature	Sand: $d_p = 200$-800 μm; Spherical immersed body
27.	Zabrodsky, Antonishin, and Parnas [50]	$$Nu_{p,max} = 0.88\,Ar^{0.213}$$ Physical characteristics of gas taken at $T_{sr} = \dfrac{T_w + T_b}{2}$	$D_T = 20$-60 mm, $d_p = 280$-2440 μm; $\rho_p = 2000$-4000 kg/m³, $80 < Ar < 10^6$; $T < 1200$ °C

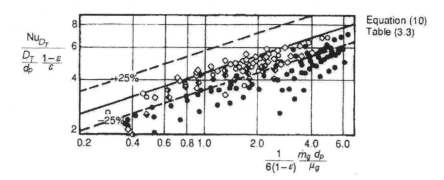

Figure 3.15. *Comparison of experimental data obtained by Grawel and Saxena [65] with correlation No. 10, Table 3.3*

Grawel and Saxena [65] have, based on their numerous experimental results, suggested correlation No. 8 from table 3.3, which is significant since it takes into account particle heat capacity. At the same time, they have compared their experimental results with other authors' correlations as may be seen from the example in Fig. 3.15, where the comparison has been done with correlation No. 10, Table 3.3, which shows great differences.

Figure 3.16 shows the correlation of Grawel and Saxena (No. 8) compared with the same experimental data. Agreement with the measurements may be regarded as very good, despite the ±25% scattering, which is for measurement of such processes a very good agreement.

The correlation of Grawel and Saxena (No. 8) may be recommended for use in calculations, as the most modern one, taking into account particle heat capacity,

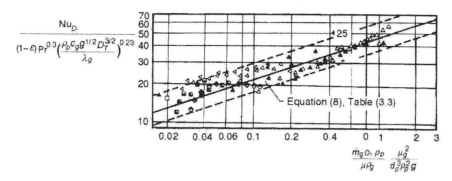

Figure 3.16. *Comparison of experimental data obtained by Grawel and Saxena [65] with correlation No. 8, Table 3.3*

and was obtained by analysis of many authors' experiments. Also the correlations of Kurochkin (Nos. 3, 4 and 5) may be recommended, since they have been obtained for a wide range of parameters and are the only ones that may be used for tubes of different cross-sectional shape.

Much more often in literature, correlations can be found for calculating the maximum heat transfer coefficient. When calculating and designing devices, one most often tends to obtain optimal heat transfer conditions. In Table 3.3 with Nos. 11 to 19 are correlations for maximum heat transfer coefficient to a single immersed tube. The value of optimum velocity is most often obtained using expression (3.35), being widely accepted in the literature.

Except formulae (3.36) of Zabrodsky, (3.37) of Khan, and (3.38)-(3.42) of Botterill, which have already been mentioned, a more recent formula of Grawel and Saxena (No. 15, Table 3.3) is also recommended. When using correlations for maximum Nusselt number for horizontal immersed tubes, accuracy of more than ±25% should not be expected.

In Table 3.3 from number 22 to 27 are empirical expressions obtained for immersed heat transfer surfaces rarely used: vertical immersed tube, horizontal finned tube and small bodies.

Heat transfer to tube bundles. For analysis of furnaces and boilers with fluidized bed combustion it is often necessary to calculate heat transfer coefficients for tube bundles immersed in the fluidized bed. Table 3.4 gives correlations most frequently seen in the literature. Correlations Nos. 1 to 4 give maximum values, and correlations Nos. 5 to 9 give the heat transfer coefficient as a function of fluidization velocity in the range from v_{mf} to v_{opt}, for a horizontal immersed tube bundle. Apart from these correlations Table 3.4 also gives the relationships for vertical bare and finned tube bundles, even though these cases are rare in application.

Experiments have shown that a key influence on heat transfer comes from the horizontal distance between tubes. Influence of vertical pitch has either not been determined or is much smaller.

In the literature, correlations Nos. 3, 4, 6, and 8 are recommended for horizontal bundles of smooth tubes. Correlation No. 6 is modified from Vreedenberg (Nos. 1 and 2, Table 3.3). For a vertical bundle the expression of Gel'perin (No. 11) is recommended, and for finned tubes expressions Nos. 12 and 13. All these expressions have been successfully used by several authors to present their experimental data. As in the case of a single immersed tube accuracy of better than ±25% should not be expected. Significant differences in different correlations also found in literature remind that all mentioned expressions should be used only for the parameter ranges used in experiments.

Table 3.4. *Empirical correlations for heat transfer to the immersed tube bundle*

No.	Author	Correlation	Conditions
		BUNDLE OF HORIZONTAL SMOOTH TUBES	
1.	Gel'perin, Ainstein, and Zaikovski [76]	$Nu_{p,max} = 0.74 Ar^{0.22}\left(1 - \dfrac{D_T}{S_h}\right)^{0.25}$	In-line tube arrangement, $D_T = 20$ mm, $d_p = 160\text{-}350$ μm; 0.380 m × 0.350 m; $S_h/D_T = 2\text{-}9$, $215 < Ar < 2200$, $H_b = 0.38\text{-}0.44$ mm
2.	Gel'perin, Kruglikov, and Ainstein [64]	$Nu_{p,max} = 0.74 Ar^{0.22}\left[1 - \dfrac{D_T}{S_h}\left(1 + \dfrac{D_T}{S_v + D_T}\right)\right]^{0.25}$	Staggered tube arrangement; $S_h/D_T = 2\text{-}9$; $S_v/D_T = 0\text{-}10$; Other conditions as in No. 1
3.	Chekansky et al. [77]	$\alpha_{max} = 28.8\, \Phi_p^{0.2}\, \lambda_g^{0.6}\, d_p^{-0.36}\left(\dfrac{x}{d_p}\right)^{0.04}\left(\dfrac{D_T}{D_{20}}\right)^{-0.12}$ x – tube pitch D_{20} – tube with $D_T = 20$ mm	Sand and corundum, $d_p = 220$ and 440 μm; $D_T = 50$ and 102 mm; Staggered tube arrangement, $16 \leq x/d_p \leq 63$, $200 < Ar < 4800$; Constant 28.8 has dimension $\left[\dfrac{W}{K}^{0.4}\dfrac{1}{kgm^{2.22}}\right]^{0.2}$
4.	Grewal and Saxena [78]	$\alpha_{max} = 0.9\, \dfrac{\lambda_g}{d_p}\left(Ar\dfrac{D_{12.7}}{D_T}\right)^{0.21}\left(\dfrac{c_p}{c_g}\right)^{45.5 Ar^{-0.7}}\left[1 - 0.21\left(\dfrac{S_h}{D_T}\right)\right]^{-1.75}$ $D_{12.7}$ – tube with $D_T = 12.7$	$300 \leq Ar \leq 10000$, Staggered and inline tube arrangement $1.75 D_T \leq S_h \leq 9 D_T$; 600×600 mm, 600×300 mm, $H_b = 600$ mm, $d_p = 250$ μm and 660 μm sand, $D_T = 14$ mm
5.	Petrie, Freeby, and Buckham [79]	$Nu_{D_T} = 14\left(\dfrac{v_f}{v_{mf}}\right)^{1/3} Pr^{1/3}\left(\dfrac{D_T}{d_p}\right)^{2/3}$	$D_T = 12.7$ and 19 mm, Staggered and in-line tube arrangement $S_h/D_T = 2.7$ and 4.4
6.	Andeen and Glicksman [80]	$Nu_{D_T} = 900(1-\varepsilon)\left(Re_{D_T}\dfrac{1}{Ar^*}\right)^{0.326}$	Modified correlation No. 2 from Table 3.2

7.	Ainstein [81]	$Nu_p = 2.92(1-\epsilon)(Re_p/\epsilon)^{0.277} Pr^{0.33}$	
8.	Zabrodsky et al. [75]	$a = 7.2\dfrac{\lambda_g(1-\epsilon)^{2/3}}{d_p} + 262\nu_f^{0.2} c_g \rho_g d_p$	$d_p = 2$ and 3 mm, $\rho_p = 1000$ and 2300 kg/m³, $D_T = 30$ mm Staggered tube arrangement: $S_h = 45\text{-}100$ mm, $S_v = 45\text{-}100$ mm In-line tube arrangement: $S_h = 50\text{-}100$ mm, $S_v = 50\text{-}100$ mm
9.	Bartel and Genety [82]	$\dfrac{aD_T}{\lambda_g} = \dfrac{11(1-\epsilon)^{0.5}}{1+\left[\dfrac{0.2512}{\dfrac{v_f d_p^{0.24}}{v_g}\left(\dfrac{d_p}{0.000203}\right)}\right]^2}\,\gamma^2\,\dfrac{D_T}{d_p}$	$D_T = 15.9$ mm, Staggered tube arrangement 0.39-0.21 m, glass beads $d_p = 203\text{-}470$ μm, $H_b = 0.58$ m

BUNDLE OF VERTICAL SMOOTH TUBES

10.	Wender and Cooper [83]	$Nu_p = 3.51\cdot10^{-4} C_R (1-\epsilon)\left(\dfrac{c_g \rho_g}{\lambda_g}\right)^{0.43} Re_p \left(\dfrac{c_p}{c_g}\right)^{0.8}\left(\dfrac{\rho_p}{\rho_g}\right)^{0.66}$ C_R – from [3], page 273, Fig. 4, $C_R = 1\text{-}1.8$	Correlated results of a number of authors $10^{-2} < Re_p < 10^2$ where constant $3.51\cdot10^{-4}$ has dimension $\left(\dfrac{s^2}{m^2}\right)^{0.43}$
11.	Gel'perin and Ainstein [8]	$Nu_{p,max} = 0.75 Ar^{0.22}\left(1 - \dfrac{D_T}{S_h}\right)^{0.14}$	$S_h / D_T = 1.25\text{-}5$

Table 3.4.   *Continued*

No.	Author	Correlation	Conditions
		BUNDLE OF HORIZONTAL FINNED PIPES	
12.	Bartel and Genetti [82]	$$Nu_p = \frac{7.34\left[1 - \dfrac{0.027 + 4.3 l^{1.5}}{S_h^{(1.12 + 3.2 l^{0.6})}}\right]}{\left[1 + \dfrac{0.00102 + 0.047 l^{0.8}}{Re_p^{(0.33 + 0.4 l^{3.3})} d_p^{(1.23 - 0.57 l^{0.13})}}\right]^2}$$ l – fin thickness	d_p = 203, 297, 470 μm (glass beads) l = 0-22 mm, S_h = 27-122 mm, D_T = 16 mm (horizontal tubes), δ = 0.6 mm (fin thickness) 315 fins/m, Gas: air
13.	Priebe and Genetti [81]	$$Nu_p = (1013 d_p - 3.8)(Q/A_t)^{0.24} (d_p/m)^{-(0.55 + 5.94_p)} \left[1 + \frac{819 + 5.88 m}{(v_f/v_{mf})^{0.164}}\right]$$ Q – total heat exchanged; A_t – total tubes surface with fins m – distance between fins	D_T = 13 mm (horizontal tubes) l ≈ 8 mm

3.5.3. Influence of geometrical parameters on heat transfer

The influence of geometry on heat transfer intensity from a single immersed tube or tube bundle should be considered bearing in mind heat transfer mechanisms described in Section 3.5.1. Due to the key role of particle heat convection the greatest influence will be of those geometrical parameters on which depend particle mixing intensity in the fluidized bed, size of bubbles and character of their motion and the duration of particle-to-surface contact. Those are: height and cross-section dimensions of the bed, position of heat transfer surface, tube diameter, tube bundle arrangement and their distance from one another, tube surface quality and dimensions of fins.

Regardless of the actual value of geometrical parameters, flow pattern of gas and particle motion around immersed tubes or tube bundles are similar.

Visual observations show that below a horizontal tube in a fluidized bed there is a region with lower particle concentration, and near the tube surface there exists a gaseous region devoid of particles. With an increase in fluidization velocity this region diminishes. Above the tube exists the region where particles are moving slowly or not at all. With the increase in fluidization velocity this region sharply diminishes too. The most intensive heat transfer is at the sides of the tube due to intensive "washing" of tube lateral sides with particles. Contrary to what may be expected, heat transfer intensity on the sides of the tube increases most slowly with the increase in fluidization velocity due to the increase in tube to bubble contact time. Despite all this, maximum local heat transfer coefficient is always at the sides of the tube, and its position moves downstream with an increase in fluidization velocity [8]. The increased side surface of tubes with elliptical cross section makes them a very good choice for heat transfer in fluidized beds.

The situation is very similar for the tube bundle, too [12]. The behavior of the bed with an immersed tube bundle is more "calm," due to more small bubbles. Bubbles "carry" particles upwards between the tubes, and they return back downwards along the bed walls. In the case of large relative pitches, bubble size is comparable to the distance between the tubes. When tube distance is small, bubble size is comparable to the tube diameter.

In most experiments it has not been noted that bed height and cross-section significantly influence heat transfer, although these parameters do influence bubble size and character and intensity of particle motion.

Exhaustive experimental research [12] has shown that in the range of $H_{mf} =$ 50-180 mm, heat transfer coefficients increase for both single tubes and tube bundles. With an increase in particle size from 500 to 1850 μm, the influence of bed height decreases due to the smaller influence of particle convection to the total heat transferred (Fig. 3.17).

In bed regions where particle motion is more intensive, heat transfer from the tubes will also be more intensive. It is expected that heat transfer should be less intensive along the walls and immediately above the distribution plate. The given formulae for calculating heat transfer coefficients for tubes also hold true only for

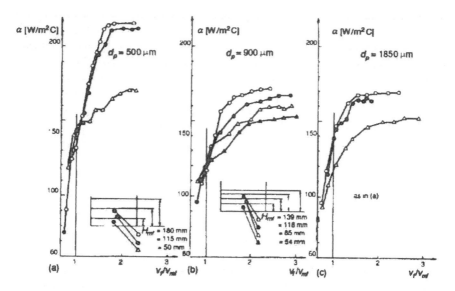

Figure 3.17. *Effect of bed height on heat transfer coefficient between the fluidized bed and immersed single tube. Measuring tube is placed at half bed height. Measurements of B. Grubor [12]*

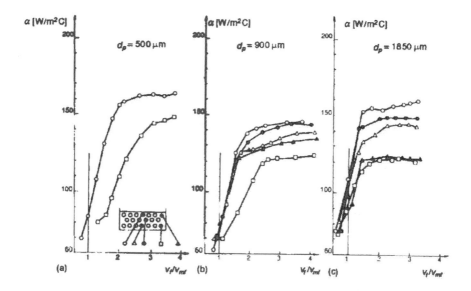

Figure 3.18. *Dependence of heat transfer coefficient for tube in tube bundle immersed in fluidized bed on tube position. Measurements of B. Grubor [12]*

heat transfer surfaces not influenced by bed walls and the distribution plate. Placement of tubes in the bundle will also influence the amount of heat transferred. Tubes in the middle of the bundle, and nearer to the bed surface have larger heat transfer coefficients. The same may be concluded for the whole tube row in the bundle (Fig. 3.18) [12]. Increase in particle size diminishes this influence.

Increasing tube diameter decreases heat transfer intensity, due to the longer particle-to-tube contact time, i. e., the decrease in temperature difference. According to the measurements of Berg and Baskakov [6] increase of tube diameter over 10 mm does not lead to a further decrease in heat transfer coefficient.

Tube inclination, its horizontal or vertical position, does not influence significantly the value of the heat transfer coefficient [8, 11, 12]. The tube pitch in the bundle has a much greater influence. Numerous experiments [8,12] have shown that the greatest influence on heat transfer comes from the horizontal pitch.

In the in-line tube bundle vertical pitch between pipes has no significant influence [9, 63, 75, 81]. The influence of horizontal pitch may be seen in Fig. 3.19 [12].

In the staggered tube bundle, the influence of horizontal pitch depends on the vertical pitch size. According to the measurements of Gel'perin, for relative horizontal pitch of $S_h/D_T > 6$, vertical pitch does not significantly influence the heat transfer (Fig. 3.20) [8]. The influence of pitch also depends on the particle size. With increase in particle size, and decrease in particle convection heat transfer intensity, influence of tube spacing on heat transfer decreases (Fig. 3.21).

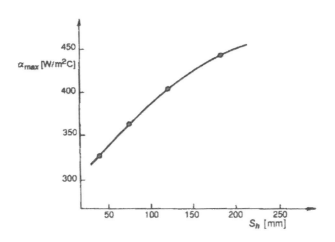

Figure 3.19.
Effect of horizontal pitch on heat transfer coefficient of immersed horizontal corridor tube bundle. According to the measurements of Gel'perin (Reproduced by kind permission of the author Prof. J. F. Davidson from [8])

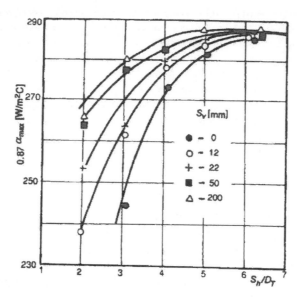

Figure 3.20. *Effect of vertical pitch on heat transfer coefficient for immersed horizontal staggered tube bundle. According to the measurements of Gel'perin (Reproduced by kind permission of the author Prof. J. F. Davidson from [8])*

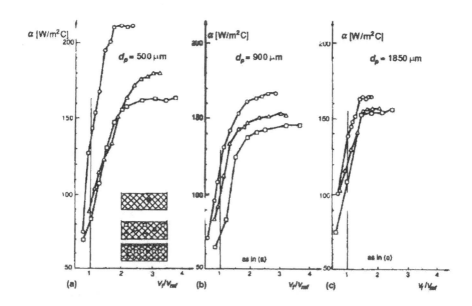

Figure 3.21. *Effect of pitch on heat transfer coefficient for immersed horizontal staggered tube bundle. Measurements of B. Grubor [12]*

The character of particle-to-surface contact and thickness of the gas gap between particles and surface depend on the smoothness of the surface. According to measurements [49], when surface roughness is comparable to the particle size, heat transfer will be more intensive due to the more intensive particle-to-surface contact. Even roughness much greater than particle size may lead to increased heat transfer, due to greater numbers of particle-to-surface collisions.

Similar influence may come from the fins on the tube surface. Fins increase the surface in contact with the bed and are also exposed to particles. Too many fins (decreased fin spacing) may cause a decrease in particle mobility and hence a decrease in heat transfer [79].

3.5.4. Radiative heat transfer in the fluidized bed

Previous sections have dealt with heat transfer processes to the immersed surfaces in the case where the bed temperature is below 600 °C, and when, according to many authors, the radiation component of heat transfer may be neglected. In fluidized bed combustion bed temperature ranges from 800 to 900 °C. At these temperatures heat transfer by radiation cannot be neglected [2, 6, 10, 84, 85]. It also should be borne in mind that temperature of the tube wall is 100-150 °C if the cooling fluid is water.

The influence of bed temperature on the heat transfer coefficient for the immersed tube bundle, according to the measurements of Draijer [2], may be seen in Fig. 3.22. Measurements have been carried out in an experimental boiler of 1.5 MW_{th} power, and with the combustion of coal. Bed particle size has been around 0.8 mm. The increase in bed temperature in the range of 700-900 °C, shows around 15% increase in heat transfer coefficient.

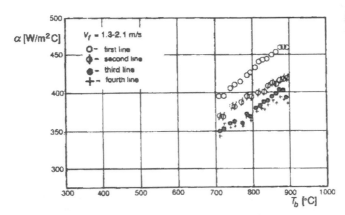

Figure 3.22.
Effect of bed temperature on heat transfer coefficient for immersed horizontal tube bundle. According to the measurements of W. Drijer (Copyright 1986 from [2]. Reproduced by permission of Routledge, inc., part of the Taylor & Fransis Group)

Based on the increase of heat transfer coefficient with the increase in bed temperature and temperature difference between the bed and the surface, it is not easy to discern the influence of heat transfer by radiation from the influence of the changed thermo-physical gas properties. At temperatures common in fluidized bed combustion boiler furnaces, these two effects are of the same order of magnitude and overlap each other, so it is hard to separate their influence in experiments with immersed tubes or tube bundles.

Figure 3.23 shows the results of measuring heat transfer to a single tube immersed in the fluidized bed during coal combustion [86]. Experiments were carried out in an experimental furnace of 0.5 MW_{th} power; inert material was sand d_p = 0.3-1.2 mm; fluidization velocity 1.5-2.5 m/s; diameter of the measured tube D_T = 18 mm; and the cooling fluid was water with inlet temperature of 10-15 °C. Experiments have been carried out without combustion at bed temperature of 350-450 °C, and with lignite combustion at bed temperature of 750-850 °C. The same figure also shows the results of other authors [2, 87] obtained for combustion at similar temperatures. Experimental results have been compared to the empirical correlations of Maskaev and Baskakov (No. 14, Table 3.3), Varrigin and Martjunshin (No. 25, Table 3.3), Broughton (Nu_p = 1.22 $Ar^{0.195}$ [2]), and Grewal and Saxena (No. 15, Table 3.3), obtained under the conditions where heat transfer by radiation may be neglected. Results of these experiments are in the accuracy range of these formulae, ±25%, so in these conditions it has not been possible to determine the influence of radiation on the total heat transfer.

According to the measurements of Baskakov [10] using the probes which measure radiation flux, radiation may account for up to 50% of the total heat transferred when the bed temperature is higher than the surface temperature and for

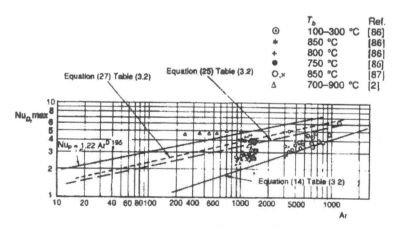

Figure 3.23. *Heat transfer for single tube immersed in fluidized bed with coal combustion. Measurements of B. Arsić [86]. Comparison with experimental data of different authors and correlations given in Table 3.2*

large particles (d_p = 5-7 mm) and dark surfaces. For smaller particles (d_p = 0.5 mm), bed temperature of 1074 K and surface temperature of 400 K, radiation accounts for around 15% of heat transfer (Fig. 3.24).

Heat transfer by radiation to a surface immersed in the fluidized bed can be discussed for two different cases:

– in the period of time when the surface is in contact with the bubble, the surface "sees" the particles of the bubble "cloud," which are at the bed temperature, and
– in the period of time when the surface is in contact with particles ("packet" of particles) the surface "sees" only the first row of particles for which temperature is close to the surface temperature. Particles farther away from the surface, for which temperature is much higher and close to the bed temperature, do not contribute much to the heat transfer. According to Chen and Churchill [88] influence of these distant particles amounts to only 6%.

The closer the temperature of the first particle row to the surface temperature (and that is often the case for small particles), the influence of radiation is smaller. So, in general, the influence of radiation heat transfer increases with larger particle size and higher surface temperature. According to Baskakov [10] for surface temperatures lower than 400 K the influence of radiation is small even when the bed temperature is higher than the surface temperature. Radiation may be neglected when the bed temperature is lower than the surface temperature. These ratios may be seen in Fig. 3.24 where the results of measuring total heat transfer coefficients for surface temperature of 400-1400 K and bed temperatures of 323-1473 K are compared with the convective component of heat transfer coefficient for two types of particles (Al_2O_3) d_p = 0.5 mm and d_p = 5-7 mm.

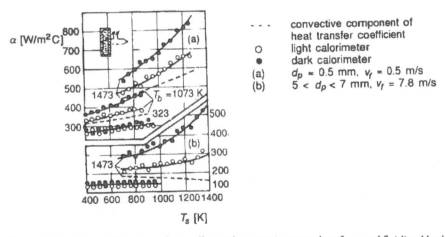

Figure 3.24. *Overall heat transfer coefficient between immersed surface and fluidized bed as a function of surface temperature and bed temperature. Measurements of Baskakov (Reproduced by kind permission of the author Prof. J. F. Davidson from [10])*

The radiation component of the heat transfer coefficient for fluidized beds and immersed surfaces, eq. (3.34) can be calculated using the expression:

$$\alpha_{rad} = \sigma \varepsilon_{bs} (T_b^2 + T_s^2)(T_b + T_s) \qquad (3.44)$$

Emissivity ε_{bs}, can be calculated as for parallel plates at small distance:

$$\varepsilon_{bs} = \frac{1}{\dfrac{1}{\varepsilon_b} + \dfrac{1}{\varepsilon_s} - 1} \approx \varepsilon_b \varepsilon_s \qquad (3.45)$$

for ε_b and $\varepsilon_s > 0.8$.

Emissivity of fluidized bed surface ε_b is higher than the emissivity of particles, ε_p, for the same reason for which the emissivity of a smooth surface is lower than that of the rough surface. Instead of the bed surface emissivity, effective bed emissivity may be used which takes into account cooling of particles near the bed surface. Figure 3.25 shows the values of these emissivities for different bed materials and particle sizes in a bed temperature range of 873-1550 K and surface temperature 540-1500 K, according to measurements of Kovensky [89]. Using these data α_{rad} can be calculated by expressions (3.44) and (3.45).

For bed materials for which $0.3 < \varepsilon_p < 0.6$, the following simplification is recommended [10]:

$$\alpha_{rad} = 7.3 \sigma \varepsilon_p \varepsilon_s T_s^3 \qquad (3.46)$$

3.5.5. Modelling of heat transfer processes to immersed surfaces

Simultaneously with the exhaustive experimental research of heat transfer processes for immersed surfaces, many researchers have tried to make models of heat transfer based on the assumptions about physical processes which take place in the fluidized bed. These models are individual attempts to include all the following processes: conduction and convection in the gas, heat transfer by particle-to-surface contact and particle-to-particle contact, and particle heating. This would enable the prediction of heat transfer intensity and the influence of particular parameters. Numerous models are known, and the most well known are models of M. Leva [108], Levenspiel and Walton [56], Mickley and Fairbanks [91], van Heerden [38], Wicke and Fetting [92], Zabrodsky [93], Ernest [94], Gabor [95], Baskakov [96, 97, 98], Gel'perin and Ainstein [99], Botterill [100, 101, 102], and H. Martin [103]. Detailed reviews and analyses of these models can be found in the books [6-9, 11, 12].

This book will not deal with the details of these models, only with their basic assumptions and stages in their development.

The first proposed models [56, 108] were based on the assumption that the only thermal resistance between the surface and the fluidized bed is the layer of gas

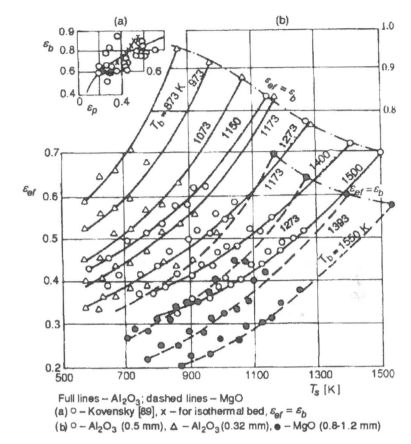

Full lines – Al$_2$O$_3$; dashed lines – MgO
(a) ○ – Kovensky [89], x – for isothermal bed, $\varepsilon_{ef} = \varepsilon_b$
(b) ○ – Al$_2$O$_3$ (0.5 mm), △ – Al$_2$O$_3$(0.32 mm), ● – MgO (0.8-1.2 mm)

Figure 3.25. *Effective emissivities of fluidized bed measured by radiometer [89]. Emissivity measured: (a) radiometer above the surface of fluidized bed, (b) radiometer behind an immersed quartz plate cooled by air flow, giving an effective emissivity (Reproduced by kind permission of the author Prof. J. F. Davidson from [10])*

at the surface. Mixing of particles in the fluidized bed is supposed to be so intensive that heat is transferred to the depth of the bed without any resistance. With these assumptions, the heat transfer coefficient between the surface and the fluidized bed may be defined by the expression:

$$\alpha \approx \frac{\lambda_g}{\delta} \tag{3.47}$$

The main problem in developing models like these is determining the thickness of the gas layer depending on other parameters, above all velocity and number of bed particles hitting the surface. One of the major drawbacks of these models is that they do not take into account thermophysical properties of solid particles. Some later models [93], as well as the models of Wicke and Fetting [92] and others, tried to remove this drawback by taking into account heating of one or more rows of particles around the exchanger surface.

A major breakthrough in development of these models is the "packet" model of Mickley and Fairbanks [91] which, in agreement with the two-phase model of Davidson, assumed that the surface alternately exchanges heat with the bubbles and the emulsion phase. The emulsion phase, in conditions of minimal fluidization, in the form of particle "packets," reaches the surface, heats (or cools) to a certain temperature and then goes to the bed volume where it exchanges heat with other bed particles. Alternating at the exchanger surface, "packets" (groups) of particles transfer heat from the surface. The total heat transfer coefficient may be expressed as:

$$\alpha = \alpha_{gB}\, f_o + (1 - f_o)\alpha_e, \quad \text{where is} \quad \alpha_e = \alpha_{pc} + \alpha_{gc} \tag{3.48}$$

Heat transfer to the particle "packets" is considered to be a non-stationary process. Mickley and Fairbanks have considered particle "packets" as homogenous half-infinite space and obtained the following expression:

$$\alpha_e = \alpha_\tau = \sqrt{\frac{\lambda_{ef}\, \rho_{mf}\, c_p}{\pi \tau}} = \alpha_{pc} \tag{3.49}$$

The basic drawback of these models is the possibility of reaching unrealistically high values for the heat transfer coefficient. It was obvious that another heat resistance should be introduced between the surface and the "packet" of particles. Baskakov [96] and Ernest [94] were the first to suggest taking into account thermal resistance of the gas layer between the surface and particle "packet":

$$\alpha_{pc} = \frac{1}{R_g + R_\tau} \tag{3.50}$$

where:

$$R_g = b\frac{d_p}{2\lambda_{ef}}, \quad R_\tau = \sqrt{\frac{\pi \tau}{\lambda_{ef}\, \rho_{mf}\, c_p}} \tag{3.51}$$

Further model development on the part of Baskakov [97, 98], Gel'perin [99] and Botterill [102] mainly takes into account different ways to "pack" particles next to the exchanger surface. Instead of the particle "packet" porosity which is equal to ε_{mf}, near the surface, greater porosity is assumed. Differences also appear due to different assumptions about the gas layer thickness, i. e., different values of the coefficient b. Xavier and Davidson [9] have used $b = 1/2$, Čatipović [49] $b = 1/3$, and Baskakov [6] has shown that one should use different values for coefficient b to achieve agreements of theory and experimental results for heat transfer in fluidized beds of different particle size.

In fluidized beds of small particles ($d_p < 1$ mm) due to small fluidization velocities heat transfer to bubbles may be neglected, so:

$$\alpha = (1 - f_o)\alpha_e,$$

and for very small particles, $d_p < 0.15$ mm:

$$\alpha_e \approx \alpha_\tau = \frac{1}{R_\tau} \qquad (3.52)$$

At incipient fluidization, there are no bubbles and particle motion, so:

$$\alpha = \alpha_{gc} = \alpha_{mf} \qquad (3.53)$$

In case of large particles ($d_p > 1$ mm) the role of convective heat transfer is the key one:

$$\alpha = (1 - f_o)\left[b\frac{2\lambda_{ef}}{d_p} + \alpha_{gc}\right] + f_o\alpha_{gB} \qquad (3.54)$$

According to Baskakov [6, 96] the difference between total heat transfer coefficient and its convective component α_{gB} may be neglected at $Ar > 10^8$ (i. e., $d_p > 1$-2 mm):

$$\alpha = f_o\,\alpha_{gB} \qquad (3.55)$$

Although earlier models that viewed the fluidized bed as the sum of single particles exchanging heat with the surface [96, 100] did not have much success, the most recent heat transfer model of H. Martin [103, 104] when dealing with particle heat transfer uses the analogy with the kinetic theory of gases. According to the detailed analysis and comparison of different models [12, 13], the model of H. Martin shows very good agreement with results of different experiments [13].

3.6. Heat transfer to the walls of the fluidized bed combustion boiler furnace

Experimental correlations for calculating heat transfer coefficients given in Section 3.5 may be used to calculate the amount of heat obtained by immersed tube bundles in fluidized bed combustion boilers. When using these expressions one of the major problems is determining fluidized bed porosity. Besides that, average particle diameter is used in these expressions and, in FBC boilers, particle size distribution of inert bed material can be very wide.

The given formulae were obtained in experiments which took care that exchangers used in the measurements were far enough (at least 100-150 mm) from the distribution plate and the free bed surface to avoid their influence on the measurement results. These fact must be taken into account when placing heat transfer surfaces in FBC boilers.

Immersed tube bundles are not the only heat transfer surfaces in FBC boilers, in most cases walls of these furnaces are water-tube walls. Water-tube walls stretch from the distribution plate to the furnace exit and convective pass. Water-tube furnace walls take a significant part of the total heat generated. These exchanger surfaces are subjected to very different conditions: a great part of the water-tube wall is in contact with the fluidized bed in which combustion takes place, part of the surface is washed by the particles of the free bed surface, and the greatest part is in the freeboard where solid particle concentration is very low.

In the first approximation, for calculating heat transfer to the part of the water-tube walls in contact with the fluidized bed, expressions given in Section 3.5 may be used, and for calculating heat transfer to the part of the surface in the freeboard, formulae and data used for pulverized coal combustion boiler calculations may be used.

For more accurate calculations one must bear in mind that processes in the fluidized bed boiler furnace are much more complex. Detailed investigations of hydrodynamics, and above all the change of particle concentration along the axial and radial dimension of the fluidized bed furnace, and heat transfer on the furnace walls have been carried out at Chalmers University at Göteborg, in a 16 MW_{th} FBC boiler. The results are shown in [85] together with the general methodology for calculating heat transfer in FBC furnaces.

In Section 3.1 it has already been mentioned that the main differences in conditions (and mechanisms) of heat transfer in the FBC boiler furnace are caused by the change in particle concentration along the bed height, from 1000 to 0.1 kg/m^3. According to the measurements of Andersson [105], heat transfer coefficients range from 650 W/m^2K in the bed to ~60 W/m^2K in the freeboard.

As these investigations have shown [85], tube bundles immersed in the bed decrease fluidized bed density (i. e., increase its porosity), and this fact may have significant influence on the heat transfer when expressions from Section 3.5 are used. Considering heat transfer to water-tube furnace walls which are in contact with the fluidized bed, significant changes in heat flux have been determined along the bed height. Maximum heat transfer coefficient has been measured at the level of fixed bed height, and at half the bed's height heat transfer coefficient drops to almost 50% of its maximum. Near the distribution plate heat flux is very small [85, 105]. The main reason for the decrease in heat flux in regions far from the bed surface is directed particle circulation downwards along the bed walls. These details are not being taken into account by any model of heat transfer in FBC boilers.

In the splash zone, due to eruptive ejection of large numbers of particles by bubbles and instability of the bed surface, heat transfer coefficients to the walls are very high and close to maximum bed values. In the splash zone particle concentration decreases exponentially, and far from the bed surface it becomes nearly constant. In the freeboard, near the walls, almost two times greater than average particle concentrations have been measured [85, 106]. Near the wall are particles with diameter greater than the average diameter bed particle size.

In such complex conditions Bengt-Ake Andersson [5, 107] has proposed the following method for calculating heat transfer to the furnace walls in FBC boilers.

Total heat transfer coefficient includes heat transfer by particles, gas and radiation:

$$\alpha = \alpha_{pc} + \alpha_{gc} + \alpha_{rad} \qquad (3.56)$$

Heat transfer due to particle convection is calculated according to the H. Martin model [103, 104]:

$$\alpha_{pc} = \frac{\lambda_g}{d_p}(1-\varepsilon)Z(1-e^{-N}) \qquad (3.57)$$

where:

$$Z = \sqrt{\frac{gd_p^3(\varepsilon - \varepsilon_{mf})}{5(1-\varepsilon_{mf})(1-\varepsilon)}} \frac{1}{6} \frac{\rho_p c_p}{\lambda_g} \qquad (3.58)$$

$$N = \frac{\mathrm{Nu}_{wp}}{C} Z \qquad (3.59)$$

$$\frac{1}{\mathrm{Nu}_{wp}} = \frac{1}{(\mathrm{Nu}_{wp})_{max}} + \frac{\lambda_g/\lambda_p}{4\left(1+\sqrt{\frac{3}{2}\frac{C}{\pi}\frac{\lambda_g}{\lambda_p}}Z\right)} \qquad (3.60)$$

$$(\mathrm{Nu}_{wp})_{max} = 4\left[(1+\mathrm{Kn})\ln(1+\frac{1}{\mathrm{Kn}})-1\right] \qquad (3.61)$$

$$\mathrm{Kn} = \frac{4}{d_p}(\frac{2}{\gamma}-1)\frac{\lambda_g\sqrt{2\pi RT_b/M_g}}{p(2c_g - \frac{R}{M_g})} \qquad (3.62)$$

$$\gamma = \{1+10^{[K_1-(1000/T_p+K_2)]/K_3]}\}^{-1} \qquad (3.63)$$

In these eqs. C is a constant in the range 1.0–4.0, and the best agreement with measurements is obtained for $2.0 < C < 2.6$, while other constants are $K_1 = 0.6$, $K_2 = 1$, and $K_3 = 2.8$ [85].

When calculating heat transfer in the fluidized bed, up to the level of the fixed bed ρ_p = const. (i. e., ε = const.) has been used, and in the splash zone:

$$\frac{\rho}{\rho_b} = \exp[-(h - H_{mf})/(\overline{H} - H_{mf})] \tag{3.64}$$

where $\overline{H}$ is the extrapolated bed height obtained when the linear change in bed pressure is extrapolated towards zero. In this way particle concentration change above the bed (i. e., high values of heat transfer coefficient) has been taken into account.

Above this zone, in the freeboard, constant particle concentration is used which is equal to the particle concentration at the furnace exit.

This way the furnace is divided into three parts: dense bed, splash zone and the freeboard.

Heat transfer by radiation is calculated using the expression:

$$\alpha_{rad} = \frac{\sigma(T_b^4 - T_s^4)}{\left(\dfrac{1}{\varepsilon_x} - \dfrac{1}{\varepsilon_{bs}} - 1\right)(T_b - T_s)} \tag{3.65}$$

where, in the dense fluidized bed:

$$\varepsilon_{bs} = \left\{\frac{\varepsilon_s}{(1 - \varepsilon_s)B}\left[\frac{\varepsilon_s}{(1 - \varepsilon_s)B} - 2\right]\right\}^{0.5} - \frac{\varepsilon_s}{(1 - \varepsilon_s)B} \tag{3.66}$$

where $B = 0.667$, and in the areas with low particle concentration, i. e., in the freeboard

$$\varepsilon_{bs} = \varepsilon_g + \varepsilon_m - \varepsilon_g \varepsilon_m \tag{3.67}$$

and also:

$$\varepsilon_m = 1 - e^{-\sigma_p L} \tag{3.68}$$

where L is the characteristic dimension of the furnace:

$$L = 3.5\frac{V_F}{A_F} \tag{3.69}$$

and σ_p, is the radiation absorption coefficient of the particles, and is calculated as:

$$\sigma_p = \varepsilon_p n \frac{\pi d_p^2}{4} \tag{3.70}$$

and

$$n = \frac{6\rho_m}{\rho_p \, \pi d_p^3} \tag{3.71}$$

Heat transfer by gas convection is calculated as in the model of H. Martin, according to the Baskakov expression [6]:

$$\mathrm{Nu}_{gc} = 0.009 \, \mathrm{Pr}^{1/3} \, \mathrm{Ar}^{1/2} \tag{3.72}$$

and in the freeboard:

$$\frac{\alpha_{gc} h}{\lambda_g} = 0.332 \mathrm{Pr}^{1/3} \mathrm{Re}_h^{1/2} \tag{3.73}$$

In the fluidized bed all three components of heat transfer are calculated and in the freeboard particle convection may be neglected. Using this model and values for particle concentration (density) measured in the furnace of the 16 MW$_{th}$ FBC boiler at Chalmers University, Andersson [85, 105] has obtained good agreement with the measured values of heat transfer coefficient along the entire height of the furnace walls (Fig. 3.26).

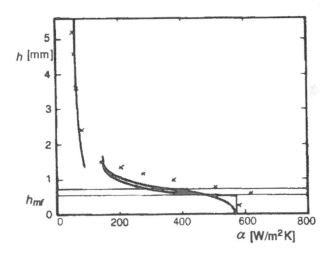

Figure 3.26. *Heat transfer coefficient along the furnace wall, from distribution plate to the furnace exit in Chalmers University demonstration FBC boiler with thermal power of 16 MW. Comparison of measured and calculated values according to Andersson [85]*

Nomenclature

a_p specific particle surface (per unit particle volume) ($= A_p/V_p = 6/\phi_s d_p$), [m²/m³]

a_{ef} effective thermal diffusivity of fluidized bed, [m²/s]

A_F cross section of the furnace, [m²]

A_p external particle surface area, [m²]

A_T cross section of the tube immersed in fluidized bed, [m²]

c_g specific heat of gas at constant pressure, [J/kgK]

c_p specific heat of particle of the inert bed material, [J/kgK]

C_{O_1} oxygen mass concentration in fluidized bed per unit volume (far from active particle), [kg/m³]

$C_{O_{1p}}$ oxygen mass concentration, per unit volume at particle surface, [kg/m³]

d diameter of active particle, diameter of fuel particle, [m] or [mm]

d_p particle diameter of the inert fluidized bed material, [m] or [mm]

D_b dimension (diameter) of fluidized bed cross section, [m]

D_B diameter of bubble in fluidized bed, [m]

D_G molecular diffusion coefficient of gas, [m²/s]

$D_h^{\bullet}$ hydraulic diameter ($= 4A_T/O_T$), [m]

D_s effective particle mixing coefficient in fluidized bed, [m²/s]

D_T diameter of the tube immersed in fluidized bed, [m]

f_T temperature coefficient defined by eq. (3.22)

f_o relative time period of bubble contact with the surface immersed in fluidized bed,

g gravity, [m/s²]

H_b height of the fluidized bed, [m]

h_{Bc} heat transfer coefficient between bubble and "cloud" based on unit bubble surface, [W/m²K]

$(H_{Be})_B$ heat transfer coefficient between bubble and emulsion based on unit bubble volume, [W/m³K]

$(H_{Bc})_B$ heat transfer coefficient between bubble and "cloud" based on unit bubble volume, [W/m³K]

$(H_{ce})_B$ heat transfer coefficient between "cloud" and emulsion based on unit bubble volume, [W/m²K]

I_p enthalpy of all particles in fluidized bed per unit bed volume, [J/m³]

i_p specific enthalpy of all particles in fluidized bed per unit mass of the bed, [J/kg]

k mass transfer coefficient, [m/s]

l length, [m]

L_{max} maximum distance from distribution plate at which temperature of gas jet issuing from bubble cap orifice becomes equal to bed particle temperature, [m]

M_g molecular mass of the gas, [kg/kmol]

O_T perimeter of the tube immersed in fluidized bed, [m]

q heat flux, [W/m²]

q_B volume gas flow rate through the bubble, [m³/s]

q_R heat energy generated during combustion, [J/kg]

r_R radius of the orifice on distribution plate, [m]

R universal gas constant, [J/molK]

R_g heat resistance of the gas, [m²K/W]

R_t heat resistance of particle "packet," [m²K/W]

S_{Bc} surface of bubble to "cloud" interface, [m²]

S_h horizontal pitch of immersed tube bundle, [m]

S_v vertical pitch of immersed tube bundle, [m]

t_b bed temperature, [°C]

T	active particle temperature (fuel particle temperature), [K]
T_b	bed temperature, [K]
T_{gB}	gas temperature in bubble, [K]
T_p	particle temperature, [K]
T_R	gas temperature at orifice on distribution plate, [K]
T_S	temperature of the surface immersed in fluidized bed, [K]
u_t	free fall (terminal velocity) of an isolated particle, [m/s]
v_B	rise velocity of group of bubbles in fluidized bed, [m/s]
v_f	superficial velocity of fluidizing gas (fluidization velocity), [m/s]
v_{mf}	superficial gas velocity at incipient fluidization (minimum fluidization velocity), calculated for solids free bed cross section, [m/s]
v_{opt}	fluidization velocity for which heat transfer coefficient for surface immersed in fluidized bed has maximum value, [m/s]
v_R	gas velocity in orifice on distribution plate, [m/s]
V_B	bubble volume, [m^3]
V_F	volume of the furnace, [m^3]
V_p	particle volume, [m^3]

Greek symbols

α	heat transfer coefficient for active particle (fuel particle) in fluidized bed, also average heat transfer coefficient for surface immersed in fluidized bed, [W/m^{2}K]
α_e	heat transfer coefficient between emulsion phase and tube immersed in fluidized bed, [W/m^{2}K]
α_{gB}	heat transfer coefficient between gas in bubbles and tube immersed in fluidized bed, [W/m^{2}K]
α_{gc}	gas convection heat transfer coefficient, [W/m^{2}K]
α_{gcmax}	gas convection component of the maximum heat transfer coefficient for surface immersed in fluidized bed, [W/m^{2}K]
α_{ge}	heat transfer coefficient for emulsion gas convection, [W/m^{2}K]
α_{max}	maximum heat transfer coefficient for surface immersed in fluidized bed, [W/m^{2}K]
α_{mf}	heat transfer coefficient for immersed tube at incipient fluidization (when there are no bubbles), [W/m^{2}K]
α_p	heat transfer coefficient between gas and fluidized bed particles, [W/m^{2}K]
α_{pc}	particle convection heat transfer coefficient, [W/m^{2}K]
α_{pcmax}	particle convection component of the maximum heat transfer coefficient for surface immersed in fluidized bed, [W/m^{2}K]
α_{rad}	radiation heat transfer coefficient, [W/m^{2}K]
α_T	heat transfer coefficient for tube immersed in fluidized bed, [W/m^{2}K]
α_t	heat transfer coefficient for the particle "packets" and tube immersed in fluidized bed. According to the model of Mickley and Fairbanks, [W/m^{2}K]
γ_B	particle volume fraction in bubbles
δ	thickness of gas layer at the surface of the tube immersed in fluidized bed during contact with bed particles, [m]
ε	void fraction (porosity) of fluidized bed
ε_{bs}	generalized emissivity of the system fluidized bed-immersed surface, including view factor
ε_b	emissivity of the bed surface
ε_g	emissivity of the gas
ε_m	emissivity of the gas-particle mixture in the furnace
ε_{mf}	void fraction at incipient fluidization

ε_p emissivity of the particles in the bed

ε_s emissivity of the immersed heat transfer surface

η_h factor showing possibility for particle temperature to reach gas temperature in bubble, and after that moment not to participate in heat transfer

λ_g heat conductivity of the gas, [W/mK]

λ_{ef} effective heat conductivity of fluidized bed, [W/mK]

λ_p heat conductivity of particle, [W/mK]

μ_g dynamic viscosity of the gas, [kg/ms]

ρ_g gas density, [kg/m^3]

ρ_m density of gas-particle mixture, [kg/m^3]

ρ_{mf} bed density at incipient fluidization, [kg/m^3]

ρ_p density of inert bed particles, [kg/m^3]

ϕ_s particle shape factor

σ Stefan-Boltzman constant (= $5.67 \cdot 10^{-8}$) [W/m^{2}K^4]

σ_p radiation absorption coefficient of the particles

τ contact period of particle "packet" and immersed tube, [s]

Dimensionless criterial numbers

$$\mathrm{Ar} = \frac{g d_p^3 \rho_g (\rho_p - \rho_g)}{\mu_g^2}$$ Archimedes number based on difference between particle and gas densities

$$\mathrm{Ar}_b = \frac{g d_p^3 \rho_g (\rho_b - \rho_g)}{\mu_g^2}$$ Archimedes number based on fluidized bed density

$$\mathrm{Ar}^\circ = \frac{g d_p^3 \rho_g \rho_p}{\mu_g^2}$$ Archimedes number based on bed particle density

$$\mathrm{Ga} = \frac{g d_p^3}{v_g^2}$$ Galileo number

$$(j_D)_{mf} = \frac{k}{v_{mf}} \left(\frac{v_g}{D_G} \right)^{3/2}$$ Dimensionless mass transfer factor, (= $k\, \mathrm{Sc}^{2/3}/v_{mf}$) according W. Prins, eq. (3.16)

$$\mathrm{Kn} = \frac{4}{d_p} \left(\frac{2}{\gamma} - 1 \right) \frac{\lambda_g \sqrt{2\pi R T_b / M_g}}{p \left(2 c_g - \dfrac{R}{M_g} \right)}$$ Knudsen number

$$\mathrm{Nu} = \frac{\alpha d}{\lambda_g}$$ Nusselt number for heat transfer between active particle (fuel particle) and fluidized bed

$$\mathrm{Nu}_{D_r} = \frac{D_T \alpha_T}{\lambda_g}$$ Nusselt number based on diameter of the tube immersed in fluidized bed

$$\mathrm{Nu}_{gc} = \frac{\alpha_{gc} d_p}{\lambda_g}$$ Nusselt number for immersed tube defined by gas convection and bed particle diameter

$Nu_{max} = \dfrac{\alpha_{max} D_T}{\lambda_g}$ Maximum Nusselt number of immersed tube based on tube diameter

$Nu_p = \dfrac{\alpha_p d_p}{\lambda_g}$ Nusselt number for bed particle

$Nu_p = \dfrac{\alpha_T d_p}{\lambda_g}$ Nusselt number of immersed tube based on bed particle diameter, Nu_{pmax} is its maximum value

Nu_{wp} Nusselt number according to the model of H. Martin, eq. (3.60)

$Pr = \dfrac{c_g \mu_g}{\lambda_g}$ Prandtl number

$Re = \dfrac{\rho_g v_f d}{\mu_g}$ Reynolds number based on active particle diameter

$Re_{D_t} = \dfrac{\rho_g D_T v_f}{\mu_g}$ Reynolds number based on immersed tube diameter

$Re_h = \dfrac{\rho_g v_f h_f}{\mu_g}$ Reynolds number based on the height of the furnace

$Re_{mf} = \dfrac{\rho_g v_{mf} d_p}{(1 - \varepsilon_{mf}) \mu_g}$ Reynolds number in eq. (3.16)

$Re_{opt} = \dfrac{\rho_g v_{opt} d_p}{\mu_g}$ Reynolds number for which intensity of heat transfer in fluidized bed has maximum value

$Re_p = \dfrac{\rho_g v_f d_p}{\mu_g}$ Reynolds number for fluidized bed particle

$Sc = \dfrac{\mu_g}{\rho_g D_G}$ Schmidt number

$Sh = \dfrac{kd}{D_G}$ Sherwood number, for active particle in fluidized bed, based on active particle diameter

References

[1] F Johnson, BA Andersson, B Leckner. Heat flow measurements in fluidized bed boilers. Proceedings of 9th International Conference on FBC, Boston, 1987, Vol. I, pp. 592-597.

[2] W Draijer. Heat transfer in FB-boilers. In: M Radovanović, ed. Fluidized Bed Combustion. New York: Hemisphere Publ. Co., 1986.

[3] D Kunii, O Levenspiel. Fluidization Engineering. New York: R. E. Krieger Publ. Co., 1977.

[4] NP Muhlenov, BS Sazhin, VF Frolov, eds. Calculation of Bubbling Fluidized Bed Apparatuses (in Russian). Leningrad: Khimiya, 1986.

[5] OM Todes, OB Citovitch. Apparatuses with Bubbling Fluidized Beds (in Russian). Leningrad: Khimiya, 1981.

[6] VA Baskakov, BV Berg, AF Rizhkov, NF Filipovskij. Heat and Mass Transfer Processes in Bubbling Fluidized Beds (in Russian). Moscow: Metalurgiya, 1978.

[7] JSM Botterill. Fluid-Bed Heat Transfer. London: Academic Press, 1975.

[8] NI Gel'perin, VG Ainstein. Heat transfer in fluidized beds. In: JF Davidson, D Harrison eds. Fluidization, 1st ed. London: Academic Press, 1971. pp. 471–540.

[9] AM Xavier, JF Davidson. Heat transfer in fluidized beds: Convective heat transfer in fluidized beds. In: JF Davidson, D Harrison, eds. Fluidization, 2nd ed. London: Academic Press, 1985; pp. 437-463.

[10] AP Baskakov. Heat transfer in fluidized beds: Radiative heat transfer in fluidized beds. In: JF Davidson, D Harrison, eds. Fluidization, 2nd ed. London: Academic Press, 1985, pp. 465–472.

[11] B Grubor. Heat transfer in fluidized beds – Literature survey (in Serbian). Report of the Institute of Nuclear Sciences Boris Kidrič, Vinča, Belgrade, IBK-ITE-342, 1982.

[12] B Grubor. Heat transfer between fluidized bed and immersed horizontal tubes (in Serbian). MSc thesis, Mechanical Engineering Faculty, University of Belgrade, 1982.

[13] B Arsić. Heat transfer between fluidized beds and immersed tubes – Literature review (in Serbian). Report of the Institute of Nuclear Sciences Boris Kidrč, Vinča, Belgrade, IBK-ITE-327, 1982.

[14] SDj Nemoda. Heat transfer mechanisms and temperature field in fluidized beds (in Serbian). MSc thesis, Mechanical Engineering Faculty, University of Belgrade, 1987.

[15] M Stojiljković, D Milojević, M Stefanović. Transient Heat Transfer in Fluidized Bed (in Serbian). Report of the Institute of Nuclear Sciences Boris Kidrič, Vinča, Belgrade, IBK-ITE-624, 1987.

[16] DJ Gun. Transfer of heat & mass to particles and a gas in non-uniformly aggregated fluidized bed. Int. J. of Heat and Mass Transfer 4:467-492, 1978.

[17] JF Davidson, D Harrison, eds. Fluidization, 2nd ed. London: Academic Press, 1985.

[18] N Wakao, S Kagnei, T Funazakri. Effect of fluid dispersion coefficients on particle-to-fluid heat transfer coefficients in packed beds. Correlation of Nusselt number. Chem. Eng. Sci. 3:325-336, 1979.

[19] R Toei, R Matsuno, H Kojima. Behavior of bubbles in the gas-solid fluidized bed. Kakagu Kogaku 11:851-857, 1966.

[20] AI Tamarin, DM Galerstein, VM Shuklina. Investigation of heat transfer and burning coke particle temperature in bubbling fluidized bed (in Russian). Journal of Engineering Physics 1:14-19, 1982.

[21] AI Tamarin, DM Galerstein, VM Shuklina, SS Zabrodsky. Investigation of convective transfer processes between burning particle and bubbling fluidized bed (in Russian). Proceedings of 6th All-Union Heat and Mass Transfer Conference, Minsk, 1980, Vol. VI, Part 1, pp. 44-49

[22] RD La Nauze. Fundamentals of coal combustion in fluidized beds. Chem. Eng. Res. Des. 1:3-33, 1985.

[23] RD La Nauze, K Jung, J Kastl. Mass transfer to large particles in fluidized beds of smaller particles. Chem. Eng. Sci. 11:1623-1633, 1984.

[24] W Prins. Fluidized bed combustion of a single carbon particle. PhD dissertation, University Twente, Enschede (the Netherlands), 1987.

[25] BV Berg, AP Baskakov. Experimental investigation of heat transfer between bubbling fluidized bed and cylindrical surface (in Russian). Journal of Engineering Physics 1:27-30, 1966.

[26] S Nemoda. Heat transfer coefficients between bubbling fluidized bed and spherical probe with small Biot-number (in Russian). Proceedings of Minsk International Forum Heat and Mas Transfer, Minsk, 1988, Additional volume, pp. 55-59; Termotekhnika, 2:89-100, 1988.

[27] MM Avedesian, JF Davidson. Combustion of carbon particles in a fluidized bed. Trans. Inst. Chem. Engrs. 2:121-131, 1973.

[28] KK Pillai. The effective radial conductivity of fluidized beds. J. Inst. of Energy 419:100-102, 1981.

[29] P Basu, J Broughton, DE Elliott. Combustion of single coal particles in fluidized beds. Institute of Fuel, Symp. Series No. 1, Fluidized Combustion, Vol. 1, pp. A3-1-A3-10, 1975.

[30] AL Gordon, NR Amundson. Modelling of fluidized bed reactors. Part IV: Combustion of carbon particles. Chem. Eng. Sci. 12:1163-1178, 1976.

[31] S Yagi, D Kunii. Studies on effective thermal conductivities in packed beds. AIChE Journal 3:373-381, 1957.

[32] LS Leung, IW Smith. The role of fuel reactivity in fluidized-bed combustion. Fuel 2:354-360, 1979.

[33] JP Congalidis, C Georgakis. Multiplicity patterns in atmospheric fluidized bed coal combustors. Chem. Eng. Sci. 8:1529-1546, 1981.

[34] RK Chakraborty, JR Howard. Combustion of char in shallow fluidized bed combustor: Influence of some design and operating parameters. J. Inst. of Energy 418:48-54, 1981.

[35] P Basu. Measurement of burning rates of carbon particles in turbulent fluidized bed. Proceedings of 2nd International FBC Conference Fluidized Combustion: Is it Achieving Its Promise?, London, 1984, DISC/3/17-DISC/3/25.

[36] RD La Nauze, K Jung. The kinetics of combustion of petroleum coke particles in a fluidized bed combustor. Proceedings of 19th Symposium /Int./ on Combustion. 1983, Pittsburgh: The Combustion Institute, pp. 1087-1092.

[37] RD La Nauze, K Jung. Mass transfer to large particles in fluidized beds of smaller particles. Chem. Eng. Sci. 11:1623-1633, 1984.

[38] C van Heerden, APP Nobel, DW van Krevelen. Mechanism of heat transfer in fluidized beds. Ind. Eng. Chem. 6:1237-1242, 1953.

[39] PK Agarwal, WE Genetti, YY Lee. Model for devolatilization of coal particles in fluidized beds. Fuel 8:1157-1165, 1984.

[40] JF Stubington, Sumaryono. Release of volatiles from large particles in a hot fluidized bed. Fuel 7:1013-1019, 1984.

[41] RK Chakraborty, JR Howard. Burning rates and temperatures of carbon particles in a shallow fluidized-bed combustor. J. Inst. of Fuel 12:220-224, 1978.

[42] IB Ross, MS Patel, JF Davidson. The temperature of burning carbon particles in fluidized beds. Trans. Inst. Chem. Eng. 2:83-88, 1981.

[43] ME Aerov, OM Todes. Hydraulical and Thermodynamical Principles of Operation of Apparatuses with Fixed and Fluidized Beds (in Russian). Moskva: Khimiya, 1968.

[44] AK Bondareva. Measurements of thermal conductivity in bubbling fluidized beds (in Russian). Reports of Soviet Academy of Sciences 4:768-770, 1957.

[45] S Nemoda, G Kanevče. Temperature field in bubbling fluidized beds used for thermal treatment of metallic parts, Proceedings of Minsk International Forum Heat and Mas Transfer, Minsk, 1988, Additional volume, pp. 60-62.

[46] JSM Botterill. Fluidized bed behavior. In: JR Howard, ed. Fluidized Bed Combustion and Applications. London: Applied Sci. Publishers, 1983. pp. 1-36.

[47] AP Baskakov, BV Berg, OK Vitt. et al.. Heat transfer to objects immersed in fluidized beds. Powder Technology 5-6:273-282, 1973.

[48] HS Mickley, DF Fairbanks, RD Hautorn. The relation between the transfer coefficient and thermal fluctuations in fluidized-bed heat transfer. Chem. Eng. Progr. Symp. Series 32:51-60, 1961.

[49] NH Čatipović. Heat transfer to horizontal tubes in fluidized beds. Experiment and theory. PhD dissertation, Oregon State University, 1979.

[50] SS Zabrodsky, NV Antonishin, AL Parnas. On the fluidized bed to surface heat transfer. Canadian Journal of Chemical Eng. 1/2:52-60, 1976.

[51] AR Khan, JF Richardson, KJ Shakiri. Heat transfer between fluidized bed and a small immersed surface. In: JF Davidson, DL Keairns, eds. Fluidization, Proceedings of 2nd Enginering Foundation Conference. Cambridge: Cambridge Univ. Press, 1987, pp. 351-356

[52] AOO Denloye, JSM Botterill. Bed to surface heat transfer in a fluidized bed of large particles. Powder Technology 2:197-215, 1978.

[53] AP Baskakov, VM Suprum. Determination of the convective component of the heat-transfer coefficient to a gas in a fluidized bed. Int. Chem. Eng. 2:119–125, 1972.

[54] WM Dow, M Jakob. Heat transfer between a vertical tube and a fluidized air-solid flow. Chem. Eng. Progress. Symp. Series 637-648, 1951.

[55] RD Toomey, HF Johnstone. Heat transfer between beds of fluidized solids and the walls of the container. Chem. Eng. Progr. Symp. Series 5:51-63, 1953.

[56] O Levenspiel, JS Walton. Bed wall heat transfer in fluidized system. Chem. Eng. Progr. Symp. Series 9:1-13, 1954.

[57] CY Wen, M Leva. Fluidized-bed heat transfer. A generalized dense-phase correlation. AIChE Journal 3:482-488, 1956.

[58] HA Vreedenberg. Heat transfer between fluidized bed and a horizontal tube. Chem. Eng. Sci. 1:52-60, 1958.

[59] HA Vreedenberg. Heat transfer between a fluidized bed and a vertical tube. Chem. Eng. Sci. 2:274-285, 1960.

[60] YuP Kurochkin. Heat transfer between tubes with different cross sections and two-phase flow of granulated materials (in Russian). Journal of Engineering Physics 6:759-763, 1966.

[61] AN Ternovskaya, YuG Korenberg. Pyrite Kilning in a Fluidized Bed. Moskva: Khimiya, 1971.

[62] VG Ainstein. An investigation of heat transfer process between fluidized beds and a single tube. In: SS Zabrodsky. Hydrodynamics and Heat Transfer in Fluidized Beds. Cambridge: M.I.T. Press, 1966.

[63] SC Saxena et al.. Heat transfer between a gas fluidized bed and immersed tubes. In: Advances in Heat Transfer, Vol. 14. London: Academic Press, 1978.

[64] NI Gel'perin, VYa Kruglikov, VG Ainstein. In: VG Ainstein, NI Gel'perin. Heat transfer between a fluidized bed and a surface. Int. Chem. Eng. 1: 67-73, 1966.

[65] NS Grewal, SC Saxena. Heat transfer between a horizontal tube and a gas-solid fluidized bed. Int. J. Heat Mass Transfer 11:1505-1512, 1980.

[66] DB Traber, VM Pomerantsev, IP Mukhlenov, VB Sarkits. Heat transfer between bubbling fluidized bed of catalyzer and heat transfer surface (in Russian). Journal of Applied Chemistry 11:2386-2393, 1962.

[67] AV Chechetkin. High temperature heat carriers. New York: McMillan Co., 1963.

[68] VK Maskaev, AP Baskakov. Characteristics of heat transfer in fluidized bed of coarse particles (in Russian). Journal of Engineering Physics 6:589-593, 1973.

[69] SS Zabrodsky. Hydrodynamics and Heat Transfer in Fluidized Beds. Chambridge: M.I.T. Press, 1966.

[70] VG Ainstein, NI Gel'perin. Heat transfer in fluidized beds. Int. Chem. Eng. 1:61-66, 1966.

[71] A Baerg, J Klassen, PE Gishler. Heat transfer in a fluidized solids bed. Can. J. Research 28:287-307, 1950.

[72] HS Mickley, CA Trilling. Heat transfer characteristics of fluidized beds. Ind. Eng. Chem. 1135-1147, 1949.

[73] VMKo Sastri, MR Vijayarghavan. Effects of surface roughness on heat transfer in fluidized beds. Indian Inst. of Technology, Madras. New York: Hemisphere Publ. Co. 1975.

[74] NN Varrigin, IG Martjushin. A calculation of heat transfer surface area in fluidized bed equipment (in Russian). Journal of Chemical and Mechanical Engineering. 5:6-9, 1959.

[75] SS Zabrodsky, et al.. Heat transfer in a large-particle fluidized bed with immersed in-line and staggered bundles of horizontal smooth tubes. Int. J. of Heat Mass Transfer 4:571-579, 1981.

[76] NI Gel'perin, VG Ainstein, Zaikovski. Hydraulic and heat transfer characteristics of bubbling fluidized bed with immersed horizontal tube bundles (in Russian). Journal of Chemical and Oil-Treatment Mechanical Engineering 3:17-20, 1968.

[77] VV Chekansky, BS Sheindlin, DM Galerstein, KS Antonyuk. Investigation of heat transfer in bubbling fluidized bed with narrow-spaced horizontal staggered tube bundle (in Russian). Reports of the special design bureau of automatization and oil treatment. Oil Chemistry 3:143-152, 1970.

[78] VA Borodulya, VL Gnazha, AI Zheltov, SN Upadhyay, SC Saxena. Heat transfer between gas-solid fluidized beds and horizontal tube bundles. Letters in Heat and Mass Transfer, 1980.

[79] JC Petrie, WA Freeby, JA Buckham. In-bed heat exchangers, finned surfaces on the fluidized-bed side of the tubes increases overall heat transfer rates. Chem. Eng. Progr. 7:45-57, 1968.

[80] BR Andeen, LR Glicksman. Heat transfer to horizontal tubes in shallow fluidized beds. Presented at ASME-AIChE Heat Transfer Conference, St. Louis, U.S.A., 1976, Paper 76-HT-67.

[81] SC Saxena, et al.. Modelling of a Fluidized Bed Combustor with Immersed Tubes. Chicago: University of Illinois, 1976.

[82] WE Genetti, R Schmall, ES Grimett. The effect of tube orientation on heat transfer with bare and finned tubes in fluidized beds. Chem. Eng. Progr. Symp. Ser. 67:90-96, 1971.

[83] L Wender, GT Cooper. Heat transfer between fluidized-solids beds and boundary surfaces – Correlation of data. AIChE Journal 1:15-23, 1958.

[84] WPM van Swaaij, FJ Zuiderweg. The design of gas-solids fluidized beds-prediction of chemical conversion. Proceedings of International Symposium on Fluidization, Toulouse (France), 1973, pp. 454-467

[85] BA Andersson. Heat transfer in stationary fluidized bed boilers, PhD dissertation, Chalmers University of Tech., Göteborg (Sweden), 1988.

[86] B Arsić, S Oka. Heat transfer of the horizontal tube immersed in fluidized bed during coal combustion (in Serbian). Report of the Institute of Nuclear Sciences Boris Kidrič, Vinča, Belgrade, IBK-ITE-478, 1984.

[87] CF Morck, J Olafsson, B Leckner. Operating experiences form a 16 MW atmospheric fluidized bed. Report No. A81–105, Chalmers University of Tech., Göteborg (Sweden), 1981.

[88] TC Chen, SW Churchill. Radiant heat transfer in packed beds. AIChE Journal 1:35-41, 1963.

[89]   VI Kovensky. About calculation of radiative heat transfer in disperse systems (in Russian). Journal of Engineering Physics 6:983-988, 1980.

[90]   OM Todes, OB Citovich. Apparatuses with Bubbling Fluidized Beds (in Russian). Leningrad: Khimiya, 1981.

[91]   HS Mickley, DE Fairbanks. Mechanism of heat transfer to fluidized beds. AIChE Journal 9:374-348, 1955.

[92]   E Wicke, F Fetting. Wärmeübertragung in Gas-Wirbel-Schichten. Chem. Ind. Technik 6:301-309, 1954.

[93]   SS Zabrodsky. Analysis of experimental data on the maximum heat-transfer coefficient of fluidized beds with the surrounding walls (in Russian). Journal of Engineering Physics 4:22-30, 1958.

[94]   R Ernest. Der Mechanizmus des Wärmeüberganges an Wärmeaustauschern in Fliesbetten. Chem. Ind. Technik 3:166-173, 1959.

[95]   JD Gabor. Wall to bed heat transfer in fluidized and packed beds. Chem. Eng. Progress. Symp. Series 105:76-86, 1970.

[96]   AP Baskakov. Mechanism of heat transfer between bubbling fluidized bed and immersed surface (in Russian). Journal of Engineering Physics 11:20-25, 1963.

[97]   AP Baskakov. Heat and Mass Transfer Processes in Bubbling Fluidized Beds (in Russian). Moscow: Metalurgiya, 1978.

[98]   AP Baskakov. The mechanism of heat transfer between fluidized bed and a surface. Int. Chem. Eng. 2, 1964.

[99]   NT Gel'perin, VG Ainstein, AV Zaikovskij. About mechanics of heat transfer between immersed surface and heterogeneous bubbling fluidized bed of coarse particles (in Russian). Chemical Industry 6:418-426, 1966.

[100] JSM Botterill, JR Williams. The mechanism of heat transfer to gas fluidized beds. Trans. Inst. Chem. Eng. 5:217-230, 1963.

[101] JSM Botterill, MHD Butt, GL Cain, KK Redish. Effect of gas and solid thermal properties on the rate of heat transfer to gas-fluidized beds. AAH Drinkenburg, ed. Proceedings of International Symposium on Fluidization, Amsterdam: Univ. Press, 1967, pp. 442-454.

[102] JSM Botterill, AOO Denloye. A theoretical model of heat transfer to a packed or quiescent fluidized bed. Chem. Eng. Sci. 4:509-515, 1978.

[103] H Martin. Heat transfer between gas fluidized beds of solid particles and the surfaces of immersed heat exchanger elements. Chem. Eng. Progr. 1:157-223, 1984.

[104] H Martin. Wärme- und Stoffübertragung in der Wirbelschicht. Chem. Ing. Tech. 199-209, 1980.

[105] BA Andersson, F Johnsson, B Leckner. Heat flow measurements in fluidized bed boilers, Proceedings of 9th International Conference on FBC, Boston, 1987, Vol. 1, pp. 592-597.

[106] BA Andersson, B Leckner. Particle mass flux in the freeboard of a fluidized bed boiler, Powder Technology, 1:25-37, 1989.

[107]   F Johnsson, BA Andersson, B Leckner. Heat transfer in FBB-discussion on models. Presented at the IEA Meeting on Mathematical Modelling on AFBC, Boston, 1987.

[108]   KK Pillai. Heat transfer to a sphere immersed in a shallow fluidized bed. Letters in Heat and Mass Transfer, 1976.

[109]   M Leva, M Weintraub, M Grummer. Chem. Eng. Progress. 9:563-575, 1949.

4.

FUNDAMENTAL PROCESSES DURING COAL COMBUSTION IN FLUIDIZED BEDS

Combustion processes, or better, the set of chemical reactions with oxygen that releases heat from the fuel, depend heavily on the flow regime, heat and mass transfer, basic hydrodynamic and thermodynamic parameters, and fuel properties. Especially, when burning coal, which is physically and chemically a very complex material, the influence of coal properties is important. In this and following chapters due attention will be paid to this and other related problems. The fluidized bed, as a medium in which combustion takes place is also a complex and not yet fully understood system. The previous chapters dealing with hydrodynamics of the fluidized bed, gas and particle mixing processes and heat transfer, provide enough data to properly envisage the complexity of the conditions under which combustion takes place, and it also provides the necessary basis to analyze the influence of these processes on coal combustion and its behavior in the fluidized bed.

The aims of this chapter are: (1) to consider specific conditions under which coal combustion takes place in a fluidized bed and, at the same time, to point out interconnections to other processes taking place in the fluidized bed, (2) to show the important complexities of using coal as a fuel, and especially those properties which can have decisive influence on the combustion processes, (3) to individually consider the major processes and transformations of the coal particle in a fluidized bed, (4) to consider which data and quantities are characteristic of the combustion process, and necessary for further research and theoretical analysis, (5) to point out which data on coal and combustion conditions are necessary for the proper organization of combustion in a fluidized bed and for the design and calculation of FBC boilers for different coals, and (6) to show basic approaches to mathematical modelling of combustion processes in fluidized beds.

This chapter provides the basic knowledge necessary to understand the properties of FBC boilers, their specific design characteristics, and the influence of coal characteristics on the concept and design of FBC boiler furnaces (which is the main issue in the following chapters). This knowledge is also necessary for developing calculation methods suitable for FBC boiler furnaces and for mathematical modelling of combustion processes in general, which are the basis of engineering calculation methods.

This, and following chapters are devoted to coal combustion. However, many of the basic assumptions and conclusions are applicable to the combustion of solid fuels in general, above all to combustion of biomass (wood waste, agricultural waste) and oil shale.

4.1.   Characteristic features of combustion in fluidized beds

In the introductory chapter, the specific properties of FBC boilers have been described as compared to other combustion technologies. When comparing these technologies, it has been stated more than once that the main cause of all differences lies in the presence in the furnace of solid particles of inert material moving chaotically. If, on the other hand, we consider a single coal particle, and the processes and transformations taking place during its combustion, we can rightfully say: the main difference that coal particle "sees" is the presence of inert material particles and the interaction with them.

4.1.1.   Combustion conditions in fluidized beds

In fluidized beds, combustion takes place at temperatures of 800-900 °C, which is much lower than in conventional boilers, for which combustion takes place at 1000-1200 °C. Therefore, the limiting factor for the combustion rate is not only the intensity of O_2 diffusion towards the coal particle surface, but also the rate of chemical reaction.

The temperature field in a fluidized bed is uniform, due to intense inert material particle mixing, and the intense heat transfer between particles and between gas and particles. This ensures favorable and similar conditions for combustion in the entire fluidized bed.

When fuel particles enter the fluidized bed of inert material, due to the chaotic and directed movements of bed particles, they should spread uniformly across the bed. Depending on fuel particle density and size, bed porosity (i. e., its density), and the uniformity of fluidization, the distribution of fuel particles in the bed volume will be more or less uniform. In real bubbling FBC boiler furnaces, however, the fact is that some types of fuel will mostly burn in the upper half of the bed, or at the bed surface with periodical immersions into the bed. Nonetheless, fuel particles will be in continuous motion and in collisions with inert material particles, and also in constant contact with air.

At the moment when a cold fuel particle enters a fluidized bed of hot inert material particles (temperature 800-900 °C), the heating process begins due to continuous collisions with hot particles. From the data given in Chapter 3, the heat transfer coefficient from the fluidized bed to fuel particle of 5-10 mm in diameter is around 300 W/m²K, i. e., at temperature differences between bed and particle of 800 °C, the heat flux at the particle surface is 240 kW/m².

In bubbling fluidized bed boilers coal with a particle size of 5-10 mm and up to 50 mm is used. Therefore, despite the intense heat transfer process, the fuel particle in a fluidized bed is heated at a rate of around 100 °C/s. Devolatilization lasts from 10 to 100 s, depending on the type of coal and particle size used, while residual char burning takes much longer – 100 to 2000 s. This is an important difference from the behavior of particles in pulverized coal combustion boilers, since compared to pulverized coal combustion, these characteristic times are much longer. Thus for example, the devolatilization and burning of char for particles 90 μm in diameter lasts 0.1 and 1-2 s respectively.

The relatively long duration of the combustion processes when burning large fuel particles, and the various interactions of these processes, became of interest as research topics with the development of fluidized bed combustion boilers.

The mass content of coal (i. e., char) as a percentage of the total bed mass in stationary bubbling bed combustion conditions is typically about 2-5%, depending on the coal type, fuel particle size and the char burning rate. This mass content is small enough to make the assumption that one fuel particle is not influenced by other fuel particles in the bed. Any possible fuel particle collisions are likely to be of sufficiently short duration so as not to contradict this assumption. Therefore it is justifiable to consider one lone fuel particle in the "sea" of inert material particles when investigating and modelling fuel combustion processes. The influence of other fuel particles is taken into account only by employing the overall O_2, CO and CO_2 concentrations.

When considering the approach of oxygen molecules to the fuel particle surface, i. e., the process of O_2 diffusion, it must be borne in mind that the inert material particles interfere with this process. The mass transfer process towards the fuel particle is hindered in the fluidized bed, as opposed to heat transfer for which the conditions in a fluidized bed and the presence of inert material particles are favorable. While the temperature field is homogenous in fluidized beds, the concentration field, i. e., oxygen distribution, is very heterogeneous. Fuel particles inhabit the emulsion phase of the bed and consume all or most of the oxygen available in the emulsion phase. Further, a large amount of air, i. e., oxygen, is moving in bubbles and can pass through the bed without reacting with fuel particles. The concentration differences of O_2, CO and CO_2 between bubbles and emulsion can be very great indeed, and the process of mass transfer (i. e., the process of O_2 diffusion) from bubbles to emulsion can be a limiting factor for combustion processes in the fluidized bed.

Obviously, the conditions under which the combustion of coal particles in a fluidized bed take place, are significantly different from the combustion conditions in other types of boilers. Research on these combustion processes requires an understanding of processes in fluidized beds which were the subject in previous chapters – namely, the processes of particle motion and mixing, motion of bubbles and processes of mass transfer between bubbles and emulsion phase, heat transfer between fuel particles and fluidized bed, and gas diffusion to and from the fuel particle.

4.1.2. Physical processes during coal particle combustion in fluidized beds

From the moment it enters the fluidized bed of hot inert material particles, a coal particle (its combustible and mineral components) suffers an array of physical, physicochemical, and chemical transformations. Analysis of these transformations and processes is made difficult by the fact that they normally occur simultaneously, so it is difficult to follow their separate development and properties.

Due to the intense heat transfer, the coal particle temperature rises rapidly after entering the fluidized bed. When the temperature of 100 °C is reached, an intense process of evaporation and drying begins.

Further temperature increase causes thermal degradation of hydrocarbons, i. e., the release of volatiles. The process of volatile evolution begins at temperatures around 450 °C. Moisture and volatile release may cause particle fragmentation into several pieces as the intense heat transfer to the fuel particle causes rapid evaporation and formation of volatiles inside the particle. If the pores in the coal particle are not capable of conducting all of the evolved gaseous matter to the particle surface, the inner pressure will break the particle into several pieces. Fuel particle disintegration (or fragmentation) in a fluidized bed is a very important process that dramatically changes the conditions for the subsequent combustion process, as the combusting surface, an important parameter for heat generation, changes dramatically.

Volatiles form an upward elongated cloud around the fuel particle, which diffuses and penetrates the emulsion phase and enters the bubbles, and also if favorable conditions for mixing with oxygen exist, ignites, burning around the particle, in the emulsion phase but also in the bubble phase. Very often, the conditions are such that there isn't enough time for volatiles to burn completely in the bed, so their burning continues in the freeboard. This is why creating favorable conditions for mixing with the rest of the oxygen in the freeboard is very important. Volatile evolution, its ignition and combustion, precedes or overlaps the process of igniting and combustion of the char itself.

Char formed after the volatile evolution has an inner structure which is very different from that of the initial coal. Combustion of char can take place at the surface or in the pores of the particle. For oxygen to reach the carbon with which it must react, it must overcome a number of obstructions to mass transfer: (1) resistance to mass transfer between the bubble and emulsion phase, (2) resistance to oxygen diffusion through the emulsion phase to the coal particle surface, and (3) resistance to penetrating open pores in the char particle. At the same time a burning char particle can behave in at least two different ways: (1) simultaneously with carbon combustion, mineral components (ash) in the coal particle separate, husk or peel off, so that the coal particle dimension decreases steadily, and (2) burning inside the particle, so that the carbon gets used up leaving the mineral skeleton of the particle intact – particle dimensions remains the same until the end of combustion.

During char combustion, due to collisions with inert material particles, the process of char particle wear (or attrition) is also inevitable. Small particles of char separate from the main particle due to the mechanical influence of bed particles. This process, called attrition, depends on the coal type, and may produce a significant amount of very small combustible particles. These particles, and the final residue of large particle combustion, are removed from the bed by the combustion products, and usually they do not even burn in the freeboard, representing the main source of combustion losses (unburned carbon in the fly ash).

Release and combustion of volatiles can have a major influence on coal combustion characteristics in the fluidized bed. The sudden release of a large quantity of volatiles, most of which will not be able to burn near the particle, and the movement of unburned gaseous matter through the bed, its mixing with oxygen, ignition and burning conditions, demand serious investigation. Despite the very intense mixing, some gaseous combustible matter leaves the bed unburned and providing the conditions for its combustion above the bed is a very difficult challenge, which has to be met by proper design of the combustion process in fluidized bed boilers.

Figure 4.1 shows all of the processes mentioned above and outlines their complexities when combusting a coal particle in a fluidized bed. What cannot be shown in a figure like this, and what in real life makes following and analyzing these processes difficult, is the constant chaotic movement of fuel and inert bed material particles.

In the following sections the processes experienced by the fuel particle will be described in more detail: heating, drying, evolution and combustion of volatiles, particle fragmentation, char combustion and attrition. However, a more detailed account of processes taking place above the bed is not the subject of this book. In

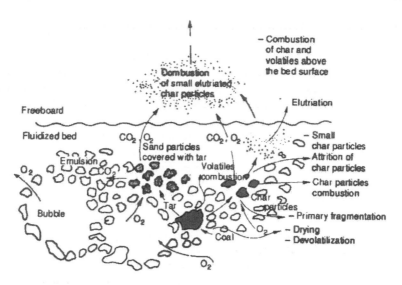

Figure 4.1. *Coal particle burning in the fluidized bed – schematic of the different processes*

this part of the furnace, the processes of volatile combustion, and the environment for unburned, small char particle combustion and the heat and mass transfer processes themselves, are similar to those in pulverized coal combustion, so the reader is directed to the literature on conventional combustion for such problems [1, 2].

Finally, it is necessary to mention one process specific to fluidized bed combustion. When fuel is supplied to the surface of the bed, during its passage through the high-temperature combustion products in the freeboard, such coal particles can suffer significant changes even before they reach the bed surface. Depending on the particle size, coal characteristics, temperature in the freeboard and the time necessary to reach the fluidized bed surface, the fuel particle can totally or partially lose its moisture and release volatiles and char, and burning can start. Some particles will not even reach the surface of the bed, but will be carried out of the furnace, regardless of their state of combustion. Thus, for overbed feeding, one must take into account the fact that the properties of coal particles that reach the bed surface and continue burning are no longer the same as before entering the furnace.

The behavior of shale and the mineral part of coal particles (ash) is a significant problem that has to be considered along with the combustion processes in fluidized bed boilers. Ash can stay in the bed and replace inert material particles, it can be carried out of the bed as fly ash, or it may "sink" to the bottom of the fluidized bed. Investigating ash behavior is important for optimizing coal feeding and combustion, and the overall concept and design of FBC boiler furnaces. In this chapter, we will not further take into account whether the inert material is sand or coal ash, or consider further what happens to that ash; instead, later chapters will deal with ash behavior.

4.1.3. Parameters influencing combustion in the fluidized bed

A discussion of parameters which influence coal combustion in the fluidized bed, as well as combustion in all other conditions, must begin with a consideration of the temperature and excess air. The bed temperature in the furnace stationary regime has to be chosen and maintained under very narrow boundaries. An upper limit is provided by the ash sintering temperature and optimum temperature for the reaction of CaO and SO_2, when high-sulfur fuels are burned, and the lower temperature limit is set by the minimum temperature necessary for complete combustion. It is usual to select bed temperature between 800 and 900 °C. At these combustion temperatures it is most often assumed that burning takes place in the diffusion regime, but depending on particle size, porosity and reactivity of coal, kinetics of the chemical reaction can also be a limiting factor for combustion rate. Measurements show [3] that in char combustion the particle temperature is 50-250 °C higher than the bed temperature (depending on the fuel particle size, fluidization velocity and bed particle size), which justifies the assumption that conditions may exist at which the bed temperature, i. e., kinetics of chemical reaction, is not a significant factor for coal combustion processes in the fluidized bed in a stationary regime. In experimental research on coal combustion in fluidized beds, the influence of bed temperature is nevertheless regularly explored for the following reasons:

- the influence of temperature on other processes – particle heating, drying and release of volatiles,
- the need to know the devolatilization rate, rate of volatiles and char burning at low temperatures which occur at boiler start-up,
- the boiler needs to operate in a stationary regime at lower power and at temperatures lower than 800 °C (sometimes as low as 700 °C), and
- the determination of the char burning rate at different temperatures allows estimation of the activation energy.

When considering excess air, bearing in mind the specific conditions of combustion in the fluidized bed due to the existence of bubbles, one must first address the ability of oxygen to reach the combustible material, either volatile, or carbon in char.

The fact that a significant amount of air, i. e., O_2, can bypass the fluidized bed in bubbles that do not react with the fuel, means that a simple increase of excess air cannot ensure complete combustion, and typically excess air is set at 120 to 130 percent in a FBC boiler. Optimum conditions for efficient combustion are also provided by: proper and vigorous fluidization, proper, and regular fuel distribution, good mixing of volatiles in the bed, and intensive "washing" of bubbles by gas from the emulsion phase. These conditions are achieved by employing higher fluidization velocity and by good organization of the fluidization, by producing smaller bubbles, and in a fluidized bed of larger particles by ensuring that one is in the "slow" bubble regime.

Fluidization velocity and the size of inert material particles also influence the processes of heat and mass transfer between fuel particles and the bed, and hence the rate of volatiles release, the conditions for their ignition, and the ignition of char. An increase of fluidization velocity and decrease of inert material particle size intensify these processes, while the char particle attrition process also depends on the fluidization velocity and the intensity of chaotic movement of the bed particles.

The type and properties of coal have a decisive influence on the combustion process, and all other processes occurring in a FBC. At this point we will mention: fuel particle size, volatiles content and the properties of the mineral part of the fuel, as well as the porosity of fuel particles. These properties influence the processes of particle fragmentation during volatiles evolution, the devolatilization rate, fuel particle temperature, intensity of kinetic and diffusion regimes of char combustion, manner of combustion and char attrition process.

Research on the combustion process and coal particle transformation has potentially very practical design and fundamental goals. Data acquired from both applied and fundamental research contribute to optimizing combustion in FBC boilers and more accurate prediction of their properties and behavior during operation.

To be able to choose appropriate calculation methods to design FBC boilers, it is necessary to have the following data: optimum temperature at which to start feeding fuel at the boiler start-up, the optimum temperature for steady state operation, the ratio of heat generated in the bed and above it, the burning rate of coal particles and coal mass content in the bed, expected losses due to unburned carbon in fly ash due to the incomplete burning process, how to choose the method for coal feeding, and the appropriate coal particle size.

To predict properties, or for the mathematical modelling of processes in the furnace, it is necessary to know the characteristic temperatures, process rates and durations, for all processes.

The characteristic temperatures are: temperature at which evolution of combustible volatiles begins, the volatiles ignition temperature and the char ignition temperature. For most applications knowledge of the temperature at which coal particle ignition occurs is sufficient.

The characteristic process rates are: the devolatilization rate, volatiles burning rate and char burning rate, i. e., coal particle burn-out rate (or fixed carbon consumption rate).

Characteristic times are: the time for heating and drying of coal particles (these can be calculated very precisely), devolatilization time and volatiles burn-out time, char burn-out time, or for most practical applications the overall coal particle burn-out time.

For processes involving particle fragmentation and char attrition, the characteristic features are: number and size of particles after fragmentation and mass flow rate of char particles which are the result of attrition.

Most of the features mentioned above can currently only be determined experimentally for a given coal, typically by performing experiments in which the various dominant parameters are varied over a suitably wide range.

By determining some fundamental characteristic quantities for a process, for example the activation energy for volatiles evolution, activation energy and heat of reaction for volatiles and char combustion, and by postulating the physical and mathematical model for all processes that the coal particle undergoes, however, it is possible to reduce the required level of experimental research necessary to characterize a particular coal.

4.2. Coal as combustible matter

Engineering experience gained over more than 200 years clearly demonstrates that combustion process properties depend strongly on the type of coal used. However, despite numerous experimental studies and theoretical analyses and much long-term experience gained in furnaces designed for a wide range of different types of coal, "hunches" and intuitive deduction are probably more important than the use of "proven and known" relationships in the combustion process and coal properties.

A good example of this is the relation between coal flammability (i. e., coal ignition temperature) and the content of volatiles in coal. Coals with higher volatile content ignite at lower temperatures, and have higher char burning rates – they are "more reactive." Experiments show that, for example, Kolubara lignite ignites in fluidized beds at temperatures of 300-350 °C [4], although volatile matter (VM) does not ignite below 400 °C, which is somewhat contradictory. It can be, however, that although the general relation of some combustion properties to VM content in coal is known, it is still not clear whether it is the influence of VM alone, or that the VM content is just a general measure of coal quality, its age or some other quality. What can be said is that knowing the coal properties is essential to understand their behavior during combustion.

4.2.1. Classification of coals

Coal is organic sediment rock, created by the carbonification of large layers of plant matter [5, 6]. The degree of carbonification shows how much of the original plant matter has changed its structure and how closely it approaches the properties of pure carbon. From the carbonification degree the quality (rank) of coal is determined, i. e., its type. Different approaches exist for classification of coals. Internationally acknowledged classification ranks coals by the percent of volatiles and coking properties. By American standards (ASTM), older coals are ranked by

% VM, and younger coals by calorific value. Table 4.1 gives the ASTM coal classification.

Table 4.1.   *ASTM coal classification*

No.	Name of coal according to ASTM	Volatile content* [%]	Calorific value** [MJ/kg]
1.	Meta anthracite	0-2	>35.4
2.	Anthracite	2-8	>35.4
3.	Semi-anthracite	8-14	>35.4
4.	Low-volatile bituminous coal	14-22	>35.4
5.	Medium-volatile bituminous coal	0-2	>35.4
6.	High-volatile bituminous coal A	31-40	32.6-35.4
7.	High-volatile bituminous coal B	31-45	30.2-32.6
8.	High-volatile bituminous coal C and subbituminous coal A	35-45	25.6-32.6
9.	Subbituminous coal B	35-50	19.4-25.6
10.	Lignite	40-50	<19.4

* dry mineral matter free (dmmf)
** with analytically bound water
Comment: -- from 1. to 5. classification is by volatile content
           -- from 6. to 10. classification is by volatile content and calorific value

International classification divides coals into classes, and introduces class numbers 0-9, with slightly different ranges for VM content and calorific value. Table 4.2 shows international classification, along with the coal names by American standards, and Table 4.3 shows coal classification by Yugoslav standards.

Table 4.2.   *International coal classification*

No.	Volatile content* [%]	Calorific value** [MJ/kg]	ASTM coal name
0.	0-3	>34.4	Meta anthracite
1.	3-10	>34.4	Anthracite
2.	10-14	>34.4	Semi-anthracite
3.	14-20	>34.4	Low-volatile bituminous coal
4	20-28	>34.4	Medium-volatile bituminous coal
5.	28-33	>34.4	
6.	33-40	32.6-34.4	High-volatile A bituminous coal
7.	33-45	30.2-32.6	High-volatile B bituminous coal
8.	33-50	25.6-30.2	High-volatile C bituminous coal and subbituminous A coal
9.	40-50	19.4-25.6	Subbituminous B coal

*   dry mineral matter free (dmmf),
** with analytically bound water

Table 4.3. *Yugoslav coal classification according to Yugoslav standards (YUS B.HO.001)*

No.	YUS coal	Total moisture [%]	Lower calorific value* [MJ/kg]	VM* [%]
1.	Anthracite	<5	>30	<15
2.	Stein coal	<10	>30	<40
3.	Brown coal	10-30	26-30	-
4.	Brown-lignite	30-40	25-26	-
5.	Lignite	>40	23-25	-

* daf – dry, ash free

Chemical composition of the combustible matter, and above all carbon content, follows coal division for all such classifications. As the carbonification degree increase, the carbon content also increases, and the hydrogen and oxygen contents decrease, as can be seen in Table 4.4.

Table 4.4. *Average chemical composition of coal combustible matter*

Coal type	C* [%]	H* [%]	O* [%]	H_2O* [%]
Lignite	65-72	4.5	30	>15
Subbituminous coal (brown coal)	72-76	5.0	18	10-15
High volatile bituminous coal C (brown coal)	76-78	5.5	13	5-10
High volatile bituminous coal B (gas coal)	78-80	5.5	10	3-5
High volatile bituminous coal A (gas steincoal)	80-87	5.5	4-10	1-2
Medium volatile bituminous coal (fat coal)	89	4.5	3-4	<1
Low volatile bituminous coal	90	3.5	3	<1
Anthracite	93	2.5	2	<1

* daf – dry ash free

For classification of coals into classes and subclasses, it is also necessary to know a few basic properties: volatiles content, calorific value, carbon, hydrogen and water content, etc. These coal properties along with some others, are routinely determined by the so-called proximate and ultimate analyses.

From the standardized proximate analysis the following properties are determined: the water content as received, the content of analytical water, ash content, combustible matter content, volatiles content, fixed carbon content, char content, total sulphur content (sulphate, sulphide, and organic), carbon dioxide, higher and lower calorific value, and carbon dioxide in char. These quantities are determined for the following bases: with water content as received, with analytical water content, dry, and dry ash free (daf).

Standardized ultimate analysis of combustible matter also gives percentage of carbon, hydrogen, oxygen, nitrogen and combustible sulphur. Data are provided below for typical types of coal.

Along with proximate and ultimate analyses, the particle size distribution of coal can be determined, and for fluidized bed combustion it is especially important to determine the content of particles smaller than 1 mm in diameter. Coal classification by mean particle size is given in Table 4.5.

Table 4.5.   *Coal classification by particle size according to YUS*

Assortment [mm]	Steincoal and anthracite	Brown coal	Brown-lignite and lignite	Dried coal
Piece	60-250	60-250	80-350	>60
Cube	30-60	30-65	40-120	30-60
Nut	15-30	15-40	20-65	12-30
Pea	5-20	5-20	10-35	5-15
Small	0-10	0-15	0-20	–
Groats	0-5	3-15	5-20	–
Dust	0-3	0-5	0-10	0-5

Besides the above mentioned data, in coal analyses, characteristics of the mineral part of the coal, so-called chemical analysis of laboratory ash, as well as characteristic temperatures of ash – temperature of beginning of sintering, half sphere softening temperature and melting temperature, are determined. Table 4.6 shows the standard way of presentation for the proximate and ultimate analyses, chemical ash analysis and characteristic ash temperatures. The basic constituent parts of coal and definitions of quantities determined by proximate and ultimate analysis are shown in Fig. 4.2.

Due to the thermal decomposition of coal taking place before and during the combustion process, the combustible matter is reduced into its volatile and nonvolatile components. The solid residue is called char. In addition, the mineral part of the coal also suffers changes and we call such newly formed components – ash. The characteristic times of evolution and combustion of VM are much shorter (10-100 s depending on coal particle size) than char combustion time (100-2000 s). The slower process, in this case reaction of heterogeneous oxidation of carbon and other hydrocarbon compounds in char, has a decisive influence on the combustion process and the time coal particles must spend in the furnace to ensure efficient combustion. To envisage all the characteristics of coal mass transformation during combustion it is necessary to know and analyze the basic data on chemical structure of organic matter, structure and composition of the mineral components and the porosity of coal and char particles. Data obtained by proximate and ultimate coal analyses demonstrate that engineering problems also demand some basic knowledge of these coal properties.

Table 4.6.   *Proximate and ultimate coal analysis*

VINČA INSTITUTE OF NUCLEAR SCIENCES
Laboratory for thermal engineering and energy

COAL AND ASH ANALYSIS

MINE: KOLUBARA
FIELD:
COAL NAME:
DATE: 01. 02. 1989.
NOTICE: AS RECEIVED

COAL ANALYSIS

	COMPONENT	Unit	BASIS			
			With moisture as received	With chemically bounded moisture	Dry	Dry, ash free
	Moisture as received	%	58.34			
	Chemically bound moisture	%	9.09	17.91		
Proximate anal.	Ash	%	7.99	15.74	19.18	
	Combustible matter	%	33.67	66.35	80.82	100.00
	Volatiles	%	19.13	37.69	45.91	56.80
	Fixed carbon	%	14.31	28.20	34.35	42.50
	Char	%	22.54	44.41	54.09	43.20
	Total sulphur	%	0.53	1.05	1.28	1.58
	Sulphur in sulphates	%	0.30	0.59	0.79	
	Sulphur in sulphides	%	0.00	0.01	0.01	0.01
	Organic sulphur	%	0.23	0.45	0.55	0.68
	Higher calorific value	kJ/kg	8521	16790	20452	25305
	Lower calorific value	kJ/kg	6769	15573	19472	24092
Ultimate an.	Carbon	%	21.52	42.41	51.66	63.92
	Hydrogen	%	1.98	3.91	4.76	5.89
	Combustible sulphur	%	0.23	0.46	0.56	0.69
	Nitrogen	%	0.45	0.89	1.09	1.35
	Oxygen	%	9.48	18.68	22.75	28.15

CHEMICAL ASH ANALYSIS

Component	SiO_2	Al_2O	Fe_2O_3	CaO	MgO	Na_2O	K_2O	TiO_2	P_2O_3	SO_3
[%]	53.85	21.40	9.50	7.47	2.83	0.22	0.84	1.03	0.08	2.33

ASH MELTING (FUSIBILITY)

Characteristic temperatures	Unit	Oxidation conditions	Reduction conditions
Sintering temperature	°C	890	1040
Softening temperature	°C	1180	1110
Halfsphere temperature	°C	1260	1205
Melting temperature	°C	1298	1265

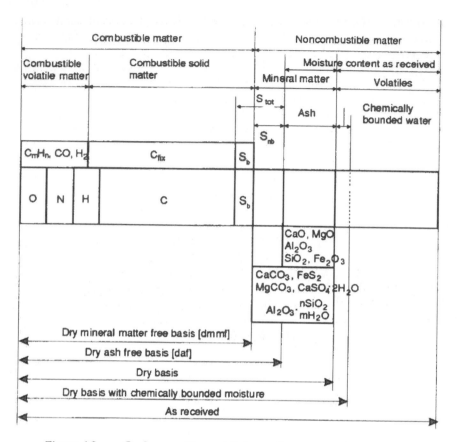

Figure 4.2. *Coal composition – definition of components determined by standard ultimate and proximate analyses: S_{tot} – total sulphur, S_b – sulphur burnable, S_{nb} – sulphur nonburnable*

In this section we will consider only the most basic properties of the structure of coal as a sediment organic rock [7, 8]. For more details the reader should see the literature [5, 6, 9].

4.2.2. Coal petrography

The coal classification given in the previous section can show complex heterogeneous coal structure only in a most approximate but still simple and useful way. Coal petrography gives more detailed data on the coal structure in a way usual for the investigation and classification of other rocks in the earth's core.

Petrography is based on visual observations of thin coal slices (rocks) as well as observing their behavior when illuminated, their color and consistency. Unequivocal relations cannot be established between this type of coal classification and coal classification by the degree of carbonification, and it is impossible to explain coal behavior in combustion by petrographic analysis alone.

The basic coal constituents are called macerals. Macerals originate from parts of the plant material from which the coal originated and they change their properties during the carbonification process [5, 9].

Three basic types of macerals exist in coal: vitrinite, exinite, and inertinite. Vitrinite is the basic coal constituent (60-90% of mass), it is transparent when obtained in thin slices, and its chemical composition and properties depend on the degree of carbonification. Exinite is a part of coal which is rich in hydrogen. It originates from parts of plants resistant to decomposition and it accumulates through decomposition of "secondary" parts of plants – spores, pollen, resin, cork, algae and fungi. Inertinite is non-transparent in thin slices, and often resembles wood-coal, it always has a pronounced wooden structure. Carbon content (mass % of C) increases, and H/C ratio decreases in macerals in the order: exinite, vitrinite, inertinite. The older the coal is, the smaller the differences between the macerals are. Maceral behavior during devolatilization (pyrolysis) shows a great influence of petrographic structure (origin). Vitrinite is the part of coal from which char is made, exinite becomes liquid and decomposes to gases and tar, and inertinite neither evaporates nor softens. It has been determined that combustion efficiency is inversely proportional to inertinite content, which influences carbon content in ash particles [7].

4.2.3. Chemical structure of organic matter in coal

The basic chemical structure seen in coal is that of aromatic hydrocarbons for which the benzene ring is the fundamental unit. This ring consists of six carbon atoms bonded by three resonance stabilized single and three double alternating bonds. Hydrogen atoms occupy the remaining free bonds of carbon atoms (Fig. 4.3). This basic unit can be combined into higher, polynuclear aromatic and naphthene hydrocarbons, or one or more hydrogen atoms may be replaced by other hydrocarbons or groups ($-OH$), ($-COOH$), ($=CO$) or by sulphur atoms.

Figure 4.4 shows a few more complex combinations which form a "coal molecule." Polynuclear aromatic groups (Fig. 4.4a) are the basic structure of the coal molecule.

As can be seen in Fig. 4.4, in a "coal molecule" these rings are usually connected by methylene bonds ($=CH_2$). The hydroaromatic group (Fig. 4.4b) contains hydroxyl group ($-OH$), aliphatic groups contain hydrocarbon ($-CH_3$), ($-C_2H_5$) or a methylene bond ($=CH_2$), (Fig. 4.4c, and 4.4d). These combinations contain most of the hydrogen in coal. Oxygen in coal is bonded in groups in which the hydrogen atom is replaced by a hydroxyl group ($-OH$), carboxyl group ($-COOH$) or carbonyl

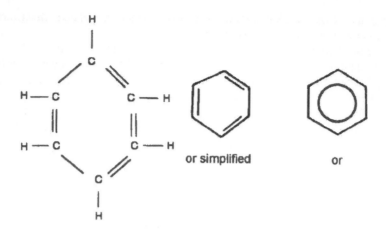

Figure 4.3. *Benzene ring*

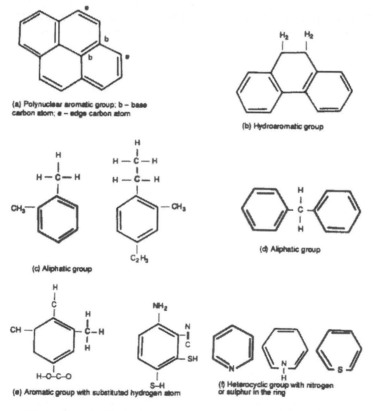

(a) Polynuclear aromatic group; b – base carbon atom; e – edge carbon atom

(b) Hydroaromatic group

(c) Aliphatic group

(d) Aliphatic group

(e) Aromatic group with substituted hydrogen atom

(f) Heterocyclic group with nitrogen or sulphur in the ring

Figure 4.4. *Different functional groups in coal chemical structure*

group (=CO), Fig. 4.4e. Sulphur and nitrogen are bonded in heterocyclic groups – pyridine, pyrol, and thiophene (Fig. 4.4f).

The chemical structure of a typical bituminous coal [7] is given in Fig. 4.5. This "coal molecule" consists of an array of aromatic/hydroaromatic clusters containing 2-5 benzene rings, mutually connected by CH_2, C_2H_4 or C_3H_6 groups, sulphide bonds (S or S–S) or eteric bonds (over oxygen).

Aliphatic bonds are weak, they allow clusters to form and position in multiple planes, which allows for their interconnections, and ability to form porous structures. Aliphatic, hydroaromatic and heterocyclic bonds are also more susceptible to breaking and allow for evolution of volatile matter. Char consists of polynuclear aromatic structures, rich in carbon. In these structures, edge carbon atoms are at least an order of magnitude more reactive than the carbon atom.

Since the rank (age) of coal is a measure of its structural similarity to the structure of graphite, it is logical to expect that older coals have less aliphatic and hydrocyclic structures. This also produces a lower porosity. Here, the aromatic bond ratio is the lowest in exinite, and the highest in inertinite. In the pyrolysis and carbonification process at high temperatures, the carbon in aromatic bonds stays in the char (as fixed carbon), tar originates from hydrocarbons in hydroaromatic structures, and volatile matter (CH_4, CO and CO_2) is the product of hydrocarbons in aliphatic structures.

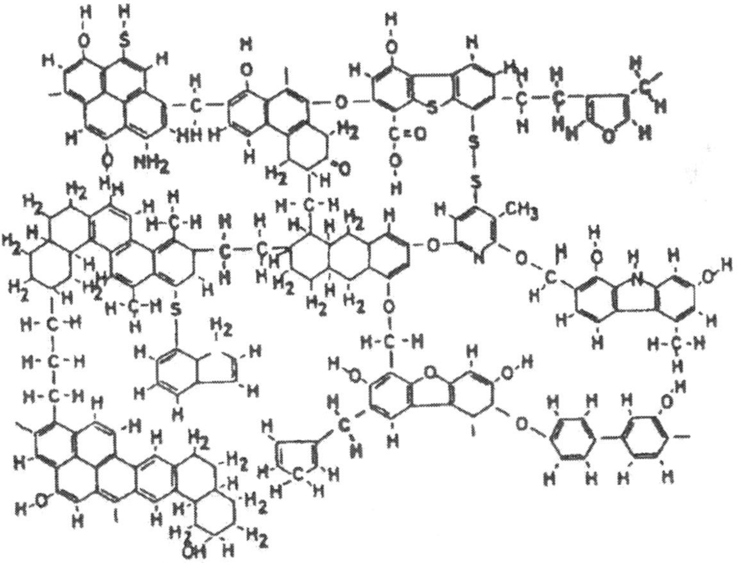

Figure 4.5.   *Chemical structure of bituminous coal (Reprinted from [7]. Copyright 1978, with permission from Elsevier Science)*

4.2.4. Chemical structure of mineral matter in coal

The mineral matter in coal has most probably been formed by the effect of "geological water flows." The following minerals are among those most often found in coal:

(1) aluminosilicates (clays) – kaolinite, $Al_2SiO_3O_5(OH)_4$ and illite $KAl_3Si_3O_{10}(OH)_2$, make about 50% of mineral matter in coal,

(2) oxides – silicates, SiO_2, hematite, Fe_2O_3, about 11%,

(3) carbonates – calcium carbonates, $CaCO_3$, siderite, $FeCO_3$, dolomite $CaCO_3 \cdot MgCO_3$, and ankerite, $2CaCO_3 \cdot MgCO_3 \cdot FeCO_3$, about 10%, and

(4) sulfides and sulfates, as pyrite, FeS_2 and gypsum $CaSO_4 \cdot 2H_2O$, make the remaining 25%.

Mineral matter is most often dispersed in coal as particles of around 2 μm in diameter. During combustion, mineral matter due to heating undergoes significant changes and turns into ash, which typically has composition ranges of: SiO_2 (20-60%), Al_2O_3 (10-35%), Fe_2O_3 (5-35%), CaO (1-20%), and MgO (1-5%). At high temperatures, depending on the composition of mineral matter, the ash melts and leaves the coal or char particles through the pores.

Along with the above mentioned elements and compounds, coal contains around 20-30 different trace metals. Some metal atoms (e. g., boron) are bonded to the "coal molecule" while others (zirconium, manganese) are found in the inorganic part – in mineral matter. Copper can be found bonded to both organic and inorganic molecules. Elements such as boron, barium, manganese, antimony and zirconium appear in concentrations of 500-1000 ppm, and other trace metals are typically found in concentrations of 5-500 ppm.

4.2.5. Porosity of coal and char particles

The chemical reactivity of the char, except from the type and structure of coal, also depends on porosity and pore structure, i. e., real specific surface available for heterogeneous chemical reaction. Coal and char particle porosity is defined by the following quantities: specific internal volume of pores, ζ_p; specific internal surface of pores, A_g; internal volume or surface distribution depending on the pore diameter, δ_p

The particle pore volume for different coals ranges from 0.01 to 0.12 cm^3/g, and specific surface is 100-600 m^2/g.

Pores are classified into three large groups: micropores with diameter less than 20 Å, mezopores with diameter from 20 to 500 Å, and macropores with diameters larger than 500 Å. Although it is usually assumed that pores are cylindrical, they can also be cones or "cavernous" in structure.

The greatest part of the specific pore surface is found in the form of micropores, but this surface is very often inaccessible to oxygen, because the bandwidth of pore size near the surface is not enough to let oxygen penetrate into the particle. This is why pore size distribution is important parameter for coal reactivity. Macropores prevail in younger coals and micropores in high quality coals.

Pore structure can be determined by three main methods: using a picnometer, mercury porosimetry and gas adsorption.

A better knowledge of the causes of higher or lower coal reactivity, determining real char combustion rate, and postulating mathematical models of char particle burning, demands a detailed knowledge of pore structure, specific internal surface of pores and the process by which oxygen diffuses through the pores of char particles. Further, one must also take into account the fact that the pore structure changes during pyrolysis and devolatilization of the coal particle, as well as during gasification and combustion. Depending on the heating rate, temperature and type of coal (composition of volatiles) during devolatilization, the average pore size may increase or decrease and micropores can completely close. The pore structure of the char particle after devolatilization differs greatly from that of the original coal. During combustion the average char particle pore size increases. The coal and char particle porosity will not be considered in more detail in this book and interested readers should consult the literature on this subject [7, 8, 10].

4.2.6. Coal characteristics that influence the combustion process

The characteristics that make coals different are numerous. In the previous sections, the most important characteristics have been mentioned, but not all of them are equally significant for the combustion process itself. The influence of so many different parameters (coal particle density, porosity, pore structure, specific surface of pores, chemical composition, volatile matter content, fixed carbon, ash structure, etc.) can not be investigated individually. Thus it is natural to presume that connections exist between different coal properties, and thus reduce the number of parameters to be considered. In [8], it is pointed out that chemical composition of coal (carbon, hydrogen and oxygen content), as an independent variable has a major influence on all other coal properties and on the coal behavior during combustion.

Figure 4.6, partially taken from [8], shows that all coals in coordinate system C, H, fall into a relatively narrow hatched region (data for numerous Yugoslav coals have been added). A greater scattering of data in the lignite area is obvious. To the left of this region high-quality coals (anthracite) reside, and to the right are shown lignites from different countries. Since oxygen content is determined by the two other major elements – carbon and hydrogen – Fig. 4.6 shows that coal properties can be represented by only one parameter – carbon content.

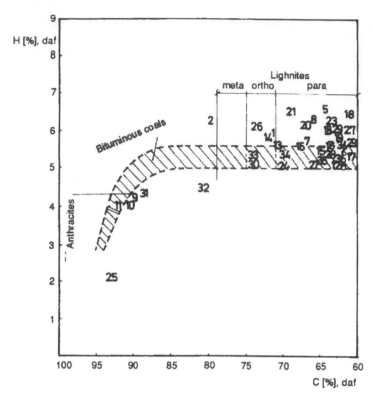

Figure 4.6. *Relation between hydrogen and carbon contents in coals of different rank and origin – basis for coal classification. Hatched region – coals from U.S.A., according to [8]. Numbers – data for various Yugoslav coals, as denoted in legend: 1 – Bogovina (10-30 mm), 2 – Bogovina (0-10 mm), 3 – Kolubara (5 mm), 4 – Kolubara (10 mm), 5 – Kolubara (15 mm), 6 – Aleksinac (5 mm), 7 – Aleksinac (10 mm), 8 – Aleksinac (15 mm), 9 – Vrška Čuka (5 mm), 10 – Vrška Čuka (10 mm), 11 – Vrška Čuka (15 mm), 12 – Djurdjevik (coarse), 13 – Djurdjevik (fine), 14 – Djurdjevik (coarse), 15 – Kolubara (dried), 16 – Czech coal, 17 – Piskupština, 18 – Ugljevik, 19 – Bogutovo selo, 20 – Miljevina, 21 – Albanian coal, 22 – Drmno, 23 – Bela Stena , 24 – Kakanj, 25 – Metalurgical coal, 26 – Raša, 27 – Kamengrad, 28 – Velenje, 29 – Bitolj, 30 – Stranjani, 31 – Jerma, 32 – Rtanj, 33 – Dobra sreća, 34 – Podvis – Tresibaba, 35 – Timok basin, 36 – Kreka*

Figures 4.7 and 4.8 support these relationships between the oxygen and carbon content (Fig. 4.7) and the relationship between coal calorific value and fixed carbon content (Fig. 4.8). Data for numerous Yugoslav coals have been added to

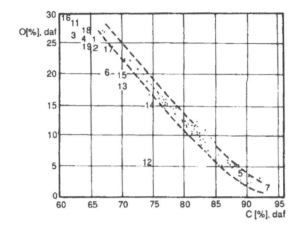

Figure 4.7. *Relation between oxygen content and carbon content in different coals. Points between dashed lines – coals from different countries; Numbers – Yugoslav coals as in Table 6.1, Chapter 6*

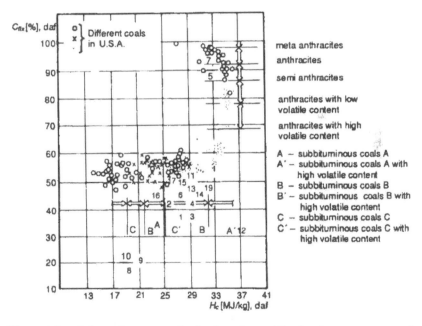

Figure 4.8. *Relation between coal calorific value and fixed carbon content in coals of different rank and origin. Basis for ASTM classification. Points of different shape – coals from different countries; Numbers – Yugoslav coals as in Table 6.1, Chapter 6*

these figures, and these data are also presented in Tables 6.1, 6.2, and 6.3 in Chapter 6.

It can also be demonstrated that, though with greater scatter, volatile content can be related to carbon content in coal [8].

Bearing in mind these relationships between coal properties, it is logical to base coal classification on carbon content or volatiles content.

It can be expected that any combustion process property depends on the volatiles content. Since straightforward relationships between volatiles content and other coal properties do not exist, the influence of other properties must be considered separately.

4.3.   Fragmentation of coal particles in fluidized beds

Coal is subjected to intense thermal shocks at the beginning of its residence in the fluidized bed, and to the continuous mechanical action of inert material particles during its entire residence time (lifetime), where the coal particle goes through major mechanical changes – in size and shape. These changes lead to an increase of fuel combustion surface, a decrease of fuel mass inventory in the bed, and contribute to the increase of losses caused by elutriated unburned particles.

Initially, research on fragmentation and the size and shape change of fuel particles in fluidized beds was motivated by interest in determining the losses caused by elutriation of unburned char particles. These are particles whose free fall velocity is lower than the velocity of gases in the furnace, so that they can be removed from the bed and furnace before complete combustion occurs. More recently, research has been aimed at determining the real size of coal and char particles that burn in the bed. This datum is necessary for postulating physically correct mathematical models of combustion.

The mechanical changes the coal particle undergoes in the fluidized bed have different causes and happen either sequentially or simultaneously depending on the process under consideration. According to their physical nature they are:

(1) coal particle fragmentation due to thermal stress and pressure increase in particles during devolatilization (primary fragmentation),

(2) char particle fragmentation during combustion due to internal burning in the particle and increase in porosity (secondary fragmentation),

(3) attrition of char particles during combustion due to mechanical interactions with inert material particles, and the formation of very small particles which are elutriated but not yet burnt (attrition), and

(4) char particle fragmentation to produce smaller particles at the end of burn-out (percolation). These particles are also elutriated.

Figure 4.9 schematically shows the nature and order of mechanical changes of a coal particle during its life and combustion in a fluidized bed. Much detailed re-

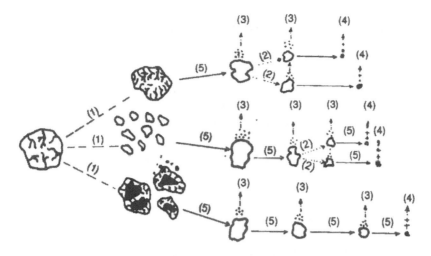

Figure 4.9. *Schematic of the order and nature of mechanical coal particle changes during residence and combustion in fluidized beds*
(1) Primary fragmentation – – –, *(2) Secondary fragmentation* . . . . . .,
(3) Attrition – · – · – , *(4) Percolation* + + +,
(5) Combustion ——

search of the attrition process has been carried out by Massimilla et al. [11-15], and experimental data can be found in [16-19]. Secondary fragmentation has rarely been investigated [20], but primary fragmentation research has been gaining attention [21, 22, 95].

4.3.1. Primary fragmentation

During devolatilization, and after moisture evaporation, depending on the type, structure and properties of the coal and conditions in the fluidized bed, three situations are possible: (1) the number and size of coal particles is unchanged, (2) the particles fragment, but total volume of the newly formed particles is the same as total volume of the parent particles, and (3) particles fragment and they also expand (swell) (Fig. 4.9). All three processes have been observed in experiments made by Dakić et al. [21], depending on coal type.

Due to very rapid coal particle heating in fluidized beds, after the moisture evaporation, the evolution of volatiles formed by thermal decomposition of coal is very violent. Inside the particle pressure increases due to the dramatic increase of volatiles volume. Depending on particle porosity, volatiles will or will not easily escape the coal particle.

If the particle is very porous, and open pores lead to the particle surface, volatiles will easily escape the particle, not changing its size or shape. If however, the mobility of volatiles is impeded by either small porosity, the large size of particles or blocking of pores, particle fragmentation and/or swelling will occur. Particle swelling, without fragmentation, with formation of "cavernous" (bubbles) in the particle center is most commonly observed in brown (bituminous) coals.

Dakić et al. [21] have shown that a coal's tendency to fragmentation can be determined by the ratio:

$$PRN(\textit{pore resistance number}) = \frac{VM \ content}{equilibrium \ moisture \ content} \qquad (4.1)$$

which represents the ratio of released volatiles and volume of coal particle pores.

By analysis of a number of brown coals and lignites in an experimental furnace (Fig. 6.2, Chapter 6), the relationship can be demonstrated between the coal's tendency to swelling or fragmentation (*crucible swelling number*) and proposed PRN ratio. Intensive particle fragmentation has been observed for coals with PRN≈ 15-17. Coals with smaller PRN have enough opened pores for easy passage of volatiles. Coals with larger PRN swell, and their particles deform but they rarely fragment.

An important result of this research is that for every coal there exists a critical particle size, i. e., the biggest particle that will rarely fragment during devolatilization.

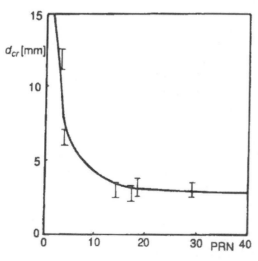

Figure 4.10.
Relationship between critical coal particle fragmentation diameter and volatile content to equilibrium moisture content ratio (PRN). According to measurements of Dakić [95]

Figure 4.10 shows the relation between critical fragmentation diameter and PRN. The diagram is very useful in engineering practice as it allows one to use only the results of ultimate analysis, to predict if the particle will fragment in a fluidized bed.

Coal particles of sizes which fall above the curve in Fig. 4.10 will most probably fragment. The critical fragmentation diameter obviously also depends on conditions in the fluidized bed (temperature, fluidization velocity, inert material particle size), so further research is needed to more fully determine the influence of these parameters.

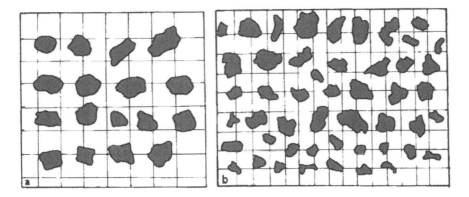

Figure 4.11. *Photo of Polish bituminous coal particles; (a) before and (b) after primary fragmentation. According to measurements of Dakić [95]*

Depending on coal type and size of parent particles, the number of particles after fragmentation and during devolatilization can be 2-3 times the number of parent particles. The particle shape before and after fragmentation is shown in Fig. 4.11 for one Polish coal.

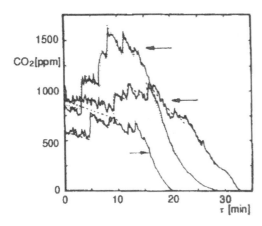

Figure 4.12.
Change in CO_2 concentration during combustion of a single particle of Kentucky No. 9 coal in fluidized bed. According to measurements of Sundback et al. Abrupt CO_2 concentration changes denote secondary fragmentation (Reprinted with kind permission of the Combustion Institute, Pittsburgh, from [20]. Copyright 1984)

4.3.2. Secondary fragmentation

Secondary fragmentation, fragmentation of char particles during combustion, has long been an area with many uncertainties, because experiments with large numbers of particles [15] are difficult to perform in such a manner as to obtain the appropriate data. Only in experiments with single particles [20] has it been determined that for certain coals, particularly bituminous ones, due to their increased porosity the char particles also fragment.

Figure 4.12 shows the measured change of CO_2 concentration for single particle burning for Kentucky No. 9 coal (PRN = 12.5, 41.5% volatiles, 3.3% moisture). Repeated abrupt concentration changes are clearly seen, as a result of repeated char particle fragmentation, at the point of critical porosity. By measuring the pore structure of char residue Kerstein and Niksa [23] have determined that critical porosity at which char particles break is around 80%.

Based on 36 experiments with single particles of diameter 2.2-6.2 mm, and temperatures of 1023-1123 K, it has been found that after devolatilization two types of particles form: very small ones formed by fragmentation, and larger ones formed by swelling. In char combustion, based on analysis from Fig. 4.12, for a particle with starting diameter of 7.8 mm, it has been found that during combustion it goes through 6 fragmentations, forming, in all, 22 particles.

The process of secondary fragmentation requires further research, especially because it has been observed for only one coal in only one experimental investigation [20, 24].

4.3.3. Attrition of char particles

The term fragmentation discussed in the previous sections is used for fuel particle disintegration – "parent" coal or its char residue – into several particles whose size is such that they cannot be elutriated out of the bed. Char particle disintegration to several small particles at the very end of burnout, because of pore structure burnout (*percolation*) will only be briefly mentioned here. Investigating percolation is difficult because it is hard to differentiate percolation from attrition [15]. The formation of very fine particles from the fine particles "glued" to larger particles, or by coal attrition before devolatilization is less important and has not been independently investigated.

The process of attrition (wearing) is the constant formation of very small particles of char, due to abrasion by inert material particles. Newly formed particles are small enough to be elutriated out of the bed and furnace.

As already stated in earlier sections, not all coals suffer either primary or secondary fragmentation, but throughout its entire life in the bed, a fuel particle is exposed to abrasion by inert material particles.

The detailed experimental research of Massimilla et al. [11-15] has shown that char particle attrition during combustion is much more intense than pure "mechanical" attrition in cold fluidized beds. This is the reason why constants obtained in experiments without combustion [16, 19] cannot be used to calculate losses due to unburned fixed carbon.

Figure 4.13 compares mass "production" of fine char particles of a South African coal, by purely "mechanical" attrition and attrition during combustion [14, 19]. "Mechanical" attrition has a peak at the beginning, when char particles are still rough, irregular and jagged. If a particle is not burning, the abrasive action of inert

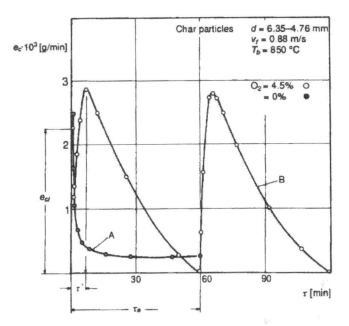

Figure 4.13. *Comparison of char attrition rate for one South African coal, without combustion (A) and during combustion (B). According to measurements of Massimilla (Reproduced with permission of the American Institute of Chemical Engineers from [14]. Copyright 1985 AIChE. All rights reserved)*

material particles (sand, ash, limestone) causes it to become smooth and attrition intensity rapidly decreases almost 10 times and remains constant through the rest of its life, until the particle becomes small enough to be elutriated itself.

If the char particle is burning, attrition intensity at first sharply decreases, and then again reaches high values and stays high. Average attrition intensity during burnout of particle 6.3-4.76 mm in size is around 5 times higher than that due to purely mechanical attrition.

Sudden increases of attrition intensity during combustion, and the subsequent constant high values, are explained by the continuous formation of crests, peaks and bulges at the burning particle surface, due to irregular and unequal burning of char particle porous structure. Two opposing processes take place simultaneously: abrasion which tends to remove surface irregularities and heterogeneous combustion which roughens the surface again.

Research carried out in Naples, on experimental furnaces of different scale (diameter 40 mm, 140 mm and 370 mm), by continuous fuel feeding and fuel feeding in batches, over a very wide range of regime parameters (d = 9.0-6.35 mm; 6.35-4.76 mm; 4.76-4.0 mm; v_f= 0.5-1.6 m/s; T_b = 750, 850, 950 °C; O_2 concentration = 0-21%; d_p = 0.2-0.4 mm) have shown that the mass of fine particles formed by attrition process can be expressed as:

$$e_c = k_c \left(v_f - v_{mf} \right) \frac{m_c}{d} \quad [\text{g/min}] \tag{4.2}$$

where m_c is in grams, velocities in m/s and average char particle diameter is in metres. For more detail see eq. (7), in Table 2.4, Chapter 2. The attrition constant k_c depends on coal type, bed temperature and oxygen concentration in bed, i. e., the combustion intensity.

According to Fig. 4.14, (T_b = 850 °C, O_2 = 4.5%, v_f = 0.8 m/s), for the coals analyzed (anthracite, coke, brown coal), attrition constant ranges from 0.77·10^{-7} to 6.3·10^{-7} [14, 19]. In Table 4.7 these values are given along with constants for coals analyzed in [25].

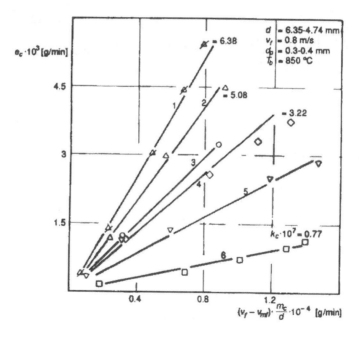

Figure 4.14. *Attrition constant for char of different coals. According to measurements of Chirone et al. (Reproduced with permission of AIChE from [14]. Copyright 1985 AIChE. All rights reserved):*
1 – NBAG anthracite char, 2 – Marine coal, 3 – Metallurgical coke, 4 – Petroleum coke, 5 – South African coal, 6 – Snibston coal

The final value in Table 4.7 is recommended based on 121 data points, from the experiments carried out with parameters: $T_p = 650$-$950\,°C$, $v_f = 0.5$-$1.6\,m/s$, $d_p = 0.2$-$0.4\,mm$, $d = 0.4$-$1\,mm$ to $d = 6$-$9\,mm$, $\lambda = 1$-1.4.

Table 4.7.   *Attrition constants of different coals*

Ref.	Coal	Attrition constant
19	NBAG Anthracite-char	$6.36 \cdot 10^{-7}$
	South African coal	$2.07 \cdot 10^{-7}$
	Marine coal	$5.08 \cdot 10^{-7}$
	Snibston coal	$0.77 \cdot 10^{-7}$
	Petroleum coke	$3.22 \cdot 10^{-7}$
	Metallurgical coke	$3.70 \cdot 10^{-7}$
25	Steam slack	$(2.13 \pm 0.36) \cdot 10^{-7}$
	Queensland coal	$(3.32 \pm 0.80) \cdot 10^{-7}$
	Petroleum coke	$(1.09 \pm 0.08) \cdot 10^{-7}$
	Polish coal No. 10	$(2.88 \pm 0.76) \cdot 10^{-7}$
	West Kentucky coal	$(2.57 \pm 0.66) \cdot 10^{-7}$
12	South African coal	$(1.86 \pm 0.33) \cdot 10^{-7}$

The influence of temperature and inert material particle size on the value of attrition constant can be seen in Fig. 4.15.

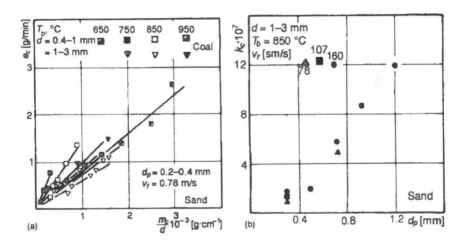

Figure 4.15. *Influence of temperature and inert bed material particle size on char attrition. (a) Attrition rate in dependence of total char particle surface; (b) Attrition constant depending on bed particle diameter. According to measurement of Arena et al. (Reproduced with permission of AIChE form [12]. Copyright 1985 AIChE. All rights reserved)*

4.4.   Devolatilization and combustion of volatile matter

The complex conditions under which coal combustion takes place in a fluidized bed, the complexity of coal as a fuel and the numerous simultaneous processes that take place from the moment a coal particle enters the fluidized bed to its burnout, can only be investigated independently, and then an overall physical and mathematical model can be produced from those investigations. This approach has been used here. In Section 4.1 the overall complexity of conditions and processes for coal burning in fluidized beds are described but without going into too much detail. Section 4.2 gives the basics of coal as a fuel and its most important properties. Section 4.3 provides a discussion of the first of many changes that the coal particle goes through in the fluidized bed. If the fragmentation process is simplified, then initially the coal doesn't change its properties – only its geometry. The first process in which significant changes of coal properties take place is devolatilization, i. e., the release of volatile matter from the coal particles. When this process finishes the remaining solid matrix is now called char, and has a completely different set of chemical and physical properties. Indeed, it is possible to regard coal combustion in fluidized beds as the combustion of two fuels, namely: combustible volatiles (homogenous) and char (heterogeneous). In reality mutual influences exist between these two processes and they also take place in parallel, even in a single coal particle.

Coals that are most often used in fluidized beds are of low quality and without preprocessing (as mined) – lignites and subbituminous coals. These coals contain 30-50% of volatiles (on dry and ash free basis), and if we take into account the moisture, the volatiles content is much higher. A significant amount of the energy produced when burning these types of coal comes from combustion of volatiles (up to 50%). That is why for the engineering design calculation on fluidized bed boilers it is important to be familiar with the processes of devolatilization and the combustion of volatiles.

Research of these processes is also important for emissions control and reduction of harmful combustion products. Volatiles in their initial composition contain high percentages of CO, for which combustion into CO_2 needs careful planning. Despite all measures, CO originating from volatiles makes a significant part of flue gases. NO_x compounds also originate from nitrogen compounds found in the volatiles. Several investigations have shown that the greatest amount of SO_2 is also released in the process of devolatilization.

Besides its influence on processes near, or at the surface and in the coal particle, as well as the influence on the final composition of combustion products, release and combustion of volatiles affect local properties and events in fluidized bed furnaces. Intense local devolatilization in some bed regions leads to a lack of oxygen, incomplete combustion and local overheating. Location of the most intense devolatilization, mixing of volatiles with gasses already in the bed, place of combustion and heat generation from volatiles (around the coal feeding points, in the bed and above it), all have great effect on construction and properties of FBC boilers.

The problems and processes tied to volatile production can obviously be divided into two groups: processes tied to fuel particles which under certain conditions may be considered independently from other fluidized bed characteristics and processes; and those dependent on fluidized bed properties, which are mostly concerned with the already-released volatiles. For both processes it is necessary to have the following information: the amount and composition of volatiles and the kinetics of devolatilization – the duration and rate of devolatilization depending on the coal type and conditions in fluidized bed, ignition temperature, duration and rate of volatiles combustion.

Research on noncombustibles, i. e., devolatilization (moisture, nitrogen compounds), has not received great attention. Information on particle drying processes during heating in the fluidized bed is almost nonexistent in the literature. Except for [26] where a model of coal particle drying in a fluidized bed has been proposed, the reader must be directed to literature on drying of solid porous materials in general and moisture transport in porous media [27]. Devolatilization processes are usually considered with the assumption that release of moisture (drying), which is not chemically bonded, has already happened and therefore a completely "dry" particle is assumed.

4.4.1. Volatile matter yield and composition

Release of volatile matter, devolatilization, has been thoroughly investigated, especially in conditions existing in pulverized coal combustion boilers. This research, especially popular in the 1950s and 1960s, has been summarized in several reviews [8, 28, 29]. Experimental research for fluidized beds is less prolific and has only been carried out more recently. One of the most important early works is probably [30], and an exhaustive review of experimental data on coal devolatilization in fluidized bed combustion is given in [3].

Contemporary research of processes during coal heating, coal chemical structure and coal chemistry [31, 32], shows that the long accepted view that "coal contains a fixed amount of matter from which volatiles are produced, and that devolatilization rate depends on concentration of this matter in coal" was incorrect. In fact there are a number of complex physico-chemical reactions occurring, some of which occur simultaneously and some of which are sequential. Due to heating of organic matter in the coal, numerous chemical bonds which make up the complex coal molecule (see Fig. 4.5) break and produce a large number of compounds of smaller molecular weight, which can be either liquid or gaseous.

The process of thermal coal decomposition – devolatilization, according to many investigations [28, 30-32], begins at temperatures around 700 K (427 °C), and consists of two basic stages: first very rapid reactions at lower temperatures take place and form the largest part of the volatiles; and then at higher temperatures slower reactions occur which produce mainly hydrogen. It is also known that in some coals (particularly bituminous coals) liquid products of thermal decomposition (tar) may also be present.

During heating, coal passes through the phase when its structure may be regarded as plastic. This may happen at temperatures as low as 350 °C due to coal matter decomposition into molecular structures which are liquid at these temperatures. As temperature increases coal plasticity increases and then sharply decreases when the liquid components start to evaporate or go through further decomposition into gaseous and solid components. Coal matter again solidifies at around 500 °C. By the time this temperature is reached most of the hydrocarbons have already been released [30]. These various processes can be explained by the changes in chemical structure of coal and a number of possible pathways are given in [8, 30-32].

Aromatic-hydroaromatic coal structure with numerous methylene and ethylene bonds, as well as peripheral groups (see Section 4.2, and Figs. 4.3, 4.4 and 4.5), allows the possibility for many different reactions to occur due to the high temperatures: breaking of bonds, recombination of newly formed groups and release of peripheral groups.

The breaking of bonds between aromatic and hydroaromatic groups gradually degrades macromolecules into smaller parts, which may become small enough to form liquid or gaseous products. Not every bond breaking leads to tar. Tar formation rate depends on the number of bonds which must be broken to get molecules of appropriate size. Different analyses have shown [31] that tar consists of one primary aromatic group with different numbers of side chains or groups. The processes responsible for coal plasticity in the first stage of devolatilization occur throughout the coal mass and are not limited to its surface. Recombination of aromatic and hydroaromatic groups also leads to very stable aromatic structures and forming of char. These processes take place during coal mass solidification.

The release of peripheral or side groups leads directly to the production of light, gaseous products. However, light gases may also originate form broken bonds between carbon groups. The gaseous products produced from peripheral groups resembles the old view of volatiles release and depends on the number, i. e., concentration of these groups in coal matter. However, gaseous products that originate from broken methylene and ethylene bonds depend on the probability of spatial distribution of atoms in coal molecules.

Given the facts noted above, during devolatilization, carbon groups break apart and recombine, simultaneously and in succession, and many of the original molecules will never become small enough to become volatile. Any carbon group may either become tar, or char, depending on overall conditions. The thermal process of bond breaking and recombination of carbon groups also explains formation of the resulting very porous char structure.

The end products of thermal decomposition of coal, i. e., devolatilization, are: gaseous products with a very complex composition, liquid, but evaporable products (tar) and char, which is almost pure carbon (ignoring the mineral matter content).

For research and modelling of the combustion process, it is also important to know the chemical and mass composition of volatiles. In order of importance, the following factors influence the yield, chemical and mass composition of volatiles: chemical composition of coal; temperature history of coal (heating rate and maximal temperature); pressure and composition of surrounding gases; and particle size.

Volatiles usually consist of the following gases: CH_4, C_2H_6, CO, CO_2, H_2, H_2O, NH_3 and H_2S, but also include many other hydrocarbons. An example of volatiles composition and its change with the increase of temperature is given in Table 4.8, for a Yugoslav lignite, according to [33].

Table 4.8.   *Volatiles content and composition of lignites*

Volume [%]	Temperature [°C]			
	500	700	800	900
H_2	7.94	26.30	31.72	31.29
CH_4	20.00	24.59	22.27	20.30
CO	35.97	28.66	33.11	39.79
CO_2	31.00	14.86	9.03	5.98
C_2H_6	0.57	1.74	2.11	1.86
C_2H_4	0.62	6.75	0.20	0.02
H_2S	2.04	0.50	0.82	0.42
C_3H_8	0.96	2.03	0.67	0.22
C_3H_6	0.35	0.37	–	0.026
C_4H_{10}	0.01	0.05	0.06	0.014
C_4H_8	0.52	0.07	0.015	0.003
Total volume [ml]	74.03	222.75	297.34	438.18

For this example for one particular type of coal the combustible volatiles mostly consist of CH_4, CO, and H_2 along with noncombustible CO_2 It is also evident that with temperature increase the volatiles content also increases.

Analyzing numerous experimental data on volatiles composition for different coals Essenhigh and Suuberg [8] have concluded:

(1) in devolatilization products of lignite, oxygen compounds are dominant, while for bituminous coals hydrocarbons dominate,

(2) a simple relationship cannot be proposed between proximate coal analysis and volatiles composition,

(3) many research data for volatiles composition and volatiles yield are contradictory, and in many experiments mass balance has not been satisfied,

(4) the maximum temperature at which devolatilization takes place affects composition and amount of released gases, and

(5) but the coal particle heat rate, from 2500 K/s through 10^4 K/s, apparently does not affect the composition of gases formed by devolatilization.

Although the above conclusions are the result of analyzing many different studies on this subject [8], the problem of determining and predicting the composition of devolatilization products for a particular coal is still open and insufficiently investigated at this time. Experimental results are obviously influenced by the manner and conditions of experiment and typically three experimental routines have been used: (a) heating of the coal specimen (usually particles of 40-1000 μm) on an electrically heated grid, with heating rate of 100-12000 °C/s and at temperatures of up to 1200 °C; (b) heating of gas entrained coal particles (particle size of 20-300 μm) in an electrically heated tube, with heating rate up to 50000 °C/s and maximum temperature up to 2000 °C; (c) heating of coal particles in a fluidized bed (size of 200-15000 μm), with heating rate of 100-1000 °C/s, and maximum temperature of 400-1000 °C. Experimental conditions for methods (a) and (b) correspond to heating rate, maximum temperature and particle size in pulverized coal combustion. For fluidized bed combustion, method (c) yields more realistic results, especially because tests with particles of 1-20 mm in size are also possible.

One of the first experiments with conditions similar to those in fluidized beds was performed by Morris and Keairns [30]. In order to provide data for the design and construction of fluidized bed gasifiers, they thoroughly investigated volatile content of three different types of coal (subbituminous coal C from Wyoming, bituminous coal C with high volatiles content from Indiana, and coking coal with high content of volatiles from Pittsburgh), at temperatures of 760, 872, 982 °C and with coal particle size of 0.5-4.0 mm. In this study, the fluidization was achieved with nitrogen in order to suppress any combustion. The composition and amount of volatiles has also been investigated later by other authors [32, 34-37] and reviews of most of this research are given in [8, 28]. Results obtained in [30] very clearly show the basic principles of devolatilization, which agree well with subsequent research.

Figure 4.16 shows the change of composition and amount of volatiles for two coals with high volatile content (Pittsburgh – 41.2% VM_o, 76.5% C, 4.8% H_2O, 6.9% O_2, 0.95% N_2, 1.85% S and 8.9% ash, and Wyoming – 46.1% of volatiles, 73.3% C, 5.0% H_2, 16.5% O_2, 1.23% N_2, 0.49% S_2 and 3.4% of ash). The area under the curves represent the total quantity of released gas.

Despite large differences between the investigated coals, the volatile content and time of devolatilization are very similar. The greatest differences are in quantity of CO and CO_2 produced. Coal with higher percent of O_2 (Wyoming, close to lignite) releases a higher percent of oxides: CO, CO_2 and H_2O. A similar general conclusion (1) is reached by Suuberg and Essenhigh [8] based on analysis of data available to them. With the exception of H_2, other gases were released in less than 10 s (particle size was 1.55 mm). The maximum H_2 content is reached somewhat

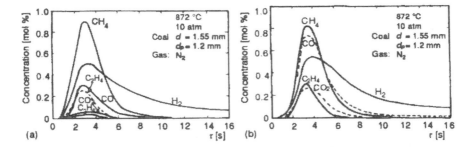

(a)   (b)

Figure 4.16. *Change of composition and yield of volatiles for two American coals: (a) Pittsburgh, (b) Wyoming. According to the measurements of Morris and Keairns (Reprinted from Fuel [30]. Copyright 1979, with permission from Elservier Science)*

later than for other gases, and thereafter hydrogen content production dominates in the mixture of devolatilization products until the end of the devolatilization process.

These results confirm that devolatilization consists of two basic stages: in the beginning, rapid low-temperature reactions release the largest part of the volatiles, and then at higher temperatures, slow reactions release mostly H_2. Due to the temperature gradient inside the coal particle, these two processes overlap, especially in larger particles. These investigations also show that major devolatilization products are: CH_4, CO, CO_2, H_2 and C_2H_4.

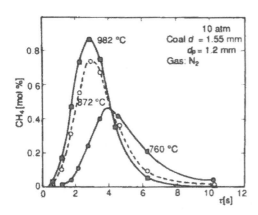

Figure 4.17.
Effect of bed temperature on methane evolution from American bituminous coal (Indiana). According to the measurements of Morris and Keairns (Reprinted from Fuel [30]. Copyright 1979, with permission from Elsevier Science)

The influence of the maximum devolatilization temperature is illustrated by the case of methane (CH_4) release (Fig. 4.17). At lower temperatures (760 °C) the mass of released methane from Indiana coal – bituminous coal C with high volatile content (brown coal) – decreases significantly and devolatilization time increases. It is interesting to compare the change of volatile composition of lignite (Table 4.8) [33], for which with an increase of temperature, the percentage of CH_4, CO and CO_2 decreases, and hydrogen content increases in accordance with conclusions (3) and (4) of Suuberg and Essenhigh [8].

Except for the gaseous products of coal pyrolysis, mentioned before, during heating of some types of coals, a significant amount of tar (liquid products) is created, and these represent compounds with the smallest molecular weight that are still liquid. In [32], it is suggested that tar is created by the decomposition of monomers into two tar molecules. A monomer is defined as the smallest nonvolatile component of coal. The pyrolysis of lignite and younger coals yields mostly gaseous products, with a tar content of 15-30% of the devolatilization products [8]. Bituminous coals can have 50-70% of tar in the devolatilization products. Data on the quantity and percentage of tar in pyrolysis products are contradictory through the literature due to the possible secondary reactions inside the coal particle occurring in the experimental apparatus, in which further decomposition of relatively large molecules and further reactions with carbon or with gaseous combustion products can occur.

The significance and effects of these secondary reactions have also been observed in the experiments of Stubington and Sumaryono [35], which were carried out in a fluidized bed at temperatures of 750, 850 and 950 °C. For this study, three Australian coals with different volatile contents 20-43.5%, particle size 3-11 mm, were investigated. Figure 4.18 [35] shows the characteristic graph of H_2 and CH_4 release for one Australian coal (36% volatiles, 64% C_{fix}, bituminous coal with low volatile content) for particles larger than 6 mm.

The characteristic second peak of H_2 release can be seen, and is especially pronounced for larger particles and higher temperatures. With temperature increase, however, the tar content is virtually constant, and the quantity of gaseous products of pyrolysis increases significantly for three types of coal. The tar-to-gaseous products content ratio is in the range of 70:30 at 750 °C and 20:80 at 950 °C.

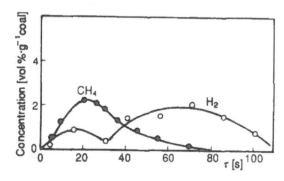

One possible explanation of these results is that the process of primary devolatilization begins at 300-400 °C and continues for temperatures of over 1000 °C when coal particles are burning rapidly. During these processes hydrogen from coal can bond with oxygen from coal forming H_2O, or with carbon, forming C_nH_m. If hydrogen forms mostly hydrocarbons, oxygen is free to react with carbon forming CO and CO_2 and can decrease the char content. With the tempera-

Figure 4.18.
Hydrogen and methane evolution from Great Northern coal in different stages of devolatilization at 950 °C. According to the measurements of Stubington and Sumaryono (Reprinted from Fuel [35] Copyright 1979, with permission from Elsevier Science)

ture increase, increase of carbon oxides has been shown along with a corresponding decrease of char content (a similar tendency is observed in Table 4.8, for lignites in the temperature range 700-900 °C). A decrease of particle size (i. e., increase in heating rate) works in the same direction. Therefore, increasing the temperature and heating rate is favorable for increased carbon oxide production, while lower temperatures and heating rates favor higher H_2O production. Similar tendencies have also been observed for lignites in [38].

Pyrolysis products primarily pass through the porous coal structure to the surface of the particle. More reactive compounds of pyrolysis products suffer secondary reactions – cracking, condensation and polymerization. Due to cracking of larger hydrocarbon molecules and lighter hydrocarbons, hydrogen and carbon are released, and subsequently adsorbed by the pore surface, increasing the char mass. The intensity of secondary reactions depends on temperature and the residence time pyrolysis products spend inside the pores. With temperature increase, as the result of secondary reactions, the yield of higher hydrocarbons, methane and ethylene also increases. With the increase in particle size, the char content increases at the expense of tar and gaseous products of pyrolysis.

The characteristic two peaks of hydrogen content in products of pyrolysis can also be explained by secondary reactions. The first peak is caused by the decomposition in the initial phase – primary devolatilization at high temperature in the region near the particle surface. The second peak is the result of secondary reactions of tar and other unstable hydrocarbons decomposing inside the particle. These reactions happen later due to the slow propagation of the temperature wave towards the inside of the particle. Therefore, the second maximum is more pronounced in larger particles and at higher temperatures.

The results described here show that the composition of products of coal pyrolysis is not simply due to the coal properties, i. e., it cannot be explained by the nature and chemical composition of organic matter in the coal alone. The content of a particular mixture of components depends also on the experimental conditions: maximum temperature, heating rate and particle size, as well as on experimental procedures. Secondary reactions in experimental apparatuses cannot be prevented, so it is not always possible to deduce whether primary or secondary products of pyrolysis have been detected and the contradictory results of some experiments are most often the result of secondary reactions.

The total amount of volatiles released from coal heated to a particular temperature also cannot be regarded as the characteristic coal property. Although for engineering purposes volatiles content is determined by standardized methodology (heating to 900 °C of pulverized coal specimen of size up to 90 μm in a laboratory oven) and is used as a characteristic property of coal in standard classification, it is clear that the real quantity of volatiles released under particular conditions also depends on temperature, heating rate and particle size. If the volatiles content obtained by a standardized procedure can be used as a realistic parameter for pulverized coal combustion, this is certainly not true for fluidized bed combustion where temperatures and heating rates are lower, and coal particles much larger.

Attempts to find a relationship between volatiles content and coal proximate analysis, according to Suuberg and Essenhigh [8] following an exhaustive review of the literature, have failed. Proposed correlations have at best ±20% precision. Among the numerous proposed correlations Suuberg and Essenhigh point out one [39] giving the volatiles content of coal which is otherwise determined using standardized procedures:

$$VM_0 \, [\%] = -0.408 \, [C] + 11.25 \, [H] + [O] + 1.3 \, [S] \qquad (4.3)$$

In the above equation concentrations are in % of mass, on dry basis, and oxygen concentration is on an ash free dry basis.

It is claimed that this correlation has a standard deviation of ±1.5%, which with 95% statistical confidence gives an error of 8% for $VM_0 = 40\%$. Despite these good results, volatiles content based on proximate coal analysis used for coal classification is not reliable for coal combustion analysis.

Numerous experimental results exist, showing that the volatiles content when heating coal in an inert atmosphere is very different from the volatiles content determined in the standardized way at 900 °C. A review of experiments carried out under conditions characteristic for pulverized coal combustion (particle size 20-90 μm, temperatures 1000-1200 °C, heating rates 10^4-10^5 °C/s) has been given by Anthony and Howard [28], unifying the results with the Badzioch and Hawksley relation [40]:

$$VM^* = Q_{VM} \, (1 - VM_c) VM_0 \qquad (4.4)$$

The ratio VM^*/VM_0 for the investigated American, English and French coals ranges from 0.75 to 1.36, and coefficient Q_{VM} from 1.3 to 1.8 depending on coal type.

Badzioch and Hawksley [40] have carried out extensive experimental research, the results of which illustrate very clearly the behavior of coal during devolatilization in pulverized coal combustion, and fill in most of the gaps in previous investigations [31, 41]. Ten different bituminous coals with volatiles content $VM_0 = 17.7$-42.0% (on a dry ash free basis) have been tested, as well as one semi-anthracite, $VM_0 = 11.5\%$. Four coals did not readily swell (crucible swelling number 0.5-2) and others were very susceptible to swell (factor 6-8). The average particle size was 20-60 μm, temperature 400-1000 °C, and heating rate (2.5-5)·10^4 °C/s. All tests were carried out in a nitrogen atmosphere.

It has been determined that volatiles content strongly depends on temperature. The influence of particle size and heating rate was minimal, due to the narrow range in which these parameters have been changed. The results of these measurements could be described by the formula (4.4) where VM_c, content of volatiles left in char, can be represented by the exponential formula:

$$VM_c = \exp[-K_1(T - K_2)] \qquad (4.5)$$

For $T = K_2$, $VM_c = 1$ which agrees with the physical fact that at low temperatures devolatilization rate is very low, and that volatile content may be neglected (i. e., $VM^* \approx 0$). For $T > K_2$, $VM_c \approx 0$, meaning that volatile content approaches maximal asymptotic value independent of temperature.

For bituminous coals susceptible to swelling, the temperature dependence is more pronounced, and constant K_2 is physically explainable and varies in the ranges of 438-746 K, meaning that devolatilization begins at temperatures of 200-480 °C. Constant $K_1 = (2.77-3.97) \cdot 10^{-3}$ K^{-1}. With these constants, the decrease of volatiles content was investigated over a range of temperatures and $(1 - VM_c)$ has been shown to vary from 0.5 to 0.9.

Coals not susceptible to swelling demonstrated much less dependence of volatiles content on temperature, $(1 - VM_c) = 0.80\text{-}0.88$. Constants obtained are $K_1 = 1.23 \cdot 10^{-3}$ K^{-1}, and $K_2 = -485$ K. For all types of coal investigated, the Q_{VM} coefficient ranged between 1.3 and 1.8.

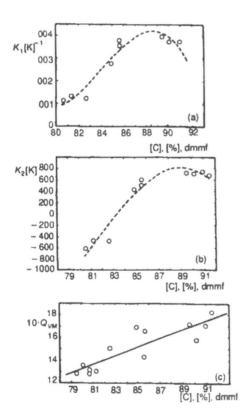

Badzioch and Hawksley managed to relate the constants obtained to the coal type, using the coal carbon content on a dry and ash free basis as a characteristic feature. Figures 4.19a, b and c show these relations for constants K_1, K_2, and Q_{VM}. However, it has been pointed out [40] that any results and constants obtained cannot be used out of the range of parameters explored or for other types of coal.

Having recognized the weakness of previous experiments, Anthony et al. [44] in their experiments used heating rates between 10^2 and 10^4 °C/s, particle size 50-1000 µm, and ambient pressure of devolatilization process 0.001-100 bar. Particle heating was performed on an electrically heated grid. Temperature range was the same as in [40], 400-1000 °C. A bituminous coal (Pittsburgh, $VM_o = 41.5\%$), and one lignite (Montana, $VM_o = 44.7\%$) were investigated. Typical mass change of the samples investigated are shown in Fig. 4.20, for the lignite for pressure of 1 bar.

Figure 4.19.
Variation of constants K_1, K_2 and Q_{VM} in expressions (4.4) and (4.5) for kinetics of volatiles evolution from coal. According to the measurements of Badzioch and Hawksley (Reprinted with kind permission from [40]. Copyright 1970 American Chemical Society)

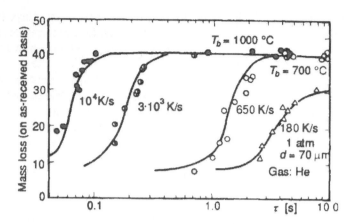

Figure 4.20. *Effect of time on devolatilization mass loss from Montana lignite at different heating rates and final temperatures. According to the measurements of Anthony et al. (Reprinted with kind permission of the Combustion Institute, Pittsburgh, from [44].Copyright 1975)*

Maximum mass loss (i. e., volatiles mass released) does not depend on the heating rate (in the range 650-10^4 °C/s), and at temperature of 1000 °C is around 41%, which is less than the volatiles content determined using standardized procedures. The volatiles content at 700 °C was much lower, 31% and the increase of volatiles content was negligible at temperatures of 900-950 °C.

It is important to point out that the wide range of external pressures (0.001-100 bar) does not influence the quantity of volatiles released from lignite. Negligible changes of volatile content from lignite when heating at temperatures higher than 900 °C have been obtained in the investigation of Mississippi lignite [42] with particle size d = 3-5 mm in a fluidized bed (Fig. 4.21). Similarly, negligible changes have been observed at temperatures as low as 600 °C, while the devolatilization itself begins at 200 °C.

Preliminary research for some Yugoslav coals (lignite, brown coal, and anthracite) shows similar tendencies [43]. Devolatilization has been carried out in a laboratory oven, in nitrogen atmosphere, by measuring the volatiles left over in char when the devolatilization process was continued until there was no measurable specimen mass change.

A high percent of released volatiles is observed from lignite (Fig. 4.22) even at low temperatures (500 °C), which agrees well with the investigations discussed above for American lignites [34, 44] shown in Fig. 4.20 and 4.21. In the temperature range of 700-900 °C, the volatiles content reaches that obtained by

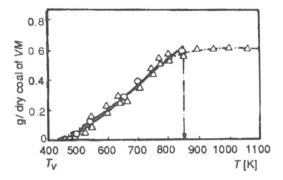

Figure 4.21.
Total devolatilization mass loss from coal obtained at different temperatures. According to the measurements of Agarwall et al. (Reprinted from Fuel [42]. Copyright 1979, with permission from Elsevier Science)

standardized analysis, while the influence of particle size is not pronounced, and a release of volatiles greater than VM_o has not been observed.

Tests with brown coal and anthracite show different behavior (Fig. 4.22). Here, the influence of particle size is more pronounced, and volatile contents greater than VM_o have been measured at 900 °C.

Investigation of some Chinese coals [45] in the high-temperature range (900-1500 °C) yielded similar results (Fig. 4.23). Measurements were carried out with coal particles of 3-9 mm in diameter, in high-temperature argon plasma. Similarly, a higher volatile content than determined by proximate analysis (VM_o) has also been observed in [40] and analyzed in detail for bituminous (brown) coals in [28, 44].

It has been established that at pressures less than 5 bar, larger quantities of volatiles are released if the specimen is heated rapidly to the desired temperature. A

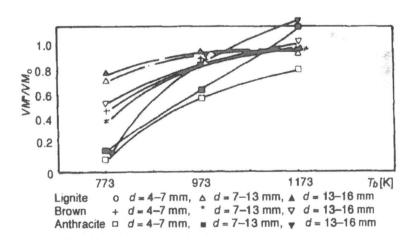

Figure 4.22. *Total devolatilization mass loss from three Yugoslav coals, obtained at different temperatures (lignite $VM_o = 41\%$, brown coal, bituminous $VM_o = 34.5\%$, $VM_o = 7\%$), [43]*

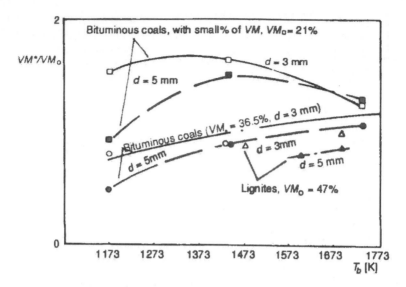

Figure 4.23.
Effect of temperature and coal particle size on devolatilization mass loss for three Chinese coals. According to the measurements of W. Fu et al. [45]

pressure influence at heating temperature of 1000 °C may also be seen in Fig. 4.24, for bituminous coal (Pittsburgh, VM_0 = 41.5%). A heating rate in the range of 650 to 10^4 °C/s has a negligible influence. At high pressures, as well as at very low ones, the volatile content asymptotically reaches constant value.

This type of behavior of bituminous coals is explained by the secondary reactions of primary devolatilization products [44]. Many primary devolatilization products are chemically very reactive. Secondary reactions (polymerization, cracking) of hot volatiles at hot surfaces of coal particle porous structure lead to formation of new solid compounds inside the particle.

Increased pressure or increased particle diameter prolongs the time volatiles spend in pores, and decreases the quantity of released volatiles. In the investigated particle size range, 50-1000 μm, increase of volatile content of around 20% [44] has been observed for smaller particles. Similar tendencies in the particle size range characteristic for fluidized beds, 3-9 mm, have been observed in experiments [45], and are shown in Fig. 4.23.

Finally, although the temperature range higher than 1000 °C is not interesting for fluidized bed combustion, we will mention that Kobayashi et al. [46], reported significant increase of released volatiles in the temperature range of 1000-2100 °C, for bituminous coal with high volatile content (Pittsburgh), with particle size of 20-500 μm and heating rates of 10^4-10^5 °C/s.

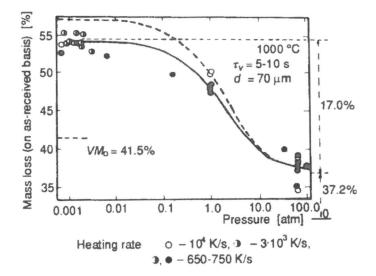

Heating rate $o - 10^4$ K/s, $\mathrm{\oplus} - 3 \cdot 10^3$ K/s,
$\mathrm{\oplus}, \bullet - 650\text{-}750$ K/s

Figure 4.24. *Effect of pressure on devolatilization mass loss from Pittsburgh seam bituminous coal. According to the measurements of Anthony et al. (Reprinted with kind permission of the Combustion Institute, Pittsburgh, from [44]. Copyright 1975)*

Results of current research are dedicated mainly to pulverized coal combustion conditions, but point to several conclusions of interest for fluidized bed combustion. The quantity of volatiles released from the coal during devolatilization:

–   depends very much on the temperature in the range 400-900 °C, and increases with temperature increase,
–   is smaller for lower heating rates, under 600 °C/s,
–   for lignites is less than VM_o, and for bituminous coals may be greater than VM_o, and
–   for lignites does not depend on pressure, and for bituminous coals decreases with the increase of pressure, and at atmospheric pressure can be greater than VM_o.

A very small number of experiments has been carried out in fluidized beds, or under conditions similar to those in fluidized bed combustion, so it is hard to tell if the above conclusions can always be applied to fluidized bed combustion. That is particularly true for the influence of coal particle size and heating rate.

Practically all experimental research on the devolatilization process in fluidized beds was intended to determine devolatilization time and/or combustion time of volatiles [3, 47-52], and not to obtain data about the volatiles content, ex-

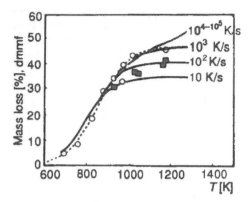

Figure 4.25.
Effect of heating rate on devolatilization mass loss from coal. According to the measurements of D. Merrick (Reprinted by permission of Elsevier Science from [36]. Copyright 1986 by the Combustion Institute)

cept perhaps for the studies of Morris and Keairns [30] and Stubington and Sumaryono [52]. These researchers investigated the composition and quantity of pyrolysis products, but did not give data on total volatile content. Vural et al. [53] examined tar quantity in fluidized beds during devolatization and the influence of the type of inert material, but their reported data did not include the secondary reactions during volatiles passage through the fluidized bed of inert material.

Certain insight into the behavior of coals during devolatilization in fluidized bed may be obtained from experimental results of Merrick [54] and Niksa [36]. These experiments were carried out under conditions very close to conditions in fluidized bed combustion. Merrick carried out measurements in an experimental furnace for five U.S.A. coals used for coking ($d = 3$ mm, $VM_o = 16$-38%), with very low heating rate (3-8 K/min), at a temperature of 950 °C, and in all cases obtained volatiles contents less than those determined by standardized procedures, and was able to establish the empirical relation: $VM^* = VM_o - 0.36\ VM_o^2$. The results of Niksa, obtained by heating on an electrically heated grid, for bituminous coal with high volatile content are shown in Fig. 4.25, and compared with his mathematical model of the devolatilization process. In the range of temperatures found in fluidized bed combustion, 600-900 °C, a significant influence of heating rate in the range 10^2-10^3 K/s is observed.

The lack of data about devolatilization in fluidized beds can be illustrated by comments of La Nauze [3], who after an exhaustive review of the literature concluded that all the authors of fluidized bed coal combustion models used models for volatiles release based on experiments carried out under conditions characteristic for pulverized coal combustion [8, 28].

4.4.2. Control processes and kinetics of devolatilization

The results of research on the processes that occur during coal heating (pyrolysis, devolatilization), described in the previous section, show that the total mass of volatiles released depends on: coal type, final temperature, heating rate and pressure. Some of the diagrams (Figs. 4.16, 4.17, 4.18) also show that total quantity of volatiles and particular compounds in such mixtures depend very much on time, and over a very long time period they asymptotically approach constant values for given temperatures. A knowledge of devolatilization kinetics, mass flux of volatiles and total time needed for devolatilization, is important for coal combustion analysis just as is the knowledge of the quantity and composition of volatiles. In mathematical modelling of coal combustion processes considering kinetics of devolatilization is inevitable, whether this process is neglected (based on the argument that devolatilization time is short compared with times of other characteristic processes – e. g., mixing time or char combustion time), or it is necessary to include in the model.

Current research provides much more detailed insights into the coal molecule chemical structure (see Section 4.2) and chemical composition of volatiles for different coals. These insights confirm the thesis that the devolatilization process consists of a number of chemical reactions produced by coal heating, which break the bonds between aromatic/hydroaromatic rings that make up the coal molecule, bonds between peripheral groups and these rings, and further decomposition or bonding of newly created groups (secondary reactions). Results of these reactions are: char residue, tar and gaseous compounds.

There are many chemical reactions which may occur simultaneously (in parallel) or sequentially, and which may have different directions – increasing or decreasing the quantity of volatiles, i. e., the devolatilization process is also influenced by three physical processes: (1) intensity of heat transfer from surroundings to coal particle, (2) heat transmission rate (propagation velocity of temperature wave), i. e., temperature gradient in coal particle, and (3) filtration process (flow) of gaseous products through the porous coal structure to the particle surface. The conditions in which the process of pyrolysis takes place determine which of these three processes will be the "controlling" process, having the greatest influence on rate, quantity and composition of released volatiles and devolatilization time. Modelling the pyrolysis process in order to predict these quantities for a particular coal, has been the aim of many authors in the past [26, 28, 32, 36, 40, 42, 44, 50, 54-57], and always starts with an analysis of which of these three processes is dominant [8, 28, 34, 58]. It should be noted that in the past several years due to the increased use of biomass as a renewable energy source, a number of models for biomass pyrolysis have also been produced. Since coal also has plant origins, these models may be useful in the research of coal pyrolysis [59-62].

The breaking point in analysis of the controlling processes is the question: does devolatilization time depend on the particle size? The intensity of these three transfer processes depends on the particle size, while the rate of chemical reactions depends only on temperature. If the heat transfer from the surroundings to the particle surface is the dominant process (i. e., the other processes are much faster), then inside the particle there will be no temperature gradient and the devolatilization process takes place in isothermal conditions in the entire particle volume. The heat transfer intensity will be inversely proportional to the particle size. In pulverized coal combustion and in fluidized bed combustion, although particle sizes are very different (20-90 μm, compared to 1-25 mm, respectively), the heat transfer to the coal particle is very high. In the first case the radiation contribution is very large, and in the second case heat transfer by contact with inert material particles is very strong. This is why it is believed [26, 28, 50] that the heat transfer process is not predominant for devolatilization.

Heat transmission through the coal particle, being a nonstationary process of body heating, depends on the Fourier number $= a_t\, t/d^2$. If this process is dominant (i. e., the slowest one) then the total devolatilization time will be proportional to d^2 [26]. It can also be shown [26, 28, 50] that devolatilization time is also proportional to d^2 if the mass transfer through the porous coal particle is dominant.

The first investigation of devolatilization kinetics was performed under conditions similar to those in pulverized coal combustion (particle size 20-90 μm, temperature 1000-2000 °C). It is not surprising that, under these conditions, it has been shown that devolatilization time does not depend on the particle size [28, 31, 40, 41, 44]. Analysis of the results available [28, 40, 41, 58] shows that a critical particle size exists under which the devolatilization process can be considered to take place in isothermal conditions. If Biot's number Bi < 0.02, i. e., particle size is less that 100-200 μm, then the temperature distribution in the entire particle volume will be uniform, and the devolatilization rate will depend solely on the rate of chemical reaction.

By analyzing the available data on coal particle devolatilization time in fluidized bed conditions ([30, 35, 42] and others who will be mentioned in the next section), La Nauze [50] concluded that for particles greater than 300 μm the devolatilization time is proportional to d^N, where N is between 1.5 and 2.2 — meaning that, in modelling the processes in fluidized beds, one must take into account the transport processes inside the particle. For now it is not possible to determine whether heat transfer through the particle is more important than the diffusion of volatiles through the porous particle structure, or vice versa.

The first mathematical models of the devolatilization process assumed that the kinetics of chemical reactions controlled the whole process, because of the great interest in pulverized coal combustion, and experimental results that were available in such conditions. Many authors (see [28]), and among them also [40, 41], assumed that devolatilization rate could be represented by the first order reaction:

$$\frac{dVM}{dt} = k_R\,(VM^* - VM) \tag{4.6}$$

in which the rate of reaction can be expressed in the form of the Arrhenius law:

$$k_R = k_{Ro}\,\exp\left(-\frac{E}{R_g T}\right) \tag{4.7}$$

This model is based on the assumption that all the complexity of chemical processes during devolatilization may be represented by one reaction with an average,

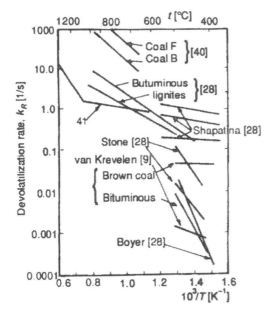

Figure 4.26.
Arrhenius plot of simple first order coal devolatilization rate according to different investigators. From Anthony and Howard (Reproduced with permission of the AIChE from [28]. Copyright 1976 AIChE. All rights reserved)

common activation energy. As a result of this simplification, the model was able to reproduce experimental data only if, for every particular set of experimental conditions, the optimum kinetics parameter set is determined: VM^*, k_{Ro}, E. Depending on the experimental conditions and coal type, reaction rate k_R can differ by 10^4 times, which may be clearly seen in Fig. 4.26 taken from [28, 44]. Although it was possible to represent some experimental data by (4.6) and (4.7), using appropriate sets of constants k_{Ro}, E, and VM^*, the inability to encompass the influence of experimental conditions, and especially the fact that VM^* depends on temperature, persuaded Anthony and Howard [44], to further develop the existing model.

The basic assumption of the new model was that thermal decomposition of coal can be described as the sum of a great number of simultaneous independent chemical reactions. Alternative models have also been developed, with the basic assumption that pyrolysis is an array of consecutive chemical reactions [9, 46, 63, 64], but the model of parallel reactions is used more often, since the empirical constants it requires can easily be determined by experiment.

The quantity of a particular component of released volatiles (CH_4, CO, CO_2, etc.) may be represented by an expression of the form (4.6):

$$\frac{dVM_i}{d\tau} = k_{Ri} \left(VM_i^* - VM_i \right) \tag{4.8}$$

where, also:

$$k_{Ri} = k_{Roi} \exp\left(-\frac{E_i}{R_g T} \right) \tag{4.9}$$

Index i represents one of the many independent reactions, which, since different bonds exist in a coal molecule, have different activation energies, i. e., reaction rates.

The quantity of volatiles left in char, at the moment t, is obtained by integrating expression (4.8):

$$VM_i^* - VM_t = VM_i^* \exp(\int_0^t k_{Ri} dt) \qquad (4.10)$$

Kinetic constants for each reaction (for each component of volatiles), VM_i^*, k_{Roi}, and E_i, must be determined experimentally. In a more recently published review of research in devolatilization, Suuberg [8] gives the recommended values of k_{Roi} and E_i for the following volatile components: CO_2, H_2O, CO, HCN, CH_4, H_2, tar, and NH_3. Knowing these values, as well as VM_i^* for a particular coal and temperature, in principle, it is possible, as was done using a somewhat modified model by Merrik [54], to determine the kinetics of release of each gas and change of volatile composition during devolatilization.

To simplify the procedure, Anthony and Howard [28] have adopted the following assumptions:

– for all reactions the same coefficient k_{Roi} may be used, and
– the number of reactions is sufficiently large so that the activation energy E may be represented by the continual distribution function $f(E)$.

With these assumptions, part of released volatiles with activation energy between E and $E + \Delta E$, may be expressed as $f(E)dE$. The quantity of released i-th component, VM_i^* is then a differentially small fraction of total released quantity:

$$dVM^* = VM^* f(E)\, dE = VM^* dF \qquad (4.11)$$

with:

$$\int_0^\infty f(E)dE \approx 1 \qquad (4.12)$$

The total quantity of volatiles left in char is obtained by summing contributions of every reaction, i. e., by integrating the expression (4.10) in the whole interval of activation energies:

$$VM^* - VM = VM^* \int_0^\infty \exp\left[-\int_0^\infty k(E)dt\right] f(E)dE \qquad (4.13)$$

It is also necessary to introduce the assumption about the shape of the $f(E)$ distribution. Gaussian distribution is usually adopted [26, 44, 55], but there have been other functions proposed (Merrik [54] adopts Rosin-Rammler's distribution):

$$f(E) = [\sigma(2\pi)^{1/2}]^{-1} \exp\left[-\frac{(E - E_o)^2}{2\sigma^2}\right] \qquad (4.14)$$

If (4.14) is used in the integral in (4.13), and integration limits of $-\infty$ to $+\infty$ are adopted, we obtain:

$$VM^* - VM = \frac{VM^*}{\sigma(2\pi)^{1/2}} \cdot$$

$$\cdot \int_{-\infty}^{+\infty} \exp\left[-k_{Ro}\int_0^{\tau} \exp\left(-\frac{E}{R_g T} \right) d\tau - \frac{(E-E_o)^2}{2\sigma^2} \right] \qquad (4.15)$$

Expression (4.15) describes the devolatilization process as a set of infinite number of simultaneous reactions, and requires only one constant more than the one-reaction model: VM^*, k_{Roi}, E_i, and σ. For integration in (4.15) one needs to know or propose coal particle temperature change in time, assuming that there is no temperature gradient in the coal particle. At very high heating rates (coal particles $<100\ \mu m$) the assumption that temperature is constant over time is justified [44].

Using their own experimental data (e. g., see Fig. 4.20) and data on temperature change during devolatilization, Anthony and Howard [44] managed to describe their experiments very successfully using just two data sets for VM^*, k_{Roi}, E_i, and σ. Solid lines in Fig. 4.20 represent calculation results using the constants given in Table 4.9.

Table 4.9. *Kinetic devolatilization parameters used in devolatilization models*

Kinetic parameters	Lignite for all pressures [44]	Bituminous coal, 70 bar [44]	Indiana coal constants used for CH_4 [34]	Mississippi lignite [42]
Pre-exponential coefficient k_{Ro} [s⁻¹]	$1.07\cdot10^{10}$	$2.91\cdot10^9$	–	–
Average activation energy E_o [kJ/mol]	204	154.5	–	–
Standard deviation σ [kJ/mol]	39.3	17.5	–	–
Total volatiles content VM^* [%]	40.63	37.18	–	–
Pre-exponential coefficient k_{Ro} [s⁻¹]	$1.67\cdot10^{13}$	–	–	$1.67\cdot10^{13}$
Average activation energy E_o [kJ/mol]	235.75	212.2	268	192
Standard deviation σ [kJ/mol]	45.7	29.4	15	40

Using a physically more appropriate value for $k_{Roi} = 1.67\cdot10^{13}$ s⁻¹, equally good agreement with experimental data has been obtained, with somewhat higher and physically more appropriate activation energies. Using the same or similar models, other authors, regardless of coal type, have adopted $k_{Roi} = 1.67 \cdot 10^{13}$ s^{-1}, with good results. The devolatilization model described here is the most often used of models of numerous authors for describing experimental data, and has also been used for developing models for other conditions (fluidized bed combustion [26, 56], coking [54]), in which the rate of chemical reaction is not the only or the most important factor. This model removed many shortcomings of the simple model with one chemical reaction:

- it has enabled prediction of particular volatile component kinetics, using separate sets for each component (see [8]),
- activation energies of $E_0 \pm 2\sigma = 142\text{-}326$ kJ/mol have been obtained, which agree with typical values for organic compound decomposition, and for lignites and bituminous coals are very close to real values,
- the model predicts that the time needed to reach VM^* for a given temperature increases with the increase of temperature, and
- the model predicts that release of volatiles will continue for higher temperatures, irrespective of the fact that for the preceding temperature level an asymptotic value of volatiles yield was already established.

. However, the model of many parallel reactions still has a number of drawbacks:

- it does not include the influence of ambient pressure and particle size, and
- it cannot predict total quantity of volatiles for a given temperature and heating rate, VM^*, but requires this value as input, experimental datum.

Merrik [54] has, for coking conditions and very slow heating rates, 3-8 K/min, modified the model using Rosin-Rammler's distribution instead of Gaussian distribution, and applied it to describing the kinetics of the following volatile components: CH_4, C_2H_6, CO, CO_2, tar, H_2, H_2O, NH_3, and H_2S and for determining physical characteristics and chemical composition of coke [54, 65, 66].

The application of the Anthony-Howard model to fluidized bed combustion has been considered by La Nauze [50], Agarwal et al. [34, 42] and Borghi et al. [56].

Agarwal and La Nauze have concluded that chemical reaction kinetics cannot be the only control process for devolatilization in fluidized bed combustion, because in burning large particles (usually > 1 mm) it is hard to believe that the temperature gradient inside the particle is negligible. For typical conditions in fluidized beds and estimated thermal characteristics of coal, it can be shown that Biot's number is 1-20, while isothermal conditions in particles exist only for Bi ≤ 0.02.

Using expression (4.13) in the case of different temperatures along the radius of spherical coal particle (no primary fragmentation, all other factors constant in particle volume), yields:

$$\left[\frac{VM^*-VM}{VM^*}\right]_{sr}=\frac{24}{d^3}\cdot$$

$$\cdot\int_0^{d/2}\left\{\int_0^\infty\exp\left[-k_{Ro}\int_0^\tau\exp\left(-\frac{E}{R_gT}\right)dt\right]f(E)dE\right\}r^2dr \qquad (4.15a)$$

where $f(E)$ is given by expression (4.14).

Temperature as a function of time and position inside a coal particle is obtained using the analytical solution of the well-known differential equation for transient heat conduction in spherical particles. By parametric analysis, and using kinetic parameters close to those given in Table 4.9, the significant influence of Biot's number has been shown, i. e., the influence of the heat transfer process inside the particle. Results given in Fig. 4.27 show that the kinetics of chemical reaction prevail as the control process of devolatilization for particles smaller than 100-200 μm, which is close to the experimentally obtained values.

Figure 4.28 shows the comparison of the calculated CH_4 release time for a bed temperature of 872 °C and experimental data of Morris and Keairns [30], and in Fig. 4.29, the kinetics of CH_4 release at different temperatures are compared to the results of the same authors obtained for Indiana (U.S.A.) coal (bituminous C coal,

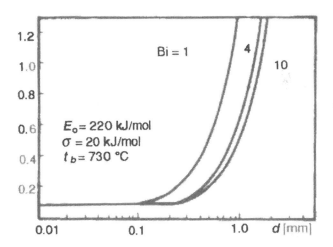

Figure 4.27. *Effect of coal particle size on time necessary for evolution of 95% of total amount of volatiles. Calculation according to Agarwall's model taking into account particle heating – expression (4.15a) (Reprinted from Fuel [34]. Copyright 1984, with permission from Elsevier Science)*

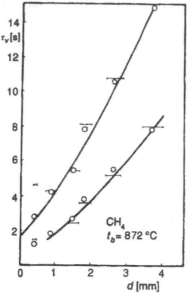

Figure 4.28.
Evolution time for 50% and 95% of total amount of volatiles from coal. Lines according to the model of Agarwall et al. points – measurements of Morris and Keairns [30] for Indiana coal (Reprinted from Fuel [34]. Copyright 1984, with permission from Elsevier Science)

with high volatile content). Kinetic constants used in the calculations are given in Table 4.9.

An important contribution of this model is that it encompassed dependence of volatiles release kinetics on coal particle size, which is an experimental fact determined by numerous experiments, and especially for fluidized bed combustion of large coal particles (>1 mm). The model also showed good agreement with experimental data for Mississippi coal [42], using constants from Table 4.9.

Borghi et al. [56] use the original Anthony-Howard model, claiming that for particles of around 1 mm, and when the volatile content is over 50% of asymptotic value, there are no significant differences between the isothermal model [10] and nonisothermal model [34].

One of the more recent models of this type has been developed by Zhang [68] in his PhD dissertation. This model represents the enhanced Agarwal model, which considers a more complete equation of heat conduction through the particle, with energy sources and sinks added. For very large particles (up to 50 mm) a simplified variant of this model (assuming infinitely fast chemical reaction of thermal decomposition of coal) gives better agreement with experimental data than the mass transfer model considered to be more appropriate for larger particles [69, 70]. The simplified Zhang model in

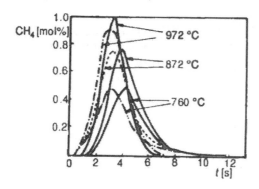

Figure 4.29.
Effect of temperature on kinetics of methane evolution and total evolution time for Indiana coal, according to the model of Agarwall et al. and measurements of Morris and Keairns [30] (Reprinted from Fuel [34] Copyright 1984, with permission from Elsevier Science)

fact assumes that the devolatilization rate is equal to the propagation velocity of the temperature wave at which devolatilization takes place. For the model to be accurate it is necessary to determine this characteristic temperature.

There were no attempts to encompass, in fluidized bed conditions, the influence of volatiles filtration through the porous coal structure, although a number of attempts have been made to upgrade the Anthony-Howard model with inclusion of the mass transfer process [69, 70]. Article [55] which gives the review of models taking into account the mass transfer process, states as their major drawback that they do not simultaneously take into account heat conduction in the coal particle.

Trying to overcome this drawback, Phuoc and Durbetaki [55] have, based on the Anthony and Howard model, postulated a model which takes into account the following processes: heat conduction through the particle, mass transfer from the particle center to the surface, and kinetics of chemical reactions, also taking into account secondary reactions of volatiles in char pores. This complex model has not been compared to experimental data, but gives interesting qualitative conclusions:

- the increase of particle inner pressure depends on the ratio of the rate of chemical reaction which "produces" volatiles to the mass transfer rate by molecular diffusion through the micropores,
- in pores smaller than 10^{-5}cm, molecular diffusion is the dominant process, leading to high pressures inside the coal particle (3-12 bar above the pressure around the particle),
- if convective flow of volatiles prevails (pores are larger than 10^{-5} cm) there is no significant increase of pressure inside the particle, and
- the ratio of volatiles "production" to convective flow rate in pores depends on the heating rate, particle size and pressure. The model takes into account the influence of these quantities.

Model [55] still has a shortcoming of being unable to predict the finite asymptotic quantity of volatiles VM^*, but requires it as input, experimental datum.

Realizing that the basic drawback of the widely used models, based on the ideas of Anthony and Howard, is that they do not take into account current knowledge about the structure of the coal macromolecule (see Section 4.2), which may be seen from the requirement that the finite asymptotic value of VM^* must be given as input datum for particular conditions, Niksa developed a model based on entirely new assumptions [32]:

- the coal molecule is made of complex and intertwined aromatic, hydroaromatic and aliphatic nuclei (see Fig. 4.4), which consist of several (for bituminous coals 2-4) rings for each nucleus (see Fig. 4.5), with different types of bonds and different peripheral groups, and
- devolatilization is a set of numerous chemical reactions:

    - bond dissociation is between nucleus and macromolecule decomposition, while each newly formed part may be further decomposed,

  – recombination of nuclei and forming of new, solid, aromatic structure – char residue, and
  – release of peripheral groups and forming of gaseous products.

Coal is represented as having three components: aromatic rings, fragile bonds between them and peripheral groups. These basic components not only make up the "coal molecule" but also the products of pyrolysis – free monomers (mobile aromatic rings), tar, char, and gaseous products.

Four chemical reactions are taken into account: breaking of chemical bonds between rings, release of peripheral groups, forming of tar by dissociation of polymers and forming of char residue.

Each reaction is represented in a manner accepted in the Anthony-Howard model – expressions (4.13) and (4.14), with appropriate activation energies:

  (a) mass of the char residue is the result of a number of conflicting processes: dissociation of molecular bonds, recombination, and formation of tar by decomposition of polymers, and depends on the number of free aromatic rings, number of bonds available for recombination, number of formed solid aromatic chains,
  (b) mass of newly formed tar depends on the number of free monomers,
  (c) mass of gas depends on the two-stage reaction – forming of new peripheral groups during the breaking of intermolecular bonds and release of peripheral groups, and
  (d) the number of free monomers, above all, depends on the probability that breaking of chemical bonds in a "coal molecule" will produce monomers and not some other part of the ring chain, and on the probability that newly formed monomers will recombine into larger chains, and finally it depends on the rate at which monomers degrade into tar.

The approach to modelling of the devolatilization process, applied by Niksa, is theoretically closer to the physical and chemical nature of coal and the devolatilization process, and is based on the most recent findings about the nature and structure of the coal macromolecule. This model removes two significant drawbacks of the models based on the unidirectional parallel reactions: it is sensitive to the influence of heating rate and does not require the finite asymptotic quantity of released volatiles as an input datum.

Comparing this model with the results from two independent experiments for bituminous coal with high volatile content [46, 71], Niksa obtained very good agreement (see Fig. 4.25 and Fig. 4.30) [36, 57].

Further development of the model of devolatilization will probably be based on this type of model. It is necessary to improve the model to encompass cases in which heat and mass transfer are the control processes, in order to take into account the influence of pressure and particle size. Models of this type require data on the physical and chemical structure of the coal molecule, which are still very

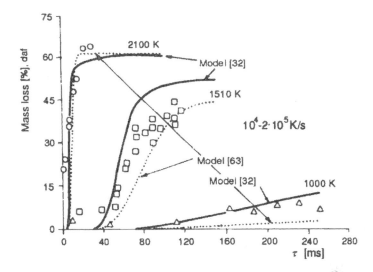

Figure 4.30. *Effect of time and temperature on mass loss from a Pittsburgh seam HVA bituminous coal. Calculation according to the model of S. Niksa, compared with the measurements and model of Kobayashi et al. (Reprinted by permission of Elsevier Science from [57]. Copyright 1986 by The Combustion Institute)*

rare and available from specialized laboratories. This is the reason why models of the Anthony-Howard type are still being used in combustion models, as can be seen in the exhaustive review of La Nauze [3].

4.4.3. Devolatilization time in fluidized beds – experimental results

Measurements of devolatilization time in the available literature show great differences as can be seen in reviews of these experiments in Table 4.10, together with the list of the most important experimental conditions. Review papers of La Nauze [50] and Pillai [49, 51], which analyze other authors' results should also be taken into account.

These differences may be due to different experimental methods used for measuring the duration of devolatilization:

– in experiments carried out in inert atmospheres, completion of the devolatilization process can be determined accurately enough by gas analysis
– experiments carried out in oxidative atmosphere, at temperatures below 700 °C include char combustion, and at temperatures above 700 °C include combus-

Table 4.10. *Review of experimental research in which devolatilization time has been measured*

Author (year) [Ref]	Coal type, origin	Experimental		Fluidized bed				Coal particle size [mm]	Devolatilization time [s]	Comment
		Atmosphere	Measurement method	Material d_p [mm]	Dimension [mm]	Temperature [°C]	Velocity [m/s]			
Morris Keairs (1979) [30]	High volatile bituminous Wyoming (U.S.A.) Indiana (U.S.A.) Pittsburgh No. 8	Nitrogen 10 bar	Gas analysis	Char 1-1.4	$D_b = 30$ $H_b = 35$	760-982	0.335	0.5-4	1 to 20	– no relationship between t_v and d^N has been given according to [34, 50] $N = 2$ for Pittsburgh No. 8 according to analysis in [68] $N = 1.04$
Pillai (1981) [48]	12 coals (lignite, bituminous anthracite) – bituminous U.S.A. Rexco-coke Coalit-briquette Pittsburgh No. 8 Glen Brook F.R.G Lohberg Osterfeld U.K. Nostell Agrina – lignite Texas (U.S.A.) Hot Creek (Canada) – anthracite Niederberg (F.R.G.) – graphite rock (Sweden)	Air 12-13 % O_2	Visual	Chamote 0.6	$D_b = 100$ $H_b = 75$	770-1010	1.2-1.5	0.25-8	1 to 25	– for all coals and all temperatures relationship $t_v = a_v d^N$ has been obtained – $N = 0.34$-1.76 – for Pittsburgh No. 8 coal $N = 0.34$-0.65 – $a_v = T_s^{1.8}$
Oka (1983, 1985) [74, 75, 76]	Dried lignite Kolubara (Yugoslavia)	Air	Measurement of combustion products temperature above the bed	Sand 0.63-1.0	$D_b = 147$ $H_b = 220$	550-700	1.0	5-25	5 to 40	– relationship between t_v and d^2 given by diagram – for particles $d = 5$-15 mm, $N = 2$ – for particles $d = 15$-25 mm $N = 0$ possible fragmentation determining the end of devolatilization is uncertain

Author (year) [ref]	Coal	Atmosphere	Method	Bed material (mm)	Dimensions (mm)	Temp.				Remarks
Agarwal (1984) [42]	Mississippi lignite (U.S.A.)	Air	Char mass measurement	Glass spheres 0.6	$D_b = 76$, $H_b = 127$	250-900	0.51	3-5	10-50	– results and discussion suggest $N = 2$ – relationship between t_c and d^A is not given directly
Stubington Sumaryono (1984) [35]	Bituminous Bulli, Greta, Great Northern (U.S.A.)	Nitrogen	Gas analysis	Sand 0.5-0.6	$D_b = 35$, $H_b = 80$	750-950	1.5 v_{mf}	1-15	30-200	– $N = 2$ according to analysis in [50]
Pillai (1985) [51]	Bituminous Rufford (UK) Pittsburgh No 8 (U.S.A.)	Air 13% O_2	Visual	Chamote 0.62	$D_b = 100$, $H_b = 175$	900	1 2	6-17	50-300	– $N = 0.83$ for Pittsburgh No 8 – $N = 1.83$ for Rufford coal
Andrei (1985) [73]	Lignite Montana (U.S.A.)	N_2– O_2, mixture 5 and 8% O_2	Char mass measurement	Sand 0.2	$D_b = 100$, $H_b = 120$	750-900	0.067	1.67-3.28	4-13	– relationship between t_c and d^A has not been given – results show $N = 1$-1.2
Prins (1987) [72]	– lignite, – bituminous – anthracite dried, wet	N_2–O_2 mixture 3, 6 and 21% O_2	Visual	Al_2O_3 0.5-0.6	15×188 mm^2 $H_b = 125$	200-800	0.25-0.8	4-9	7-54	– relationship between t_c and d^A has not been given, results show $N = 1.7$ – influence of fluidization velocity has not been determined – influence of O_2 concentration has not been determined – influence of number of particles has not been determined (1-35)
Zhang (1987) [68]	– bituminous Evans, Minto (Canada)	Air, nitrogen	Visual char mass measurement	Sand 0.425-0.5	$D_b = 100$, $H_b = $ -	750-900	0.21-0.33	5-50	20-800	– for Minto coal $N = 1.54$ – for Evans coal $N = 1.61$ – influence of fluidization velocity and number of particles has not been determined (5-35)
Ekinci (1988) [52]	6 coals from Turkey – lignite, bituminous, anthracite	Air 13% O_2	Visual	Sand 0.71	$D_b = 110$, $H_b = 100$	700-850	–	1-10	1040	– $N = 0.53$ to 1.49

tion of both volatiles and char. These processes, as well as higher particle temperatures, may affect devolatilization, and
— methods by which the end of devolatilization is determined can also affect the results:

   — visual detection of the disappearance of a visible flame around coal particles occurring at the bed surface may be erroneous due to different particle residence times in the bed. The duration of devolatilization in these experiments is equivalent to the duration of volatiles combustion. It has been shown [72], that this method yields longer devolatilization times in experiments with more particles than in experiments with one particle, because measurement of the statistically improbable longest time is possible,
   — the measurement of particle mass, by interrupting the process, can also lead to the wrong results because the process continues up to the moment of actual measurement, and
   — when the temperature of combustion products is measured, estimates of the end of the volatiles combustion process can be seriously in error.

In contemporary measurements most attention has been paid to the influence of the following parameters: coal type (lignites to anthracite), coal particle size (0.5 to 25 mm), temperature (200–1010 °C) and oxygen concentration (2-21%). Measurements have also been carried out in inert atmosphere. However, the influence of inert material particle size and fluidization velocity has not been sufficiently investigated.

Almost all authors have tried to express their results in the form:

$$\tau_V = a_V d^N \qquad (4.16)$$

Several authors determined $N = 2$, corresponding to the devolatilization process controlled by heat transfer through the coal particle or by mass transfer through the porous structure of the particle (which is impossible to differentiate in these experiments). In most cases this exponent has values of 0.5-2 which can be seen from Fig. 4.31 based on the analysis of Zhang [68]. According to Prins' analysis [72], an exponent value of 1.3 corresponds to the case when the control process is heat transfer from the fluidized bed to the coal particle. In the range of $N = 1.3$-2, for both processes external and internal heat transfer must be taken into account. Values $N < 1.3$ are likely the result of the controlling influence of kinetics of chemical reactions.

A broader analysis of the devolatilization process has been given by Zhang [68], whose experiments are the largest set to date, carried out for very large particles of 5-50 mm. He stated his doubts about La Nauze's conclusions [50] that Moriss' and Keairns' and Stubington's and Sunamaryon's experiments yield $N = 2$.

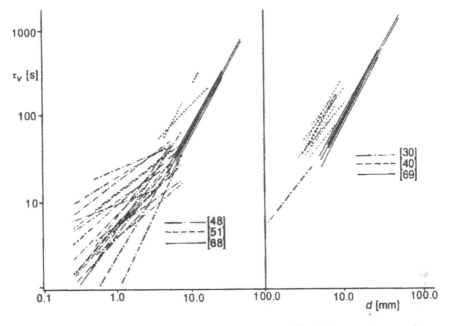

Figure 4.31. *Effect of particle size on devolatilization time for different coals according to the measurements of several authors. Analysis and comparison given by Zhang [68]*

Based on repeated analysis and on his own results, Zhang, as well as Prins [72], claims that dominant processes are heat transfer between particle and bed, and heat conduction through the particle, not mass filtration through the particle (which gives $N = 2$). By direct comparison of results obtained in the same experimental apparatus (using three experimental methods) Zhang concluded that the devolatilization process does not depend on fluidization velocity, number of particles and O_2 concentration (i. e., if combustion is going on or not) which is contrary to the claims of Prins.

Other results also suggest a small influence of the fluidization velocity [68, 72] and O_2 concentration [48, 68, 73]. The influence of inert material particle size has not been investigated. Prins [72] claims that combustion of volatiles reduces time τ_v, and Andrei [73] claims that existence of flame does not affect the devolatilization process (this is also claimed by Zhang [68]). Analyzing the results obtained in inert atmosphere, La Nauze [50] obtained $N = 2$, and in experiments with combustion $N = 0.5$-2. Further, the influence of coal type has not been clarified. In all experiments, the increase of temperature reduces devolatilization time.

The general conclusion of all these authors is that the devolatilization time in conditions characteristic for fluidized bed combustion (particle size, bed temperature) which ranged between 5 and 200 s, is not short enough to allow one to treat devolatilization as instantaneous. According to Stubington's [77], Prins' [72], and La Nauze's [3] analyses, the devolatilization time is comparable to the characteristic time of particle mixing in fluidized beds. Hence, in combustion models it can be assumed that volatiles are released from the fuel particles uniformly over the bed volume in cases of good mixing and uniform fluidization. However, due to the changing devolatilization rate of coal particles (with maximum at the beginning of the process), kinetics of devolatilization must be taken into account, and this leads in practice to higher concentration of volatiles in the vicinity of coal feeding points in the case of small coal particles and poor radial mixing.

Further research on the devolatilization process and devolatilization times is required to clarify the influence of the above-mentioned parameters, the determining control processes and verification of mathematical models. The devolatilization process in fluidized bed combustion is still not sufficiently well understood.

4.4.4. Ignition and combustion kinetics of volatile matter

The fact of releasing volatiles means that coal in a fluidized bed can be divided into two fuels – gaseous and solid, whose further behavior and combustion is completely different. In assessing furnace design concepts, and performing engineering calculations, it is usual to base them on the energy and mass balance and averaging over bed volume. However, in creating rigorous mathematical models of processes in FBC boilers, one must take into account that these two fuels ignite at different temperatures, burn at different rates, one by homogeneous, the other by heterogeneous chemical reactions and at different places in the furnace. In the combustion of high volatile coals and biomass, the problem of combustion of volatiles is much more complicated. In particular, the quantity of heat generated by combustion of volatiles is over 50% of the total heat generated and the combustion of volatiles at unusual places, and local heat generation due to their combustion can create serious problems in the operation of FBC boilers and furnaces. To date in general engineering practice the problem of volatiles combustion has been rarely investigated or has even been neglected for the following reasons: (1) fluidized bed combustion technology appeared in countries with access to high-quality coals; the tendency was primarily to burn coal wastes from coal washing and separation plants, and/or coals with high sulphur content; (2) an inadequate knowledge of physical and other characteristics of volatiles also played an important role in its neglect in such modelling. Even today it is often assumed that it is most important to know the distribution of volatiles in fluidized beds, and that if there is enough oxygen, such volatiles will burn instantaneously. These questions will be considered in more detail in the next section, and this section gives a review of the rather scanty data available on combustible volatiles as fuel.

There are three basic properties of every fuel: ignition temperature, calorific value and combustion rate. When considering these properties, one must not forget that volatiles are a mixture of several different gases.

In the case of the ignition temperature of volatiles, the literature provides very contradictory data and controversial opinions still exist. Further, there are no direct measurements of ignition temperature by standard and accepted methods [8]. Due to the different composition of volatiles, which depend on coal type and thermal decomposition conditions, many questions may be raised about how reasonable it is at all to try to determine volatile ignition temperature, by any standard methods for gaseous fuels.

Based on much indirect evidence, it is reasonable to conclude [8, 72] that volatiles do not ignite below 600 °C, and that they certainly ignite above 700 °C. The only direct measurement of volatiles ignition temperature in fluidized beds has been carried out by Prins [72]. According to his data, on the coals he investigated (lignites, brown coals, anthracite), volatiles do not ignite at temperatures below 680 °C.

Some indirect evidence for this belief are as follows:

- carbon monoxide ignition temperature is 644-657 °C,
- hydrogen ignition temperature is 530-590 °C,
- hydrocarbons ignition temperatures are: CH_4 650-663 °C, C_2H_6 520-630 °C, C_2H_4 around 542-547 °C, and C_2H_2 406-580 °C,
- C_nH_m content in combustion products of different coals sharply decreases at temperatures of about 560 °C, according to measurements in [78], meaning that ignition and combustion of hydrocarbons took place,
- in modelling combustion processes in fluidized beds, many authors assume that volatiles ignite at temperatures higher than 600-650 °C, e. g., Rayan and Wen [79],
- experiments set up for investigation of devolatilization are carried out at temperatures (≥ 700 °C, if combustion of volatiles is desired and at temperatures <700 °C if not, and
- the temperature difference between bed and combustion products in the freeboard is small or negligible at steady combustion temperatures below 600 °C and above 800 °C, but reaches a maximum at 600 °C when ignition and combustion of volatiles takes place above the bed, according to [80].

On the other hand, it has been experimentally shown that:

- coals with high volatile content ignite at lower temperatures [4, 8, 93], and
- ignition temperature and combustion rate of "parent" coals is much lower than ignition temperature and combustion rate of char of the same coal [168].

In analyzing these contradictory data one must take into account the fact that ignition temperatures of the above-mentioned volatile component-gases are determined by standard methods, in large volume, without solid walls, heated particles and other ignition initiators. In real-life conditions, when igniting volatiles released from the coal particle, catalytic influences of mineral matter and other components of coal may also be present. The problem of ignition of volatiles as well as coal particles with high volatiles content requires further research.

There is very little data on calorific value of volatiles for different coals. The calorific value of volatiles has most often been measured indirectly, based on the difference of calorific value of coal and calorific value of laboratory char, determined by standard proximate analysis. Data obtained in this manner should be used with caution in any calculations or modelling, bearing in mind the differences between the volatiles content obtained by standard proximate analysis and real quantity and composition of volatiles released (or remaining in char) in real life. Based on the data from [33], the ratio of calorific value of volatiles and coal is approximately equal to the ratio of volatiles and fixed carbon contents in coal. For dried Kolubara lignite , volatiles calorific value of 12.15 MJ/kg (dry fuel) has been measured, as opposed to a total of 22.25 MJ/kg, where the ratio of volatiles to carbon content was 60:40, dry ash free basis [81].

In more detailed analyses, it should be borne in mind that during devolatilization, the released quantity in a unit of time (mass flux) also changes with time, as does its composition, so the calorific value also changes. Data given by Suuberg [8] indicates a great dependence of calorific value of volatiles on temperature (see Fig. 4.32 for lignite) and also on time.

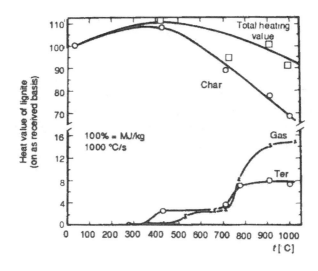

Figure 4.32. *Distribution of calorific values of products from lignite pyrolysis to peak temperatures. According to measurements of Suuberg (Reproduced with kind permission from Kluwer Academic Publishers from [8])*

According to these data, the calorific value of volatiles is, for lignites, only 25% of total calorific value, and for bituminous coals around 50%, and clearly depends on the content of heavier hydrocarbons (tar) with greater calorific value. To fully determine the properties of particular coals (especially of lignites) requires the knowledge of composition and calorific value of volatiles.

The calorific value of volatiles can be determined knowing that volatiles are a mixture of known gases (CO, H_2, CH_4, C_2H_6, C_nH_m, tar, etc.). The reactions of these gases with oxygen are well known, as well as the quantity of heat released [18, 82]. For this approach it is necessary to know (or assume) the type and order of chemical reactions. Depending on these, as a result, complete or incomplete oxides may be obtained (e. g., CO_2 or CO).

Detailed mathematical modelling of these processes also requires taking into account the rates of chemical reactions. Assuming one-step oxidation:

$$\text{Fuel} + n_1 O_2 = n_2 CO_2 + n_3 H_2 O + Q_l \qquad (4.17)$$

in which molar number n_1, n_2, and n_3 depends on the fuel composition. The corresponding rate of chemical reaction will be:

$$k_H = k_{H_o} T^n \exp(-\frac{E}{R_g T})[\text{fuel}]^a [O_2]^b \qquad (4.18)$$

where concentration in mol/cm^3, E [kJ/mol K], R_g [kJ/mol K], and T [K] are introduced.

In some models volatiles are treated as homogeneous, unique fuel, and the rate of chemical reaction is chosen for oxidation of CO, since it is the slowest reaction, and CO is in all cases an important constituent part of mixtures of volatiles (see [79, 80]). Reaction heat Q_l is taken to be equal to the calorific value of volatiles.

At higher temperatures and for substoichiometric conditions, oxidation will not proceed to CO_2 and H_2O, so CO and H_2 will also be present in the mixture. Most often, a two-step reaction is assumed:

$$C_n H_m + \left(\frac{n}{2} + \frac{m}{2}\right) O_2 = nCO + \frac{m}{2} H_2 O \qquad (4.19)$$

followed by:

$$CO + \frac{1}{2} O_2 = CO_2 \qquad (4.20)$$

Even more complex chains of reactions can be obtained if the first reaction is assumed to be:

$$C_nH_m + \frac{n}{2}O_2 = nCO + \frac{m}{2}H_2 \qquad (4.21)$$

followed by the numerous possible reactions of CO and H_2 with O and O_2. In the case of two-step oxidation – (4.19), (4.20) and (4.21), the rate of chemical reactions is calculated from expression (4.18), for which some constants are given in Table 4.11.

Table 4.11. *Kinetic constants of some important chemical reactions*

Compound	One-step			Two-step			Quasi-global		
	k_{H_a}	E	a	k_{H_a}	E	b	k_{H_a}	E	n
CH_4	$1.3 \cdot 10^8$	202.75	-0.3	$2.8 \cdot 10^9$	202.7	1.3	$4.0 \cdot 10^9$	202.75	–
CH_4	$8.3 \cdot 10^5$	125.7	-0.3	$1.5 \cdot 10^7$	125.7	1.3	$2.3 \cdot 10^7$	125.7	–
C_2H_6	$1.1 \cdot 10^{12}$	125.7	0.1	$1.3 \cdot 10^{12}$	125.7	1.65	$2.0 \cdot 10^{12}$	125.7	–
C_3H_8	$8.6 \cdot 10^8$	125.7	0.1	$1.0 \cdot 10^{12}$	125.7	1.65	$1.5 \cdot 10^{12}$	125.7	–
C_2H_6	$2.0 \cdot 10^{12}$	125.7	0.1	$2.4 \cdot 10^{12}$	125.7	1.65	$4.3 \cdot 10^{12}$	125.7	–
CO	$4.0 \cdot 10^{14}$	167.6	1	–	–	0.25	–	–	–
CO	$1.3 \cdot 10^{14}$	125.7	1	–	–	0.50	–	–	–
H_2	$3.2 \cdot 10^{15}$	167.6	1	–	–	0.25	–	–	–
NH_3	$4.9 \cdot 10^{14}$	163.4	0.86		–	1.04	–	–	–
$CO + O_2 = CO_2 + O$	–	–	–	–	–	–	$3.1 \cdot 10^{11}$	157.8	0.0
$CO + OH = CO_2 + H$	–	–	–	–	–	–	$1.5 \cdot 10^7$	-3.35	1.3
$H + O_2 = O + OH$	–	–	–	–	–	–	$2.2 \cdot 10^{14}$	70.40	0.0
$H_2 + O = H + OH$	–	–	–	–	–	–	$1.8 \cdot 10^{10}$	37.30	1.0

All units are in cm, s, mol, and K, except E which is in kJ/mol·K

Reaction rate for the quasi-global approach (4.21), can be calculated from expression [8]:

$$\frac{d[C_nH_m]}{d\tau} = -6 \cdot 10^4 \, T_p^{0.3} [C_nH_m]^{1/2} [O_2] \exp\left[-\frac{102.2}{R_gT}\right] \qquad (4.22)$$

for aliphatic hydrocarbons, and:

$$\frac{d[C_nH_m]}{d\tau} = -2.08 \cdot 10^7 \, T_p^{0.3} [C_nH_m]^{1/2} [O_2] \exp\left[-\frac{163.4}{R_gT}\right] \qquad (4.23)$$

for cyclic hydrocarbons. Concentrations are in mol/cm³, pressure in atm. where 1 atm = 0.981 bar = $0.981 \cdot 10^5$ N/m², T [K], and R_g [kJ/molK].

Detailed data on chemical reactions and their rates can be found in [18, 82] and other such specialized books.

4.5.   Volatile matter combustion in fluidized beds

In the previous section, the process of coal particle devolatilization was considered independently of coal particle surroundings. The influence of surroundings has been considered by exploring the effect of temperature, heating rate, particle size and O_2 concentration. Special attention has been paid to values of these properties that are appropriate for fluidized bed combustion. What happens to the volatiles released from the coal particle and the influence of other processes taking place in the fluidized bed have, however, not been considered. The existence of inert material particles, bubbles, mixing of gas and particles in the fluidized bed certainly influence the devolatilization process and the destiny of volatiles until their final burn-out. The behavior of coal particles during devolatilization, movement, mixing and burning of volatiles in fluidized beds will be discussed below.

4.5.1.   High volatile coal combustion in real conditions

Numerous data, obtained by following these processes in industrial-scale FBC boilers and furnaces under real operating conditions, show that for the combustion of coal (and biomass) with high volatiles content, a number of undesirable processes and some specific problems arise [80-87].

Any concept or design of an FBC boiler burning coals with high volatiles content must take into account the specific issues associated with burning high volatile fuels and the design of such boilers must be suitable to combust two fuels with very different properties:

– one fuel is gaseous (volatiles from coal) and the other is solid (char),
– the release and combustion of volatiles lasts seconds, while char combustion lasts minutes,
– volatiles ignition temperature is between 650 and 750 °C, and char ignition temperature can even be under 300 °C,
– the combustion of volatiles is homogeneous (takes place in gaseous phase), while char combustion is heterogeneous, and the burning rates of the two components are very different,
– volatiles residence time in fluidized beds is at most several seconds and it is possible that they will leave the bed without coming into contact with oxygen. Hence, the combustion of these volatiles must be able to occur in the freeboard, and

276

Chapter 4Chapter 4

- the char stays in the fluidized bed a long time, until it either "burns out" to the critical size which can be elutriated out of the bed, or is withdrawn by the bed withdrawal system. Further, char will rarely stay only in oxygen-deficient parts of the bed, i. e., the char particle will certainly see oxygen rich zones during its combustion history.

If these differences are not taken into account, combustion will be inefficient and major problems may arise in practice.

The following phenomena can occur leading to problems in volatiles combustion in fluidized beds:

- even at optimum combustion conditions, not all volatiles will burn out in the fluidized bed, and some volatiles will burn in the freeboard. For this combustion to occur properly, good mixing must be provided, with an "intensification" of turbulence in combustion products flow, and there must be ample oxygen. In the case of adiabatic furnace walls, the temperature of combustion products above the bed may exceed bed temperature by up to 200 °C [80, 81],
- when supplying large coal particles to the bed surface, large flames can be seen above the bed, due to the combustion of volatiles mainly originating from the coal particles near or at the bed surface [88]. When the supply of such fuel is temporarily stopped, after some given period in which volatiles burn out completely, the flame will disappear, indicating that only char is burning,
- in overbed feeding, volatiles are released around and under the coal feeding point, during the coal particle fall to the bed surface, and burn in the freeboard [88],
- when small coal particles (<3 mm) are fed below the bed surface, if there is non-uniform fluidization and inadequate number and choice of locations for the feeding point, local overheating occurs, due to the high concentration of heat generation – combustion of volatiles and char in the immediate vicinity of the feeding point,
- local hot spots in fluidized bed may cause ash softening and sintering, which can cause defluidization and possibly installation shut-down, and
- during the start-up period and coal feeding at bed temperatures lower than 600 °C, a sudden rise of temperature will occur when the bed temperature exceeds 600 °C and volatiles combustion begins. Heat generation in the furnace abruptly rises almost two-fold, possibly leading to uncontrolled rise of temperature exceeding the allowed limits [80].

4.5.2. Coal particle behavior during the devolatilization process in fluidized beds

In experiments described in Section 4.4.3, for one or several particles, many authors have reported yellow flame around the coal particles occurring at the fluidized bed surface. The disappearance of this flame has been taken as the end of devolatilization. Pillai [48, 49] was the first to observe that deep in the fluidized bed, during devolatilization, a coal particle is "enclosed" in a bubble of volatiles. As a result, the coal particle's ability to move is reduced during devolatilization. A bubble of volatiles and the flame existing at its edge due to the volatiles combustion, produce upward drag on the coal particle, dragging it towards the bed surface, disturbing its chaotic movement. The data of Yates [89], on volatiles release in the form of small discrete bubbles, observed by X-ray photography, have been explained by Pillai [11] as the result of limited coal particle movement. In this experiment, the coal particle was fixed to one place by a wire holder.

Detailed observation of coal particles during volatiles release and combustion, and char combustion, have been carried out in two-dimensional planar fluidized beds with transparent walls, for example by Radovanović [169], and has answered many questions about this process. Some of the more important results from such studies are described in detail below.

Coal particles, influenced by combustible volatiles' bubbles and surrounding flames, tend to float and stay in the upper regions of the bed. During devolatilization their degree of movement is restricted. After devolatilization, the degree of movement of char particles is significantly higher.

At temperatures sufficiently high for volatiles ignition, three possible flame shapes have been observed (Fig. 4.33):

(a) if the coal particle is (floating) at the surface of the bed, volatiles burn on the particle surface with a typical yellow diffusion flame (Fig. 4.33a),

(b) if the coal particle is at the bubble boundary inside the fluidized bed, a similar yellow diffusion flame is formed (Fig. 4.33b), and

(c) sometimes, at the bed surface there appears a small blue premixed flame, when the coal particle resides in the emulsion phase, several centimeters below the bed surface (Fig. 4.33c).

Flames of volatiles combustion have not been observed in the emulsion phase at all. Explanations for such behavior should be sought in the intense heat transfer and high heat capacity of inert material particles, which prevent the formation of flame, as well as tend-

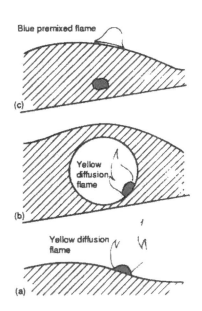

Figure 4.33.
Position and type of the flame during volatile combustion: (a) coal particle on the bed surface, (b) coal particle at the edge of bubble, (c) coal particle in emulsion phase, flame on the bed surface. According to Radovanović [169]

ing to quench radical processes. In the bubble phase of the bed such a high heat sink does not exist, and there is enough oxygen to enable the formation and persistence of a diffusion flame at the coal particle surface. Combustion of volatiles in a bubble takes place at temperatures higher than the average bed temperature, meaning that the average gas temperature in such bubbles during that period (before char combustion) is higher.

Another interesting question, which is important for modelling the processes in fluidized bed has been clarified by experiments of Prins [72, 90]: namely, are combustion of volatiles and char combustion simultaneous, or do these two processes occur sequentially? Figure 4.34 [72] shows three possible combinations observed in experiments. At temperatures lower than the volatiles ignition temperature, when there is no volatiles flame around the fuel particle, ignition of char particles is observed after an ignition time lag, as a bright spot on the fuel particle (Fig. 4.34a). Eventually, the whole particle surface ignites, and the particle becomes very bright, and is evidently at a higher temperature than the inert material particles.

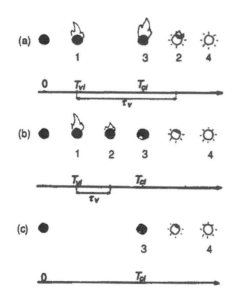

At higher temperatures, when burning bituminous coals with high volatile content, char ignition starts only after the volatiles combustion finishes, and when the yellow diffusion flame around the particle disappears (Fig. 4.34b).

When burning brown coal and lignite, due to the high reactivity of char from these coals, char ignition occurs locally, while the volatiles combustion is still occurring. Yellow diffusion flame can also be seen at the moments when the char particle is very bright, i. e., ignited (Fig. 4.34c). This figure also shows the relationships between these processes and their characteristic times.

As has been mentioned earlier, these experiments have shown that volatiles ignition starts at temperatures around 680 °C, while char residue ignites at lower temperatures (for lignite as low as 250 °C).

Figure 3.34.
Three possible sequences of the processes during coal particle combustion in a fluidized bed. According to Prins. (Reproduced by kind permission of the author Dr. W. Prins from [72])
(a) coal with small amount of volatiles, (b) char ignition after the end of volatiles combustion, (c) char ignition before the end of volatiles combution, 1 – ignition of volatiles, 2 – flame extinction, 3 – char ignition, 4 – char combustion

At temperatures which do not lead to ignition of volatiles, the time lag before char ignition in high-quality (older) coals is much longer than for younger coals. However, when volatiles are burning, younger coals show a much greater time lag for char ignition. These facts provide some interesting insights, which are useful in modelling of coal combustion in fluidized beds.

If there is no combustion of volatiles, then oxygen can freely reach the coal particle surface by diffusion. Char ignites sooner for more reactive (younger) coals.

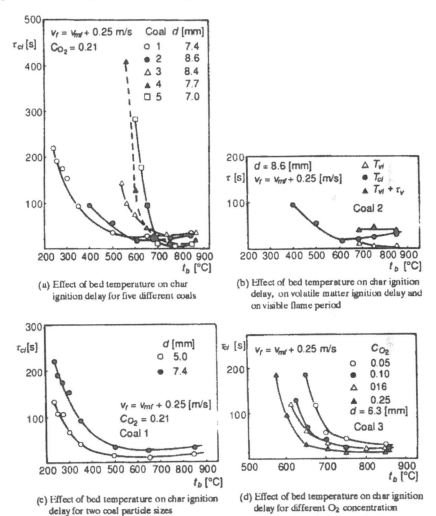

(a) Effect of bed temperature on char ignition delay for five different coals

(b) Effect of bed temperature on char ignition delay, on volatile matter ignition delay and on visible flame period

(c) Effect of bed temperature on char ignition delay for two coal particle sizes

(d) Effect of bed temperature on char ignition delay for different O₂ concentration

Figure 4.35. *Char ignition delay and volatile matter ignition delay for coal particle combustion in fluidized bed (Reproduced by kind permission of the author Dr. W. Prins from [72]*
Coal: 1 – lignite, 2 – subbituminous coal (brown coal), 3 – bituminous coal, 4 – char of coal 3, 5 – anthracite

During the continuous release and combustion of volatiles, oxygen consumption in the area around the particle is so great that there is no excess oxygen to react with fixed carbon of the char. In the combustion of younger coals, due to the greater reactivity of their char at the end of devolatilization process, the small amount of O_2 that reaches the particle surface is only capable of igniting the char on the bottom part of the particle where there is no flame.

Due to necessary delay time for coal particle heating, and moisture evaporation, a period of delayed ignition (and even a delay of devolatilization) also exists. Hence, the lifetime of a coal particle in a fluidized bed may be divided into the following characteristic periods – volatiles ignition time lag, devolatilization (and volatiles combustion) period, char ignition time lag, char combustion period, and characteristic periods: (1) – volatiles ignition, (2) – volatiles burn-out, and (3) – char ignition. Qualitative estimates of these times and of the influence of particular parameters are shown in Fig. 4.35.

4.5.3. Distribution and combustion of volatile matter in fluidized beds

The construction and design of FBC furnaces, to achieve optimum combustion and realize the advantages of fluidized bed combustion, should among other things provide combustion of the maximum quantity of volatiles in bed and at the same time prevent "hot" spots in the bed, i. e., provide a uniform temperature field. Achieving these goals by proper furnace design requires one to carry out proper engineering calculations of the furnace and ensure the correct placement of heat transfer surfaces. Further, in order to have detailed and reliable mathematical models of the processes that take place in the bed, one must have a knowledge of the volatiles distribution in the bed and the conditions for their in-bed ignition and combustion. In Chapter 2, which deals with the hydrodynamics of fluidized bed, and in Section 2.3.6 which deals with mixing of gas and particles in the fluidized bed, the consequences of poor radial mixing in terms of its ability to create inadequate combustion conditions have been described. Design measures to enhance mixing have also been described.

The discussion of the practical consequences of inadequate mixing provided in that section has been directed mostly to the mixing of solid coal particles. However, this consideration is even more important for the gaseous part of solid fuel, i. e., volatiles (see Fig. 2.37). Data given in Section 2.3.6, about the value of the apparent gas diffusion coefficient (in the axial direction 0.1-1 m^2/s, and in the radial direction 10^{-3}-10^{-4} m^2/s) speak strongly about the problems of proper, uniform distribution of volatiles in fluidized beds. Gas mixing processes in fluidized beds have simply not been sufficiently investigated, and the very small number of research studies devoted to these problems are still inadequate to the demand for better knowledge in this area. Worse still, not only are data scarce, but some of it is contradictory. Unfortunately, even less work has been devoted to the problems of volatiles distribution [91, 92], and good direct measurement of volatiles distribution in fluidized beds does not at this time exist. Analysis of practical problems is made even more difficult due to the fact (see Fig. 2.36) that gas mixing conditions depend heavily on the installation size.

The behavior of single fuel particles during devolatilization and the release, ignition and combustion of volatiles described in the previous section cannot as yet give answer to the following questions. What happens when there are large assemblies of fuel particles close to each other releasing volatiles? Does a continuous field of volatiles form in the bed? Will flames of each particle join into one big flame? Do volatiles burn also in the emulsion phase, or only in the bubbles? What conditions are necessary for ignition and combustion of volatiles?

Direct answers to these questions, based on measurements, still don't exist. For the purpose of design of boilers and furnaces overall measurements of heat generation distribution in the bed and above it, for combustion of different coals, are used [93-95]. The problem of distribution of volatiles, their ignition and combustion, except for design and calculation of boilers, is most often considered in designing mathematical models of processes in fluidized bed furnaces. In the rest of this section and the next section of this chapter, more will be said about mathematical models of the processes in the furnace.

The distribution of volatiles and distribution of heat generated by their combustion is the result of a number of processes taking place in fluidized beds, and it depends on the type and properties of fuel, and geometric and design parameters of furnaces. Except about processes of release and combustion of volatiles for each particular coal particle, several other things must be taken into account:

(a) is coal fed above the bed or under its surface?
(b) where under the bed's surface is coal fed?
(c) it is inevitable that the number of coal feeding points must be finite, and as small as possible – which in some points leads to high fuel concentration,
(d) devolatilization time is 5-50 s, and char combustion time is up to 10 min. and depends heavily on the particle size,
(e) distance particle traverses in axial direction is an order of magnitude larger than in radial direction, considering the ratio of apparent diffusion coefficients of particles in axial and radial directions,
(f) time needed for coal particle to travel from the bottom to the surface of the bed is comparable, if not shorter than devolatilization time,
(g) released volatiles reach the surface of the bed in less than 10 s (bed height of 1.0 m), and in that time by diffusion in radial direction traverse a mere 10 cm,
(h) ignition temperature of volatiles is 680 °C, and ignition temperature of char is 250-650 °C,

(i) burning rate of volatiles is much higher than that of char,

(j) sufficient concentration of oxygen is required for ignition and combustion of volatiles,

(k) regardless of appropriate conditions in the bed, part of volatiles will leave the bed unburnt, and will burn above the bed,

(l) due to increased oxygen concentration, volatiles will more probably burn in bubbles, and in experiments with one particle flame of volatiles in emulsion phase has not been observed, and

(m) different coals have different densities, and tendencies to float at the bed's surface.

Bearing in mind these facts, several basic conclusions may be drawn:

(1) When feeding coal above the bed it is more probable that volatiles will burn above the bed. Usually large particles (0-50 mm) of more reactive, younger coals are fed above the bed. Coal particle size prolongs devolatilization time and enables uniform distribution of volatiles, but the quantity of volatiles is large so the problems with their combustion are greater. Younger coals are lighter and tend to stay in the upper half of the bed.

(2) When feeding coal under the bed's surface, usually near the distribution grid, smaller particles (< 3 mm) are used, devolatilization time is shorter and conditions are favorable for incomplete combustion of volatiles and generation of hot zones. However, these coals usually have a low volatile content, and the path up to the bed's surface is long enough to allow for ignition and burn-out of volatiles.

(3) Depending on the ratio of characteristic time of devolatilization (τ_v) and particle residence time in the fluidized bed, i. e., the distance the particle traverses in time τ_v (see eqs. 2.93, 2.95 and 2.96, and Figs. 2.27-2.29), two extreme cases can be distinguished: (1) all volatiles are released instantaneously in immediate vicinity of coal feeding points, and (2) volatiles are uniformly generated in the whole bed volume. Depending on the conditions, a whole spectrum of possibilities exists between these two extreme cases.

It is usually supposed that the conditions for ignition and combustion of volatiles will be met if volatiles come in contact with oxygen (at $T_b > T_w$), and an infinitely rapid oxidation process is assumed.

One of the rare attempts, based on the observation of the devolatilization process and gas mixing in fluidized bed, to predict how much volatiles will burn in the bed, and how much above the bed, is the so-called "plume" model proposed by Park [96-98].

The "plume" model assumes that volatiles are released instantaneously at the coal feeding point. Released volatiles move upwards with fluidization velocity v_f, and in the radial direction expand according to the laws of molecular diffusion with apparent diffusion coefficient, D_{gr}. At the boundary of this "plume" and surrounding oxygen, combustion controlled by diffusion rate takes place (Fig. 4.36).

Based on the Park model, values of the non-dimensional number $H_b D_{gr}/v_f L^2$ (where L is distance between coal feeding points), determines whether volatiles will burn out in the bed or not. The value of this number is around 10^{-1} and suggest that most of the volatiles burn out in bed. Bywater [99] improved this model by including diffusion of solid particles. However, there are no data on experimental verification of this model.

In [77] the estimate of volatile "plume" expansion is based on the distance a particle traverses during devolatilization time, τ_v:
in the vertical direction

$$\overline{\Delta l_a} = \tau_v \, \overline{u}_p = \tau_v \, 1.9 (\, v_f - v_{mf} \,)^{0.5} \tag{4.24}$$

and in the radial (horizontal) direction:

$$\overline{\Delta l_r^2} = 2 D_{gr} \tau_v \tag{4.25}$$

In one of the most recent models of fluidized bed combustion, developed by international cooperation in International Energy Agency of OECD [100-105], the distribution of volatiles is based on the ratio of devolatilization time τ_v, and the time

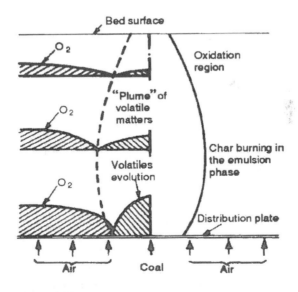

Figure 4.36.
"Plume" model of Park et al. – combustion of volatile matter in "plume" above the under-bed coal feeding point (Reproduced with permission of the AIChE from [97]. Copyright 1980 AIChE. All rights reserved)

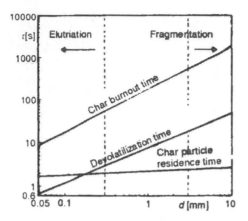

Figure 4.37.
Effect of particle size on time scales of the various subprocesses of coal combustion in fluidized bed – char burn-out time, devolatilization time, and char particle residence (turnover) time (Reproduced with kind permission of the author Dr. G. Brem from [104])

needed for particles to reach the bed's surface [106], (i. e., the shortest time the particle spends in the bed):

$$\tau_v = \frac{H_{mf}}{\delta_1 (v_f - v_{mf})(\delta_2 + \delta_3)} \quad (4.26)$$

According to [100], coefficients are chosen in the range of δ_1 = 0.35-1.0, δ_2 = 0.35, and δ_3 = 0.6-1.0, and depend on the particle diameter.

For standard conditions in fluidized bed combustion, a qualitative notion about the relationship of characteristic times depending on the fuel particle diameter, is given in Fig. 4.37.

The devolatilization time is much longer than the shortest time the particle spends in the bed, for particles larger than 1mm, justifying the assumption that the volatiles are uniformly distributed in the bed. Volatiles released from particles smaller than 1mm will partially burn above the bed.

4.5.4. Modelling of volatile matter distribution and combustion in fluidized beds

Researchers developing the first models of fluidized bed combustion were primarily interested in the combustion of geologically older coals with low volatiles content. Hence, early models described only char combustion, and did not take into account the existence and influence of volatiles. More recently, interest has grown in the combustion of geologically younger coals (lignites, brown coals) with high volatiles content, and several attempts exist to encompass both char and volatiles combustion in fluidized bed combustion models. However, due to inadequate knowledge on the devolatilization process, ignition and combustion of volatiles, and also on the mixing and distribution of volatiles, current models are based on very simplified process mechanism.

In Section 4.7, which deals with mathematical models describing the processes in fluidized bed combustion, a systematization of existing models is given (in Tables 4.15 and 4.16), which shows all the assumptions used. Assumptions about generation, distribution and combustion of volatiles are shown in column D, with lower-case letters (a) through (g) and numbers (1) through (4) of the classification shown below. Assumptions about the combustion in the freeboard are in column G, denoted by the letters (a) through (d) in the list below. Tables are based on the review in [3] and data have been added for more recent models.

We will briefly list the assumptions used in modelling:

D. Release and combustion of volatiles

> *Devolatilization*
> (a) release of volatiles is instantaneous at the level of the coal feeding plane,
> (b) the devolatilization time is much longer than the characteristic time of coal particle mixing – distribution of volatiles is uniform throughout the bed volume,
> (c) devolatilization is instantaneous at the coal feeding point,
> (d) some arbitrary volatiles generation distribution is adopted,
> (e) devolatilization is tied to coal particle and its movement through the bed,
> (f) kinetics of devolatilization are taken into account, and
> (g) combustion of volatiles is not taken into account.

> *Combustion of volatiles*
> (1) volatiles burn with a diffusion-controlled flame, combustion is instantaneous if there is enough oxygen,
> (2) kinetics of the chemical reaction (usually CO oxidation) control the rate of volatiles combustion,
> (3) combustion is controlled by both chemical kinetics and diffusion, and
> (4) combustion is instantaneous and controlled by the devolatilization rate.

G. Combustion above the bed

> (a) is not taken into account at all,
> (b) combustion of char particles leaving bed is included,
> (c) volatiles burn instantaneously when they leave the bed, and
> (d) volatiles burn in the whole region above the bed.

A review of models given in Table 4.15 shows that, only since 1985, did the devolatilization process gain real attention, but that even recent models do not adequately include this process. Most often, based on the estimate of characteristic times, distribution of volatiles generation is assumed [107], the kinetics of volatiles release is not taken into account and instantaneous combustion is also assumed when there is enough oxygen. Processes taking place above the bed are also simplified and rarely included in models.

4.6.   Char combustion

In the stationary combustion regime, a rather small amount of coal, i. e., char, actually exists in the fluidized bed. The mass ratio of fuel, compared to the inert material mass is of the order of 2-5% in the fluidized bed, depending on the char particle size and reactivity. Without deviating much from real conditions it is possible to investigate fluidized bed combustion by experiments with one char particle surrounded by the inert material particles and gas with appropriate oxygen concentration. Therefore, this section will primarily deal with the combustion of single particles, most often of spherical shape. Virtually all experimental research on the nature and kinetics of combustion processes is carried out on single particles or a batch (portion) of a small number of particles of the same size and shape.

Fragmentation and devolatilization processes drastically change shape, size, composition and characteristics of coal particles fed into the boiler furnaces and reaching the fluidized bed. These processes, lasting 1-100 s will help determine the fuel particle size, and char combustion takes place in the fluidized bed. The duration of the char combustion process is an order of magnitude larger than the duration of devolatilization and combustion of volatiles, and hence primarily influences many characteristics of combustion and the furnace itself.

Char is combustible matter from the coal particle left over after devolatilization. Bearing in mind that the composition and quantity of volatiles depend on the coal type, heating rate and maximal temperature at which decomposition of hydrocarbons takes place, it is obvious that the composition and characteristics of char may be very different. That is why it is necessary to investigate the char combustion process for every type of coal independently, and include the conditions under which char was formed.

When investigating char combustion in a fluidized bed, it should be noted that char may be formed in two ways: in a laboratory oven under standard conditions, at different temperatures and with different heating rates; and in a fluidized bed in inert atmosphere or in air, at different bed temperatures. It has been widely accepted that to reproduce conditions sufficiently similar to those in the real fluidized bed, char must be prepared in a fluidized bed. Preparation of char in a fluidized bed in a nitrogen atmosphere is easier and often used, although it should be noted that combustion of volatiles changes the temperature conditions of devolatilization and hence the final composition of char.

Roughly speaking, char consists of left-over hydrocarbons, mineral matter and elemental carbon. Besides the composition, which may be different for the same type of coal if prepared under different conditions, the basic characteristic of a char particle is its porosity (see Section 4.2.5). The total chemical reactivity of char depends on the type (composition) of coal and its porosity. Porosity influence occurs not only because of the large pore specific surface, but because it also determines the local concentration of oxygen inside the char particle. In principle, the O_2 concentration inside the particle is always lower than at its surface. Only inside large, open pores may the O_2 concentration be close to the concentration on the surface. The depth of O_2 penetration determines the char particle combustion rate, the character of combustion, i. e., will the combustion take place at the particle's surface only, or will it also take place inside the particle.

Regardless of the fact that combustion process characteristics depend on the char composition and porosity, many experimental studies have been carried out with particles of pure carbon. These investigations discovered many characteristics of combustion in fluidized beds and permitted exploration of the influence of many factors (particle size, O_2 concentration, etc.). However, real data on the character and the rate of combustion of a particular coal can be obtained only by experiments with its char particles.

Char combustion in fluidized beds is a chain of processes: oxygen diffusion from the bubbles to the emulsion phase, oxygen diffusion in the emulsion phase to the char particle surface, oxygen diffusion through the particle's porous structure to the carbon molecules at the surface of the pores, heterogeneous carbon oxidation occurring at the surface, and diffusion of oxidation products occurring in the reverse direction to the bubbles with simultaneous homogeneous reaction in the gaseous phase. The total combustion rate of char particles in a fluidized bed depends on the character and the ratio of the rates of these processes.

Experimental research and mathematical modelling of char combustion are aimed at determining ignition temperature, combustion rate, burn-out time and temperature of char particles.

Knowing these parameters and the character and pathways of the above-mentioned processes on the char combustion process, significantly influences the ability to predict the efficiency of coal combustion in fluidized bed, and the emission of carbon monoxide and nitrogen oxides from fluidized bed boilers. In real combustion conditions, ash behavior (its elutriation and possible sintering) is strongly related to the character of char particle combustion.

Before development and research of fluidized bed combustion, char combustion had been thoroughly investigated under conditions characteristic for pulverized coal combustion, with particle size less than 200 μm. Among the first studies were those of Essenhigh and Beer [107, 108]. Thorough reviews of coal and char combustion research under pulverized coal combustion conditions have been given in [1, 7, 8, 109]. Regardless of very thorough research on char combustion under these conditions, which discovered many of the aspects of processes that also applied to fluidized bed combustion, differences in particle size, combustion rate, and conditions for oxygen diffusion, must be investigated under the conditions similar to those in fluidized bed combustion. Data presented in the following sections are based mostly on research on char combustion in fluidized beds, which follows the pioneering work of Avedesian and Davidson [110]. Basic knowledge about the nature and chemistry of the char combustion process accumulated in this research [1, 7, 109] formed the basis for most of the subsequent experiments in fluidized beds.

4.6.1. Kinetics of heterogeneous chemical reactions on the surface of carbon (char) particles

Char particle combustion rate is defined as the mass of carbon burned per unit time and unit area. Since the inner surface of porous particle structure is usually unknown, combustion rate is most often expressed using the outer particle surface, and is called apparent combustion rate. Physically it is more appropriate to express combustion rate in terms of inner, true particle surface. However, bearing in mind that the total surface of porous particle is not always accessible to oxygen due to obstructed O_2 diffusion through the pores, the question of the most appropriate definition of combustion rate remains open, and it is a matter of consensus, and when considering any study, it is essential to know on which surface area the combustion rate is based.

Based on the mass balance, combustion rate per unit of outer surface area can be expressed as the rate of oxygen transfer to the outer particle surface:

$$R = \Lambda k (C_{O_2} - C_{O_{2P}}) \qquad (4.27)$$

The combustion rate can also be expressed as the apparent rate of chemical reaction R_c of the nth order:

$$R = R_c (C_{O_{2P}})^n \qquad (4.28)$$

The concentration of oxygen at the particle surface is usually unknown and may be eliminated from expressions (4.27) and (4.28), yielding:

$$R = R_c (1 - \chi)^n C_{O_2}{}^n \qquad (4.29)$$

where:

$$\chi = \frac{R}{R_m} \quad \text{and} \quad R_m = \lambda k C_{O_2} \qquad (4.30)$$

Parameter R_m is the maximum possible combustion rate in the case when the rate of chemical reaction is so great that $C_{O_{2P}} = 0$, and combustion rate depends only on the mass transfer towards the particle.

An expression for χ can also be obtained [109]:

$$\frac{\chi}{(1 - \chi)^n} = R_c \frac{C_{O_2}{}^n}{R_m} = R_c \frac{C_{O_2}{}^{n-1}}{\Lambda k} \qquad (4.31)$$

This expression shows that the influence of mass transfer process on the overall combustion rate depends on n, the order of chemical reaction. With the usual assumption that $n = 1$, i. e., if carbon oxidation is a chemical reaction of first order, ratio χ does not depend on the oxygen concentration. Equation (4.31) can be used for determining the order of chemical reaction n, based on the experimental determination of combustion rate [109]. Although it has been suggested [72, 111] that in fluidized bed combustion, at temperatures of 900-1500 K, an order of chemical reaction $n = 1/2$ is more appropriate, $n = 1$ is used more frequently. In that case from (4.27) the usual expression for combustion rate may be obtained:

$$R = \frac{C_{O_2}}{\left(\dfrac{1}{\Lambda k} + \dfrac{1}{R_c} \right)} \tag{4.32}$$

In the case where $n = 1/2$, a more complex expression is obtained:

$$R = -\frac{R_c^2}{2\Lambda k} + \frac{1}{2}\left(\frac{R_c^4}{\Lambda^2 k^2} + R_c^2 C_{O_2} \right)^{1/2} \tag{4.33}$$

and for $n = 0$

$$(R - R_m)(R - R_c) = 0 \tag{4.34}$$

From these relationships, it follows that for finite oxygen concentration at the outer particle surface, combustion rate is determined by the rate of chemical reaction $R = R_c$, and that it jumps to $R = R_m$ when $C_{O_{2P}} = 0$ [3, 109].

The chemical reaction rate R_c is called apparent, because besides chemical reaction it includes also the influence of other processes and parameters: (1) true (inner) rate of chemical reaction of oxygen with inner particle surface; (2) inner surface area size; and (3) oxygen diffusion through the pores in the particle. The relationship between combustion rate R (or R_c) and inner (true) rate of chemical reaction R_i, per unit area of pore surface and unit concentration, and in the case when mass transfer is not the limiting process, is given by the expression [109]:

$$R = \eta \gamma \rho_P a_P R_i [C_{O_2} (1-\chi)]^m \tag{4.35}$$

Reaction efficiency coefficient η is a function of Thiele's module Φ [111]:

$$\left[\eta \Phi^2 \frac{(m+1)}{2} \right] = \frac{\gamma R (m+1)}{[8 D_e C_{O_2} (1-\chi)]} \tag{4.36}$$

All right-hand-side values in (4.36) can be measured enabling calculation of coefficient η and then R_i from (4.35).

If the chemical reaction is slow enough, so that pore diffusion manages to supply enough oxygen to the surface (oxygen concentration inside the particle is equal to the concentration at the surface) it will be $\eta = 1$:

$$R = \gamma \rho_P a_P R_i [C_{O_2} (1-\chi)]^m \tag{4.37}$$

Under these conditions, combustion takes place in kinetic regime I, the particle burns in all of its volume, and its size remains constant. Then $n = m$, is the true order of chemical reaction, R_c and R_i differ only in surface area they have been defined for, and the ratio $R_c = (A_{pore}/A_p)R_i$ holds.

When pore diffusion and chemical reaction both control the combustion rate, the left hand side of equation (4.36) tends to $1/\eta$, so from (4.35) and (4.36) follows:

$$R = 2 \left\{ \frac{2 a_P \rho_P D_e R_i [C_{O_2} (1-\chi)]^{m+1}}{m+1} \right\}^{0.5} \tag{4.38}$$

Combustion under these conditions is called kinetic regime II, combustion rate R is independent of the particle size, and apparent order of chemical reaction is:

$$n = \frac{m+1}{2} \tag{4.39}$$

During combustion the particle gets smaller and does not change density.

Combustion rate of coal particles depends on, besides the values seen in (4.29), (4.32) or (4.33), (4.35), and (4.37), three other important parameters which express the intensity of the three processes: (1) mass transfer coefficient, k, determines the oxygen transfer to the outer particle surface, (2) the effective O_2 diffusion coefficient in particle pores, D_e, and (3) the rate of chemical reaction, R_i (or R_c).

The coefficient of mass transfer from the emulsion phase to the fuel particle, in fluidized bed combustion is obtained from the correlations described in Section 3.3.

According to Smith [109], an effective pore diffusion coefficient can be obtained from the expression:

$$D_e = \frac{D_{pore} \xi_P}{l_{pore}} \tag{4.40}$$

For pores larger than 1-3 μm, $D_{pore} = D_G$, and for small pores (<1 μm), the Knudsen diffusion coefficient is used:

$$D_{pore} = 9.7 \cdot 10^3 \, r_{pore} \left(\frac{T_P}{M} \right)^{0.5} \tag{4.41}$$

where the numerical constant has the dimensions [(cm/s)(g/Kmol)] and where:

$$r_{pore} = \frac{2 \xi_P l_{pore}^{0.5}}{a_P \rho_P} \tag{4.42}$$

Apparent chemical reaction rate R_c or true (inner) rate of chemical reaction at the pore walls inside the particle, is usually represented in Arrhenius form:

$$R_c \text{ (or } R_i \text{)} = A \exp\left[-\frac{E}{(R_g T_P)}\right]^k \qquad (4.43)$$

where E is apparent or true activation energy, and k is apparent (n) or true (m) order of chemical reaction, depending on whether it is an apparent or true rate of chemical reaction.

In an exhaustive review and analysis of the experimental results of many authors, Smith [109] concludes that investigations of combustion rate R_c, for chars of different origin and for particles <100 μm show:

-   apparent order of chemical reaction ranges from $n = 0.2$ to $n = 1$. Experimental results of Field [1] assuming $n = 0.5$ and $n = 1.0$ do not show great differences in scattering around the relationship obtained by least-squares method.
-   combustion rate does not depend on the particle size in this range, but shows that char burns in kinetic regime II. Very fine particles, $d_p < 20$ μm, have lower combustion rate than larger particles, speaking of combustion in kinetic regime I,
-   apparent activation energy for char of different origin (from petroleum coke to brown coal) ranges from 34 kcal/mol (142 kJ/mol) to 16 kcal/mol (67 kJ/mol), and the average can be taken as 20 kcal/mol (84 kJ/mol), and
-   apparent combustion rate R_c, for chars of different origin differs by a factor of 10 and more, which is manifested in the difference in pre-exponential coefficient and apparent activation energy. Different conditions, and different particle sizes, after devolatilization give char of different chemical composition and porous structure, leading to different reactivity.

Differences in combustion rate of char of different coals are understandable because R_c includes inner (true) carbon reactivity and also the value and availability of inner pore surface, which may range from 1 m^2/g for petroleum coke to around 1000 m^2/g for brown coal char.

Smith [112, 113] has collected and analyzed numerous experimental data for different kinds of porous carbon: char from different coals, nuclear graphite, various highly purified graphites. Figure 4.38 shows true reactivity of porous carbon of different origin. It may be seen that there are great differences in reactivity of carbons of different origin, although the influence of pore size and pore surface area has been eliminated. Differences reach up to four orders of magnitude. The average activation energy is ~40 kcal/mol (167.5 kJ/mol), which is around twice the value obtained for apparent char activation energy. It is thought that such great differences in the reactivity of porous carbon come form the impurities in carbon and the absorbed gases. Results obtained from experiments with highly purified carbon show little scattering and an activation energy around 60 kcal/mol (~250 kJ/mol).

The great differences in reactivity of investigated chars and porous carbon show that the carbon oxidation process is still not well understood [7, 109] and that for engineering purposes it is necessary to first determine reactivity of char of inter-

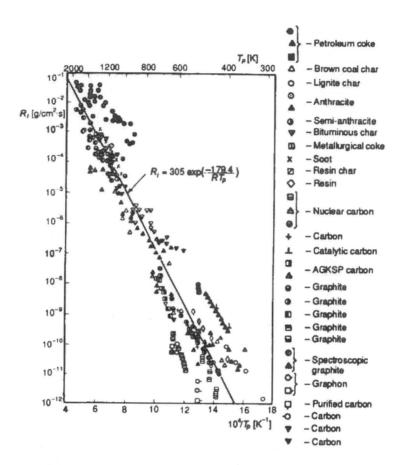

Figure 4.38. *Intrinsic reactivity of chars (carbons) of different origin. According to I. W. Smith (Reprinted with kind permission of the Combustion Institute, Pittsburgh, from [109])*

est, at different temperatures, for different particle sizes and oxygen concentrations, taking care to organize char production under conditions similar to those in real life.

4.6.2. Chemical reactions and control processes during carbon (char) particle combustion

One of the basic questions, and one of the still unsolved problems, which arises when considering the carbon (or char) combustion process, is: which chemical reactions take place? As possible reactions, the following are considered:

$$C + O_2 = CO_2 \tag{4.44}$$

$$C + \frac{1}{2}O_2 = CO \tag{4.45}$$

or jointly:

$$(1+p)C + (1+\frac{p}{2})O_2 = pCO + CO_2 \tag{4.46}$$

as well as:

$$CO_2 + C = 2CO \tag{4.47}$$

$$CO + \frac{1}{2}O_2 = CO_2 \tag{4.48}$$

Reactions with carbon are heterogeneous and take place at the solid surface of the particle (outer or inner surface of porous structure) while reaction (4.48) is homogeneous and takes place in the gaseous phase.

In pioneering work on carbon particle combustion in fluidized beds, Avedesian and Davidson [110] have assumed the following chain of reactions: oxygen does not reach the particle surface and is used up in a thin layer at some distance from the particle in carbon monoxide oxidation (reaction 4.48). Newly formed carbon dioxide diffuses to the particle and from the particle. At the particle surface CO_2 reacts with carbon (reaction 4.47) forming CO, which is transferred from the particle by diffusion. This chain of reactions has been dubbed the "two-layer carbon oxidation model." Concentration profiles of CO, CO_2, and O_2 are shown in Fig. 4.39a for the case of heterogeneous reaction at the particle's surface. Analyses and experimental data given in [114] show that temperature conditions in fluidized beds (<1000 °C) do not favor reaction (4.47). In the majority of recent models it has been assumed that carbon oxidation at the particle surface

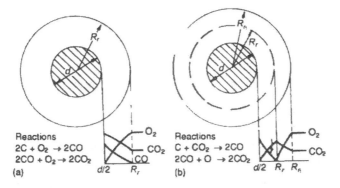

Figure 4.39. *Concentration of O_2, CO, and CO_2 around burning carbon particle in diffusion combustion regime: (a) Single-layer model, (b) two-layer model*

forms CO and CO_2, and that these gases are conveyed away by diffusion, while reaction (4.48) in the gaseous phase then transforms carbon monoxide to carbon dioxide (concentration profiles are given on Fig. 4.39b).

Determining the ratio of compounds formed around the particle, CO/CO_2, is one of the key problems in modelling carbon particle combustion. Experimental research on this problem is difficult because it is impossible to directly measure concentration of these gases at the surface or in the immediate vicinity of the particle. Measurements are possible far from the particle, when the gas phase reactions are finished.

Analysis of the results from many authors, given in [7], shows that the ratio CO/CO_2 can be expressed as:

$$p = \frac{[CO]}{[CO_2]} = A \exp\left(-\frac{E}{R_g T}\right) \tag{4.49}$$

where constants for low pressures are $A \approx 316$, $E = 25\text{-}37.5$ MJ/kmol, and for higher pressures, $A \approx 2500\text{-}3500$, $E = 50\text{-}80$ MJ/kmol. In modelling char combustion in a fluidized bed the following equation is most often used [72, 114, 115]:

$$p = 2500 \exp\left(-\frac{52000}{R_g T}\right) \tag{4.50}$$

A basic question which has to be answered by these models is the determination of the distance from the particle surface at which carbon monoxide oxidation ends. Oxygen consumption and particle temperature depend on whether CO transforms into CO_2 near the particle surface or far from it. Oxidation of CO into CO_2 depends on the particle size and temperature, i. e., on the ratio of reaction rate of CO oxidation and CO diffusion from the particle.

According to calculations of Ross et al. [116] for particles smaller than 0.5 mm, CO spreads up to 5-6 particle diameters, and particle temperature will correspond to the CO forming reaction. When burning particles larger than 5 mm, CO spreads to relatively smaller distances, (~ 1 d), and particle temperature corresponds more to the CO_2 forming reaction.

In fluidized bed combustion, in both cases, inside this zone there will exist many inert material particles (1000-3000), so the influence of these particles must be taken into account.

Combustion of higher rank coals in fluidized beds is typically carried out by feeding coal under the bed surface, with particle size between 0.1 mm and 3-5 mm, i. e., in the range where none of the above-mentioned models apply.

Caram and Amundsen [117,118] and Ross et al. [116] have proposed a so-called continuous model which takes into account all four major reactions (4.44), (4.45), (4.47) and (4.48). The concentration distribution of O_2, CO, and CO_2 is given in Fig. 4.40 for this model. At the particle surface, concentration of these three gases, CO, CO_2, and O_2, is not zero, and oxidation of CO to CO_2 takes place in the entire area around the particle.

All models described assume that heterogeneous reactions (4.44), (4.45), and (4.47) take place only at the outer surface of the char (carbon) particle. Due to the porous structure of coal, and especially char, combustion is also possible inside the particle. Only if the reactions at the particle surface are fast enough to consume all available oxygen, will reactions inside the particle become impossible. Otherwise, oxygen will diffuse through the porous structure inside the particle and react with the carbon at pore walls.

As has already been mentioned in the previous section, if the diffusion of O_2 through the porous structure is much faster than the chemical reaction at the inner surface of pores, then O_2 concentration in contact with pore surface will be equal to the concentration at the particle surface. In that case oxygen transfer through the boundary layer outside the particle is very intensive, so O_2 concentration at the particle surface is equal to the concentration far from the particle. At lower temperatures, combustion takes place in kinetic regime I, and combustion rate is defined by expression (4.37). The combustion process is controlled only by the rate of chemical reaction. Figure 4.40a shows the concentration profiles of O_2 inside and around the porous carbon or char particle.

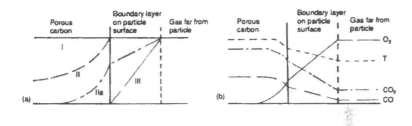

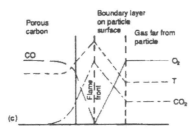

Figure 4.40. *Concentration of O_2, CO, and CO_2 around burning porous carbon particle: (a) Oxygen concentration, I – kinetically controlled combustion regime, II – combustion regime controlled by pore diffusion, IIa – combustion regime controlled by pore diffusion and external gas diffusion, III – combustion regime controlled only by external gas diffusion; (b) Single-layer model; (c) two-layer model*

If the diffusion process inside the pores is slow, it will control the penetration of O_2 through the pores, and inside the particle a gradient of O_2 concentration will be formed. The combustion process is partly controlled by diffusion through the pores and partly by the rate of chemical reaction. At somewhat higher temperatures combustion is carried out in kinetic regime II, and combustion rate is given by expression (4.35) where η is the efficiency coefficient, i. e., particle volume utilization, and it shows the decrease of combustion rate due to the diffusion through the pores. Equation (4.36) gives the relation between η and Thiele's modulus, expressing the ratio of rates of two processes, chemical reaction and diffusion.

With further increase of temperature, the combustion process becomes controlled by O_2 diffusion through the boundary layer around the particle. Oxygen concentration at the particle surface decreases (compared to the concentration far from the particle). Combustion is carried out in regime IIa, where process is controlled by chemical reaction kinetics, diffusion through the porous particle structure and diffusion through the boundary layer around the particle (curve IIa at Fig. 4.40a).

Even higher temperatures create the diffusion regime (curve III in Fig. 4.40a), where the combustion process is controlled by the diffusion rate of O_2 through the boundary layer around the particle.

In the case when combustion takes place inside the particle, we can also speak of two combustion models – single-layer and two-layer (and even continuous). Concentration profiles of O_2, CO and CO_2 for single-layer and two-layer models, analogous to those shown in Fig. 4.39a and b, but including combustion inside the particle are shown on Fig. 4.40b and c. Profiles shown in Fig. 4.40b correspond to lower temperatures when reactions (4.44), (4.45), (4.46), and (4.48) take place. Due to combustion inside the particle, O_2 concentration at the particle surface is not zero, as in Fig. 4.39a. At higher temperatures, reaction (4.47) is also possible, and concentration profiles correspond to the two-layer model (Fig. 4.40c), analogous to the Fig. 4.39b. Due to combustion inside the particle, CO_2 concentration is not zero at the outer surface of the particle.

Brem [104] has formulated a "map of combustion regimes," based on the experiments and mathematical modelling of char combustion for different coals in fluidized beds. For char of one Polish coal ($E/R = 28.15$ K) Fig. 4.41 shows this map of regimes in coordinate system bed temperature – char particle size:

(1) for $\Phi \ll 1$ combustion is in kinetic regime I,
(2) for $\Phi \gg 1$ combustion is controlled by diffusion process,
(2a) for $Bi_m \gg \Phi$ combustion is in kinetic regime IIa,
(2b) for $Bi_m \ll \Phi$ combustion is in diffusion regime III, and
(3) $1 < \Phi < 5$ combustion takes place in kinetic regime II.

Regime boundaries in Fig. 4.41 correspond to particular types of coal and combustion conditions (i. e., for particular char reactivity R, and oxygen concentration). However, it may be concluded that in most cases, under conditions of fluidized

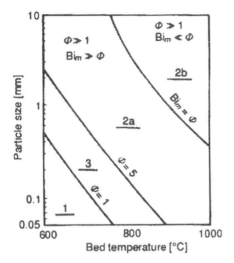

Figure 4.41.
Combustion regime map based on data obtained from combustion experiments with Polish 10 coal char particle in fluidized bed. According to the combustion model of Brem (Reproduced with kind permission of the author Dr. G. Brem from [104])
Regions: 1 – kinetic regime, 2a – pore diffusion regime, 2b – external diffusion regime, 3   transient regime

bed combustion (t_b = 700-900 °C, and d = 1-10 mm), the combustion process is controlled by pore diffusion and diffusion in the boundary layer around the char particle.

4.6.3. Carbon (char) particle burning models

Combustion of the carbon (char) particle changes the mass of carbon, particle size and particle density. The following simple relationships exist between these three values:

$$M_c = M_{co} (1-f) \qquad (4.51)$$
$$d = d_o (1-f)^\alpha \qquad (4.52)$$
$$\rho_p = \rho_{po} (1-f)^\beta \qquad (4.53)$$

where $3\alpha + \beta = 1$

Three characteristic cases are possible:

(1) particle size decreases, density is constant ($\beta = 0, \alpha = 1/3$). Combustion takes place at the outer particle surface – *unreacting shrinking core model*,

(2) particle size remains the same while density decreases, because combustion takes place inside the particle's porous structure ($\beta = 0, \alpha = 1$) – *reacting core model*, and

(3) all other combinations of constants α and β, result in the so-called *progressive conversion model*, when combustion inside the particle takes place only in the reaction zone near the surface, but the core still exists where there is no reaction between carbon and oxygen.

When postulating a physical model of carbon (char) particle combustion and its further mathematical formulation, the following basic assumptions are usually adopted:

–   the particle is spherical,
–   processes are quasi-stationary,
–   isothermal and isobaric conditions are inside the particle,

- physical properties are constant, and
- gases are ideal – the diffusion coefficient is taken as for ordinary binary diffusion.

Differences between models originate from the following different approaches:

- is combustion taking place at the inner or outer surface of the particle,
- which type of O_2 and carbon reaction is assumed,
- what order of oxidation reaction is assumed,
- is the ash presence taken into account, and
- are changes in particle's porous structure taken into account?

Unreacted (shrinking) core model (1), is the simplest model of carbon particle combustion, which does not take into account particle structure. Only diffusion in the boundary layer around the particle and the rate of chemical reaction on the particle surface are taken into account. Such a model is applicable when the rate of chemical reaction on the particle surface is higher than O_2 diffusion through the porous particle structure. Figure 4.42 shows the schematic view of the unreacted (shrinking) core model and concentration distribution of O_2 and C in and around the particle.

The carbon particle combustion rate according to this model can be obtained based on expressions describing diffusion and chemical kinetics:

- mass flow rate of O_2 towards the particle is:

$$\dot{M}_{O_2} = k(C_{O_{2\infty}} - C_{O_{2p}})\pi d^2 \tag{4.54}$$

- carbon consumption is proportional to mass flow rate of O_2:

$$\dot{M}_C = \Lambda \dot{M}_{O_2} \tag{4.55}$$

where

$$\Lambda = \frac{1+p}{1+\dfrac{p}{2}} \cdot \frac{12}{32} \tag{4.56}$$

is the stoichiometric ratio of oxidation reaction, and

- carbon consumption expressed through the rate of chemical reaction on the surface of the particle is:

$$\dot{M}_c = \pi d^2 r_c C_{O_{2p}} \tag{4.57}$$

if the reaction is of the first order.

By combining expressions (4.54) through (4.57), the combustion rate of carbon particle is obtained:

$$\dot{M}_C = \pi d^2 \left[\frac{1}{\Lambda k} + \frac{1}{R_c} \right]^{-1} C_{O_{2\infty}} \tag{4.58}$$

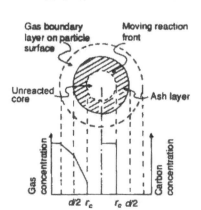

Figure 4.42.
Unreacted (shrinking) core model of char combustion

In the case of char particle combustion, if it is assumed that the particle ash skeleton remains unchanged, and that there exists resistance to O_2 diffusion through the ash layer on the particle surface (see Fig. 4.43b), the particle combustion rate is:

$$\dot{M}_c = \pi d^2 \left[\frac{1}{\Lambda k} + \frac{d_o^2}{d^2 R_c} + \frac{d_o(d_o - d)}{d\Lambda D_e'} \right]^{-1} C_{O_{2\infty}} \qquad (4.59)$$

Burn-out time for a carbon particle according to this model is obtained from:

$$\dot{M}_c = \frac{d}{dt}(\rho_p \frac{\pi}{6} d^3) = \pi d^2 R \qquad (4.60)$$

or after differentiation and integration:

$$\tau = \frac{\rho_p}{2} \int_0^{d_o} \frac{d(d)}{R_c C_{O_{2\infty}}} \qquad (4.61)$$

For two limiting cases very simple expressions can be obtained. In the kinetic regime, when combustion rate is controlled only by the rate of chemical reaction at the particle surface ($k \gg R_c$, and $D_e' \gg R_c$):

$$\tau = \frac{\rho_p}{2} \int_0^{d_o} \frac{d(d)}{R_c C_{O_{2\infty}}} = \frac{\rho_p d_o}{2 R_c C_{O_{2\infty}}} \qquad (4.62)$$

i. e., well-known linear dependence of the burn-out time on particle size.

In the case of the diffusion regime ($R_c > k$, and without ash layer, or $D_e' \gg k$):

$$\tau = \frac{\rho_p}{2} \int_0^{d_o} \frac{d(d)}{\Lambda k C_{O_{2\infty}}} \qquad (4.63)$$

If the molecular diffusion process takes place (i. e., influence of convection is negligible):

$$Sh = k \frac{d}{D_e} = 2 \qquad (4.64)$$

and it follows that:

$$\tau = \frac{\rho_p}{2} \int_0^{d_o} \frac{d \cdot d(d)}{\Lambda 2 D_e C_{O_2}} = \frac{\rho_p d_o^2}{8 D_e \Lambda C_{O_{2\infty}}} \qquad (4.65)$$

which is the well-known quadratic dependence of particle burn-out time on particle diameter.

The relative role of diffusion, or of the kinetics of chemical reaction may be judged from expressions (4.62) and (4.65) respectively.

In conditions appropriate for pulverized coal combustion, at temperatures 1200-1500 °C, according to numerous measurements of coal particle burn-out time [1, 7, 107, 109], the kinetic regime controls combustion for particles $\leq 1 \mu m$, and diffusion process for particles $>100 \mu m$.

In fluidized bed combustion, many coals lose their outer layer of ash as a consequence of mechanical interaction with inert material particles, and expression (4.58) holds true. Due to lower combustion temperatures (800-1000 °C) the influence of combustion kinetics is more pronounced, but because most of the particles are large, $d_o \geq 1$ mm, combustion usually takes place in the diffusion regime. In the section dedicated to the experimental research of combustion kinetics in fluidized bed, the character of char combustion for different coals in fluidized beds will be discussed in more detail.

For combustion of porous carbon and char particles, models which do not take into account combustion at the inner surface of the porous structure and cannot realistically describe the combustion process. Therefore, in the last several years different variants of model (3) are developed, whose extreme variant, when the diffusion process through the pores does not pose any resistance, is the model designated as (2). These models are called progressive conversion models. Figure 4.43 shows two basic and simple variants of this model – with and without porous ash layer. The basic concept of the simplest reacted core model (no ash layer), is represented by the equation describing O_2 concentration change inside the particle:

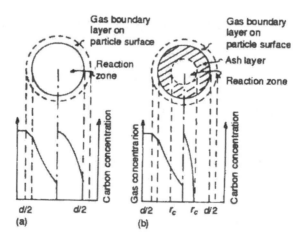

Figure 4.43. *Reacted core model of char combustion;*
(a) without ash layer, (b) with ash layer

$$\frac{1}{r^2}\frac{d}{dr}(r^2 D_e \frac{dC_{O_2}}{dr}) - r_v C_{O_2} = 0 \tag{4.66}$$

where R_v is the rate of O_2 consumption per unit volume:

$$R_v = r_v C_{O_2}^n = \frac{\rho_c}{\Lambda} R_s \tag{4.67}$$

and:

$$R_s = \frac{A_P}{M_C} R_t = \frac{6}{\rho_c d} R_t \tag{4.68}$$

while R_t is given by eq. (4.43).

Equation (4.66) can be transformed into dimensionless form by introducing values $z = r/(d/2)$, $c^* = C_{O_2}/C_{O_{2\infty}}$, assuming $r_v = const$ and $D_e = const$, to reach the following form:

$$\frac{1}{z^2}\frac{d}{dz}(z^2 \frac{dc^*}{dz}) - \frac{(d/2)^2 r_v}{D_e} c^* = 0 \tag{4.69}$$

with boundary conditions:

$$r = \rightarrow \frac{dC_{O_2}}{dr} = 0, \quad r = d/2 \rightarrow D_e \frac{dC_{O_2}}{dr} = k(C_{O_{2\infty}} - C_{O_{2p}}) \tag{4.70}$$

or in non-dimensional form:

$$z = 0 \rightarrow \frac{dc^*}{dr} = 0, \quad z = 1 \rightarrow D_e \frac{dc^*}{dz} = \frac{(d/2)k}{D_e}(1 - c^*) \tag{4.71}$$

The non-dimensional coefficient beside the second term on the left hand side of equation (4.69) is the square of the already mentioned Thiele's modulus:

$$\Phi = \sqrt{\frac{(d/2)^2 r_v}{D_e}} \tag{4.72}$$

which describes the ratio of rate of chemical reaction and diffusion through the porous structure of the particle.

Efficiency of carbon particle mass consumption, which depends on Φ, given by eq. (4.36), can be obtained after solving eq. (4.69), from the expression:

$$\eta = \frac{\int(r_v C_{O_2})dV_p}{r_v C_{O_{2p}} V_p} = \frac{R_v}{r_v C_{O_{2p}}} \tag{4.73}$$

The combustion rate can now be calculated using oxygen consumption:

$$\dot{M}_c = \pi d^2 \Lambda k(C_{O_{2\infty}} - C_{O_{2p}}) \tag{4.74}$$

where oxygen concentration at the outer particle surface is obtained from:

$$\pi d^2 k(C_{O_{2\infty}} - C_{O_{2s}}) = \frac{\pi d^3}{6} R_v = \frac{\pi d^3}{6} \eta r_v C_{O_{2s}} \qquad (4.75)$$

or:

$$c_p^* = \frac{C_{O_{2s}}}{C_{O_{2\infty}}} = \frac{1}{1 + \dfrac{\eta d r_v}{k C_{O_{2\infty}}}} \qquad (4.76)$$

In the literature, many modifications of the progressive conversion model exist. These tend to make the model more realistic, often by introducing carbon and char characteristics and information on the char particle porous structure and subsequent changes in its structure during combustion. Analyses show that for $\Phi \geq 100$, a progressive conversion model becomes the unreacting core model. Also, at high temperatures a progressive conversion model becomes similar to an unreacting core model. The reacted core model is applicable to fluidized bed combustion and combustion of pulverized coal, that is, in its simpler version – without porous ash layer. A porous ash layer must be taken into account for combustion in fixed beds (grate combustion, large particles) and in the later phases of particle burn-out.

Due to significant changes in the porous structure of char particles, more recent models [7, 72, 104] try to take these changes into account. Above all this is achieved by postulating porous structure formation and its development in the models and by taking into account the influence of temperature change on the combustion rate.

4.6.4. Mathematical modelling of single char particle combustion in a fluidized bed

A physical and mathematical description of the char (or pure carbon) particle combustion process, relies on the process and model analysis from the previous section. The major differences, which must be taken into account, are that the models just described treat single particle in infinite space filled with air (as is effectively the case for pulverized coal combustion), while in fluidized bed combustion, a fuel particle is surrounded by a large number of inert material particles – sand, ash and limestone. Hence, it is necessary to account for the fact that the access of oxygen to the char particle surface is obstructed and that there is a difference in the oxygen concentration in the emulsion and bubbling phases. The change in conditions for O_2 transfer to the particle surface can be included through the mass transfer coefficients (Section 3.3).

Mass transfer from the bubbles to the emulsion phase represents additional resistance to O_2 access to the fuel particle surface, which also lowers the combustion rate. This resistance must be taken into account along with other diffusion resistances analyzed in the previous section. Mass transfer between bubbles and the emulsion phase is taken into account in the model for the particle environment, i. e., by an appropriate fluidized bed model, which must be postulated together with the char (carbon) particle combustion model.

Mathematical models of char combustion in fluidized beds are developed for two good reasons: (a) as an auxiliary, but very important, tool for description and analysis of the processes during experimental research of coal (char or carbon) combustion kinetics in the fluidized bed, in order to determine characteristics of the coal and combustion process (combustion rate, activation energy, pre-exponential coefficient, combustion regime, etc.), or, (b) as part of any complex mathematical models describing all processes in fluidized bed combustion furnaces.

Table 4.12 gives a review of the basic characteristics of the majority of models postulated. Models 1, 2, 3, 4, 16, 17, 18, and 20 belong to (a) model group, while others are parts of more complex models.

In addition to the basic assumptions listed in the previous section, postulated mainly for pulverized coal combustion, in the case of fluidized bed combustion the following special conditions must be taken into account: (1) most often the combustion of large coal (char) particles is considered assuming that the diffusion processes can control the combustion rate, but (2) combustion takes place at temperatures lower than 1000°C, so that the kinetics of chemical reactions cannot be ignored, especially at the end of particle "life-time" when its dimensions are reduced.

The first, and subsequently most commonly used model of carbon particle combustion in a fluidized bed has been postulated by Avedesian and Davidson [110]. Key assumptions of this model are a diffusion combustion regime, combustion at the particle surface, shrinking core model and a two-layer model of a diffusion layer around the particle, with reaction (4.47) at the surface of the particle and reaction (4.48) at the boundary between the two layers. Processes in fluidized beds have been modeled by the two-phase Davidson model.

Based on the equation for molecular diffusion in gas surrounding the particle:

$$\frac{d}{dr}\left(r^2 \frac{dC_i}{dr}\right) = 0 \qquad (4.77)$$

with boundary conditions corresponding to the assumed chemical reactions and two-layer model of the diffusion layer:

$$
\begin{aligned}
&r = d/2 \quad [CO_2] = 0 \\
&r = R_r \quad\; [CO_2] = [O_2] = 0 \\
&r \to \infty \quad [CO_2] = 0, [O_2] = [O_2]_\infty
\end{aligned}
\qquad (4.78)
$$

and if at layer boundary ($r = R_r$), the equality of diffusion coefficients is assumed:

Table 4.12.   Review of char (or pure carbon) particle fluidized bed combustion models

No.	Author, year, reference	Type of particle	Combustion regime	Oxidation reaction type	Order of reaction	Diffusion through the ash layer	Change of porous structure	Fluidized bed model	Mass transfer model
1.	Avedisian, Davidson, 1973, [110]	Carbon	Diffusion, surface combustion, shrinking core model	Two-layer model (4.47), (4.48)	-	-	-	Two-phase model	$Sh = 2\ \varepsilon_{mf}$
2.	Basu et al, 1975, 114]	Carbon	Diffusion, surface combustion, shrinking core model	Single-layer model (4.46)	-	-	-	Two-phase model	$Sh = 2\ \varepsilon_{mf}$
3.	Campbel, Davidson, 1975, [119]	Carbon	Diffusion, surface combustion, shrinking core model	Two-layer model (4.47), (4.48)	-	-	-	Two-phase model	$Sh = 2\ \varepsilon_{mf}$
4.	Becker, Beer, Gibbs, 1975, [120]	Carbon	Diffusion, surface combustion, shrinking core model	Two-layer model (4.47), (4.48)	-	-	-	Two-phase model	-
5.	Gordon, Amundson, 1976, [121]	Carbon	Diffusion, surface combustion, shrinking core model	Two-layer model (4.47), (4.45)	-	-	-	Two-phase model	$Sh = 2$
6.	Horio, Wen, 1978, [122] Rajan, Krishnam, Wen, 1978, [123]	Char	Diffusion+kinetic, surface combustion, shrinking core model	Single-layer model (4.46)	-	-	-	Three-phase model	$Sh \approx 2$
7.	Chen, Saxena, 1978, [124]	Char	Diffusion+kinetic, surface combustion, shrinking core model	-	-	Taken into account	-	Three-phase model	-
8.	Saxena, Rehmat, 1980, [125]	Char	Diffusion+kinetic, surface combustion, shrinking core model	Single-layer model (4.44)	1	Taken into account	-	Two-phase model	$Sh = 2 + 0.6\ Re_{mf}^{0.5}\ Sc^{1/3}$
9.	Sarofim, Beer, 1978, [126]	Char	Diffusion+kinetic, surface combustion, shrinking core model	Single-layer model (4.46)	1	No	-	Two-phase model	-

No.	Reference		Model	Layer model					Sh
10.	Fan, Tojo, Chang, 1989, [127], Tojo, Chang, Fan, 1981, [128]	Char	Diffusion, surface combustion, shrinking core model	Single-layer model (4.44)	–	No	–	Two-phase model	$Sh = 2$
11.	Park, Levenspiel, Fitzgerald, 1978, [96]	Char	Diffusion+kinetic, surface combustion, shrinking core model	Single-layer model (4.44)	1	No	–	Exchange of O_2 between emulsion and bubbles is not taken into account	–
12.	Rajan, Wen, 1980, [9]	Char	Diffusion+kinetic, surface combustion, shrinking core model	Single-layer model (4.46) and (4.47)	–	No	–	Two-phase model	$Sh = 2$
13.	Walsh, Dutta, Beer, 1984, [129]	Char	Diffusion, surface combustion, shrinking core model	Single-layer model (4.44)	1, 0.5	No	–	Combustion above the bed	–
14.	Borghi, Sarofim, Beer, 1985, 1985, [56]	Char	Kinetic IIa, surface combustion, shrinking core model	Single-layer model (4.44)	1	No	Yes	O_2 between emulsion and bubbles is taken into account	Frossling correlation for Sh, modified for ε_{mf}
15.	Donsi, El-Sauvi, Farmisani, 1986, [130]	Char	Diffusion+kinetic, surface combustion, shrinking core model	Single-layer model (4.44)	1	No	–	Two-phase model	$Sh = 2\varepsilon + 0.6\,Re_{mf}^{0.5}\,Sc^{1/3}$
16.	Ross, Davidson, 1984, [116]	Carbon	Diffusion+kinetic, surface combustion, shrinking core model	Continuous model (4.48) and (4.44) or (4.47)	1	–	–	Two-phase model	$Sh = 2\varepsilon$
17.	Chakraborty, Howard, 1984, [134]	Char	Diffusion+kinetic, surface combustion, shrinking core model	Single-layer model (4.46)	1	No	–	Two-phase model	$Sh = 2\varepsilon + 0.69\,Re_{mf}^{0.5}\,Sc^{1/3}$
18.	Turnbull et al, 1984, [131]	Char	Diffusion+kinetic, surface combustion, shrinking core model	Continuous model (4.44) and (4.48) (4.44) or (4.47)	0.5	–	Yes	Two-phase model	$Sh = 3.5$

Table 4.12. *Continued*

No.	Author, year, reference	Type of particle	Combustion regime	Oxidation reaction type	Order of reaction	Diffusion through the ash layer	Change of porous sturcture	Fluidized bed model	Mass transfer model
19.	Zhang, 1987, [68]	Char	Diffusion, surface combustion, shrinking core model	Single-layer (4.44)	–	No	–	Emulsion phase as the filtration of the fixed layer	–
20.	Boriani et al. 1989, [132]	Char	Diffusion+kinetic, surface combustion, shrinking core model	Continuous concentration change (4.44)–(4.48)	1	No	–	Two-phase model	
21.	Prins, 1987, [72]	Carbon	Kinetic II, progressive conversion model, with combustion in porous layer and un-reacted core	Continuous change (4.46)	1	Diffusion through porous carbon layer, no ash layer	Yes	Two-phase model	Equation (3.16)
22.	Brem, 1988, 1989, 1990, [101,104,133]	Char	Kinetic IIa, combustion in porous structure and at the surface, variable particle density model with size decrease	Continuous (4.46)	1	Diffusion through porous char structure, no ash layer	Yes	Two-phase model	Equation (3.16)
23.	Groups of authors IEA-AFBC Model, 1990, [102]	Char	Kinetic IIa, combustion in porous structure and at the surface, variable particle density model with size decrease	Continuous (4.46)	1	Diffusion through porous char structure, no ash layer	Yes	Two-phase model	Equation (3.16)

$$-\frac{1}{2}\left(\frac{d[CO]}{dr}\right)_1 = \left(\frac{d[CO_2]}{dr}\right)_1 = -\left(\frac{d[CO_2]}{dr}\right)_2 = \left(\frac{d[O_2]}{dr}\right)_2 \qquad (4.79)$$

where 1 and 2 correspond to left and right hand side of the zone boundary, $r = R_r$, the following solutions are obtained:

− in zone 1:

$$[CO] = 2[O_2]_\infty\left(\frac{d}{r}-1\right)$$

$$[CO_2] = [O_2]_\infty\left(2-\frac{d}{r}\right)$$

− in zone 2:   (4.80)

$$[CO_2] = [O_2]_\infty \frac{d}{r}$$

$$[O_2] = [O_2]_\infty\left(1-\frac{d}{r}\right)$$

It can be shown that $R_r = d$, and that molar mass flow rate of oxygen is:

$$\dot{M}_{O_2} = 4\pi R_r^2 D_e\left(\frac{d[O_2]}{dr}\right)_{r=R_r} = 4\pi d D_e[O_2]_\infty \qquad (4.81)$$

From the solution of eq. (4.77), with boundary conditions:

$$r = d/2 \quad [O_2] = 0$$
$$r \to \infty \quad [O_2] = [O_2]_\infty \qquad (4.82)$$

in the case of O_2 diffusion towards the particle in absence of chemical reactions, the following is obtained:

$$\dot{M}_{O_2} = 2\pi d D_e[O_2]_\infty \qquad (4.83)$$

Using the definition of mass transfer coefficient, O_2 mass flow rate in this case may also be written as:

$$\dot{M}_{O_2} = k\pi d^2[O_2]_\infty \qquad (4.84)$$

yielding:

$$k = \frac{2D_e}{d} \qquad (4.85)$$

Avedesian and Davidson have assumed that the influence of convection is negligible, so that for the mass transfer by molecular diffusion towards the spherical particle: Sh = $kd/D_G \equiv 2$, or for the fluidized bed: Sh = $2\varepsilon_{mf}$. Using this assumption, as well as relationship (4.85), expression (4.81) may be written as:

$$\dot{M}_{O_2} = 2\pi \mathrm{Sh} D_G \, d[O_2]_\infty \qquad (4.86)$$

Different authors have used other correlations for Sherwood number, to take into account mass transfer by convection (see Tables 3.1 and 4.12).

For a more exact description of char combustion in fluidized beds it is necessary to take into account the change of O_2 concentration along the bed height, both in the emulsion and bubbling phases, and mass transfer between the phases. Most authors (Table 4.12) use a two-phase Davidson model of the fluidized bed.

Using his own fluidized bed model, Davidson calculated the change of oxygen concentration along the fluidized bed in which char combustion of total mass m_c, takes place [110, 135]:

in the emulsion:

$$\frac{[O_2]_{H_b}}{[O_2]_o} = (1 - \frac{v_{mf}}{v_f}) \exp(-X_B) +$$

$$+ \frac{[1 - (1 - \frac{v_{mf}}{v_f}) \exp(-X_B)]^2}{k' + 1 - (1 - \frac{v_{mf}}{v_f}) \exp(-X_B)} \qquad (4.87)$$

in the bubbles:

$$[O_2]_B = [O_2]_\infty + ([O_2]_o - [O_2]_\infty) \exp(-X_B \frac{y}{H_b}) \qquad (4.88)$$

where X_B is defined by expression (2.126), and k' is obtained based on the total O_2 consumption for char combustion in the bed:

$$N_p = \frac{6m_c}{\rho_{po}\pi d_o^3} \qquad (4.89)$$

or from:

$$N_p \dot{M}_{O_2} = k[O_2]_\infty A_b H_{mf} \qquad (4.90)$$

which, together with (4.86) yields:

$$k = 12 m_c \, \mathrm{Sh} D_G \left(\frac{d}{\rho_{po} d_o^3} \right) \frac{1}{A_b H_{mf}} \qquad (4.91)$$

For simpler notation k' is introduced:

$$k' = \frac{k H_{mf}}{v_f} = \frac{12 m_c \, \mathrm{Sh} D_G \, d}{\rho_{po} d_o^3 A_b v_f} \qquad (4.92)$$

Based on the O_2 balance which now can be formed for the whole bed:

- the total quantity of oxygen leaving the bed is equal to the sum of O_2 leaving the emulsion phase and O_2 leaving the bubbles:

$$(v_{mf} A_b [O_2]_\infty) + (v_f - v_{mf}) A_b \{[O_2]_\infty + ([O_2]_o - [O_2]_\infty) \exp(-X_B)\}$$

- the total quantity of O_2 entering the bed is:

$$v_f A_b [O_2]_o$$

By subtracting these two values the oxygen consumption is obtained:

$$([O_2]_o - [O_2]_\infty) A_b [v_f - (v_f - v_{mf}) \exp(-X_B)] \qquad (4.93)$$

which, expressed in moles of O_2, must be equal to the carbon consumption:

$$-\frac{N_p \pi \rho_{po} d^2}{24} \frac{d}{d\tau} \qquad (4.94)$$

From the similar equation for single particles:

$$\dot{M}_{O_2} = -\frac{1}{12} \frac{\pi \rho_{po} d^2}{2} \frac{d}{dt}(d) = 2\pi Sh D_G d[O_2]_\infty \qquad (4.95)$$

we can substitute $[O_2]_\infty$, in (4.93) and then make it equal to (4.94). After integration over time, burn-out time of all char particles in the bed, of total mass m_c, is obtained:

$$\tau_c = \frac{m_c}{12[O_2]_o A_b \{(v_f - v_{mf})[1 - \exp(-X_B)] + v_{mf}\}} + \frac{d_o^2 \rho_{po}}{96 Sh D_G [O_2]_o} \qquad (4.96)$$

The first term of this expression gives the influence of the diffusion resistance due to the gas transfer between bubbles and the emulsion phase, and the second term gives the influence of the diffusion layer around the char particle, on the burn-out time (see expression 4.65).

Almost all models mentioned so far (see Table 4.12) adopt the shrinking core model, and do not assume the existence of an ash layer at the particle surface (except in models No. 7 and No. 8, of Chen, Rehmat and Saxena). It is assumed that collisions with inert material particles prevent the formation of an ash layer at the char particle surface.

Although the question of the order of chemical reaction has not been resolved yet, for easier calculations, practically all models assume first order reactions.

A number of experiments have been carried out to determine which reactions take place at the carbon (char) particle surface [68, 72, 110, 114, 116, 119, 126, 131].

Although there are two extreme cases – that CO is formed on the particle surface by reaction (4.47) (Davidson [110]) and that CO_2 is formed by reaction (4.44) (Fan [127], Zhang [68]) the general opinion appears to be that both CO and CO_2 are formed at the particle surface, and that O_2 reaches the surface, eq. (4.46).

Basu [114] has shown experimentally, as early as 1975, that CO is the primary product of fluidized bed combustion. Further, an energy balance of a burning particle also shows that endothermic reactions of type (4.47) are not probable [68, 114]. The most recent analyses [7, 72] have shown that both CO and CO_2 form at the particle surface. The CO/CO_2 ratio changes both with combustion temperature (4.49) and with particle size. When burning particles smaller than 0.5 mm, particle temperature favors CO formation, and for particles larger than 1 mm, CO_2 formation is favored [3, 38, 116, 131]. Discussion about which chemical reactions take place at the particle surface (single- or two-layer diffusion layer), has finally been resolved with the advent of modern computers, and now all reactions (4.44) through (4.48) can be taken into account, and their role determined by the kinetics of these chemical reactions.

Changes in gas concentration around the particle are assumed to be continuous in most recent models [72, 101, 102, 104, 131-133].

The first char particle combustion models considered combustion of larger particles, with porous structure, at high temperatures, when the diffusion combustion regime prevails. From 1978 and the advent of the model of Horio and Wen [122], in most cases kinetics of chemical reaction are also taken into account, in order to model combustion of smaller particles (<1 mm), and less reactive coals.

As early as 1985, Borghi et al. [56] proposed a model that takes into account combustion inside the pores, and all modern models [72, 101, 102], bearing in mind the change in combustion conditions and relative importance of particular processes during char particle "life-time," provide for the possibility of combustion inside the particle's porous structure, in kinetic regime IIa, which encompasses all diffusion resistances, and the kinetics of chemical reaction at the pore surface. Such models require numerical solution of the equations, especially as they must take into account the change in particle porous structure.

Expression (4.96) for calculating particle burn-out time, has been obtained assuming the diffusion regime and reactions of type (4.47). Analytical solutions may also be obtained for some more complex models – kinetic regime and reactions of type (4.46) [3, 122, 124, 125]. If the expression:

$$A_b \{(v_f - v_{mf})[1 - \exp(-X_B)] + v_{mf}\}$$

is represented as Δ, one can obtain

(1) for the shrinking core model, including kinetics and diffusion and $C + O_2 = CO_2$ reaction at the particle surface:

$$\tau_c = \frac{m_c}{12\Delta[O_2]_o} + \frac{\rho_{po}d_o^2}{128\,S\,Sh D_G[O_2]_o} + \frac{\rho_{po}d_o}{64\,R_c[O_2]_o} \qquad (4.97)$$

(2) for the constant particle size model, including kinetics and diffusion:

$$\tau_c = \frac{m_c}{12\,\Delta[O_2]_o} + \frac{\rho_{po}d_o^2}{192\,S\,Sh D_G[O_2]_o} + \frac{\rho_{po}d_o}{192\,R_c[O_2]_o} \qquad (4.98)$$

In the case where an ash layer exists around the unreacted (shrinking) core of char, according to the model of Saxena and Rehmat [125], expression (4.97) needs to have the following term added:

$$\frac{\rho_{po}\,d_o^2}{96\,D_e[O_2]_o}\left(1 - \frac{1 - z^{2.3}}{1 - z}\right) \qquad (4.99)$$

where z is the ratio of ash layer volume to the burnt coal volume.

In the case of combustion at the pore surface, for kinetic regime II, in [131] the following expression for burn-out time has been obtained:

$$\tau_c = \frac{m_c}{12\,D[O_2]_o} + \frac{\rho_{po}d_o^2}{128\,S\,Sh\,Dg[O_2]_o} + \\ + \frac{\rho_{po}d_o}{64[O_2]_o[R_h(A_{pore}/V_p)D_e]^{0.5}} \qquad (4.100)$$

where:

$$R_h = 470\,T_p \exp\left(-\frac{179{,}400}{R_g T_p}\right) \qquad (4.101)$$

is the modified real chemical reaction rate at the pore inner surface, obtained from the well-known Smith expression [109] for char particle combustion rate:

$$R_c = 21 T_p \exp\left(-\frac{179{,}400}{R_g T_p}\right) \qquad (4.102)$$

where:

$E = 179{,}400$ [J/mol] and $A = 21\,T_p$ [m/sK].
In the same expression (4.102), Field proposes constants:

$$E = 149,200 \text{ [J/mol]} \quad \text{and} \quad A = 595 \ T_p \text{ [m/sK]}.$$

In mathematical modelling, one of the key problems is the choice of the reaction order and the expression for the rate of chemical reaction. From Table 4.12 it can be seen that most often first order reaction is assumed, for simplicity, although there is ample evidence against such a choice.

Most of the authors used Field's expression (4.102), models No. 6, 11-13, 16 and 17 (Table 4.12), although it has been obtained for pulverized coal combustion, i. e., particles < 500 μm, which are not characteristic for fluidized bed combustion. Several authors (No. 7-10) give no data from which activation energy and pre-exponential coefficients could be determined for expression (4.43). Other authors have used Smith's data [109].

The pre-exponential coefficients are adopted covering a wide range, depending on the coal type. The following section will deal with these differences in more detail.

4.6.5. Char combustion kinetics in fluidized beds – experimental results

Table 4.13 gives an exhaustive chronological review of experimental research on char combustion kinetics for different coals in the fluidized bed. At first experiments were aimed at determining burn-out time [110]. In later experiments, the goal was to acquire data in order to postulate mathematical models incorporating the nature of chemical reactions, and the experimentally obtained combustion rate and kinetic parameters. In some experiments, the mass transfer coefficient has also been determined, as well as the ignition temperature and particle temperature during combustion. Such changes in experimental goals have been made possible by the development of measurement methods and techniques, and this will be described later.

A large number of experimental studies were carried out with pure carbon, graphite and coke, since these are less complex materials (fuels), whose structure is better known and so that the processing and analysis of experimental data were much easier. However, based on the results of these experiments, it is not possible to accurately predict combustion behavior of different coals, or their chars. This prompted experimental studies of the behavior of many different coals (from lignites to anthracites). These comparative studies have shown that the combustion rates of various coals are very different. The influence of structure and composition of coal on the char structure, combustion rate and kinetic parameters, have not been clarified yet, and remain the subject of further research.

Experimental conditions in most cases correspond to those in industrial-scale fluidized bed combustion.

The size of particles whose combustion has been studied ranges from 0.15 to 50 mm. Hence, two groups of experiments can be recognized: (a) experiments in

Table 4.13. Review of experimental research of char (or carbon) combustion

No.	Authors, year, references	Fuel, particle diameter [mm]	Fluidized bed Material, particle diameter [mm]	Fluidized bed Height, diameter [mm]	Experimental conditions Temperature [°C]	Fluidization velocity [m/s]	Experimental method Bed heating	Organization of experiment	Measurement method	Final results of experiment
1.	Avedesian, Davidson, 1973, [110]	Coke 0.23-2.61	Ash 0.39; 0.65	300 76	900	0.171-0.353	Electrical	Combustion of particle portion (batch)	– visual, – bed temperature, – O_2, CO_2 concentration	– burn-out time
2.	Campbel, Davidson, 1975, [119]	Coal, char, coke 1.5-3.3	Ash 0.32-0.65	40-110 76	700-900	0.1-0.6	Electrical	Batch	– O_2, CO_2 concentration	– burn-out time, – burning rate
3.	Basu et al., 1975, [114]	Carbon	Sand 0.140	– 129	550-900	0.2-0.3	Electrical	Single particles	– particle mass measurement, – O_2, CO_2, CO concentration	– reaction type
4.	Basu, 1977, [136]	Carbon electrode 9-15	Sand 0.100	– 150	600-900	0.08	Electrical	Single particles	– particle mass measurement, – particle temperature	– particle temperature, – burning rate
5.	Chakraborty, Howard, 1978, [137]	Carbon electrode 3-12	Sand 0.327, 0.55, 0.78	12-50 71.5	800-900	0.25-0.71	Electrical	Batch 6 particles	– mass measurement, – particle temperature	– particle temperature, – burning rate
6.	Tamarin et. al., 1980, [138]	Coke 3-12	Sand 0.45-2.15	40 70	650-800	0.41-1.65	Electrical	Batch	– visual	– burn-out time, – mass transfer coefficient
7.	Kono, 1980, [139]	Anthracite and brown coal char 0.050-0.850	Glass beads 0.1	50-100 g	700-1000	$(15-20)v_{mf}$	Electrical	Batch	– CO_2 concentration measurement after abrupt change from inert to oxidative atmosphere	– burning rate, – kinetic constants
8.	Pillai, 1981, [48]	Lignite, steincoal, anthracite, char 0.25-8.0	Limestone	50 100	770-1010	1.2-1.5	Propane combustion	Batch	– visual	– burn-out time

Table 4.13. *Continued*

No.	Authors, year, references	Fuel, particle diameter [mm]	Fluidized bed — Material, Particle diameter [mm]	Fluidized bed — Height, diameter [mm]	Experimental conditions — Temperature [°C]	Experimental conditions — Fluidization velocity [m/s]	Bed heating	Experimental method — Organization of experiment	Experimental method — Measurement method	Final results of experiment
9.	Ross, Davidson, 1981, [116]	Char, coke, graphite 0.164-0.46 0.88-2.43	Sand 0.23; 0.55	150 101	830-900	0.45; 0.5	Electrical	Batch	– O_2, CO_2 concentration – particle temperature (optical)	– burning rate, – burn-out time, – nature of reaction, – kinetic parameters
10.	Chakraborty, 1981, [134]	Char, bituminous coal 1.84-4.375	Sand 0.3-0.85	30 71.5	800-900	0.25-0.71	Electrical	Batch	– O_2, CO_2, CO concentration – bed temperature	– burn-out time
11.	Baskakov, 1983, [140]	Coal 5-5.5	Corundum 0.12; 0.46	– 46	350-900	0.1-0.3	In oven	Batch	– mass measurement, – O_2, CO_2 concentration	– burning rate
12.	Oka, 1983, [74, 75]	Dried lignite 5-25	Sand 0.63-1.0	220 147	550-700	0.95-1.15	Electrical	Batch	– bed temperature	– burn-out time, – ignition temperature
13.	Read, Michener, 1984, [78]	Anthracite, coking coal, bituminous, lignite 0.25-1.0	Sand 0.5-1.0	150 160	300-750	0.15	–	Continuous fuel feeding	– gas concentration, – heat balance	– burning rate, – kinetic parameters, – ignition temperature
14.	Oka, 1984, [4]	Lignite anthracite 5-25	Sand 0.63-1.0	220 147	550-800	0.95-1.15	Electrical	Batch	– bed temperature	– burn-out time, – ignition temperature
15.	Basu, 1984, [141]	Carbon electrode 4-8	Sand 0.278	Turbulent bed 100	776-860	2.2-5.3	Propane combustion	Batch	– O_2, CO_2 concentration – bed temperature	– burning rate
16.	Turnbull et al, 1984, [131]	Coke, carbon char 0.15-1.0	Sand 0.15-0.7	300 100	800-900	0.15-0.3	Overheated air	Batch	– visual – CO_2 concentration	– burn-out time, – kinetic constants

No.	Reference	Fuel	Bed material		Temperature	Velocity	Heating	Operation	Measurement	Investigated parameters
17.	Oka, Arsić, 1985, [142]	Lignite, brown coal, steincoal, anthracite	Sand 0.6-1.0	220 147	550-880	0.6-1.0	Electrical	Batch	– bed temperature	– burning rate, – ignition temperature
18.	Pillai, 1985, [51]	Rufford and Pittsburgh coals 6.4-17.2	Chamotte 0.62	75 100	900	1.2	Propane combustion	Batch	– visual	– burn-out time
19.	La Nause, Jang, 1985, [143]	Coke 10	Sand	– 102	700, 900	0.45; 0.53	–	Batch	– mass measurement	– burning rate
20.	Basu, Ritche, 1985, [144]	Coke 4-14	Sand 0.130	Fast bed 120	850	4-8	Heated air	Batch	– mass measurement	– burning rate
21.	Andrei, Sarofim, Beer, 1985, [73]	Montana coal, lignite 1.65-2 2.83-3.28	Sand 0.2	120 100	800-900	2.1; 1.9 $3.2v_{mf}$	Electrical	Batch	– mass measurement	– devolatilization, – burn-out time, – burning rate
22.	Basu, Surabao, 1986, [145]	Coke, carbon electrode, coal 8-16	Sand 0.1-0.3	1000 102	776-860	0.58-2.15 $(8-30)v_{mf}$	Propane combustion	Batch	– temperature, – O_2 concentration, – mass	– burning rate, – mass transfer
23.	Durao, et al 1986, [146], [147]	Coal	Portugal anthracite U.S.A. 0.125-4	Sand 0.64, 80 100	700-900	$(3.76-6.2)v_{mf}$	Propane combustion	Batch	– temperature, – O_2 concentration, – mass	– burning rate, – devolatilization time
24.	Prins, 1987, [72]	Graphite 3-13	Alumina 0.67	115 100	712, 837	$(0.46-2)v_{mf}$	Electrical	Single particle	– particle temperature, – O_2, CO, CO_2 concentration, – bed temperature	– burning rate, – CO/CO_2 ratio, – activation energy
25.	Zhang, 1987, [68]	Char, Evans coal, petroleum coke, graphite electrode 4-50	Sand 0.328 0.655	190 100	800-900	$(1.5-5)v_{mf}$	Electrical	Batch	– visual – mass measurement, – CO, CO_2, O_2 concentration	– devolatilization time, – mass transfer, – particle temperature

Table 4.13. *Continued*

No.	Authors, year, references	Fuel, particle diameter [mm]	Fluidized bed		Experimental conditions			Experimental method		Final results of experiment
			Material, particle diameter [mm]	Height, diameter [mm]	Temperature [°C]	Fluidization velocity [m/s]	Bed heating	Organization of experiment	Measurement method	
26	Oka, Radovanović, 1987, [148]	Dried and raw lignite 5-25	Sand 0.63-1.0	220 147	300-700	0.45-0.95 $(3.5$-$4)v_{mf}$	Electrical	Batch	– bed temperature	– burn-out time – burning rate
27.	Patel, Nskala, Dixit, 1988, [149]	Coal: Texas lignite, Pittsburgh No. 8, Kentucky No.9 anthracite 4.75-6.35	Sand, ash, limestone 0.85-1.18	100-150 100	up to 840	0.91-2.74	Electrical	Batch	– bed temperature, – visual – CO, CO_2, O_2, NO_x, SO_2 concentration	– burning rate
28.	Patel, Nskala et al, 1989, [150]	Lignite, bituminous, anthracite 2.0-2.36	Sand 0.85-1.18	100-150 100	676-926	1.52	Electrical	Batch	– CO, CO_2, O_2, NO_x, SO_2 concentration	– activation energy – preexponential coefficient
29.	van Engelen, van der Honing, 1990, [151]	Char 0.4-2.0	Sand 0.63-0.8	100 100	600-950	0.76	Electrical	Batch	– gas concentration	– kinetic parameters – activation energy
30.	Brem, 1990, [194]	Char of 5 different coals 0.5-2.0	Sand 0.65-0.8	100 100	575-900	0.76	Electrical	Batch	– gas concentration	– kinetic parameters – activation energy
31.	Oka, et al, 1990, [152] Ilić, Oka et al, 1990 [153] Milosavljević, Ilić, Oka, 1990, [154] Ilić, Milosavljević, Oka, 1990, [155]	Coal, lignite, brown, anthracite 5-15	Sand 04.-0.6	220 147	500-700	$3v_{mf}$	Electrical	Batch	– CO, CO_2, O_2 concentration – bed temperature	– burn-out time – burning rate – devolatization time – ignition temperature

which combustion kinetics of high quality coals have been studied, where particles were usually pneumatically fed to the furnace below the bed surface, and (b) experiments in which combustion kinetics of geologically younger coals, usually fed to the bed surface, have been studied. In the first group particle size were typically 0.15 to 2 mm, and in the second group from 5-50 mm.

Inert material in most cases has been silica sand, with a particle size of 0.1 to 1.0 mm, i. e., the size used in industrial furnaces. However, the influence of size and type of inert material have not been studied systematically.

Fluidized bed dimensions in such studies, i. e., diameter and height, do not correspond to industrial sizes. Bed diameter ranged between 75 and 150 mm, and height from 50 to 200 mm. Height-to-diameter ratio in these experiments has been 1-2.

In experiments aiming to study combustion kinetics in conditions corresponding to furnaces in the stationary regime, bed temperature was typically 700-1000 °C, and most often 800-900 °C. In a number of experiments involving combustion kinetics under conditions corresponding to the furnace, the start-up has been determined and also the ignition temperature of the coal and char has been measured; hence experiments have been carried out at lower temperatures of 350-700 °C.

The majority of experiments have been carried out in bubbling fluidized beds (v/v_{mf} = 1-5), with fluidization velocities of 0.2-1.5 m/s.

The basic experimental technique used today was first introduced by Avedesian and Davidson in their pioneering work [110]. Typically, a heated fluidized bed was charged with a measured fuel portion (batch), and changes were followed until complete burn-out of all fuel particles occurred, i. e., so called "batch" experiments. In only two experiments [78 and 139] does it appear that a different technique was used.

In "batch"-type experiments, it is possible to measure the following parameters: existence, or disappearance of particles visually, change in mass of fuel particles, bed temperature, and composition of combustion products.

Visual observation was one of the first and simplest measurement methods in these experiments. The appearance and disappearance of a red-hot (shining), burning fuel particle is recorded, and thus the time period between ignition and burn-out is determined. This method has a number of serious shortcomings. It is only possible to observe visible red-hot (shining) particles, having temperatures above 600 °C. Due to intense mixing, particles spend most of the time in the depths of the fluidized bed, and can seldom be seen at the surface where it is possible to observe them. Measurement errors, especially in experiments with small particles, can be significant. However, the greatest shortcoming is that only total burn-out time and average combustion rate can be measured based on the starting mass of fuel particles.

Measuring change in particle mass during combustion gives a much more accurate estimate of combustion rate. However, experiments are extremely difficult to perform and potentially subject to considerable errors. Fuel particles are usually placed on a thin wire net so that they can be periodically removed from the bed and measured. However, it is very difficult to stop combustion during this phase of the experiment, and this introduces experimental error. By measuring mass change it is possible to estimate combustion rate during fuel particle burn-out.

Combustion rate and its change during burn-out can be estimated indirectly, based on the heat balance during burn-out of the particle batch, by measuring bed temperature change. In [74-76] there is a detailed analysis of potential errors and shortcomings of this method. Major causes of errors are: insufficiently accurate estimates of heat losses, non-uniform bed temperature distribution; further, it is necessary to know physical properties of the bed and fuel material (inert material mass, inert material specific heat, inert material density, etc.). This method does not allow determination of combustion time, and the combustion rate can accurately be determined only at the beginning of the combustion process. Figure 4.44 gives the characteristics of bed temperature change for "batch" combustion in a fluidized bed.

The periods for characteristic processes during particle "life-time" in the bed up to total burn-out can be clearly seen. Immediately after feeding particles into the bed ($\tau = 0$) the bed cools during the period of fuel heating, moisture evaporation and devolatilization. Next a temperature rise (point A) occurs as a result of the commencement of volatiles combustion (where coal is the fuel) or solid

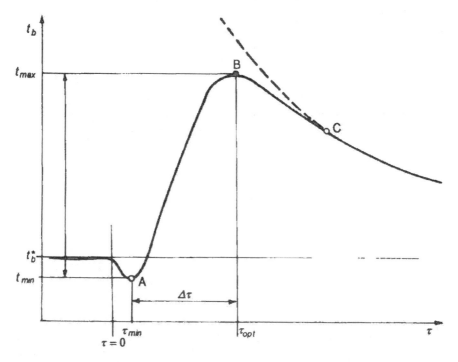

Figure 4.44. *Characteristic bed temperature change during "batch" combustion experiments in a fluidized bed*

carbon combustion (if graphite or char is the fuel) and heating of fuel particles to the bed temperature, which is maintained constant in these experiments (see Table 4.13). Based on these measurements it is not possible to determine when volatiles ignite and their combustion begins – before or after point A. Furthermore, any abrupt rise of bed temperature is the consequence of heat generation due to volatiles and char combustion. Depending on the bed temperature or fuel type, combustion of volatiles more or less influences the shape of this part of the bed temperature change curve. At the moment when the maximum bed temperature is achieved, combustion of the fuel is not finished. At that time, generated heat is equal to the heat carried away by fluidizing gas. Based on such measurements, it is not possible to determine the end of combustion which occurs after point B (point C).

If we suppose that temperature measurements are fast enough to follow real temperature changes in these experiments, the major shortcomings of this kind of experimental research are:

–   inability to accurately determine characteristic times (ignition and end of combustion),
–   unknown influence of volatiles combustion on bed temperature change, and
–   possible, unknown influence of heat losses.

An analysis of the effect of these shortcomings is given in [4, 74, 75, 142].

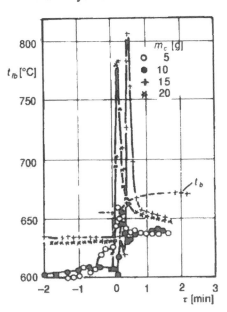

The influence of volatiles, depending on the type of coal, their quantity and fuel particle size, can be qualitatively estimated based on measurements of temperature change of gases above the bed surface. Figure 4.45 [75, 76] shows that for the period of volatiles combustion for Kolubara coal (lignite), (particle size $d = 5$ mm), temperature above the bed surface rises abruptly. However, the curve of bed temperature change does not show this rise, although the temperature of gas itself may differ from bed temperature by up to 150 °C. Due to the slower release of volatiles when larger particles ($d = 25$ mm) are burned, these changes are less sharp, and differences are typically smaller.

Figure 4.45.
Freeboard temperature during lignite "batch" combustion experiments in a fluidized bed ($d = 5$ mm, $t = 650$ °C) according to the measurements of Oka and Arsić [75,76]

The influence of volatiles combustion on bed temperature change in these "batch" experiments will be smaller if the devolatilization rate and combustion rate of volatiles are higher. Given that the heat inertia of the bed is high, its bulk temperature will not reflect such fast changes. Bed temperature change is above all the result of char combustion. A simple equation of the bed heat balance, with several simplifying assumptions [74, 75, 154], allows calculation of the char combustion rate:

$$M_b c_b \frac{d}{d\tau}(t_b - t_b^*)+$$

$$+\dot{m}_v \left. c_g \right|_0^{t_B} (t_b - t_b^*) = -H_C \frac{dm_c}{d\tau} \tag{4.103}$$

This experimental method yields more accurate data when studying kinetics of more reactive fuels and smaller particles (≤ 5 mm). It is in principle possible, by using combustion models described in Section 4.6.4, and setting bed heat balance with the corresponding fluidized bed hydrodynamics model (Chapter 2, Section 2.3.7), based on such experiment and on the measured bed temperature change over time, by optimizing model constants, to determine kinetic parameters – pre-exponential coefficient and activation energy of fuel. However, this method requires many assumptions. Hence, in recent experiments, a more accurate methodology has been used, based on measuring change of combustion products concentration. By measuring change of O_2, CO, CO_2 and other gas concentrations it is possible to directly and accurately determine the beginning of combustion, the end of combustion, the combustion rate and character of the combustion process, without additional assumptions in the previously discussed method. Using models described in Section 4.6.4 and a fluidized bed hydrodynamics model one can obtain kinetic constants. Figure 4.46 [152, 153] shows typical concentration changes for CO_2 in experiments aimed at determining combustion kinetics for three coals – Kolubara lignite, Aleksinac brown coal and Vrška Čuka anthracite (particle size 5 mm and bed temperature 700 °C). The beginning and end of combustion can be clearly seen. This type of observation was not possible using visual observations or by direct bed temperature measurement.

The release and combustion of volatiles can also be clearly seen on the concentration change curves. First, a clearly pronounced maximum originates from the release and combustion of volatiles. The end of volatiles combustion, especially for smaller particles (<5 mm), can also be determined accurately enough. When investigating the combustion kinetics of larger particles, the release and combustion of volatiles is slower and takes place simultaneously with char combustion.

The combustion rate can be easily determined directly form the change of O_2 or CO_2 and CO concentration in time [154], while:

$$\frac{dm_c}{dt} = ([CO_2] + [CO]) \frac{M_v}{M_c} \dot{m}_v \tag{4.104}$$

or

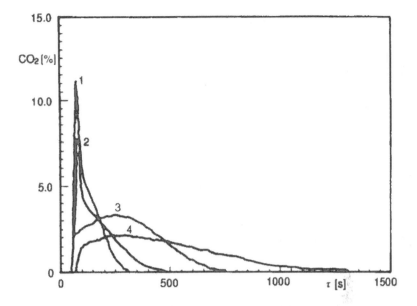

Figure 4.46. *CO_2 concentration during "batch" combustion experiments with different coals, at 700 °C, according to the measurements of Ilić and Oka [152, 153]*
Coals: 1 – Kolubara (lignite), d = 5 mm, 2 – Aleksinac (brown coal), d = 5 mm, 3 – Vrška Čuka (anthracite), d = 5 mm, 4 – Vrška Čuka, d = 15 mm

$$\frac{dm_c}{dt} = \frac{2p+1}{p+1}[O_2]\frac{M_v}{M_c}\dot{m}_v \qquad (4.105)$$

where [CO_2], [CO] and [O_2] are instantaneous measured molar concentrations of gases during the experiment.

Errors that can occur in determining combustion rate with this method are mainly caused by the following two reasons:

- if combustion of volatiles takes place simultaneously with char combustion, it is not possible to distinguish between char and volatiles combustion rates, and
- inertia of instruments for measuring gas concentration can influence accuracy of measurements when fast concentration changes are measured, and this is the case when burning more reactive fuels with smaller particles (≤5 mm), for the period of volatiles combustion. Taking into account the response time of instruments in order to correct measurements can dramatically improve measurement accuracy [155].

(a) Char burn-out time

The initial studies of coal combustion kinetics (carbon or char) in a fluidized bed carried out by Davidson et al. [110, 119] (Table 4.13 Nos. 1 and 2) dealt with determining burn-out time. Having data on the burn-out time and relying on theoretical observations, which yielded formulae (4.96) through (4.100), it has been possible to determine processes which control combustion, and also the influence of important parameters – fuel particle diameter, bed temperature, fluidization velocity, diffusion through the ash layer and inert material particle size. In later investigations as well, burn-out time measurements have been carried out mainly visually (Table 4.13, Nos. 6, 8, 16, 18 and 23) or less frequently by measuring concentration of combustion products (Nos. 9, 10, 21 and 31), or by measuring bed temperature (Nos. 12, 14, 26). In may studies, the influence of fuel particle size, temperature and coal characteristics have been investigated. Particles of 0.2 through 50 mm in size have been investigated at temperatures from 300 to 900 °C, and coal ranging from lignite to anthracite. According to the measurements of Avedesian, Campbel, and Davidson [110, 119] for coal, coke and char particles larger than 1 mm, a linear dependence of burn-out time from the square of initial fuel particle diameter has been demonstrated:

$$\tau_c \sim d_o^2 \qquad\qquad (4.106)$$

at bed temperature of 900 °C, which is in agreement with the assumption that O_2 diffusion towards the fuel particle is a controlling process, i. e., that combustion takes place in the diffusion regime (see formula 4.96). It has also been shown experimentally that Sherwood's number is Sh = 1.42, which is close to Sh = 2ε, and that the value of X_B, describing diffusion of O_2 from bubbles into the emulsion phase has a value of $X_B = 0.608$.

Later investigations [48, 73, 116, 134, 138] have shown that burn-out time must be represented as:

$$\tau_c \sim d_o^n \qquad\qquad (4.107)$$

and that exponent n depends on the particle size, bed temperature and coal characteristics. This fact is best illustrated by the results of Pillai [48] obtained for 12 different types of coal, particle sizes of 0.25-8.0 mm, in the temperature range of 770-1010 °C (Fig. 4.47). In these experiments the values of n obtained ranged from 0.6 to 1.9.

All experiments published to date have led to the widely accepted conclusion that fuel particles less than 1 mm in diameter in fluidized beds burn in the kinetic regime, and that particles larger than 1 mm (or according to some authors, Nos. 8, 10, 21 in Table 4.13, particles larger than 3 mm) burn in the diffusion regime.

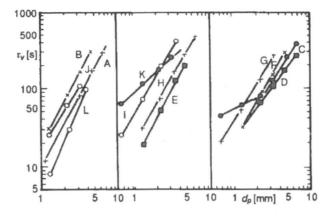

Figure 4.47. *Effect of size on particle burn-out time during combustion of different coals in fluidized bed according to the measurements of Pillai (Published with kind permission of The Institute of Energy, London, from [48]):*
A, B – coal chars, C – anthracite, D, F, E, G – brown (bituminous coals, H – high-ash bituminous coal, J, L – lignites, K – graphitic rock, I – high-ash low reactivity coal ($t_b = 1010\ °C$)

Investigation of physical processes that control combustion of particles in the fluidized bed and the experimental data of many researchers show that the boundary between the kinetic and diffusion regimes depends on the combustion temperature and coal type. In the majority of experiments, the diffusion regime has been detected for temperatures higher than 800 °C. Such low combustion temperatures at which the diffusion regime is the controlling mechanism are best explained by the fact that the temperature of burning particles is up to 200 °C higher than the bed temperature [110, 116, 146, 147].

The influence of the process of O_2 diffusion through the particle's porous structure, i. e., in what conditions will combustion follow the "shrinking core model," or "reacting core model," can be experimentally tested by measuring density of the burning particle during the combustion process, as in Prins [72]. Figure 4.48 presents his results, which show that the average particle density for coal of 9 mm in diameter is almost constant when burning at 837 °C, while at 712 °C combustion of 5 mm particles follows closely the reacting core model, where density shows linear change versus carbon conversion degree – $\alpha = 0, \beta = 1$ in eq. (4.52) and (4.53).

More reactive coals, i. e., coals whose char has a large specific pore surface, burn in the diffusion regime even at very low temperatures. Investigations of Pillai

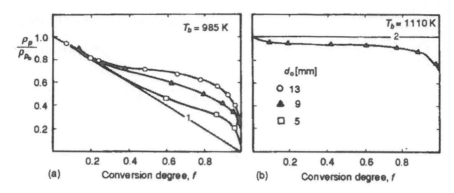

Figure 4.48. *Change of carbon particle mean density with conversion degree. According to the measurements of W. Prins (Reproduced by kind permission of the author Dr. W. Prins from [72])*

[48] show that for lignites and geologically younger coals an exponent n obtained is closer to 2, and it has been concluded that lignites even at the comparatively low temperature of 775 °C burn in the diffusion regime. According to measurements in [4, 74, 75, 152, 153] the diffusion regime for lignite particles with d = 5 mm is achieved at temperatures as low as 550 °C. Figure 4.49 shows that in the same temperature range

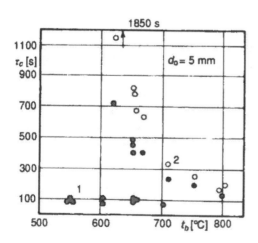

Figure 4.49.
Effect of bed temperature on burn-out times for lignite and anthracite char particles. According to the measurements of Oka and Arsić [75, 76] 1 – lignite, 2 – anthracite

550-800 °C, the burn-out time of an anthracite particle strongly depends on temperature (kinetic regime) while the lignite burn-out time does not depend on temperature (diffusion regime). Investigations in [131] show that porosity and specific surface of char pores of different coals influence the dependence of burn-out time from bed temperature (Fig. 4.50). The figure shows that char from the coal with greatest pore specific surface in the temperature range of 700-900 °C burns in the diffusion regime. With the decrease of pore specific surface, the diffusion regime is reached only at higher temperatures, and coke in these conditions burns in

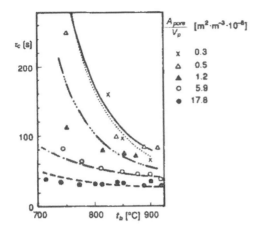

$\frac{A_{pore}}{V_p}$	$[m^2 \cdot m^{-3} \cdot 10^{-6}]$
x	0.3
△	0.5
▲	1.2
o	5.9
●	17.8

Figure 4.50.
Burn-out times for single carbon particles with different porosity at various bed temperatures. According to the measurements of Turnbull et al. (Reproduced with kind permission of the Institution of Chemical Engineers from [131])

the full kinetic regime. The influence of other parameters on burn-out time has rarely been investigated and only in isolated experiments.

The influence of fluidization velocity has been recorded only for velocities higher that 1 m/s [48]. In these experiments, higher values for Sherwood number have been determined (Sh = 4-6) than ones that apply to pure molecular diffusion (Sh = 2, or Sh = 2ε). For Re >10 transport of oxygen to the particle surface is also influenced by forced convection.

For older coals, with high ash contents (anthracite, above all) an influence of O_2 diffusion through the ash layer has been noted [73, 146, 147], in agreement with formula (4.99). In [138] it has been shown that burn-out time is inversely proportional to $d_p^{0.5}$.

In several experiments [110, 116, 119] it has been shown that the two-phase model of fluidized beds, as well as the first term on the right-hand-side of formulae (4.96) through (4.100) satisfactorily describe oxygen diffusion from bubbles into the emulsion phase.

(b) Chemical reactions in char particle combustion in fluidized beds

Possible chemical reactions of carbon particle combustion, their importance and assumed course in fluidized bed combustion have already been discussed in Section 4.6.2. These questions have also been the subject of discussion and testing in an array of experimental investigations (Nos. 1-3, 9, and 24, according to Table 4.13). In these investigations the following important questions, systematized by Prins [72], have been raised:

- does carbon combustion occur by the gasification reaction (eq. 4.47), as has first been proposed by Davidson et al. [110, 119] or by reaction (4.46), as proposed by later authors?
- where does the homogeneous reaction (4.48) take place? and

−   what is the real CO/CO_2 ratio produced by reaction (4.46)?

Closely related to these questions are the following dilemmas:

−   what is the role of O_2 diffusion through the porous char particle structure? and
−   what is the role of the kinetics of chemical reactions?

All these questions have already been addressed in previous sections, and assumptions have been stated as to the most probable course of the combustion process in fluidized beds. In this section, some specific experimental data will be discussed, which help resolve these problems, together with the resulting conclusions.

Experimental investigations of Basu [114] have clearly shown that Davidson's assumptions [110] about the existence of gasification reaction (4.47) cannot be true. Later, Davidson himself [116] retracted this hypothesis, and recently Prins [72] has directly proven experimentally that gasification reaction (4.47) is too slow, in conditions existing in fluidized beds, to significantly influence the combustion process.

However, although it can be regarded as proven that processes at the particle surface take place according to reactions (4.46) it has not yet been determined what is the primary CO/CO_2 ratio. The primary CO/CO_2 ratio, and the size of the volume where CO oxidation processes later takes place, significantly influence the estimate of true temperature of a burning particle, and hence the rate of chemical reactions.

As has already been mentioned in Section 4.6.2, according to the analysis of Ross and Davidson [116], in combustion of particles smaller than 0.5 mm, CO oxidation takes place in a relatively wide area around the particle ($\sim$5-6 d), and in combustion of particles of $\sim$5 mm in diameter, CO oxidation takes place near the particle, in the area $\sim d$ wide, leading to higher temperatures in larger particles, because in CO to CO_2 oxidation 2/3 of the total heat of CO_2 production is released. These estimates have been based on measurements of combustion rate and using equation (4.50) which defines CO/CO_2 ratio and the empirical expression for oxidation rate of carbon monoxide proposed by Howard [156]:

$$\frac{d[CO]}{d\tau}=k_{CO}[CO][O_2]^{0.5}[H_2O]^{0.5}\exp\left(-\frac{E_{CO}}{R_gT}\right) \qquad (4.108)$$

where concentrations are given in $kmol/m^3$ and constants are

$$k_{CO} = 1.3\cdot10^{11}\ m^3/kmol\ s$$
$$E_{CO} = 125.4\cdot10^6\ J/kmol$$

In an analysis of his experimental data, Prins [72] points out that the time CO spends in the vicinity of carbon particles and the oxidation rate of CO lead to the conclusion that near the particle there isn't any significant conversion of CO into CO_2, and that the measured CO/CO_2 ratios relate only to the primary production of these compounds. According to Prins' measurements in combustion of particles 3-13 mm in diameter, average CO/CO_2 ratio at 712 °C is 0.36, and at 837 °C – 0.28. Both values differ 3-10 times from those calculated using expressions (4.50) and (4.108).

Clarification of this problem obviously requires further experimental research [157], and the unreliability factor for these processes also figures in the unreliability of modelling of fluidized bed combustion.

(c) Char combustion rate

The review of experiments given in Table 4.13 shows that at the time of writing there are 18 experiments in which the combustion rate has been measured. Nevertheless it is hard to arrange and compare the results of these experiments due to their different aims, different presentation of the results and types of particles investigated. Two groups of experiments can be described:

– experiments whose aim was to study the processes and determine the influence of significant parameters, for isolated carbon or coke particles (Nos. 2, 4, 5, 19, and 24, according to Table 4.13), and
– experiments with a mainly practical aim, which were carried out by investigating large numbers of particles of different coals (Nos. 9, 12, 14, 17, 26, and 31, Table 4.13).

A separate group of experiments (Nos. 15, 20, and 22, Table 4.13) deals with the investigation of combustion rate in conditions of a turbulent and circulating fluidized bed.

The knowledge of combustion rate of char particles in fluidized beds has both theoretical and practical significance. Very different char combustion rates require theoretical explanation, and must be known also for optimum design of combustion in real furnaces. Char inventory in fluidized beds in stationary operating regimes of real furnaces, and response to the load change, depend heavily on combustion rate.

Experiments having solely theoretical goals tend to determine control processes and their relation to significant parameters. These experiments are by their character and goals similar to experiments in which burn-out time has been measured and which have been described at the beginning of this section. Measurement methods are different – measurements of change in mass, bed temperature or determining the composition of combustion products. To remove the influence of parameters which cannot be controlled – particle porosity, influence of ashes and catalytic properties of different additives – these experiments have been carried out using pure carbon or coke. The influence of diffusion from the bubbling to the emulsion phase has been eliminated by investigating single particles. The range of combustion rates of carbon particles in these experiments is shown in Table 4.14.

Table 4.14. *Comparison of combustion rates of carbon particles in fluidized beds*

Ref.	d_c [mm]	t_b [°C]	Combustion rate [mg/s]
[116]	1.5-3.3	700-900	2.4-3.6
[136]	3.0-14.0	750-800	1.0-5.0
[137]	2.0-20.0	800-900	0.1-1.5
[143]	4.0-13.0	700-900	0.2-1.6
[72]	3.0-13.0	712-837	0.006-1.3

The given values correspond to the total combustion rate of carbon particles, which depends on the experimental conditions, for which the following processes are important: O_2 diffusion through the boundary layer around the particle; the kinetics of chemical reaction; and O_2 diffusion through the particle's porous structure. Different experimental conditions influence the combustion rates obtained. The influence of experimental conditions confirms conclusions already stated in this and previous sections.

Analogous to relation (4.107), combustion rate dependence on the particle diameter can be expressed as:

$$\frac{dm_c}{dt} \sim d_c^n \sim d_c^n \qquad (4.109)$$

where $n = 1$ stands for the diffusion regime and $n = 2$ for the kinetic regime.

Basu [136] has, for particles of 3-10 mm in diameter, obtained values of $n = 1.22$-1.55 and has concluded that combustion is primarily controlled by diffusion. Chakraborthy and Howard [137] have, for particles of 3-12 mm in diameter, obtained values of $n = 1.75$-1.97 and have concluded that combustion takes place in the kinetic regime.

The characteristic diagram [143] showing qualitatively the influence of four properties on the combustion rate – carbon particle size, inert material particle size, oxygen concentration and fluidization velocity, is given in Fig. 4.51.

The detailed investigations of Prins [72] point to many peculiarities of the carbon particle burn-out process, which may be the cause of differences in experimental results. Figure 4.52 shows the results of calculations using a model for progressive conversion (model 3 in Section 4.63) which takes into account O_2 diffusion through the particle's porous structure.

Calculations have been performed for two temperatures, 712 °C, where it is believed that the rate of chemical reaction is the primary factor, and 837 °C where O_2 diffusion should have the primary influence. From the start of combustion (the moment of insertion of particles into the bed) the combustion rate increases due to increasing depth to which O_2 penetrates into the particle's porous structure. When the combustion zone inside the particle has been set up, the combustion rate reaches

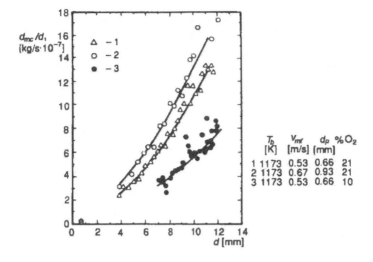

Figure 4. 51. *Char combustion rate for various particle sizes in different bed conditions. According to the measurements of La Nauze and Jung (Reproduced with kind permission of the Institution of Chemical Engineers from [3])*

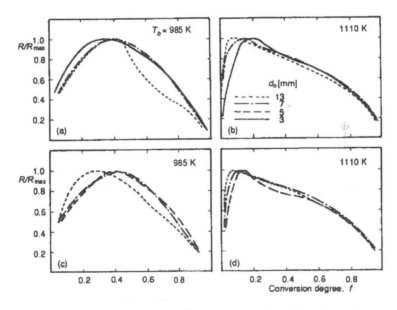

Figure 4.52. *Char combustion rate for carbon particles in dependence of conversion degree. According to the model of Prins (Reproduced by kind permission of the author Dr. W. Prins from [72])*
(a) and (b) free moving particle, (c) and (d) particle hanging on embedded thermocouple

its peak. With further particle conversion, particle size decreases, and the depth to which O_2 penetrates (reaction zone inside the particle) stays constant for a longer period of time. The combustion rate decreases because the volume content of carbon taking part in combustion decreases. The course of the process is characteristic for large particles, inside which O_2 cannot easily reach the particle center due to high resistance to diffusion inside the pores. At higher temperatures (Fig. 4.52b), when the outer diffusion of O_2 is the major combustion controlling process, the depth to which oxygen penetrates into the particle is small and the maximum combustion rate is reached quickly. Later the occurrence of a combustion rate maximum corresponds to the kinetic regime and the greater depth of reaction zone inside the particle (Fig. 4.52a). The conversion degree, f, at which maximum combustion rate is reached can be a measure of the influence of external diffusion.

Experimental investigations of combustion kinetics in circulating and turbulent fluidized beds have been sparse, due to the relatively late introduction of boilers using these two fluidization regimes. Investigations carried out by Basu et al. [141, 144, 145] with carbon particles of 4-16 mm in diameter and with fluidization velocities of up to 8 m/s (i. e., with fluidization number of up to 30 v_{mf}) have shown that in conditions of intense fluidization, the combustion rate is much higher than in bubbling fluidized beds and can be up to 40 mg/s. The main conclusion of these experiments is that combustion takes place in the regime which is mainly controlled by the rate of chemical reaction, because for circulating and turbulent fluidized beds mass transfer to the particle surface is increased.

Numerous investigations on combustion kinetics of different coals have been carried out in order to obtain empirical data necessary for choosing the concept and design of the boiler. Overall coal combustion rate or burn-out time as well as devolatilization rate have been measured. In the previously described works of Pillai [48, 51, 146, 147], the char burn-out time has been measured and corrected to take into account devolatilization time. During the course of several years, using the same methodology, comparative measurements have been made for Yugoslav coals: lignite, brown coal, steincoal and anthracite, to determine overall, apparent, combustion rate [4, 74, 75, 142, 148, 152-155]. To calculate the combustion rate, expression (4.103) has been used, and in more recent papers [152, 153] expressions (4.104) and (4.105) have been used. This permits one to take into account the total calorific value of different coals, especially for particles larger than 5 mm; the influence of volatiles combustion and fragmentation char combustion rate could not be excluded. Investigations have yielded comparative values for combustion rates of different coals, needed for the choice of boiler start-up temperatures. Figure 4.53 shows the results of experiments with temperature ranging from 350 to 800 °C.

The values obtained range from $1 \cdot 10^{-3}$ to $18 \cdot 10^{-3}$ kg/m^{2}s for particles of 5 mm diameter. These values agree well with the values obtained in the previously mentioned studies, 0.08-1.4 mg/s. The combustion rates range from those obtained

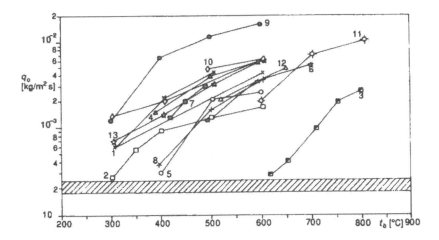

Figure 4.53. *Apparent combustion rate for different coals for various bed temperatures. According to the measurements of S. Oka et al. Shadowed region – start-up temperature criteria*

1 – Kolubara, No. 1, lignite, 5 mm; 2 – Bitolj, No. 3, lignite, 5 mm; 3 – Vrška Čuka, No. 5, anthracite, 5 mm; 4 – Miljevina, No. 6, brown (subbituminous) coal, 4.76–7 mm; 5 – Aleksinac, No. 11, brown coal, 4.76–7 mm; 6 – Raša, No. 12, bituminous coal, 5 mm; 7 – Albanian, No. 13, brown, 4.75–7 mm; 8 – Kakanj, No. 15, brown, 4.76–7 mm; 9 – Piskupština, No. 16, lignite, 4.76–7 mm; 10 – Czech, No. 17, brown, 4.76–7 mm; 11 – Ibar, bituminous coal, 4.76–7 mm; 12 – Bogovina, brown, 4.76–7 mm; 13 – Djurdjevik, brown, 4.76–7 mm

for lignites to those obtained for anthracite. The tested lignites have rather large combustion rates even at temperatures of 300 and 400 °C. Brown coals do not reach those values even at temperatures of 700-800 °C. A higher combustion rate for coals with high content of volatiles is obvious. The influence of volatiles and coal structure on the combustion rate which has been noticed, has not been clarified by these experiments and should be the subject of further investigations.

(d) Kinetic parameters

For complete mathematical modelling of combustion processes in fluidized beds it is necessary to obtain kinetic parameters A and E, for the rate of heterogeneous chemical reactions at the carbon particle surface, in expression (4.43). In many theoretical studies, the authors of mathematical models (see Section 4.6.4) have been satisfied to use kinetic constants obtained under conditions characteristic for pulverized coal combustion, using guidelines from Field [1] and Smith [109]. However, with the increased interest in mathematical modelling of processes and enhancement of measuring techniques, more experiments are aimed at obtaining these parameters in fluidized bed combustion [72, 78, 104, 116, 131, 151].

Obtaining kinetic parameters, A and E, can be rather straightforward if the combustion rate ($dm_c/d\tau$ or R_c) is determined for several temperatures, and then using Arrhenius' expression (4.43).

In principle, regardless of experimental conditions, type of carbon, char or coal, and experimental methodology, identical or close to identical results should be obtained for kinetic parameters. However, the inner surface of pore structure, actual O_2 concentration near this surface, and order of chemical reaction are not known and can be very different. Also, due to the inability to decrease or remove, by changing experimental conditions or by post-processing of experimental data, the influence of other processes controlling the combustion rate, experimental data from various authors often differ greatly and hence, their proposed values of A and E also differ strongly. These differences are clearly illustrated by Fig. 4.54.

Ross and Davidson [116] have concluded that their experimental data can best be described by the kinetic parameters of Field, expression (4.102). Turnbull et al. [131] advise, based on their experiments, the use of expression (4.101), which is obtained based on kinetic parameters of Smith, and points out that Field's expression, obtained using the assumption that combustion takes place at the particle's outer surface, also includes the diffusion processes of O_2 through the particle's porous structure.

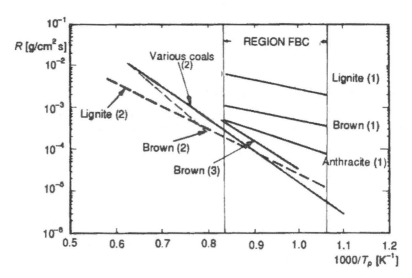

Figure 4.54. *Char combustion rate according to different authors (Reprinted with kind permission of Global Energy Group, Penn Well Corp. from [150])*
(1) – measured in fluidized bed, (2) – measured for entrained pulverized coal, (3) – measured in fixed bed

Prins failed to describe his experimental data using kinetic parameters proposed by Field and Smith. The best agreement of experimental data and mathematical model has been obtained by using the formula for combustion rate determined by his own experiments:

$$\frac{dm_c}{dt} = 2.7 \cdot 10^6 \, d^{2.6} [C_{O_2}]_p \left(\frac{[CO]}{[CO_2]} \right)^{-0.5} \exp\left(\frac{150000}{R_g T_p} \right) \qquad (4.110)$$

where the numerical constant is in $m^{0.4}/s$.

Investigating the combustion rate of different coals, ranging from lignite to anthracite, in the temperature range of 300-750 °C, Read and Minchener [78] have determined activation energy over a relatively narrow range, considering the different coal and char properties. Values obtained from this experiment were:

$$E = 91100\text{-}128000 \text{ J/mol}$$

where smaller values correspond to anthracite, which it must be noted is somewhat odd. At the same time, the pre-exponential coefficient differs by several orders of magnitude, from $3.5 \cdot 10^3$ 1/bar·s to $5 \cdot 10^6$ 1/bar·s. These results can be used only as the qualitative measure of differences between different types of coals, and not for any other purpose.

The same holds true for comprehensive investigation of van Engelen, van der Honing and Brem [104, 151], where, using numeric optimization, based on experimental data, a set of five kinetic parameters have been obtained, in which, besides activation energy there are four more specific, newly introduced parameters. For very different coals, ranging from anthracite to coke and lignite, activation energies in the range of 167000-234000 J/mol have been obtained. The results of these investigations hold true only for these coals and only if the complete set of five parameters is used.

4.6.6. Temperature of burning particles in a fluidized bed

Very early on it was noticed that the temperature of particles burning in a fluidized bed is very different from the temperature of inert material in the fluidized bed. At sufficiently high oxygen concentrations and at temperatures higher than 600 °C, after releasing and combustion of volatiles, at the bed surface char particles may be observed glowing more than inert material particles. Almost all workers who carried out experimental research on combustion kinetics of char have also dealt with the question of burning particle temperature. Most of the results obtained, given the fact that combustion takes place in the diffusion regime and at relatively low bed temperatures of 900 °C, cannot be explained without assuming that char particles are at around 200 °C higher temperature than the bed (see [110, 114]). In more recent investigations, good agreement of proposed models with experimental data has been obtained only if the rate of chemical reaction is calculated at temperatures much higher than the bed temperature (e. g., 70-150 °C higher, in [116] and [130]). These first qualitative treatments have provided good general agreement that at bed temperatures of 800-900 °C, and for carbon (char) particles of 3-15 mm in size, the temperature of burning particles may be 50-200 °C higher than the bed temperature and that this difference increases with the decrease of particle diameter.

The knowledge of burning particle temperature in fluidized bed combustion is important for several reasons:

- the change of particle temperature before combustion determines the rate of drying and devolatilization,
- particle temperature is important for determining physical properties of gases near the particle,
- particle temperature (combustion temperature) determines the conditions for creation of NO_x compounds, and
- particle temperature is important for assessing the danger of ash melting in the bed.

It can be easily concluded that the temperature of burning particles depends on the balance of heat energy generated by combustion and the heat transferred from the particle to the bed mass. However, bearing in mind all the parameters on which these processes depend, it is clear that particle temperature is determined by a large number of parameters.

Heat generated by combustion depends on: the type of chemical reactions taking place at the particle surface, CO/CO_2 ratio, and total carbon surface taking part in the reaction (i. e., combustion rate) and O_2 concentration near the particle. The importance of CO/CO_2 ratio and CO combustion zone is clear if we have in mind that oxidation of C to CO generates only 1/3 of the total heat generated in CO_2 formation [72, 116].

From the burning particle heat is transferred through the gas, by gas and particle convection and by radiation. These processes depend on a large number of parameters which have been discussed in detail in Chapter 3 (Section 3.3). The amount of heat transferred from the particle depends on the fuel particle size and inert particle size, fluidization velocity and bed temperature. It is thought that heat transfer by radiation does not greatly influence this process [38].

Due to the importance of knowing the burning particle temperature, there are numerous experimental investigations in which the sole aim, or major aim, was its measurement. Also, mathematical models now always include the energy balance equation for fuel particles [72, 103-105], and hence a determination of the coefficient of heat and mass transfer to the fuel particle and emissivity of burning particle surface and inert material, is becoming more important.

Direct measurements of particle temperature have been carried out in [38, 51, 72, 110, 136, 137, 158]. A detailed review of these and other results is given in [3, 38].

Two methods of measuring burning particle temperature have been used to date:

- photographing or optical pyrometry of particles visible at the bed surface [38, 110], and
- inserting a thermocouple in a carbon particle [51, 72, 136, 137].

In [158] a modification of the optical method is proposed, which comprises inserting an optical cable into the fluidized bed, with which it is then possible to optically measure burning particle temperature inside the bed.

Both methods used to date have drawbacks which determine the accuracy and even the feasibility of carrying out the measurement. Using optical methods it is possible to measure only the temperature of particles at the bed surface where heat transfer conditions are different than those inside the bed and only in the range of "visible" temperatures. Due to increased radiation it can be expected that measurement errors tend to give lower than actual temperatures.

Inserting a thermocouple changes the combustion and heat transfer conditions, prevents free movement of particles and allows measurement of the temperature of only a small number of particles. Towards the end of combustion, measured temperature can be completely wrong since the burning surface is very near the head of thermocouple [3]. The advantage of this method is, however, that the temperature can be monitored from the moment the particle enters the heated inert bed material.

Both methods are limited to measuring temperatures of large particles, so the experiments have been carried out with particles of 3-15 mm in diameter. In the literature, no data exist on the temperature of smaller particles.

A typical temperature history of the coal particle from the moment it enters the bed, as measured by a thermocouple inserted into the particles of 6-15 mm in diameter [48, 51], is shown in Fig. 4.55. At the temperature curve several regions can be seen:

(1) a period of particle heating to 250-350 °C, with characteristic point at 100 °C, where drying takes place and finishes,

(2) a period of combustion and devolatilization where besides the visible flame of volatiles combustion, particle heating continues. Releasing of the greatest part of volatiles concludes at particle temperature some 110 °C lower than the bed temperature,

(3) in the third period particle temperature sharply rises to temperatures 40-120 °C higher than the bed temperature due to ignition and combus-

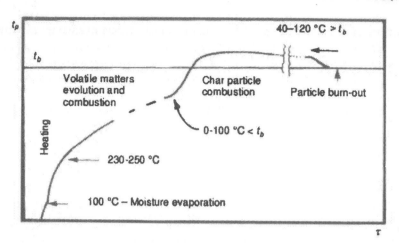

Figure 4.55. *Coal particle temperature history during combustion in a fluidized bed (Published with kind permission of The Institute of Energy, London, from [48])*

tion of char. In this period the temperature remains constant until particle size decreases to approximately 3 mm, and

(4) in the last period, until particle burn-out, measured temperature sharply decreases. The cause for such a decrease may be measurement error, mentioned earlier, but also the changed conditions for heat generation and transfer for particles smaller than 1 mm.

As has been mentioned earlier, all measurements and estimates of particle temperature in combustion agree that the temperature of a burning particle is 50-200 °C higher than the bed temperature which has, in these experiments, been 700-900 °C (for particles of 3-15 mm in size). This difference is greater for smaller particles.

Results obtained in the works of Tamarin et al. [159, 160], where the increase in particle overheating has been seen with the increase in fuel and inert material particle size, differ from these general conclusions and have been described by the empirical formula:

$$\Delta T = 670\, d_p^{0.23} d^{0.15} \qquad (4.111)$$

where diameters are in centimeters.

This result has been explained by the increased influence of particle convection on heat transfer from fuel particles with a decrease in particle size.

The increase of carbon particle overheating with increase in particle size, for graphite particles of 4-12 mm, has been also obtained experimentally by Prins [72], but in the kinetic combustion regime at bed temperatures of 712 °C. Using the experimentally obtained linear relation of carbon particle overheating, shown in Fig. 4.56a, Prins has, based on his model of progressive conversion, obtained very good agreement for the measured and calculated values of combustion rate. Figure 4.56a shows these results compared with those of La Nauze and Jung, Fig. 4.56b, for particles of petroleum coke [3], which, under similar conditions show completely different behavior. It is highly probable here that fuel particle overheating is influenced by the combustion rate and particle porosity.

The measurements of Ross et al. [38] show a clear linear relation between overheating and O_2 concentration which must be remembered when estimating particle temperature in real conditions, based on the experimental results mostly obtained at O_2 concentrations close to atmospheric.

Fluidization velocity and inert material particle size show their influence dependent on their effect on the heat and mass transfer coefficients.

Better insights into the dependence of fuel particle temperature on numerous parameters, as well as the explanation of frequently conflicting experimental results and physically based theoretical explanations and prognosis, is possible by considering the energy balance equation for fuel particles, and the hydrodynamic processes, heat and mass transfer and combustion, discussed in detail in this and previous sections. Detailed discussions of this kind have been presented in [3, 38, 72, 104, 116, 131].

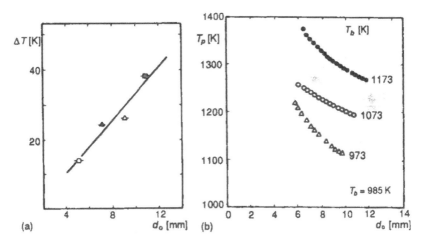

Figure 4.56. *Effect of particle size on particle overheating in a fluidized bed (Fig. 4.56a reproduced by kind permission of the author Dr. W. Prins from [72], and Fig. 4.56b reproduced with kind permission of The Institute on Chemical Engineers from [3])*
(a) According to the measurement of Prins for carbon particles, and (b) According to the measurements of La Nauze and Jung for pertoleum coke

The energy balance equation for a spherical fuel particle, assuming that the temperature is constant inside the particle, can be written as:

$$\frac{m_c c_c}{\pi d^2} \frac{dT_p}{d_\tau} = RQ_t - \alpha(T_p - T_b) - \sigma_c \varepsilon_p (T_p^4 - T_b^4) \qquad (4.112)$$

In the stationary state, if radiation is neglected since for this process there are still no reliable data, a simple equation is obtained:

$$RQ_t = \alpha(T_p - T_b) \qquad (4.113)$$

The heat of the chemical reaction depends on the CO/CO_2 ratio and the place where CO oxidation to CO_2 occurs. According to the analysis of Ross and Davidson [116], presented in Section (4.6.5), for large particles the area of CO oxidation is smaller ($r \sim d$) than for small particles. Hence for small particles there is greater probability that generated heat goes mostly to the heating of the surrounding particles of inert material. Also, the heat transfer coefficient for smaller particles is, for constant Nusselt number, greater, which leads to the higher intensity of heat transfer. This discussion can explain the fact that it has been observed that particles smaller than 1 mm burn at temperatures close to the bed temperature.

If the left hand side of (4.113) is shown like this, according to Prins [72, 112]:

$$RQ_t = R \frac{pQ_{CO} + Q_{CO_2}}{1 + p} \qquad (4.114)$$

the following expression for fuel particle temperature in stationary state is obtained:

$$T = T_b + \frac{dm_c}{d\tau} \frac{1}{\alpha \pi d^2} \frac{pQ_{CO} + Q_{CO_2}}{1 + p} \qquad (4.115)$$

where it is assumed that all the generated heat of chemical reactions is transferred to the fuel particle.

Using the model of progressive conversion, eq. (4.110) for combustion rate, taking into account diffusion through the pores, and eq. (4.115), Prins has calculated the carbon particle temperature for the diffusion regime, at 900 °C, which is shown in Fig. 4.57 (curve a). Curve b has been calculated assuming that the combustion process is controlled only by the outer diffusion of O_2. Major differences can be seen, especially for particles with smaller diameter. Calculation results have shown that in the case of particles of 13 mm in diameter, external diffusion contributes around 90% to the combustion rate, while for particles of 3 mm in diameter the contributions of external and internal diffusion are approximately equal. If O_2 diffusion in the pores is also taken into account, for particles smaller than 3 mm, the particle temperature decreases with the decrease in particle size, and is around 100 °C lower than estimated. For large particles these differences are small, and in both cases the particle temperature de-

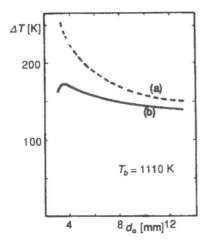

Figure 4.57.
Carbon particle overheating according to the Prins calculations (Reproduced by kind permission of the author Dr. W. Prins from [72])
(a) with pore diffusion, (b) without pore diffusion

creases with the increase in particle size. Turnbull [131] has come to similar conclusions using his model and Ross' analysis [116] on the size of the area where CO burns to CO_2. He assumed that the energy generated by CO oxidation is transferred to the burning particle and changes with the particle size, and that it depends on the ratio of fuel and inert material particle size. Starting from particle diameter of $d =$ 1.5 mm the particle temperature sharply decreases to the bed temperature with the decrease of particle diameter.

Further research is needed to clarify the relation of particle temperature to parameters significant for fluidized bed combustion. An incorrect estimate of particle temperature in modelling the processes in FBC boilers can lead to results which are far from accurate.

4.7. Mathematical modelling of processes in solid fuel combustion in fluidized bed boilers

Every researcher's ambition, or better said the ambition of the collective mind of all researchers in an wide area of science, is to mathematically describe processes in such a way as to be able to correctly and easily predict those processes. As has been already mentioned in Section 2.3.7, this research activity that we call mathematical modellin can have different goals:

(a) To describe the processes investigated in a particular experiment, and in that way, to discover the influence of certain parameters, not measured in the experiment; to enable determination of the physical and chemical properties that govern the process (e. g., rate of chemical reactions, activation energy); to give physical explanation of the observed results and enable prediction of the course of the process under different conditions. We can call these models research or phenomena models.

(b) To describe the set of single but related physical and chemical processes which take place simultaneously in any given area, in such a way as to discover their mutual relations, influence and significance of particular processes. This kind of model forms the basis and a first step towards postulating models to be used for practical, engineering purposes, for the calculation and design of real installations.

Both kinds of models, besides research goals, also have an educational character and can significantly contribute to development and dissemination of basic knowledge about the investigated phenomena. By developing and using these models, researchers and engineers are trained to think logically in terms of physical (or chemical) processes and observe the relationship between numerous processes taking place in real installations.

(c) The third group of models, so called system models, have as their immediate or future goal to describe fully all the important processes in an installation, to enable the use of knowledge, experimental data and correlations obtained experimentally, and to finally become everyday tools for the engineer, for calculations for real installations and their regimes and parameters of exploitation. These special kinds of models often are intended for use in automatic control to change the output (i. e., power of a furnace) – i. e., for guiding the operation of a full-scale device.

The first two types of models are typically mainly developed at universities and in scientific institutes. For development of the third type of model, universities and institutes usually don't have sufficient funding and/or research staff, as well perhaps the interest, to develop independently such large and complex models, so these tend mostly to be the result of international cooperation [102]. Large boiler manufacturing companies also independently develop or order the development of such models, but models obtained in that way are not usually made available to the general scientific community [104].

This classification of models has already been discussed in Section 2.3.7, related to the modelling of fluidized bed hydrodynamics. Here this discussion has been repeated but in connection to much more complex models, and more numerous processes than simply solid fuel combustion in fluidized beds. It is hard to draw a sharp line between models of type (a) and (b) because one may consider that there is no practical situation in which it is possible to model a single process. Also, the models of group (b) are often close to the models whose goal is design of a boiler, or at least its pre-design. All process models described so far in this book belong to group (a): models of fluidized bed hydrodynamics (Section 2.3.7), heat transfer models (Section 3.5.5), devolatilization models (Sections 4.4.2 and 4.5.4) and models of char particle combustion (Section 4.6.4), although a number of those encompass and describe several processes.

In this section we will deal with the development and characteristics of fluidized bed combustion models, which fall into groups (b) and (c), and describe all of the above mentioned processes. The general model that should describe processes in fluidized bed furnaces with solid fuel combustion should encompass the following:

- bed hydrodynamics (gas and particle mixing),
- heating and drying of fuel particles,
- fragmentation,
- release of volatiles,
- char combustion,
- chemical reactions in gases,
- attrition and elutriation of bed particles,
- heat transfer,
- sulphur and limestone bonding,
- creation and reduction of NO_x compounds,
- reactions and phenomena in the freeboard,
- mutual influence of the bed and freeboard,
- recirculation of unburned particles into the bed, and
- introduction of secondary air.

It is obvious that to postulate such complex models, detailed knowledge on all of the processes discussed in this book is needed, and so such modelling assumes rational decisions about the relative importance of particular processes in characteristic cases. Model simplification is carried out depending on its intended use, the FBC technology in question and the type of fuel, in order to keep such models from being overly complex and thus hard to handle. Such complex models are usually built in modules, each module being devoted to one phenomenon (e. g., fragmentation, char combustion, hydrodynamics, etc.) which enables easy modification of models for particular phenomena and upgrading with data from new research.

A great number of processes that need to be modeled, as well as the great number of proposed models for particular phenomena (see Sections 2.3.7, 3.5.5, 4.4.2, 4.5.4, and 4.6.4), the great number of proposed empirical correlations that can replace models (as in heat and mass transfer) and the significant influence of fuel type on processes, pose a set of difficult choices and dilemmas to the researcher who actually produces such a model. These dilemmas have been continuously pointed out in this book, especially in the sections dedicated to modelling.

For the above mentioned reasons, and due to the constant influx of new data, a large number of models have been created, with the goal of enabling the prediction and calculation of processes in fluidized bed combustion of coal, in accordance with the most recent knowledge base, as well as with each author's personal judgment about the significance of particular processes. Many, if not most of these models, have been listed in Table 4.15, which also presents their basic characteristics.

When postulating and developing a model the possibility of testing it must also be taken into account. The possibility of directly testing the model is usually the limiting factor in model development, and it affects the way most dilemmas are solved. Models of group (a) are tested on a small-scale, often bench experiments,

Table 4.15. *Mathematical models of fluidized bed combustion*

No.	Author	Year	A	B	C	D	E	F	G	H	Ref.
1.	Yagi et al.	1957	IV	d	I	g	II&IV	ab	a	–	162,163
2.	Avedesian	1973	I	a	I	g	I	c	a	–	110
3.	Becker et al.	1975	III	c	I	a&b1	I	a	a	–	120
4.	Campbel et al.	1975	I	a	I	g	I	c	a	–	116
5.	Gibbs	1975	I	a	I	a2&4	I	ab	a	–	164
6.	Gordon et al.	1976	I	a	I	g	I&III	c	a	–	121
7.	Saxena et al.	1976	II	cd	I	g	I	a	a	–	165
8.	Horio et al.	1977	I	a	I	g	II	a	a	–	166
9.	Gordon et al.	1978	I	a	I	g	III	a	a	–	117
10.	Horio & W	1978	III	b	II	g	II	a	a	–	122
11.	Rajan et al.	1978	III	b	II	a2	II	a	a	–	123
12.	Chen & S.	1978	II	a	II,IV	g	I	a	a	–	124
13.	Sarofim & B.	1978	I	b	II	f4	I	a	a	–	126
14.	Fan et al.	1979	I	a	III	g	I	c	a	–	127
15.	Park et al.	1979-1981	IV	d	I	c1	II	a	c	–	95, 97, 98
16.	Bywater	1980	IV	d	III	e1	II	c	a	–	99
17.	Rajan & Wen	1980	III	b	II	d2	II	ab	b	–	79
18.	Saxena & R.	1980	I	a	I, IV	g	II, IV	c	a	–	125
19.	Ch.& Howard	1981	IV	d	I	g	II, III	c	a	–	134
20.	Congalidis	1981	III	a	II	b1	III	ab	a	–	167
21.	Tojo et al.	1981	I	a	I	g	I	c	–	–	128
22.	Ross & D.	1981	I	a	I	g	II, III	c	a	–	116
23.	Walsh et al.	1984	–	–	–	–	III	a	b	–	129
24.	Turnbull et al.	1984	I	a	I	g	II, IV	c	a	–	131
25.	Borghi et al.	1985	III	c	II	b, f1	II, IV	a	b	–	56
26.	Andrei et al.	1985	III	c	II	b, f1	II, IV	a	b	–	73
27.	Donsi et al.	1986	I	a	II	g	II	ab	a	–	130
28.	Preto	1986	III	c	II	d1	II	ab	bc	+	100
29.	Zhang	1987	I	d	I	g	I	c	a	–	68
30.	Prins	1987	I	c	I, IV	g	II, IV	c	a	–	72
31.	Kostić	1987	II, III	a	II	d1	II	a	bd	–	161
32.	IEA-AFBC.	1988	III	c	II, V	a, d1	II, IV	ab	bc	+	102
33.	Boriani et al.	1989	III	c	II	e, f2	II	ab	bd	+	132
34.	Brem	1990	III	c	II, V	a, d1	II, IV	ab	bc	+	104
35.	Grubor	1992	III	c	II, V	a, d1	II, V	ab	bd	+	105

which are run so as to enable investigation of only one phenomenon and/or the influence of several characteristic parameters.

Models of group (b) are tested based on the results of investigations of real fuel combustion in experimental furnaces and pilot-plants of simple geometry and simplified construction, which enable measuring of local parameters.

Models of group (c) are tested in the same way as models of group (b) but their final verification must consist of a comparison with measurements at demonstration and industrial-size facilities, usually by measuring overall operating parameters.

However, the basis of every model must be its sub-models, which have previously been thoroughly tested on laboratory experimental devices.

The review and systematization of solid fuel fluidized bed combustion models have been presented by several authors, among them La Nauze [3], Preto [100], Kostić [161], Grubor, Oka [103], and Brem [104]. Using systematization methods given in [3] with certain extensions, and adding more recent models, Table 4.15 has been compiled. It contains perhaps almost all of the known models sorted chronologically, which describe coal combustion in a fluidized bed, and which, at their time of introduction, added something new to the description of these complex processes.

In reference to the systematization presented in earlier reviews, column D has been presented in a new way, and now contains several possible models of devolatilization, but also the possible variants of volatiles combustion. Column G has also been augmented by several possible ways to describe processes in the freeboard, and column H has been added dealing with the fuel particle fragmentation.

Table 4.16 contains further explanations that help in reading Table 4.15 and gives descriptions and characteristics of given models.

Even the detailed description of models in Tables 4.15 and 4.16 does not encompass all of the characteristics of a particular model. Mostly this is in the description of processes not directly linked to the fuel combustion, or for some parameter or feature which has been introduced in more recent models as characteristics of particular boiler or furnace designs. Some of these important characteristics are:

- does the model relate to the fuel feeding above the bed or under the bed surface,
- is the recirculation of unburned particles described,
- is the introduction of secondary air into the furnace described,
- have the processes of SO_2 capture with limestone been included,
- is reduction of NO_x compounds included,
- are the chemical reactions predicted in bubbles or only in the emulsion phase,
- is the actual size distribution of the burning fuel taken into account, and
- does the model describe steady or unsteady behavior of the furnace.

Table 4.16 *Description of models from Table 4.15. and key for reading the table*

A *Fluidized bed model*
    I Two-phase model of bubbling fluidized bed, bubble with cloud: bubbling and emulsion phase
    II Three-phase model of bubbling fluidized bed: bubble, cloud and emulsion phase
    III Fluidized bed divided into sections: two phases in each section
    IV Two-phase model of bubbling fluidized bed, bubble without the cloud: no difference between bubbling and emulsion phase
B *Gas flow through the bed*
    (a) Piston-like flow in bubbling phase, ideally mixed gas in emulsion phase. Mass exchange between phases takes place with finite speed
    (b) Ideally mixed gases in both phases
    (c) Piston-like flow in both phases. Gas exchange between phases takes place with finite speed
    (d) Piston-like flow through the bed, no exchange between phases
C *Mixing of particles in the bed*
    I Intense (ideal mixing)
    II Intense mixing in each section
    III Particle mixing with finite speed. A mixing coefficient is used when postulating mass balance
D *Release and combustion of combustible volatiles*
    – Devolatilization
    (a) release of volatiles is instantaneous in the coal feeding plane
    (b) devolatilization time is much greater than the characteristic time of coal particles mixing – volatiles distribution is uniform in whole bed volume
    (c) release of volatiles is instantaneous at the coal feeding point
    (d) arbitrary distribution of volatiles generation in the bed
    (e) devolatilization is tied to the coal particle and its movement through the bed
    (f) devolatilization kinetics are taken into account
    (g) volatiles combustion is not taken into account
    – Volatiles combustion
    (1) volatiles burn in the diffusion controlled flame, combustion is instantaneous if there is enough oxygen
    (2) chemical reaction kinetics (usually CO oxidation) control the volatiles combustion rate
    (3) combustion is controlled by both chemical reaction kinetics and diffusion
    (4) combustion is instantaneous, and it is controlled by the devolatilization rate
E *Char combustion kinetics*
    I char combustion is controlled by external diffusion
    II char combustion is controlled by both external diffusion and chemical reaction rate at the particle's external surface. Homogeneous reaction of CO to CO_2 oxidation is assumed to be very fast
    III CO to CO_2 reaction rate is comparable to the rate of other chemical reactions
    IV diffusion of O_2 reaction rate is comparable to the rate of other chemical reactions
    V O_2 diffusion through the particle's porous structure is taken into account
F *Elutriation*
    (a) char burns to the size which may be elutriated
    (b) attrition is taken into account
    (c) attrition is not taken into account
G *Combustion above the bed*
    (a) is not considered at all
    (b) combustion of elutriated char particles is taken into account
    (c) volatiles burn out instantaneously after leaving the bed
    (d) volatiles burn in the whole volume above the bed
H *Fragmentation*
    + is taken into account
    – is not taken into account

Table 4.15 includes all three model categories. Earliest models fall into group (a) (e. g., models 1-4), most of the models fall into group (b), while several most recent models (e. g., 28, 32-37) have all characteristics of models from group (c) including most of the mentioned processes not mentioned in systematization.

It is not possible, and would not fit with the character and goals of this book, to describe in detail experimental correlations and sub-models of all the most important processes mentioned in such models. All the knowledge presented up to now, experimental data, experimental correlations and models for the most important processes in fluidized bed combustion of solid fuels form a basis for description of these models and for postulating new ones. Based on Table 4.15, and relying on the above, it is possible to study in much greater detail each such model.

The goal of this section is to point to the directions and causes for development of modern fluidized bed models, to critically look at their drawbacks and possibilities and to point out possible enhancements. Hence, attention will mostly be paid to the processes directly linked to fuel particle combustion, i. e., columns C, D, E, G, and H.

Sections in which particular models have been individually discussed (2.3.7, 3.5.5, 4.4.2, 4.5.4, 4.6.4) should be regarded (and read) as the part of this analysis.

A discussion and critical analysis of the model of fluidized bed hydrodynamics has been already carried out in Section 2.3.7, and will not be repeated here. In simpler models of fluidized bed combustion, the heat transfer to the walls or immersed heat transfer surfaces is not taken into account, while in the more recent, system models, the heat transfer is calculated based on experimental correlations. Heat transfer per se is not modeled. The same applies to the heat and mass exchange between fuel and fluidized bed. Hence, heat transfer models will not be discussed. The ways in which devolatilization and volatiles combustion processes are included into the models have been discussed in Section 4.5.4 and here it will be only briefly evaluated.

With rare exceptions, when describing hydrodynamics of a fluidized bed, two-phase models with finite mass exchange rate between bubbles and emulsion phase and average bubble size has been adopted. For correct modelling of processes in large-scale deep beds, these two parameters are very important. Hence, many modern models predict the change in bubble size with increasing distance from the distribution plate.

The greatest obstacle for transferring calculation results and experimental data from small-scale experiments to industrial-scale units is the actual size and number of bubbles in the bed, i. e., their surface. Further research should be directed at this problem, but it is hard to expect more realistic results, at least as long as the bubble decay and coalescence processes are not taken into account. Inevitably the one-dimensionality of modern models of fluidized bed combustion also limits the possibility of obtaining more realistic results.

The fact that many modern models are one-dimensional also dictates the assumption that particle mixing is total, either in the whole bed volume or in particular sections into which the bed is divided (C.I or C.II).

Development of fluidized bed combustion models can best be reviewed if one analyzes columns D and E, i. e., the way of modelling the processes directly pertaining to the fuel particle.

It can be noted that until 1985 almost all models treated only carbon or char particle combustion. There are several reasons for that: (a) at the time only a small amount of data and knowledge existed about the kinetics of release and combustion of volatiles, (b) the first models tended to describe major processes, without unnecessarily complicating the model, (c) the models have been developed in countries which have high-rank coals with low volatiles content so that the major process was char combustion, and (d) this type of coal is fed with particle size of <3 mm, and below the bed surface, so release and combustion of volatiles are almost instantaneous. That was the reason for developing models 15 and 16 in Table 4.15.

For application of fluidized bed combustion for burning younger coals with high content of volatiles, more recent models have been developed to encompass sub-models of release and combustion of volatiles (models 25, 26, 28, 31-35, Table 4.15). Geologically younger coals are fed above the bed surface with particle size ranging from dust to 50 mm in diameter. Thus, more recent models contain the combustion of volatiles and combustion in the freeboard (models 25, 26, 28, 31-35) and models 32 and 35 also contain sub-models for the case of coal feeding onto the bed surface.

Combustion and release of volatiles is nevertheless a problem which still has to be solved.

In the system models (25, 31-35) and other more complex models, devolatilization models described in Section 4.4.2 have not been included. The kinetics of devolatilization has been taken into account only in models 13, 16, 25, 26, and 33. In other models, more or less realistically, the distribution of volatiles depending on the ratio of characteristic times of devolatilization and fuel mixing is assumed. Almost all models adopt instantaneous combustion of volatiles, if there is enough oxygen. Many authors think, and that comes from discussion in the research groups in the Netherlands, Sweden, Germany, Portugal, Denmark, and Yugoslavia, who were jointly developing model No. 32 [102], that realistic results cannot be obtained for combustion of fuels with high volatiles content, unless the model includes sub-models describing the kinetics of release and combustion of volatiles.

Gradual development of knowledge about combustion processes, and simultaneous development of models of fluidized bed combustion, can best be seen in column E, which describes the char combustion model used. Details of used models are given in Table 4.12 and in analysis in Section 4.6.4.

The first models (Nos. 1-9, Table 4.15) assume a diffusion model and combustion according to the model of non-reacting core. Not until the models of Horio, Rajan, Saxena, and Beer (Nos. 10-13) is the possibility of chemical reaction kinetics at the particle surface influencing the combustion rate taken into account. The more recent models of Prins, Brem, IEA Group and Grubor (No. 30, 32, 34, and 35,

Table 4.15) also assume that depending on the conditions all processes can be important in char combustion, and take into account the diffusion of O_2 through the particle's porous structure. Since more recently the importance of the primary fragmentation process has been understood, it is always included in the models of combustion which are to be experimentally tested against measurements in real installations and thus form the basis for engineering calculations.

Diffusion through the ash layer has not until more recently been taken into account. This process has been included in models only by Chen and Saxena [124] and Saxena and Rehmat [125]. It is probably more important for the combustion of younger coals for which experiments show that the ash is "peeled off" from the particle surface by mechanical action of inert material particles.

The biggest unsolved problem in modelling char combustion, as has been detailed in Section 4.6.4, is to determine the kinetic parameters, mass transfer intensity at the particle and through the particle's porous structure, determining of the CO/CO_2 ratio and the rate of CO oxidation into CO_2. It is still necessary to adjust some constants to get good agreement with the data for particular coal combustion.

Models developed with practical goals (e. g., Nos. 28, 32-35, Table 4.15) to obtain realistic values for combustion efficiency, must include fragmentation processes, attrition, elutriation, processes in the freeboard and recirculation of unburned particles. Very little attention has been paid to modelling processes in the freeboard (Nos. 17, 23, 25, 26, 28, 31-35, Table 4.15). In these models processes above the bed are treated in a very simplified (and it must be admitted) primitive manner, so this is their major drawback when they are used to calculate combustion of coals with high volatiles content.

The requirement to control SO_2 and NO_x emissions when burning fossil fuels have made system models pay greater attention to the processes of SO_2 capture by CaO derived from limestone and to NO_x reduction (Nos. 28, 32-35, Table 4.15).

Modern system models demand the following input data:

(a) *furnace data – geometrical and conceptual*
   - dimensions,
   - feeding point, and
   - shape and number of orifices on the distribution plate;
(b) *data on the bed and inert material particles*
   - bed height,
   - bulk density of the fixed bed, and
   - physical properties of inert material and limestone (particle size, density, particle size distribution);
(c) *data on coal*
   - proximate and ultimate analyses,
   - granulometric composition, and
   - char characteristics;
(d) *data on heat transfer surfaces in bed and around it;*

(e) *data on particle recirculation*
  – cyclone and filter efficiency, and
  – average size of recirculated particles.

Depending on the intended use for a model, as input parameters the following values may be used: the overall capacity of the furnace, the bed temperature or fuel mass flow rate. One of these parameters can be chosen to be the result of calculation, while the other two will be input data. With respect to the question of the character and properties of fluidized bed combustion, combustion temperature is always chosen to be between 800 and 900 °C, so the power and fuel mass flow rate remain as alternatives when choosing input data. In models intended for use in research and testing of experimental and pilot installations, the fuel mass flow rate is chosen as input datum.

Depending on models of particular processes that are used, numerous experimental correlations and constants may be needed as input data.

Using the model it is possible to calculate all important local and integral properties of fluidized bed combustion furnaces:

(a) concentration of all gases along the height of the bed and the freeboard (modern models consider reaction of up to 10 gases – CO, CO_2, H_2, H_2O, NO, NO_2, N_2O, O_2, SO_2, and CH_4),

(b) temperature of gases along the height of the bed and the freeboard (it must be borne in mind that all of the models discussed here are one-dimensional),

(c) coal mass flow rate (or furnace power),

(d) limestone mass flow rate,

(e) mass flow rate of ash or char before and after cyclone and filter,

(f) particle size distribution of all solid particles in bed in stationary state and all particle flows into and out of the furnace,

(g) heat introduced by fuel and exchanged in the bed,

(h) combustion efficiency,

(i) mass content of fuel in bed in stationary state, and

(j) concentration of all gases at the furnace exit.

The usual structure of one system model organized in a modular way (using sub-models) is shown in Fig. 4.58 according to [102, 104, 105].

Rating the drawbacks and possible enhancements of such general, system models has already been discussed above. If a "single mark or score" should be awarded then it probably mostly depends on the views and personal attitude of the evaluator. It is certain that these models still cannot provide reliable results and that their further development is required, together with further development of sub-models and their detailed testing. The complexity of models themselves has caused more time to be spent on their development than for their testing. Over the next phase, the main effort should be directed to organizing and conducting

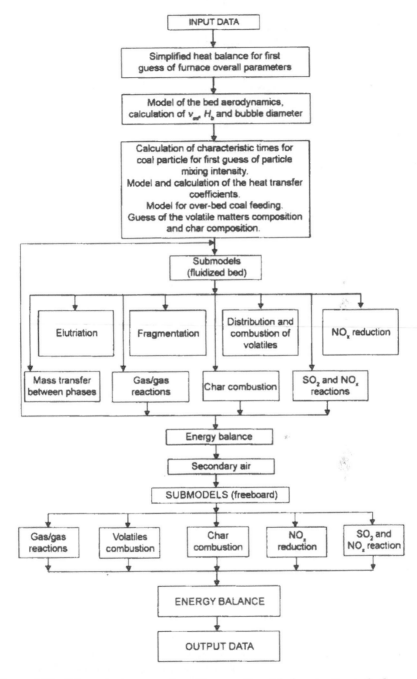

Figure 4.58. *Schematic representation of the overall model of combustion in the furnace of a fluidized bed combustion boiler*

well-planned experiments, on experimental furnaces, and pilot- and industrial-scale installations, with detailed local measurements and burning different types of coals. The results of these measurements are essential for further testing and future developments of such models.

However, it is also clear that there is no other way for detailed calculation of complex and mutually related processes in fluidized bed combustion and that the future lies in further development of such mathematical models. There is no doubt that this kind of model will increasingly become irreplaceable as an every-day engineering tool for calculation for fluidized bed boilers. This conclusion is also backed up by the data from literature, discussing the large, complex models whose development has been financed by large companies: such as the MIT-model, developed at Massachusetts Institute of Technology, TVA-model, developed at Oak Ridge Laboratories and the model developed at the University of Siegen (Germany). The exact structure and capabilities of these models are not known, since data on them is not available in the general scientific literature, but it is known that they are used by many commercial companies.

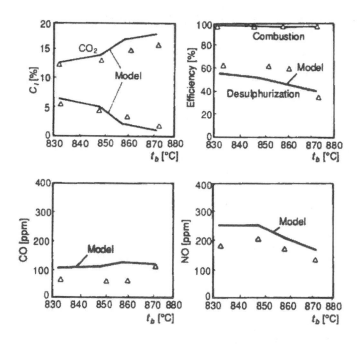

Figure 4.59. *Comparison of the model results (solid line) with data obtained from the 90 MW_{th} AKZO-AFBC (the Netherlands) boiler at various bed temperature (Reproduced with kind permission of the author Dr. G. Brem from [104])*

To become more familiar with capabilities and achievements of the models discussed, we finally give a comparison of the Brem model [104] with the measurements resulting at an industry-scale boiler with fluidized bed combustion, power 90 MWt built by Stork & Co. mounted at AKZO company in Hengelo (Holland). This boiler used a high-quality coal with calorific value of 28.2 kJ/kg, a volatile content of 28%, 10% of ash and 3.5% of moisture, particle size of 0.1 to 10 mm and with around 60% of particles smaller than 1 mm in diameter. Figure 4.59 shows some comparison results given in [104] with bed temperature in the range of 832 to 873 °C. Except for the CO concentration, qualitative and quantitative agreement is seen in all of major output parameters (CO, SO_2, and NO_x emission) and the agreement with overall parameters is most satisfactory.

Nomenclature

a, b	reaction order, eq. (4.18),
a_p	specific surface area of particle pores (area per unit mass of the particle), [m^2/kg]
a_t	coefficient of thermal diffusivity, [m^2/s]
a_v	constant in the relation between devolatilization time and coal particle diameter, [s/mm^N]
A	pre-exponential factor, [m/s]
A_b	cross section of the fluidized bed, [m^2]
A_g	specific internal pore surface area of solid particle per unit mass (intrinsic reaction surface area), [m^2/kg]
A_p	external particle surface area, [m^2]
A_{pore}	surface area of the pore walls, [m^2]
c^e	relative oxygen concentration, (=$CO_2/CO_{2\infty}$),
c_p^*	relative gas concentration at particle surface
c_b	specific heat of the inert bed material, [J/kgK]
c_c	specific heat of fuel particle, [J/kgK]
c_g	specific heat of gas (air), [J/mK]
C_{fix}	fixed carbon content in coal, [%]
C_{O_2}	oxygen mass concentration, per unit volume in fluidized bed, far from fuel particle, [kg/m^3]
$C_{O_{2p}}$	oxygen mass concentration, per unit volume at fuel particle surface, [kg/m^3]
$C_{O_{2\infty}}$	oxygen mass concentration in emulsion phase, far from fuel particle, [kg/m^3]
daf	"dry, ash free basis"
d	mean fuel particle diameter, [m] or [mm]
d_c	char particle diameter, [m] or [mm]
d_{cr}	critical fragmentation diameter, [mm]
d_{mmf}	dry, mineral matter free basis
d_o	initial fuel particle diameter, [m] or [mm]
d_p	mean particle diameter of the inert bed material, [m] or [mm]
D_{gr}	effective gas diffusion coefficient in fluidized bed in radial direction, [m_2/s]
D_G	molecular diffusion coefficient of gas, [m^2/s]
D_e	effective gas diffusion coefficient through the porous structure of coal particle, [m^2/s]

D_e effective gas diffusion coefficient through the porous structure of ash layer, [m²/s]

D_{pore} gas diffusion coefficient in particle pores, [cm²/s]

e_c elutriation rate of fine carbon particles formed by attrition, [g/min.]

E activation energy of the devolatilization process, volatiles combustion process or char combustion process, [J/kmol]

E_{CO} activation energy of the oxidation of carbon monoxide, [J/kmol]

E_o mean value of Gaussian activation energy distribution for devolatilization, [J/kmol]

f fuel particle conversion degree

H_b fluidized bed height, [m]

H_c calorific value of the fuel, [J/kg]

H_{daf} calorific value of the fuel on dry and ash free basis, [J/kg]

H_{mf} bed height at minimum fluidization, [m]

k mass transfer coefficient, [m/s]

k' coefficient defined by expression (4.92), [m/s]

k_c attrition constant

k_{CO} pre-exponential factor for CO oxidation reaction, [m³/kmol s]

k_H rate of chemical reaction, [1/s]

k_{Ho} pre-exponential coefficient in the expression for chemical reaction rate, [1/s]

k_{Ro} pre-exponential factor in the expression for coal devolatilization rate, [1/s]

k_R reaction rate of the coal devolatilization process, [1/s]

K_1, K_2 constants in eq. (4.5), [K and K⁻¹]

l_{pore} coefficient of particle pore porosity

m true order of oxidation reaction on pore surface

m_c total mass of char particles in the bed (bed inventory) during stationary combustion process, [kg]

$\dot{m}_v$ air mass flow rate, [kg/s]

M molecular mass of the gas diffusing through the pores, [kg/kmol]

M_b mass of the inert material in fluidized bed, [kg]

$\dot{M}_c$ carbon consumption, [kg/s]

M_c molecular mass of the carbon, [kg/kmol]

M_C mass of carbon in char particle in time t during combustion, [kg]

M_{co} initial mass of carbon in char particle, [kg]

$\dot{M}_{O_2}$ oxygen mass flow rate toward fuel particle, [kg/s]

M_v molecular mass of the air, [kg/kmol]

n exponent in expression for chemical reaction (4.18)

n order of the carbon oxidation reaction

n_1, n_2, n_3 number of moles O_2, CO_2, and H_2O participating in fuel oxidation reaction, eq. (4.17)

N exponent in the relation between devolatilization time and coal particle diameter

N_p number of char particles in the bed during stationary combustion process

p pressure, [N/m²], but also ratio CO/CO_2 during oxidation of carbon or char particle,

PRN ratio of the volatile content and equilibrium moisture content (pore resistance number)

q_o apparent combustion rate of coal particle, [kg/s]

Q_{CO}, Q_{CO_2} heat of carbon oxidation reaction to CO and CO_2, respectively

Q_t heat of reaction, [J/kg]

Q_{VM} ratio of total mass loss of coal during devolatilization to proximate volatile matter content

r	particle radius, [m] or [mm]
r_c	radius of unreacted core, [m] or [mm]
r_{pore}	mean pore radius, [cm]
r_v	specific volumetric rate of oxygen consumption, [1/s]
R	combustion rate defined on external particle surface area, [kg/m²s]
R_c	apparent rate of the chemical reaction of the order n, kg^{1-n}/m^{2-3n}s, for $n = 1$, [m/s]
R_g	universal gas constant, [J/molK]
R_H	modified intrinsic rate of chemical reaction on the pore surface, according to [131], eq. (4.101), [m/s]
R_i	linear rate of chemical reaction calculated per unit of the pore surface area, [kg^{1-m}/m^{2-3m}s], or for $m = 1$, [m/s]
R_m	maximum possible combustion rate, for $CO_{2p} = 0$, or when rate of chemical reaction is infinite, and mass transfer to the particle is controlling the combustion process, [kg/m²s]
R_r	radius of the region in which oxidation of carbon monoxide is taking place, [m]
R_{r1}	radius of the region around the particle in which constant CO_2 concentration is established, [m]
R_S	combustion rate per unit mass of carbon, [kg/kgCs]
R_v	rate of oxygen consumption per unit volume, [kg/m²s]
$\acute{s}$	mass ratio = 12/32, for $C + O_2 = CO_2$ reaction
t_b	bed temperature expressed in Celsius degrees
t_{fb}	freeboard temperature, [°C]
t_b^*	initial constant bed temperature in "batch" experiments for investigation of the coal combustion kinetic in fluidized beds, [°C]
T	temperature in Kelvin degrees
T_b	fluidized bed temperature, [K]
T_{ci}	char ignition temperature, [K]
T_p	particle temperature, [K]
T_v	temperature at which devolatilization starts, [K]
T_{vi}	ignition temperature of volatiles, [K]
$\bar{u}_p$	mean particle velocity in fluidized bed in vertical direction, [m/s]
v_f	superficial velocity of fluidizing gas (fluidization velocity), calculated for solids free bed cross section, [m/s]
v_{mf}	superficial gas velocity at incipient fluidization (minimum fluidization velocity) calculated for solids free bed cross section, [m/s]
V_p	volume of the particle, [m³]
VM	volatiles lost from particle up to time τ, fraction of the original coal mass, [kg/kg], or [%]
VM^*	volatiles lost from particle up to time $\tau = \infty$, in given temperature conditions, kg/kg or [%]
VM_c	proximate volatile matter in char at $\tau = \infty$, in given temperature conditions, [kg/kg]
VM_o	volatile matter content in coal as measured by proximate analysis, [kg/kg] or [%]
$VM_i, VM_i^*,$	
k_{Ri}, k_{Roi}	volatile matter content and kinetic parameters of devolatilization for the different components of volatile matter mixture released from the coal
X_B	nondimensional coefficient of gas interchange between bubble and emulsion phase, calculated per bubble volume (see Chapter 2)
y	Decart's coordinate in vertical direction
z	relative particle radius [$= r/(d/2)$]

[C], [H],
[S], [O]   mass content of C, H, S and O in coal on dry basis (in eq. 4.3.), [kg/kg]
[*fuel*]   molar concentration of fuel, [kmol/m^3]
[ ]   denotes concentration in [kg/m^3], if it is not otherwise

Greek symbols

α heat transfer coefficient for fuel particle, [W/m^{2}K]
δ_p mean pore diameter of solid particle, [Å],
$\delta_1, \delta_2, \delta_3$ factors in eq. (4.26),
Δ factor defined by the following parameters [$=A_b(v_f - v_{mf})(1- \exp(-X_B) + v_{mf}$] [m^3/s]
γ characteristic particle size (= V_p/A_p), [m^3/m^2]
ε_p emissivity of fuel particle
η ratio of the real combustion rate to combustion rate in case when pore diffusion is not the controlling process
λ excess air, [kg/kg]
λ_g heat conductivity of the gas, [J/mK]
λ_p heat conductivity of the fuel particle, [J/mK]
Λ stoichiometric coefficient of the carbon oxidation reaction, with CO and CO$_2$ as reaction products, [kgC/kgO$_2$]
ς_p total pore volume per unit mass of the solid particle, [m^3/kg]
ξ total pore volume per unit volume of the solid particle, [m^3/m^3]
ρ_c carbon density, [kg/m^3]
ρ_p particle density, [kg/m^3]
ρ_{po} initial density of fuel particle, [kg/m^3]
σ standard deviation of the Gaussian activation energy distribution for devolatilization, [J/kmol]
s_c Stefan-Boltzman constant, [W/m^{2}K^4]
τ time, [s]
τ_c char burn-out time (life time of coal particle), [s]
$\Delta\tau_{cd}$ delay period of char ignition, [s]
τ_{ci} char particle ignition time, [s]
τ_e end of moisture evaporation time period, [s]
τ_h end of fuel particle heating time period up to 100 °C, [s]
τ_l average time period needed for particle to reach fluidized bed surface, [s]
τ_v devolatilization time, [s]
τ_w time period up to the onset of devolatilization (or time period up to the ignition of volatiles at temperatures higher than T_w), [s]
Φ Thiele module,
χ = R/R_m, ratio of the combustion rate and maximum combustion rate for C_{O_2} = 0

Dimensionless criterial numbers

Bi = $\alpha\delta/\lambda_p$ Biot number for convective heat transfer for fuel particle
Bi$_m$ = kd/D_e Biot number for mass transfer for fuel particle
Fo = $a_t t/d^2$ Fourier number for fuel particle
Re$_{mf}$ = $\rho v_{mf}d/\mu_f$ Particle Reynolds number
Sc = $\mu_f/\rho_f D_G$ Schmidt number
Sh = kd/D_e Sherwood number

References

[1] MA Field, DW Gill, BB Morgan, PGW Hawksley. Combustion of Pulverized Coal. Leatherhead (GB): BCURA, 1967.

[2] R Dolezal. Large Boiler Furnaces. New York, London, Amsterdam: Elsevier Publ. Co.,1967.

[3] RD La Nauze. Fundamentals of coal combustion in fluidized beds. Chem. Eng. Res. Des. 1:3-33, 1985.

[4] S Oka. Kinetics of coal combustion in fluidized beds. Proceedings of International Seminar Heat and Mass Transfer in Fluidized Bed, Dubrovnik (Yugoslavia), 1984. New York: Hemisphere Publ. Co., 1986, pp. 371-383.

[5] N Pantić, P Nikolić. Coal (in Serbian). Belgrade: Naučna knjiga, 1973.

[6] P Nikolić, D Dimitrijević. Coal in Yugoslavia (in Serbian). Belgrade, 1981.

[7] NM Laurendal. Heterogenous kinetics of coal char gasification and combustion. In: Progress in Energy and Combustion Science. London: Pergamon Press, 1978. Vol. 4, pp. 224-270.

[8] RH Essenhigh, EM Suuberg. The role of volatiles in coal combustion. In: J. Lahaye, G. Prado, ed. Fundamentals of the Physical Chemistry of Pulverized Coal Combustion. Dordrecht (the Netherlands): Martinus Nijhoff Publishers, 1987.

[9] DW van Krevelen. Coal. New York: Elsevier Publ. Co., 1961.

[10] H March, T Siemieniewska . The surface areas of coals as evaluated from the adsorption isotherms of carbon dioxide using the Dubinin-Polanyi equation. Fuel 2:355-367, 1965.

[11] G Donsi, M Massimilla, M Miccio. Carbon fines production and elutriation from the bed of a fluidized coal combustion. Combustion and Flame Vol. 41, 1:57-69, 1981.

[12] V Arena, M D'Amore, L Massimilla. Carbon attrition during the fluidized combustion of a coal. AIChE Journal 1:40-49, 1983.

[13] P Salatino, L Massimilla. A descriptive model of carbon attrition in the fluidized combustion of a coal char. Chem. Eng. Sci. 10:1905-1916, 1985.

[14] R Chirone, M D'Amore, L Massimilla, A Mazza. Char attrition during the batch FBC of a coal. AIChE Journal 5:812-820, 1985.

[15] L Massimilla, R Chirone, M D'Amore, P Salatino. Carbon attrition during the fluidized combustion and gasification of coal. Final report – DOE Grant 1981-1985, CUEN, Depart. di Ing. Chimica, Universita di Napoli, 1985.

[16] L Lin, JT Sears, CY Wen. Elutriation and attrition of char from a large fluidized bed. Powder Technology Vol. 27, 1:105-115, 1980.

[17] P Pecanha, BM Gibbs. The importance of coal fragmentation and swelling on coal burning rates in a FBC. Proceedings of 3rd International Fluidization Conference, London, 1984, DISC/9/65-70.

[18] MA Andrei, AF Sarofim, JM Beer. Time-resolved burnout of coal particles in a fluidized bed. Combustion and Flame Vol. 61, 1:17-27, 1985.

[19]   AB Fuertes, JJ Pis, A Suarez, V Artos, F Rubeira. Coal ash attrition in a fluidized
        bed. Proceedings of 10th International FBC Conference, San Francisco, 1989, Vol. 2,
        pp. 1225-1230.

[20]   CA Sundback, JM Beer, AF Sarofim. Fragmentation behavior of single coal parti-
        cles in a fluidized bed. Proceedings of 20th Symposium /Int./ on Combustion, 1984.
        Pittsburgh: The Combustion Institute, 1985, pp. 1495-1503.

[21]   D Dakić, G van den Honing, M Valk. Fragmentation and swelling of various coals
        during devolatilization in a fluidized bed. Fuel 7:911-916, 1989.

[22]   D Dakić, G van den Honing, B Grubor. Mathematical simulation of coal fragmenta-
        tion during devolatilization in fluidized bed. Presented at the 20th IEA-AFBC Tech-
        nical Meeting, Göteborg (Sweden), 1989.

[23]   AR Kerstein, S Niksa. The distributed-energy chain model for rapid coal
        devolatilization kinetics. Part 1: Formulation. Proceedings of 20th Symposium /Int./
        on Combustion, 1984. Pittsburgh: The Combustion Institute, 1985, pp. 941-949.

[24]   P Walsh, A Dutta, R Cox, A Sarofim, J Beer. Fragmentation of a bituminous coal
        char during bubbling atmospheric pressure FBC: Effects on bed carbon load and car-
        bon conversion. Presented at 9th International Conference on FBC, Boston, 1987.

[25]   G Brem. Mathematical modeling of coal conversion processes, PhD dissertation,
        University Twente, Enschede (the Netherlands), 1990.

[26]   PK Agarwal, WE Genetti, YY Lee, SN Prasad. Model for drying during
        fluidized-bed combustion of wet low-rank coals. Fuel 7:1020-1026, 1984.

[27]   AV Likov, A Mikhailov. Theory of Energy and Mass Transfer. Englewood Cliffs,
        New York: Prentice-Hall, 1961.

[28]   DB Anthony, JB Howard. Coal devolatilization and hydrogasification. AIChE
        Journal 4:625-656, 1976.

[29]   JB Howard. In: M.A. Elliot, ed. Chemistry of Coal Utilization, second
        supplementary volume. New York: John Wiley, 1981, pp. 625-784.

[30]   JP Morris, DL Keairns. Coal devolatilization studies in support of the Westinghouse
        fluidized-bed coal gasification process. Fuel 6:465-471, 1979.

[31]   WH Wiser, GR Hill, NJ Kertamus. Kinetic study of the pyrolysis of a high-volatile
        bituminous coal. I&EC Process Design and Development 1:133-138, 1967.

[32]   S Niksa, AR Kerstein. The distributed-energy chain model for rapid coal
        devolatilization kinetics. Part I: Formulation. Combustion and Flame Vol. 66,
        2:95-109, 1986.

[33]   M Radovanović, ed. Fluidized Bed Combustion. New York: Hemisphere Publ. Co.,
        1986.

[34]   PK Agarwal, WE Genetti, YY Lee. Model for devolatilization of coal particles in
        fluidized bed. Fuel 8:1157-1165, 1984.

[35]   JF Stubington, Sumaryono. Release of volatiles from large coal particles in a hot
        fluidized bed. Fuel 7:1013-1019, 1984.

[36] S Niksa. The distributed-energy chain model for rapid coal devolatilization kinetics. Part 2: Transient weight loss correlations. Combustion and Flame Vol. 66, 2:111-119, 1986

[37] EM Suuberg, WA Peters, JB Howard. Product composition and kinetics of lignite pyrolysis. Ind. Eng. Chem. Process Des. Dev. 1:37-45, 1978.

[38] IB Ross, MS Patel, JF Davidson. The temperature of burning carbon particles in fluidized beds. Trans. Inst. Chem. Eng. 2:83-88, 1981.

[39] RC Neavel, SE Smith, EJ Hippo, RN Miller. Inter-relationships between coal compositional parameters. Fuel 3:312-320, 1986.

[40] S Badzioch, PGW Hawskley. Kinetics of thermal decomposition of pulverized coal particles. I&EC Process Design and Development 4:521-530, 1970.

[41] JB Howard, RH Essenhigh. Pyrolysis of coal particles in pulverized fuel flames, I&EC Process Design and Development 1:74-84, 1967.

[42] PK Agarwal, WE Genetti, YY Lee. Devolatilization of large coal particles in fluidized beds. Fuel 12:1748-1751, 1984.

[43] M Ilić, B Arsić, S Oka. Influence of temperature on coal devolatilization (in Serbian). Proceedings of 8th Yugoslav Symposium of Thermal Engineers YUTERM 90, Neum (Yugoslavia), 1990, pp. 443-449.

[44] DB Anthony, JB Howard, HC Hottel, HP Meisner. Rapid devolatilization of pulverized coal. Proceedings of 15th International Symposium on Combustion, 1974. Pittsburgh: The Combustion Institute, 1975. pp. 1303-1315, also Fuel 1:121-132, 1976.

[45] W Fu, Y Zhang, H Han, Y Duan. A study on devolatilization of large coal particles. Combustion and Flame Vol. 70, 3:253-266, 1987.

[46] H Kobayashi, JB Howard, AF Sarofim. Coal devolatilization at high temperatures. Proceedings of 16th Symposium /Int./ on Combustion,1976. Pittsburgh: The Combustion Institute, 1977, pp. 411-425.

[47] JF Stubington. The role of coal volatiles in fluidized bed combustion. J. of the Institute of Energy 417:191-195,1980.

[48] KK Pillai. The Influence of coal type on devolatilization and combustion in fluidized beds. J. of the Institute of Energy 420:142-150, 1981.

[49] KK Pillai. A schematic for coal devolatilization in fluidized bed combustors. J. of the Institute of Energy 424:132-133,1982.

[50] RD La Nauze. Coal devolatilization in fluidized-bed combustors. Fuel 8:771-773, 1982.

[51] KK Pillai. Devolatilization and combustion of large coal particles in fluidized bed. J. of the Institute of Energy 434:3-7, 1985.

[52] E Ekinci, G Yalkin, H Atakul, A Erdem-Senatalar. The combustion of volatiles from some Turkish coals in a fluidized bed. J. of the Institute of Energy 449:189-191, 1988.

[53] H Vural, PM Walsh, AF Sarofim, JM Beer. Destruction of tar during oxidative and
      nonoxidative pyrolysis of bituminous coal in a fluidized bed. Presented at 9th Inter-
      national Conference on FBC, Boston, 1987.

[54] D Merrick. Mathematical models of the thermal decomposition of coal. 1. The evo-
      lution of volatile matter. Fuel 5:534-539, 1983. 2. Specific heats and heats of reac-
      tion. Fuel 5:540-546, 1983. 3. Density, porosity and contraction behavior. Fuel
      5:547-552, 1983.

[55] TX Phuoc, P Durbetaki. Heat and mass transfer analysis of a coal particle undergo-
      ing pyrolysis. Int. J. of Heat and Mass Transfer 11:2331-2339,1987.

[56] G Borghi, AF Sarofim, JM Beer. A model of coal devolatilization and combustion in
      fluidized beds. Combustion and Flame Vol. 61, 1:1-16,1985.

[57] S Niksa, AR Kerstein, TH Fletcher. Predicting devolatilization at typical coal com-
      bustion conditions with the distributed-energy chain model. Combustion and Flame
      Vol. 69, 2:221-228, 1987.

[58] EM Suuberg. Significance of heat transfer effects in determining coal pyrolysis
      rates. J. of Energy and Fuels 4:593-595, 1988.

[59] J Lede, J Panagopoules, H Zhili, J Villermaux. Fast pyrolysis of wood: direct mea-
      surement and study of ablation rate. Fuel 11:1514-1520, 1983.

[60] R Chan Wai-Chun, M Kelbon, BB Krieger. Modelling and experimental verification
      of physical and chemical processes during pyrolysis of a large biomass particle. Fuel
      11:1505-1513, 1985.

[61] V Kothary, JM Antal. Numerical studies of the flash pyrolysis of cellulose. Fuel
      11:1487-1494, 1985.

[62] E Arni, RW Coughlin, PR Solomon, HH King. Mathematical modeling of lignin py-
      rolysis. Fuel 9:1495-1501, 1985.

[63] H Kobayashi. Kinetics of rapid devolatilization of pulverized coal. PhD
      dissertation, MIT, Cambridge (U.S.A.), 1976.

[64] H Reidelbach, M Summerfield. Kinetic model for coal pyrolysis optimization. Am.
      Chem. Soc., Division Fuel Chem. Preprints 1:161-183, 1975.

[65] B Atkinson, DU Merrick. Mathematical models of the thermal decomposition of
      coal. 4. Heat transfer and temperature profiles in a coke-oven charge. Fuel
      5:553-561, 1984.

[66] VR Voller, M Gross, D Merrick. Mathematical models of the thermal decomposition
      of coal. 5. Distribution of gas flow in a coke-oven charge. Fuel 5:562-570, 1983.

[67] AP Baskakov, BV Berg, OK Vitt, et al. Heat transfer to objects immersed in fluidized
      beds. Powder Technology 5-6:273-282, 1973.

[68] JQ Zhang. Devolatilization and combustion of large coal particles in fluidized sand
      bed. PhD dissertation, Queens University, Kingston, 1987. Tech. Rep. QFBC TR.
      87.2, Queens University, Kingston (Canada).

[69] WB Russel, DA Saville, MI Greene. A model for short residence time
      hydropyrolysis of single coal particle. AIChE Journal 1:65-80, 1979.

[70] GR Gavalas, KA Wilks. Intraparticle mass transfer in coal pyrolysis. AIChE Journal 3:201-212,1980.

[71] S Niksa, LE Heyd, WG Russel, D Saville. On the role of heating rate in rapid coal devolatilization. Proceedings of 20th Symposium /Int./ on Combustion, 1984. Pittsburgh: The Combustion Institute, 1985, pp.1445-1453.

[72] W Prins. Fluidized bed combustion of a single carbon particle. PhD dissertation, University Twente, Enschede (the Netherlands), 1987.

[73] MA Andrei, AF Sarofim, JM Beer. Time-resolved burnout of coal particles in a fluidized bed. Combustion and Flame Vol. 61, 1:17-27, 1985.

[74] S Oka. Investigation of coal FBC (in Serbian). Report of the Institute of Nuclear Sciences Boris Kidrič, Vinča, Belgrade, IBK-ITE-391, 1983.

[75] S Oka. Kinetics of coal combustion in fluidized bed (in Serbian). Termotekhnika, 3-4:99-123, 1983.

[76] S Oka. Kinetics of coal combustion in fluidized bed. Proceedings of International Seminar Heat and Mass Transfer in Fluidized bed, Dubrovnik (Yugoslavia), 1984, New York: Hemisphere Publ. Co., 1986, pp. 371-383.

[77] JF Stubington. The role of coal volatiles in fluidized bed combustion. J. of the Institute of Energy 417:191-195, 1980.

[78] DC Read, AJ Minchener. The low temperature combustion characteristics of various coals in a fluidized bed combustor. Proceedings of 3rd International FBC Conference Fluidized Combustion: Is it Achieving Its Promise? London, 1984, Vol. 1, DISC/26/224-231.

[79] RR Rajan, CY Wen. A comprehensive model for fluidized bed coal combustors. AIChE Journal 4:642-655, 1980.

[80] S Oka, B Arsić, D Dakić, M Urošević. Experience and problems of the lignite fluidized bed combustion. Proceedings of 3rd European Coal Utilization Conference, Amsterdam, 1983, Vol. 2, pp. 87-97.

[81] S Oka, B Arsić. Biomass combustion in fluidized bed (in Russian). In: SS Kutateladze, S Oka, eds. Transport Processes in High Temperature and Chemically Reacting Flows. Novosibirsk: Siberian Branch of the U.S.S.R. Academi of Sciences, 1982, pp. 40-49.

[82] S. Joksimović Tjapkin. Combustion Processes (in Serbian). Belgrade: Faculty for Chemical Engineering and Metallurgy, 1981.

[83] AM Manaker. Status of utility fluidized bed commercial development in the United States. Presented at ASME/IEEE Power Generation Conference, Milwaukee, 1985.

[84] S Oka. Fluidized bed combustion, a new technology for coal and other solid fuel combustion (in Serbian). In: Energy and Development, Belgrade: Society Nikola Tesla, 1986, pp.147-156.

[85] S Oka, B Grubor. State-of-the-art in circulating fluidized bed combustion (Monograph in Serbian). Report of the Institute for Nuclear Sciences Boris Kidrič, Vinča, Belgrade, IBK-ITE-645, 1987.

[86] U Renz. State-of-the-art in fluidized bed coal combustion in Germany. Yugo-
      slav-German Colloquium Low-Pollution and Efficient Combustion of Low-Grade
      Coals, Sarajevo (Yugoslavia), 1986. Sarajevo: Academy of Science of Bosnia and
      Hercegovina, Special Publications, Vol. LXXXI, Dept. of Techn. Sci., Vol. 15, 1987,
      pp. 185-214.

[87] RP Krishnan, CS Daw, JE Jones. A review of fluidized bed combustion technology
      in the United States. Proceeding of 16th ICHMT Symposium on Heat and Mass
      Transfer in Fixed and Fluidized Beds, Dubrovnik (Yugoslavia), 1984. In: WPM. van
      Swaaij, NH Afgan, eds. Heat and Mass Transfer in Fixed and Fluidized Beds. New
      York: Hemisphere Publ. Co. 1985, pp. 433-456.

[88] B Grubor, D Dakić, S Oka. Analysis of operation of retrofitted 9.3 MW$_{th}$ bubbling
      FBC boiler (in Serbian). Report of the Institute of Nuclear Sciences Boris Kidrič,
      Vinča, Belgrade, IBK-ITE-639, 1987.

[89] JG Yates, M MacGillivary, DJ Cheesman. Coal devolatilization in FBC. Chem. Eng.
      Sci. 11:2360-2361, 1980.

[90] M Radovanović, W Prins, RW Siemons. Characteristics of lignite combustion in
      fluidized bed (in Serbian). Proceedings of Symposium on Use of Lignites in
      Thermoenergetics. Belgrade: Mechanical Engineering Faculty, 1984, Vol. 2, pp.
      93-102.

[91] G Jovanović. Gas flow in fluidized beds of large particles. Experiment and Theory,
      PhD dissertation, Oregon State University, 1979.

[92] G Jovanović, NM Čatipović, TJ Fitzgerald, O Levenspiel. In: JR Grace, JM Matsen,
      eds. Fluidization. New York: Plenum Press, 1980, pp. 225-237.

[93] S Oka, B Grubor, B Arsić, D Dakić. The methodology for the investigation of fuel
      suitability for FBC and the results of comparative study of different coals. Proceed-
      ings of 4th International FBC Conference, London, 1988, I/8/1-19.

[94] S Oka, D Dakić, B Grubor, M Trifunović, B Arsić. Comparative analysis of
      fluidized bed combustion of different coals (in Serbian). Proceedings of 2nd Yugo-
      slav Congress of Chemical Eng. Dubrovnik (Yugoslavia), 1987, Vol. II, pp.
      355-358.

[95] D Dakić. Influence of solid fuel characteristics and thermodinamic parameters on
      combustion in fluidized bed (in Serbian). MSc thesis, Mechanical Engineering Fac-
      ulty, University of Belgrade, 1990.

[96] D Park, O Levenspiel, TJ Fitzgerald. A model for large scale atmospheric fluidized
      bed combustors. Presented at the 72nd Annual AIChE Meeting, San Francisco, 1979.

[97] D Park, O Levenspiel, TJ Fitzgerlad. A comparison of the plume model with cur-
      rently used models for atmospheric FBC. Chem. Eng. Sci. 1/2:295-301, 1980.

[98] D Park, O Levenspiel, TJ Fitzgerald. Plume model for large particle FBC. Fuel
      4:295-306, 1981.

[99] RJ Bywater. The effect of devolatilization kinetics on the injector region of the FBC.
      Proceedings of 6th International Conference on FBC, Atlanta, 1980, Vol. 3, pp.
      1092-1101.

[100] F Preto. Studies and modeling of atmospheric FBC of coal. PhD dissertation.
      Queen's University, Kingston (Canada), 1986.

[101] G Brem. A mathematical model as a tool for simulation and optimization of coal combustion in an AFBC. Proceedings of 4th International FBC Conferenc Fluidized Combustion in Practice: Clean, Versatile, Economic? London, 1988, pp. II/11 – 1-14.

[102] CM van den Bleek, et. al. Documentation of the IEA-AFBC Model: a) Version 1.0, April 1989, b) Version 1.1, Apeldoorn (the Netherlands): TNO, 1990.

[103] B Grubor, S Oka. Mathematical model for calculation processes in the furnace of a FBC Boiler (in Serbian). Termotekhnika 4:209-220, 1990.

[104] G Brem. Mathematical modeling of coal conversion processes, PhD dissertation, University of Twente, Enshede (the Netherlands), 1990.

[105] B Grubor. Investigation and mathematical modeling of coal combustion in bubbling fluidized bed (in Serbian). PhD dissertation, Mechanical Engineering Faculty, University of Belgrade, 1992.

[106] JB Grace. Fluidized bed hydrodynamics. In: Handbook of Multiphase Systems. Toronto: Hemisphere Publ. Co., 1982.

[107] JM Beer, RH Essenhigh. Control reaction rate in dust flames. Nature Vol. 187, 1106-1107, 1960.

[108] RH Essenhigh. The influence of coal rank on the burning times of single captive particles. J. of the Eng. for Power 7:183-190, 1963.

[109] IW Smith. The combustion rates of coal chars: A review. Proceeding of 19th Symposium /Int./ on Combustion 1982. Pittsburgh: The Combustion Institute, 1983, pp. 1045-1065.

[110] MM Avedesian, JF Davidson. Combustion of carbon particles in a fluidized bed. Trans. Inst. Chem. Eng. 2:121-131, 1973.

[111] EW Thiele. Relation between catalytic activity and size of particle. Ind. Eng. Chem. 916-920, 1939.

[112] W Prins, WPM van Swaaij. Fluidized bed combustion of a single carbon particle. In: M Radovanović, ed. Fluidized Bed Combustion. New York: Hemisphere Publ. Co., 1986, pp. 165-184.

[113] IW Smith. The intrinsic reactivity of carbon to oxygen. Fuel 7:409-414, 1978.

[114] P Basu, J Brouhgton, DE Elliott. Combustion of single coal particle in fluidized beds. Institute for Fuel, Symposium series, No. 1, Fluidized Combustion, 1975, A3.1-A3.10.

[115] J Broughton, JR Howard. Combustion of coal in fluidized beds. In: J R Howard, ed. Fluidized Beds. London: Applied Sciences Publishers, 1983, pp. 37-76.

[116] IB Ross, JF Davidson. The combustion of carbon particles in a fluidized bed. Trans. Inst. Chem. Eng. 108-114, 1981.

[117] AL Gordon, HS Caram, NR Amundson. Modeling of fluidized bed reactors-V: Combustion of carbon particles – An extension. Chem. Eng. Sci. 6:713-732, 1978.

[118] HS Caram, NR Amundsen. Diffusion and reaction in a stagnant boundary layer about a carbon particle. Ind. Eng. Chem. Fund. 2:171-181, 1977.

[119]   EK Campbel, JF Davidson. The combustion of coal in fluidized beds. Institute for
        Fuel, Symposium series No.1, Fluidized Combustion, 1975, A2-1-A2.9.

[120]   HA Becker, JM Beer, BM Gibbs. A model for fluidized-bed combustion of coal. In-
        stitute for Fuel, Symposium series No.1, Fluidized Combustion, 1975,
        A1-1-A.1.10.

[121]   AL Gordon, NR Amundson. Modeling of fluidized bed reactors-IV: Combustion of
        carbon particles. Chem. Eng. Sci. 12:1163-1178, 1976.

[122]   M Horrio, CY Wen. Simulation of fluidized bed combustors: Part 1. Combustion ef-
        ficiency and temperature profile. AIChE Symposium Series 176:101-111, 1978.

[123]   R Rajan, R Krishnan, CY Wen. Simulation of fluidized bed combustors: Part II. Coal
        devolatilization and sulfur oxides retention. AIChE Symposium Series
        176:112-119, 1978.

[124]   TP Chen, SC Saxena. A mechanistic model applicable to coal combustion in
        fluidized beds. AIChE Symposium Series 176:149-161, 1978.

[125]   SC Saxena, A Rehmat. A mathematical model for char combustion in a fluidized
        bed. Proceedings of 6th International Conference on FBC, Atlanta, 1980, Vol. III, pp.
        1138-1149.

[126]   AF Sarofim, JM Beer. Modeling of FBC. Proceedings of 17th Symposium /Int./, on
        Combustion, 1978. Pittsburgh: The Combustion Institute, 1979, pp. 189-204.

[127]   LT Fan, K Tojo, CC Chang. Modeling of shallow fluidized bed combustion of coal
        particles. Ind. Eng. Chem. Process Design and Development 2:333-337, 1979.

[128]   K Tojo, CC Chang, LT Fan. Modeling of dynamics and steady-state shallow FBC
        combustors. Effects of feeder distribution. Ind. Eng. Chem. Process Design and De-
        velopment 3:411-416, 1981.

[129]   PM Walsh, A Dutta, JM Beer. Effect of hydrocarbons, char and bed solids on oxida-
        tion of carbon monoxide in the freeboard of a fluidized bed coal combustor. Pro-
        ceedings of 20th Symposium /Int./ on Combustion,1984. Pittsburgh: The
        Combustion Institute, 1985, pp.1487-1493.

[130]   G Donsi, M El-Sawi, B Formisani. On the simulation of fluidized bed coal combus-
        tion. Combustion and flame Vol. 64, 1:33-41, 1986.

[131]   E Turnbull, ER Kossakowski, JF Davidson, RB Hopes, HW Blackshow, PTY Good-
        year. Effect of pressure on combustion of char in fluidized beds. Chem. Eng. Re-
        search & Development 7:223-233, 1984.

[132]   I Boriani, A De Marco, W Prandoni, C Grazzini, M Miccio. A system model for the
        simulation of a fluidized bed unit. Report ENEL, N.341.260-2, Rapp. 047/89, Pres-
        ented at the 9th IEA-AFBC Technical Meeting, Göteborg (Sweden), 1989.

[133]   G Brem, JJH Brouners. A complete analytical model for the combustion of a single
        porous char particle incorporated in an AFBC system model. Proceedings of
        International Conference in FBC, San Francisco, 1989, Vol.1, pp.37-46.

[134]   RK Chakraborty, JR Howard. Combustion of char in shallow fluidized bed
        combustors: Influence of some design and operating parameters. J. Inst. Energy
        418:48-58 and 55-58, 1981.

[135]   JF Davidson, D Harrisson. Fluidized Particles, Cambridge: Cambridge University Press, 1963.

[136]   P Basu. Burning rate of carbon in fluidized beds, Fuel 10, 1977.

[137]   RK Chakraborty, JR Howard. Burning rates and temperatures of carbon particles in a shallow FBC. J. Inst. Fuel 12:220-224, 1978.

[138]   AI Tamarin, DM Galerstein, VM Shuklina, SS Zabrodsky. Investigation of convective heat transfer between burning particle and bubbling fluidized bed (in Russian), Proceedings of 6th All-Union Heat-Mass Transfer Conference, Minsk, 1980, Vol. VI, Part 1, pp .44-49.

[139]   H Kono. Direct measurements of effective combustion rate constants of char particles in fluidized bed by step response curve. Presented at AIChE 73rd Annual Meeting, Chicago, 1980.

[140]   AP Baskakov, AA Ashikhmin, VA Munz. Investigation of combustion kinetics of Irscha-Borodinsky coal in low temperature bubbling fluidized bed (in Russian). Inzhenerno-fizicheskij zhurnal 1:123-126, 1983.

[141]   P Basu. Measurement of burning rates of carbon particles in a turbulent fluidized bed. Proceedings of 2nd International FBC Conference Fluidized Combustion: Is it achieving its promise? London, 1984, DISC/3/17-DISC/3/25.

[142]   S Oka, B Arsić. Comparative analysis of combustion kinetics of different coals in fluidized bed (in Serbian). Reports of the Institute of Nuclear Sciences Boris Kidrič, Vinča, Belgrade, IBK-ITE-479 and 505, 1985.

[143]   RD La Nauze, K Jung. Further studies of combustion kinetics in fluidized beds. Proceedings of 8th Australian Fluid. Mechanical Conference. Newcastle (Australia): University of Newcastle, 1983, pp. 5c.1-5c.4.

[144]   P Basu, ChD Ritchie. Burning rate of suncor coke in a CFB. Proceedings of 8th International Conference of FBC, Houston, 1985, Vol.1, pp. 28-31.

[145]   P Basu, D Subbarao. An experimental investigation of burning rate and mass transfer in a turbulent fluidized bed. Combustion and Flame Vol. 66, 3:261-269, 1986.

[146]   DF Durao, P Ferao, I Gulyurthu, MV Heitor. Combustion kinetics in shallow fluidized beds. Proceeding of International Conference on FBC, London, 1986, pp. 6.3/1-8

[147]   DFG Durao, P Ferrao, I Gulyurtlu, MV Heitor. Combustion kinetics of high-ash coals in fluidized beds. Combustion and Flame Vol. 79, 2:162-174, 1990.

[148]   S Oka, M Radovanović. Combustion kinetics of different high volatile coals in fluidized bed (in Serbian). Proceedings of 2nd Yugoslav Congress for Chemical Engineering, Dubrovnik (Yugoslavia), 1987, Vol. II, pp. 358-361.

[149]   RL Patel, N Nsakala ya, BV Dixit, JL Hodges. Reactivity characterization of solid fuels in an atmospheric bench-scale FBC. Presented at the Joint Power Gen. Conference, Philadelphia, 1988.

[150]   RL Patel, N Nsakala ya, DG Turek, DR Raymond. In-bed and freeboard combustion kinetic behavior of solid fuels in atmospheric bench-scale reactors. Presented at Power Gen. '89, New Orleans, 1989.

[151]   AB van Engelen, G van den Honing. Char characterization by combustion experiments in a small scale fluidized bed interpreted by an analytical single porous particle conversion model. Presented at the IEA-AFBC Math. Modeling Meeting, Lisbon, 1990.

[152]   S Oka, M Ilić, I Milosavljević. Kinetics of coal combustion in fluidized bed at different temperatures (in Serbian). Proceedings of 8th Yugoslav Symposium of Thermal Engineers YUTERM 90, Neum (Yugoslavia), 1990, pp. 409-416.

[153]   M Ilić, S Oka, I Milosavljević, B Arsić. Influence of volatile matter content of coal reactivity (in Serbian). Report of the Institute of Nuclear Sciences Boris Kidrič, Vinča, Belgrade, 1990, IBK-ITE-800.

[154]   I Milosavljević, M Ilić, S Oka. Experimental investigation of coal combustion kinetics in fluidized bed (in Serbian). Termotekhnika 3-4:151-165, 1990.

[155]   M Ilić, I Milosavljević, S Oka. Gas concentration measurements during transient combustion processes in fluidized bed (in Serbian). Termotekhnika 3-4:165-176, 1990.

[156]   JB Howard, GC Williams, DH Fine. Kinetics of carbon monoxide oxidation in postflame gases. Proceedings of 14th Symposium /Int./ on Combustion, 1972. Pittsburgh: Institute of Combustion, 1973, pp. 975-986.

[157]   M Ilić, S Oka, M Radovanović. Experimental investigation of char combustion kinetics – CO/CO₂ Ratio During Combustion. Proceedings of 10th Heat Transfer Conference, Brighton (G.B.), 1994, Vol. 2, pp. 81-86.

[158]   A Machek. Optical measurements of char particle temperatures and radiant emissivities in FBC. Proceedings of 3rd International Fluidization Conference Fluidized Combustion: Has it achieved its promise?, London, 1984, DISC/18/151-157.

[159]   AI Tamarin, DM Galerstein, VA Borodulya, LI Levental. Calculation of coal combustion temperature and combustion rate in shallow bubbling fluidized bed (in Russian). In: Heat and Mass Transfer Problems in Modern Solid Fuel Combustion and Gasification Technologies. Minsk: Academy of Sciences of Belarus, 1984, Vol. 1, pp. 40-49.

[160]   AI Tamarin, DM Galerstein, VM Shuklina. Investigation of heat transfer and coke particle temperature during combustion in bubbling fluidized bed (in Russian). Inzhenerno-fizicheskij zhurnal 1:21-27, 1982.

[161]   Z Kostić. Biomass combustion in fluidized bed (in Macedonian). PhD dissertation, Mechanical Engineering Faculty, University of Skopje (Yugoslavia), 1987.

[162]   D Kunii, O Lewenspiel. Fluidization Engineering. New York: R. E. Krieger Publ. Co., 1977.

[163]   S Yagi, D Kunii. Studies on effective thermal conductivities in packed beds. AIChE Journal 3:373-381, 1957.

[164]   BM Gibbs. A mechanistic model for predicting the performance of a fluidized bed coal combustor. Proceedings of 1st International Conference on FBC, Institute for Fuel, Symposium series No. 1, Fluidized Combustion, 1975, A1.1-A1.10.

[165]   SC Saxena, et al. Modeling of a fluidized bed combustor with immersed tubes. University of Illinois, Chicago, Report FE-1787-6, 1976.

[166]   M Horrio, S Mori, I Muchi. A model study for the development of low NO_x FBC. Proceedings of 5th International Conference FBC, Washington, 1977, Vol. II, pp. 605-623.

[167]   JP Congalidis, C Georgakis. Multiplicity patterns in atmospheric fluidized bed coal combustors. Chem. Eng. Sci. 8:1529-1546, 1981.

[168]   M Ilić. Investigation of fluidized bed coal combustion kinetics (in Serbian). MSc thesis, Mechanical Engineering Faculty, University of Belgrade, 1992.

[169]   M Radovanović. Combustion in fluidized beds. In: M Radovanović, ed. Fluidized Bed Combustion. New York: Hemisphere Publ. Co., 1986, pp. 127-165.

5.

FLUIDIZED BED COMBUSTION
IN PRACTICE

5.1.   Purpose and basic concept of devices for solid
fuel combustion in fluidized bed

The concept, basic parameters and dimensions of any power-generating device, depends on the nature of the fuel and the purpose of the plant. In industrial and commercial settings, energy is consumed either in the form of electric power or heat (super-heated steam, saturated steam, hot water, hot combustion products and hot air). In all of these situations, combustion in a bubbling fluidized bed has become competitive with conventional combustion technologies used in solid, liquid and gas fuel combustion, as well as waste fuel and biomass consumption. To respond to the demands of power consumers, numerous companies, worldwide, have developed and marketed FBC boilers. Those boilers can be divided into the following categories: furnaces for production of hot gases and hot air, industrial boilers producing steam or hot water for the process industries or district heating or for electricity production, and boilers for electric power generation at the utility scale. The characteristics, concept and the parameters for each of these categories are discussed below.

5.1.1.   Furnaces for production of hot gases or hot air

(a) Purpose and applications

In the process industries and agriculture, hot gases are needed in various processes – roasting, heating and most frequently for drying. Such processes can often be performed directly with combustion products the temperature of which is most commonly between 800 and 1000 °C. In agriculture, however, the requirement is typically for clean air at temperatures ranging from 40 to 150 °C, depending on the sensitivity of the product that is being dried. Drying grass serves as an exception to that rule, since here it is possible to use combustion products directly with temperatures ranging from 800 to 1000 °C.

The most commonly used fuels for these activities are liquid or gas fuels. However, various energy crises, and a rise of liquid fuel prices, together with difficulties with foreign currency, have encouraged many countries to economize by reducing the use of imported fuel. Such a need to rely on domestic power sources, primarily coal and biomass, have resulted in the development of fluidized bed combustion in many countries. Other technologies for solid fuel combustion demand high-quality coal, and pretreatment of fuel, and as such they are not appropriate for small power generation units. Fluidized bed combustion boilers meet the requirements for low air emissions and also allow the automation of plant operation.

The basic advantages for these types of applications are as follows: simplicity of operation, the possibility of combustion of various low-quality coals without special preliminary treatment, and the combustion of wastes, and in particular biomass. The disadvantages include restricted load following and change of exit gas temperature, the potential for high unburned char in the solid residue streams, significant ash and sand elutriation due to high gas velocities and a requirement for a relatively large amount of excess air.

(b) Basic parameters and requirements

Furnaces used in agriculture are typically rather small (1-5 MW_{th}), while in the process industries requirements may go up to 15 MW_{th} and above.

For drying, clean air at about 150 °C is required, and in drying highly sensitive plant products and seeds, even lower temperatures are required, typically about 40 °C. Heat transfer between combustion products and air is a difficult engineering problem, since heat exchanger surfaces are usually very large due to the low heat transfer coefficients of gases. An interesting and efficient solution is to only heat a portion of air by heat exchange surfaces immersed into fluidized bed, as in the case of furnaces constructed in the factory CER, Čačak, Yugoslavia.

When combustion products are used directly, temperatures can ranging from 800 to 900 °C, and in the case of coals with high ash melting points sometimes even as high 1000 °C. In such cases the FBC furances must use high temperature resistent materials.

Furnaces are usually lined by firebricks and have simple auxiliary systems – overbed fuel feeding is used, and there are no systems for draining bed material. Simplified construction is possible, because highly efficient combustion is not required. Typically, the operation of such a furnace is seasonal, with numerous short interruptions for equipment control and drainage of excess bed material. Due to the high excess air requirement and the low quality and therefore cheap fuels used for these applications, the specific power of these furnaces is typically 0.5 to 1 MW_{th}/m^2, per unit cross section area.

Inert bed material particles are of average size from 0.5 to 1 mm, and for large units and higher fluidization velocities, may adopt sizes up to 1 to 2 mm. Excess air can range from 1.5 to 2.5 depending on the required temperature and whether there are heat transfer surfaces immersed in the bed. The fluidization velocity may reach as high as 4 m/s, i. e. the fluidization number is 6-7, while bed height are typically 300 to 1000 mm. Combustion efficiency tends to range from 80 to 95%, and for particularly well designed boilers, they may be even 98-99% [1, 2].

(c) Basic problems

Certain basic problems have already been mentioned: limited load following range, high particle elutriation rate – i. e. overload of the flue gas cleaning system, a large surface of gas to gas heat exchanges. For these systems, large capacity high-pressure head primary air fans are required. Change in the type of fuel requires significant variation in the quantity of combustion air, which are not easy to meet given the required power and performance of the primary air fan.

(d) Some construction solutions

Early development, manufacturing and exploitation of FBC furnaces saw perhaps the most dramatic progress in Great Britain and Yugoslavia. Development and construction of simple bubbling FBC furnaces began in the seventies in Great Britain under the direction of the NCB (National Coal Board) in cooperation with a few small companies such as G. P. Warsley Ltd., Energy Equipment Ltd., and Encomech Engineering Services Ltd. [2-4]. In Yugoslavia, the firm CER, Čačak, began developing FBC furnaces in cooperation with the Institute for Thermal Engineering and Energy, VINČA Institute of Nuclear Sciences, Belgrade. The first such 4.5 MW_{th} furnace started operation in 1982, and since

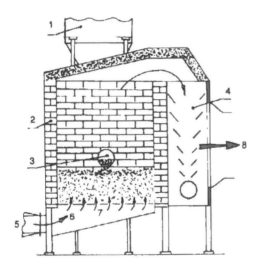

Figure 5.1.
Bubbling fluidized bed combustion hot gas generator designed by the factory CER (Čačak, Yugoslavia):
1 – coal feed hopper, 2 – refractory lined walls, 3 – coal screw feeder, 4 – particle separator, 5 – combustion air inlet, 6 – air chamber, 7 – air distribution plate, 8 – exit of the flue gases

then this company has over 100 such furnaces in operation, which range from 1 to 10 MW$_{th}$ in size, and have found applications in agriculture and the process industries [1, 5, 6]. According to 1990 data [7], about 10 to 15 companies in the world offered FBC furnaces, and only 3 to 5 companies offered hot air generators of this kind. Currently, this number is if anything lower, given the significant consolidation that has occured among manufactures of FBC equipment.

Figure 5.1 shows a simple industrial furnace for generation of hot gases of temperatures up to 800 °C made by CER, Čačak [1, 6].

The furnace is completely walled by refractory bricks, its power ranges from 0.5 to 1 MW$_{th}$, bed height is 300-500 mm, and it has overbed fuel feeding. Typically fuels include: lignite, brown coal, or anthracite, and particle sizes are in the range of 0-25 mm depending on the type of coal. There are also FBC solutions for biomass combustion for very difficult fuels such as corncobs.

Figure 5.2 shows a diagram of a FBC furnace for hot air generation with temperatures up to 150 °C manufactured by CER, Čačak [1, 6]. Furnaces are manufactured with unit power ranging from 1.5 to 4.5 MW$_{th}$. Both inbed and overbed fuel feeding is possible, depending on the coal type. Bed height is 300-500 mm, and the system uses immersed stainless steel heat exchangers as well as heat transfer surfaces in the furnace walls. Fuel and coals of various quality and particle size from 0 to 25 mm are used in this system.

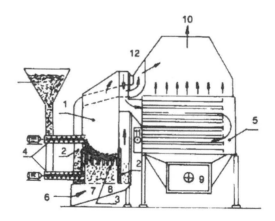

Figure 5.2.
Bubbling fluidized bed combustion hot air generator designed by the factory CER (Čačak, Yugoslavia):
1 – FBC furnace, 2 – side air fluidized beds for furnace cooling, 3 – heat exchanger immersed in fluidized bed, 4 – coal screw feeder, 5 – air heater, 6 – combustion air inlet, 7 – air chamber, 8 – air distribution plate, 9 – air inlet of the air heater, 10 – air exit from air heater, 11 – exit of the flue gases from air heater, 12 – exit of the air from immersed heat exchanger

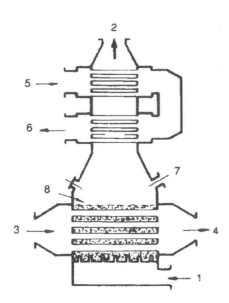

Figure 5.3.
Bubbling fluidized bed combustion hot air generator designed by Encomech Engineering [4] (Reproduced by kind permission of the author Prof. Dr. J. R. Howard from [2]):
1 – combustion air inlet, 2 – exit of the flue gases, 3 – inlet of the cold process air, 4 – exit of the hot process air, 5 – inlet of the cold process air, 6 – exit of the hot process air, 7 – opening for coal feeding, 8 – fluidized bed free surface

An interesting "construction solution" is an air heater manufactured by Encomech Engineering Services Ltd. shown in Fig. 5.3. Immersed stainless steel or ceramic pipes are used for heating air. The unit power range is from 1 to 7 MW_{ht}, and air temperatures are up to 600 °C.

5.1.2. Industrial boilers with bubbling fluidized bed combustion

(a) Purpose and applications

Industrial boilers supply process industries with heat energy in the form of superheated or saturated steam and hot water. Boilers for district heating which produce superheated steam or hot water can be included in this category. In certain cases, either occasionally or permanently, industrial boilers use surplus steam for electric power generation needed by the industries themselves or for sale to the central electric power system. Industrial boilers here, are of small or medium power up to 100 MW_{th}.

Given such purposes, one of the major requirements for this type of boilers is high reliability. An alternative, which is usually too expensive, is to have several boilers in reserve. The majority of industrial boilers use liquid or gas fuel. If they use coal, the coal choice and quality are market driven. Alternatively, it may be necessary that industrial boilers use a local fuel, e. g. coal from a local mine, waste coal left over from a previous or existing coal cleaning facility or processed coal, or waste fuel – industrial and city wastes or biomass. Industrial boilers typically have stringent requirements with respect to SO_2 and NO_x emissions due to their location near cities. This is especially the case for district heating boilers.

All of these operational requirements mesh with the characterstics and capabilities of FBC boilers, and in particular bubbling fluidized beds that are more appropriate for non-utility applications. Due to a large number of these boilers, industrial boilers represent both a large market and one with a very wide range of requirements, so there is a wide range of market niches for a large number of manufacturers. According to the data from 1990 [7] there were 67 companies worldwide that offered fluidized bed boilers. From a short review of development of FBC boilers and the more recent state of development of this technology given in Sections 1.4 and 1.6, it can be seen that boilers with bubbling fluidized bed combustion were the first to penetrate the market for industrial boilers. Given the large number of these boilers in operation it is practically impossible to provide a comprehensive review of their number and characteristics, although a number of authors have tried to make such an attempt [2, 7-18].

The characteristics of fluidized bed boilers, which permit their efficient exploitation as industrial boilers are as follows:

- fuel flexibility – simultaneously or alternative different fuels can be burnt,
- low quality fuels may be used with minimal previous fuel preparation, or waste fuel and biomass may be used, and
- low SO_2 and NO_x emissions.

As noted in the first chapter of this book, in Sections 1.4, 1.6, and 1.7, bubbling fluidized bed boilers are particularly appropriate for applications demanding small and medium unit capacity of up to 100 MW_{th}, because they are more economical in this range than conventional combustion technologies. It is therefore clear why bubbling FBC boilers first became competitive with the other combustion technologies, given the numerous advantages listed above.

This rapid breakthrough of bubbling FBC boilers into the industrial boiler market was supported by the following considerations:

- there are often boilers kept in reserve for this type of application, so investors did not find it as hard to comit to the purchase of new, still insufficiently tested technology,
- there are also a large number of old boilers with obsolete construction and characteristics so the turnover in this market is large, and
- bubbling fluidized bed combustion technology can be retrofitted thus permitting the revitalization and reconstruction of old boilers with a new, modern technology. It is also possible to retrofit liquid fuel boilers for coal or other solid fuel combustion using fluidized bed, without major problems.

In assessing the requirements and possibilities for the application of bubbling fluidized bed combustion boilers, the decisive element is the local energy situation. The energy situation typical for undeveloped countries has most frequently the following features:

- coal is the basic domestic primary energy source,
- the largest reserves are typically low quality coal or lignite, with high moisture contents and low ash melting points,
- in large systems for electric power generation, the share of coal is high,
- coal is primarily used in large facilities of unit power ≥ 200 MW$_e$,
- liquid fuels are mostly used for power generation in industry,
- in practice there are often only very limited environmental regulations, for both large and small power generation plants,
- most power generation in eastern Europe, and many other countries does not use desulphurization systems, or equipment for controlling NO$_x$ emission,
- a large number of small mines are still exploited, although they may have no market for the unwashed coal, and
- a large number of units (both "large" and "small" power generation plants) are at the end of their operational life and are facing the need for reconstruction and revitalization.

Expansion of the market for the bubbling FBC boilers in this situation can be expected in the power generation in industry, because they can best satisfy the following requirements:

- replacement of imported liquid fuels with domestic coals,
- reduction of SO$_2$ and NO$_x$ emission without requiring large investments for flue gas cleaning,
- flexibility in terms of the quality and type of fuel employed,
- efficient combustion of low-quality "as-mined" coals, biomass, industrial and city waste,
- full plant automation,
- increase of combustion efficiency, and
- the possibility of reconstruction of old plants with simultaneous increases of combustion efficiency, reduction of SO$_2$ and NO$_x$ emissions, and the possibility of low-quality fuel use.

(b) Basic parameters and requirements

These boilers have a low power range of between 5 and 15 MW$_{th}$, and a medium power range of up to 100 MW$_{th}$. Steam parameters, depending on the boiler purpose, are close to the highest parameters for electric power generation boilers. Steam pressures range from 10 to 70 bar, temperatures are most frequently between 170 and 510 °C and steam capacity is between 2 and 160 t/h.

These boilers are expected to be highly efficient and used "year round." Where they use waste fuels, requirements concerning combustion efficiency are, however, likely to be somewhat less rigid.

Nevertheless, in practice such industrial boilers with bubbling fluidized bed typically reach combustion efficiency of up to 98-99%. Numerous operational data, presented at international conferences, can verify these facts [11, 19-21].

Industrial boilers often require use of fuels with a very broad variety of characteristics – from coal (lignite) to biomass and waste fuel, as well as a considerable range of load change, even up to 1:5 in heating systems. SO_2 and NO_x emission must also be within permissible limits for the country concerned.

The basic parameters for these boilers are similar to those for large power boilers: bed height from 0.5 to 1 m, fluidization velocity from 1 to 2 m/s, fluidization number about 4, inert bed material particle size from 0.5 to 1 mm and combustion (bed) temperature between 800 to 850 °C. The specific power is 1 to 2 MW_{th}/m^2 depending on fuel type, fuel particle dimensions are from 0 to 50 mm, and excess air is around 1.2.

(c) Basic problems in design

The majority of problems in designing and constructing industrial boilers with fluidized bed combustion are the same as for FBC power boilers in general. These problems, which will be discussed in detail later, are listed below:

– stable fuel feeding can be problematic, because they often burn low quality fuels with little or no pretreatment, or they burn waste or unusual fuels,
– these units have a demand for high combustion efficiency,
– it is necessary to ensure a wide range for load turndown,
– it is necesaary to achieve a reduction of SO_2 and NO_x emission to some permitted limits, and
– it may be necessary to incorporate heat transfer surfaces immersed into fluidized bed.

(d) Basic designs of industrial bubbling FBC boilers

Depending on the power, purpose of the boiler and fuel type, there are a number of different conceptions and design solutions for industrial fluidized bed boilers. The very first boilers of this type were developed around the concept of the water shell boilers, also since many old boilers burning coal or liquid fuels had to be reconstructed, this increased the number of different concepts and design of the boiler itself as well as the auxiliary boiler systems. The very fact that there were 67 companies worldwide in 1990 manufacturing FBC boilers, indicates that specific solutions for their construction were heterogeneous, although it should be noted that in the last 10 years or so there has been a rapid consolidation of the number of companies, and the number offering bubbling bed designs internationally may now be less than 10 in the western world.

FBC boilers require: high combustion efficiency, a high overall boiler efficiency, the possibility of operating with a range of fuels, continuous and automatic load following, and maintenance with as little physical labor as possible (similar to boilers burning liquid and gas fuels), and they should include all necessary auxiliary systems.

Figure 5.4 shows an FBC boiler with all the appropriate auxiliary systems. The example chosen is a FBC boiler of 16 MW$_{th}$ capacity manufactured by Generator, Göteborg, Sweden, made in 1982. This boiler was fed with coal with a 0 to 30 mm size range, with steam capacity of 20 t/h, and is still used for heating the University in Göteborg.

Although boiler auxiliary systems will be discussed in some detail in the next section, all systems are listed here, following Fig. 5.4:

- system for storage and delivery of fuel (1, 2, 3, 6),
- system for storage and delivery of limestone (1, 2, 3, 4),

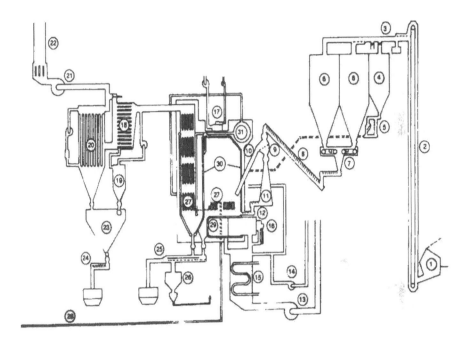

Figure 5.4. *Bubbling fluidized bed steam boiler, 20 t/h, at Chalmers University in Göteborg, Sweden, designed by Generator, Göteborg:*
1 – receiving hopper, 2 – elevator, 3 – vibrating mechanical transporter, 4 – hopper for limestone and sand, 5 – feeder for limestone and sand, 6 – coal hoppers, 7 – metering belt transporter, 8 – screw feeder, 9 – mechanical sieve, 10 – coal feeding on the bed surface, 11 – fine coal hopper, 12 – coal feeding under the bed surface, 13 – primary air fan, 14 – secondary air fan, 15 – air preheater, 16 – liquid fuel starting burner, 17 – auxiliary burner for liquid fuel, 18 – multicyclones, 19 – secondary cyclone, 20 – bag filters, 21 – forced draught fan, 22 – chimney, 23 – fly ash hopper, 24 – ash wet discharge, 25 – system for draining of the inert bed material, 26 – system for recirculation of the fly ash, 27 – superheaters, 28 – high pressure steam, 29 – water cooled distribution plate, 30 – water tube panel walls of the furnace, 31 – steam drum

- system for feeding fuel (7, 8, 9, 10), over the bed surface or under the bed surface (11, 12),
- system for feeding limestone into the bed (5),
- system for boiler start-up (16, 17),
- system for feeding and distribution of primary air (13, 15),
- system for feeding secondary air (14),
- system for storage and feeding sand into the bed (4, 5),
- system for draining material from the bed, maintenance of bed height, separation of ash and slag, and return of inert material into the furnace (25, 26),
- system for flue gas cleaning (cyclones and filters) (18, 19, 20),
- system for fly ash recirculation (26),
- water circulation system (27, 28),
- system for conveying and removal of ash and slag (23, 24, 25, 26), and
- stack (21, 22).

In reviewing the basic designs, we will begin with the solutions that have historically first developed from reconstructed fire-tube boilers.

Figure 5.5 presents a horizontal water-shell boiler developed in Great Britain in the beginning of the seventies by reconstruction of a fire-tube boiler burning liquid fuel [2-4]. A plate for primary air distribution was incorporated into the fire tube. Washed, high-quality coal with particle size 10-25 mm was fed above the bed surface. Due to small volume and small height of the space above the bed, it was impossible to use coals with high volatile content and particles below 1 mm. The boiler was manufactured by Vekos Powermaster (G.B.), of power up to 5 MW$_{th}$ or up to 10 MW$_{th}$ if there are two fire tubes. Combustion efficiency was 94-97%. Due to small space, the boiler operated with a shallow bed – 100 mm. Since there is no space for incorporation of heat transfer surfaces immersed into the fluidized bed, the boilers have high excess air and reduced boiler efficiency.

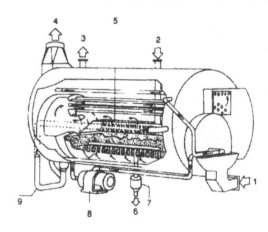

Figure 5.5.
Water shell bubbling fluidized bed combustion boiler, Vecos Powermaster (G.B.) (Reproduced by kind permission of the author Prof. Dr. J. R. Howard from [2]):
1 – fluidized air, 2 – coal feed, 3 – steam outlet, 4 – flue gases, 5 – grid for prevention of the particle elutriation, 6 – bed ash draining, 7 – aerodynamic sieve, 8 – circulating pump, 9 – recirculation of the fly ash

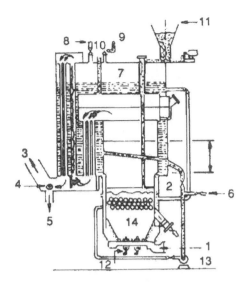

Figure 5.6.
Vertical water shell bubbling fluidized bed combustion boiler with refractory lined furnace, NEI Cochran (G.B.) (Reproduced by kind permission of the author Prof. Dr. J. R. Howard from [2]):
1 primary air inlet, 2 – bed surface, 3 – outlet of the flue gases, 4 – ash removal screw, 5 – ash discharge, 6 – feed-water inlet, 7 – steam drum, 8 – steam outlet, 9 – safety valve, 10 – measurement of the combustion products composition, 11 – coal and limestone hopper, 12 – start-up liquid fuel burner, 13 – circulating pump, 14 – immersed heat exchanger

In order to reduce the disadvantages of horizontal water shell boilers with the fire tubes, primarily in order to provide greater bed height and a greater volume and height for the freeboard, vertical water shell FBC boilers were also developed in Great Britain. Figure 5.6 shows a boiler of this type manufactured by NEI, Cohran (G.B.). The boilers are of 4 MW_{th} with a deep bed, so that it is possible to have an inbed heat transfer surface. They also have a comparatively high freeboard and three passages for the flue gases through the water shell. Furnaces are made with a water shell or covered with firebricks. Feeding of 0-12 mm coal particles takes place either underbed or abovebed surface. Combustion efficiency is 90-95%, with a moderate excess air – 1.25-1.4.

An interesting design of vertical small power, a fully automatizied bubbling FBC boiler, was proposed by Heat and Mass Transfer Institute from Minsk (Belarus), (Fig. 5.7) [59]. Basic boiler parameters are as listed:

- steam capacity 0.5-1.0 t/h,
- thermal power 0.37-0.75 kW,
- steam temperature 120 °C,
- steam pressure 1.2-1.6 bar,
- boiler efficiency of 80%,
- start-up time of 0.5 h, and
- dimensions: height 2.9 m, length 2.13 m, width 1.5 m.

Another interesting FBC boiler design developed with the aim to remove the disadvantages of horizontal water shell boilers, is a locomotive type block boiler made by Johnston Boiler Co. (G.B.), shown according to NCB technology, as presented in Fig. 5.8 [3, 4].

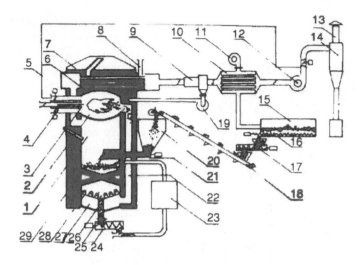

Figure 5.7. *Vertical water shell bubbling fluidized bed combustion low power boiler, designed by Heat and Mass Transfer Institute in Minsk, Belarus:*

1 – furnace, 2, 3, 4 – start-up chamber with liquid fuel burner, 5 – flue gas recirculation, 6 – heat exchanger, 7 – superheater, 8 – steam outlet, 9 – cyclone, 10 – flue gas cooler, 11 – air fan, 12 – forced draught fan, 13 – chimney, 14 – cyclone, 15 – hopper, 16 – screw feeder, 17 – coal crusher, 18 – belt transporter, 19 – fly ash recirculation system, 20 – intermediate hopper, 21 – screw feeder, 22 – primary air inlet, 23 – system for bed material return, 24 – ash removal screw, 25 – bed material discharge tube, 26 – bubble cap, 27 – air chamber, 28 – distribution plate, 29 – immersed heat exchanger

A horizontal placement for water shell boilers was used, and a screened, water-cooled pre-combustion chamber was employed, which was high enough for combustion of small particles and volatiles, with several passages for combustion products through the water shell. Coarse inert material with a maximum size of 3.2 mm was used, in order to permit higher fluidization velocities – up to 1.8 m/s and thus a greater specific power output.

Brown coal and anthracite, lignite, wood waste, waste oils and other fuels can be burned in these types of boilers. The bed temperature is 850 °C, and excess air ranges from 1.2 to 1.25 and the overall system has combustion efficiencies of the order of 97%. The capacity of these boilers can reach up to 15 MW_{th}, steam capacity may range from 15-25 t/h, and steam pressure 10-20 bar. NBC developed an interesting variety of these boilers, in which two horizontal water shell boilers were

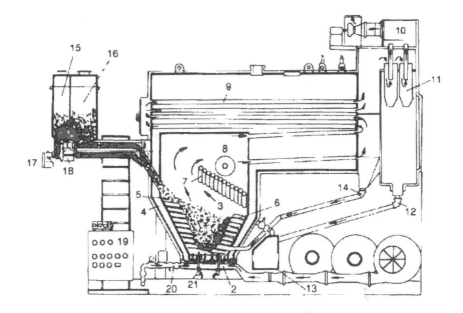

Figure 5.8. *Locomotive-type bubbling fluidized bed combustion steam boiler, Johnston Boiler Co. (G.B.) (Reproduced by kind permission of the author Prof. Dr. J. R. Howard from [2]):*
1 – primary air fan, 2 – air chamber, 3 – fluidized bed, 4 – water shell of the furnace, 5 – pilot burner, 6 – water tubes, 7 – water tubes above the bed, 8 – opening for furnace inspection, 9 – flue gas tubes, 10 – forced draught fan, 11 – particle separator, 12 – ash discharge, 13 – ash hopper, 14 – fly ash recirculation system, 15 – limestone hopper, 16 – coal hopper, 17 – limestone screw feeder, 18 – coal screw feeder, 19 – control box, 20 – flue gas recirculation, 21 – liquid fuel inlet

attached to a single pre-combustion chamber, thus allowing a considerable increase of unit power [3, 4, 20].

Further increase of unit boiler power and steam parameters for FBC boilers could not be achieved by reconstruction of shell boilers. Therefore, a new type of boilers appeared with a completely screened pre-combustion chamber and a sufficiently immersed heat exchanger surface in the bed to attain maximum specific power from 1 to 2 MW_{th}/m^2. A conventional shell boiler served only as the convective part of the boiler. Figure 5.9 shows such boiler design manufactured by Babcock Power Ltd. (G.B.) [3, 4].

Boilers were designed over a power range from 6 to 30 MW_{th} and steam pressure 30 to 40 bar. Excess air was 1.3 and the fluidization velocity was 2.3 m/s.

Use of reconstructed shell boilers was aimed at producing cheaper designs and use of already well developed design solutions. However, these boilers were

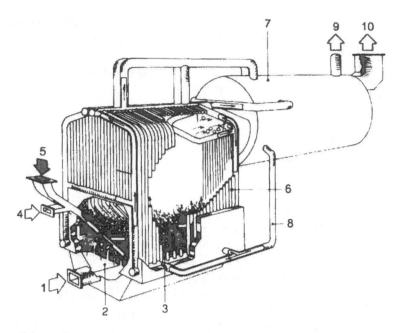

Figure 5.9.   *"Composite" bubbling fluidized bed combustion steam boiler with wa-*
ter-cooled furnace, designed by Babcock Power Ltd. (G.B.) (Reproduced by
kind permission of the author Prof. Dr. J. R. Howard from [2]):
1 – primary air inlet, 2 – air chamber, 3 – fluidized bed, 4 – secondary air in-
let, 5 – coal, 6 – membrane water cooled wall furnace, 7 – steam drum, 8 –
water down flow, 9 – steam outlet, 10 - flue gases outlet

limited to lower unit powers and lower steam pressure and temperature. Higher unit
power and higher steam pressure could be achieved only with water-tube boilers.
The first industrial boiler of this type was designed and made by Stone International Fluidfire Ltd. (G.B.) in 1980, with power ranging from 3 to 20 MW$_{th}$, but
with a steam pressure of 55 bar and steam temperature of 400 °C (Fig. 5.10). An interesting design of the distribution plate ensures better fuel mixing in the bed.

   The first "optimally" designed bubbling fulidized boiler, was probably the
water-tube boiler of Gibson Wells Ltd. 1982. Figure 5.11 [3, 4] shows the schematic diagram for this boiler. The capacity of the first boiler of this type was 22 t/h
of steam at 16.5 bar and 257 °C, and the unit was used for processing steam and
electric energy generation. The boiler had two separate fluidized beds and a multiple gas pass with steam superheaters and an evaporator. Its distribution plate was
water-cooled and was an integral part of the boiler water circulation system. Heat
transfer surfaces in the bed were inclined at an angle, enabling natural circulation
and load following by varying the bed height. Coal particles of 0-18 mm size were
fed overbed.

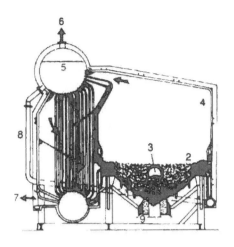

Figure 5.10.
The first industrial fluidized bed combustion boiler with water-tube furnace walls, Fluidfire Ltd. (G.B.) (Reproduced by kind permission of the author Prof. Dr. J. R. Howard from [2]):
1 – distribution plate, 2 – fluidized bed, 3 – opening for coal feed, 4 – water-tube wall, 5 – steam drum, 6 – steam outlet, 7 – flue gases outlet, 8 – water downcomer , 9 – bed material discharge

Following this historical review of the development of bubbling FBC boilers, which was provided an insight into different designs appropriate for reconstruction of old boilers burning liquid fuels, descriptions of two more modern FBC boiler designs will be provided. Modern FBC boilers can have high power and high steam parameters and successfully operate in the comercial realm, while satisfying all of the previously discussed general requirements for industrial boilers. One of the first "modern" boiler designed and constructed by Foster-Wheeler, started operation in October 1982 in Rotterdam port Europort for Royal Dutch Shell, the Netherlands, for combined steam and electric power generation.

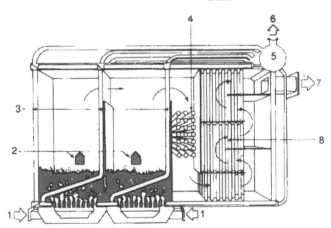

Figure 5.11. *Fluidized bed combustion boiler with water-wall furnace, Gibson Wells Ltd. (G.B.) (Reproduced by kind permission of the author Prof. Dr. J. R. Howard from [2]):*
1 – primary air, 2 – opening for coal feed, 3 – vertical water-tube panel walls, 4 – convective heat exchanger, 5 – steam drum, 6 – steam outlet, 7 – flue gases outlet to feed water heater, 8 – convective heat exchanger

The description here comes from a detailed report about the design and operation of this boiler presented at the 3rd European Coal Utilization Conference held in 1983 in Amsterdam [22]. Figure 5.12 gives the general outline and the layout of auxiliary systems of this boiler, whose basic parameters are given in Table 5.1.

Description of water and steam circulation system (Fig. 5.13). The boiler has natural circulation and a single steam drum. The basic parameters of steam and boiler at maximum power output are given on Table 5.1. The boiler self power consumption is 1850 kW.

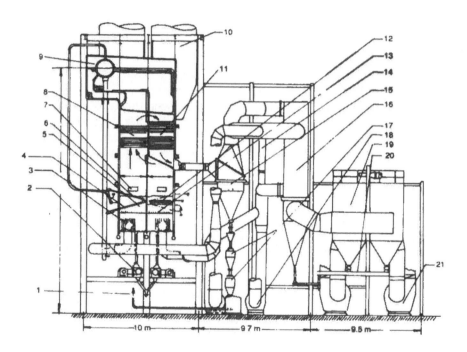

Figure 5.12. *Fluidized bed combustion boiler designed by Foster Wheeler, constructed in Rotterdam port EUROPORT (Reproduced by kind permission from Industrial Presentation Group, Rotterdam, from [22]):*
1 – fly ash recirculation system, 2 – bed material and bottom ash cooler, 3 – start-up burner, 4 – immersed heat exchanger, 5 – fluidized bed, 6 – primary superheater 1, 7 – coal feeders, 8 – final superheater, 9 – steam drum, 10 – coal hoppers, 11 – feed water heater, 12 – primary superheater 2, 13 – distribution plate, 14 – by pass of the air heater, 15 – mechanical air separator, 16 – air heater, 17 – fly ash recirculation system, 18 – primary air fan, 19 – bag filter, 20 – fly ash removal system, 21 – flue gases draft fan

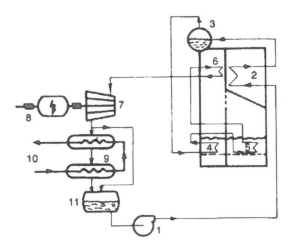

Figure 5.13.
Water and steam circulation in Foster Wheeler FBC boiler (Reproduced by kind permission from Industrial Presentation Group, Rotterdam, from [22]):
1 – feed water pump, 2 – feed water heater, 3 – steam drum, 4 – superheater 1, 5 – superheater 2, 6 – final superheater, 7 – steam turbine, 8 – electrical generator, 9 – heat exchanger, 10 – hot water for oil reservoirs, 11 – de-aerator

Table 5.1. *Basic parameters of Foster-Wheeler boiler in Rotterdam*

Steam capacity	50 t/h	Design coal:	Bituminous coal
Steam pressure	82 bar	– calorific value	25810 kJ/kg
Steam temperature	495 °C	– ash content	14%
Feed water temperature	145 °C	– moisture content	7.7%
Bed dimensions, cell A	2.44 × 4.99 m²	– volatile matter content	23%
Bed dimensions, cell B	2.29 × 4.99 m²	– fixed carbon	55.3%
Bed height in operation	1.22 m	– sulphur content	0.5%
Bed temperature	899 °C	– sulphur retention	90%
Fluidization velocity	2.74 m/s	Overall boiler efficiency	84.9%
Freeboard height	4.57 m	Electric power generation	6 MW$_e$
Ca/S molar ratio	3	Specific power per unit area	
Coal mass flow rate	6381 kg/h	of the bed cross section	1.7 MW$_{th}$/m²

Feed water enters at a temperature of 145 °C and is then heated in an economizer consisting of smooth tube bundles placed into a screened second gas pass. Water from the economizer enters the steam drum and mixes with water circulating through superheaters and evaporators. The evaporating part of the water circulation system consists of a membrane panel tube waterwall which encircles

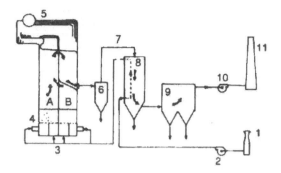

Figure 5.14.
Air and flue gases circulation system in Foster Wheeler
FBC boiler (Reproduced by kind permission from
Industrial Presentation Group, Rotterdam, from [22]):
A – bed cell A, B – bed cell B, 1 – primary air intake, 2 –
primary air fan, 3 – primary combustion air, 4 – start-up
burner, 5 – steam drum, 6 – mechanical   particle
separator, 7 – flue gases, 8 – air heater, 9 – bag filter,
10 – flue gases draft fan, 11 – chimney

the entire furnace, separates cells A and B, and forms the walls of the convective
part of the boiler. Heat transfer surfaces immersed into the beds of both cells and
are included in the water circuit. The water returns from the drum to two collectors
with several downcomers. Natural circulation is achieved due to the density difference of saturated water in downcomers and the steam-water mixture in heated water tube walls and unheated risers.

The first and the second superheaters are placed into the bed of cell A and
the bed of cell B. Final superheating takes place in an irradiated suspended
superheater placed in the freeboard of the cell A. The steam enters the turbine at a
pressure of 82 bar and exits under pressure of 7 bar. Exhaust steam is used in the
heat exchanger to heat water that heats the oil in the oil tanks. Condensate from the
heat exchangers after deaeration passes into the boiler feeding pump.

Description of the air and flue gas flow system (Fig. 5.14). The air from
the main ventilator is heated in the tubular air-heater by combustion products. During the start-up, flue gases by pass the air heater in order to achieve rapid heating of
the baghouse filters as soon as possible. From the air-heater, the air is conducted
into the plenum chambers of the bed of cell A and the bed of cell B, having been divided previously into two streams in both cases. One stream enters the starting sections A1 and B1 (see Fig. 5.15), and the other air stream, when appropriate
temperature is achieved during start-up, is introduced into the remaining two sections A2 and B2.

The distribution plate is constructed in the form of "bubble caps" with
built-in water cooled tubes, which are included into the general water circulation

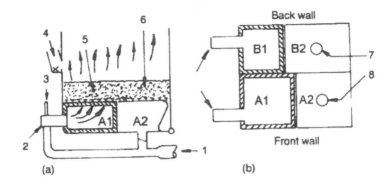

Figure 5.15. *Organization of the start-up and air distribution in the Foster Wheeler FBC boiler (Reproduced by kind permission from Industrial Presentation Group, Rotterdam, from [22])*
(a) vertical cross section of the bed cell A, (b) horizontal cross section of the air distribution chamber, 1 – primary air, 2 – start-up burner, 3 – combustion air for start-up burner, 4 – coal feeding, 5 – part of the bed in working (fluidized) state, 6 – part of the bed in quiet (defluidized) state, 7 – air inlet in bed cell B2, 8 – air inlet in bed cell A2

system of the boiler. Starting sections are lined with firebricks. Cooling of the distribution plate enables pre-heating of starting sections of the bed with combustion products from start-up, and liquid fuel burners with temperature up to 815 °C, without danger of damage due to thermal expansion.

Inside the beds of cell A and B there are no physical separations between sections A1 and A2, and B1 and B2. The water-tube panel wall separates only the active beds A and B, and the space above them. Gases from cells A and B mix at a certain height above the bed surface passing then into the second flue gas pass. A sufficient height of the space above the section A of the bed is left for burnup of small particles and volatiles. After leaving the second flue gas pass, combustion products pass through a system of cyclones with 85% efficiency, and the unburnt particles are fed into the furnace again to increase combustion efficiency. Combustion products continue to pass through the air-heater and the fine particles are separated in the baghouse filter. Combustion products is then passed into the stack by means of a flue gas draught fan, and a low underpressure is maintained in the furnace.

Starting-up the boiler (Fig. 5.15). Figure 5.15 shows that the boiler fluidized bed is divided into two parts – cells A and B. Inert material beds from these two cells are separated above the distribution plate by a water-tube panel wall, and they have a completely separated primary air supply. Primary air supply goes through four plenum chambers, of which A1 and B1 are starting chambers with built in start-up liquid fuel burners.

Air heated up to 815 °C fluidizes only the inert material in the starting sections A1 and B1. Mixing of heated and cold inert bed material and warm up of inert material up to the start-up temperature when coal feeding begins is achieved by periodical fluidizing of inert material in sections A2 and B2.

System for handling and feeding of solid materials into the furnace (Fig. 5.16). Trucks supply coal, limestone and inert material (sand), and then these are transported by a bucket elevator and discharged into the relevant bunkers. Inert material is fed into the furnace pneumatically.

A vertical bucket elevator and another belt transporter take coal and limestone into separate day hoppers by means of a common belt transporter. From the day bunker, the coal is transported by screw feeder and then fed by gravity to a coal spreader with a rotor flipper. The screw feeder regulates coal mass flow rate, and the coal spreader can operate at two speeds. Coal is spread over the surface of section A1 (i. e. B1) at a lower rotor flipper speed, and onto the entire bed surface in cell A (i. e. B) at a higher speed. Equipment for coal feeding in cells A and B are identical.

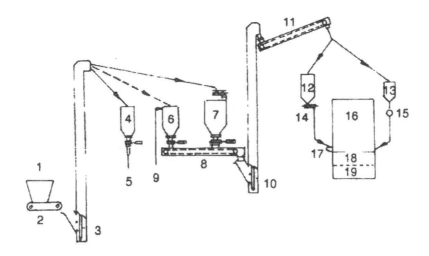

Figure 5.16. *Coal, limestone and sand transportation and feeding system for the Foster Wheeler FBC boiler (Reproduced by kind permission from Industrial Presentation Group, Rotterdam, from [22]):*
1 – receiving hopper, 2 – belt feeder, 3 – vertical transporter, 4 – sand hopper, 5 – toward the boiler, 6 – limestone hopper, 7 – coal hopper, 8 – belt transporter, 9 – pneumatic transport for the limestone, 10 – vertical transport, 11 – belt transporter, 12 – coal hopper, 13 – limestone hopper, 14 – screw feeder, 15 – rotating feeder, 16 – boiler, 17 – coal spreader, 18 – fluidized bed, 19 – air distribution chamber

Limestone from the day bunker is delivered to two small hoppers, and gravity fed to each section of fluidized bed of cells A and B via a rotary feeder.

System for removal of the excess inert bed material from the furnace (Fig. 5.17). If ash and tramp material (shale and stones) remain after coal burning in the fluidized bed, and/or if limestone is used for sulphur retention, the bed height will rise during operation. In addition, the size distribution of the bed material will change. A special system serves to maintain constant bed height and constant average particle size for the inert bed material. This system can operate either continuously or periodically. The material from the bed is drained at a single point through a vertical draining tube. Only one point is necessary to drain material from the entire furnace, if overflow openings in the water-tube panel wall which divide two cells of the furnace, are used. Inert material is cooled in screw conveyers. After grinding to a size which corresponds to the design inert bed particle size, inert material is taken into a spent bed material storage silo, by means of pneumatic transport. From the storage silo, the material can be loaded into trucks or recycled back to the furnace, or into cell A or B by pneumatic transport line.

System for separation of flying ash and recycling unburnt particles (Fig. 5.18). In order to increase combustion efficiency, a system was introduced for reinjection of fly ash back into the furnace. Only particles separated in the cyclones are returned to the furnace. These particles are injected below the bed surface by

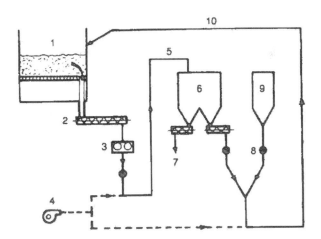

Figure 5.17.
Inert material removal system from the bed in Foster Wheeler FBC boiler (Reproduced by kind permission from Industrial Presentation Group, Rotterdam, from [22]):
1 – fluidized bed, 2 – screw transporter, 3 – crusher, 4 – conveying air fan, 5 – toward hoppers, 6 – hopper for bed material, 7 – toward material disposal, 8 – rotating feeder, 9 – fresh inert material for the bed, 10 – recirculation of the inert material

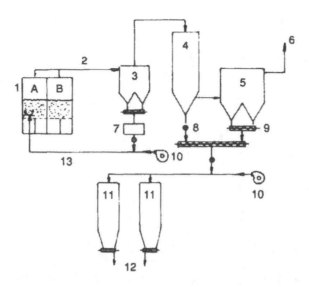

Figure 5.18.
*Flue gases cleaning system and fly ash recirculation system
for Foster Wheeler FBC boiler (Reproduced by kind
permission from Industrial Presentation Group, Rotterdam,
from [22]):
A – bed cell A, B – bed cell B, 1 – furnace, 2 – flue gases, 3 –
mechanical particle separator, 4 – air heater, 5 – bag filters, 6
– chimney, 7 – fly ash hopper, 8 – rotating feeder, 9 – screw
feeder, 10 – conveying air fan, 11 – fly ash hopper, 12 – toward
the disposal, 13 – recirculation of the unburned char*

means of a pneumatic transport line, at eight points near the distribution plate, in
section A1. Recirculation of fly ash enables an increase of efficiency from 93 to
98%.

Load control and following. In order to permit the furnace load to be varied
widely, the fluidized bed is divided into four sections with separate supply of primary air and the possibility of feeding fuel only into sections A1 and B1. Boiler
load following is provided in two ways: (a) by fluidization shutdown (slumping the
bed) in some of the sections which interrupts combustion in the section and reduces
total heat generation, and (b) by air flow rate variation and corresponding fuel flow
rate change. Using the second procedure it is possible to change bed temperature
from 770 to 900 °C, thus enabling continuous power variation within tight limits.

The following maximum continuous load changed is possible:

–   30 to 50% when only the entire cell A operates,

- 42 to 70% when three sections operate (A1, A2 and B1), and
- 60 to 100% when both cells operate (the entire furnace).

This boiler is one of the first larger fluidized bed boilers constructed [22]. The boiler met all the design requirements, but demonstrated some basic problems in terms of the systems for handling and feeding of solid materials, and also in terms of the pneumatic transport lines. However, all of these problems were satisfactorily resolved.

It is interesting to examine the design and basic parameters of a 90 MW$_{th}$ bubbling FBC boiler, manufactured by Stork, Hengelo (the Netherlands), because there is signifcant and detailed information available on its operation and its operational behavior [23-25], and in Chapter 4 of this book, Section 4.7, some results obtained from this boiler are compared with calculation results obtained using the mathematical model developed by Brem [25] (Fig. 4.59). The boiler started operation in 1986/87 in the salt industries AKZO-Hengelo in the Netherlands, and was used to generate superheated steam needed in the salt production process.

Figure 5.19 presents the basic diagram of the boiler and the layout of auxiliary systems. It is possible to follow the distribution of heat transfer surfaces on the diagram and the water and steam circulation system. The boiler is interesting, because a relatively high unit power was achieved by using a large cross section area for the furnace in the lower section where the active fluidized bed is, and then the cross section of the freeboard is reduced by more than twice some metres above the bed. The basic design parameters of the 90 MW$_{th}$ AKZO-AFBC boiler are given in Table 5.2.

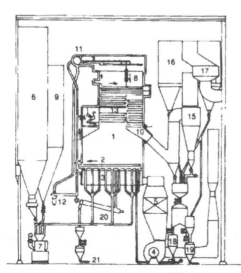

Figure 5.19.
Industrial fluidized bed boiler, 90 MW$_{th}$, designed and constructed by Stork Co. (the Netherlands) (Reproduced by kind permission of the ASME from [24]): 1 – boiler, 2 – fluidized bed, 3 – distribution plate, 4 – primary air fan, 5 – air heater, 6 – two coal hoppers, 7 – two coal mills, 8 – two air heaters for the mills, 9 – limestone hopper, 10 – feed water heater, 11 – steam drum, 12 – circulating pump for immersed heat exchanger, 13 – superheater, 14 – water injection, 15 – three cyclones, 16 – bag filters, 17 – flue gases cooler, 18 – hopper for recirculation of the unburned char particles, 19 – fly ash hopper, 20 – two water cooled screw transporters for bed material, 21 – discharge hopper for used bed material

Table 5.2.   *Basic design parameters of the 90 MW$_{th}$ boiler by Stork (the Netherlands)*

Steam capacity	115.2 t/h	Operation conditions	Base load 8300 h/year
Steam pressure	100 bar	Bed temperature	850 °C
Steam temperature	530 °C	Fluidization velocity	2.25 m/s
Feed water temperature	150 °C	Excess air	1.2
Range of load following	35-110%	Flue gas temperature	120 °C
Fuel	Coal	Boiler efficiency	90%
Design coal	Lignite	Combustion efficiency	98%
Calorific value	25 MJ/kg	Coal consumption	14.4 t/h(design coal)
Ash content	15%	Specific power per unit bed	
Moisture content	10%	cross section area	1.6 MW$_{th}$/m^2
Volatile matter content	30%	Ca/S molar ratio	2
Sulphur content	2%	SO$_2$ retention efficiency	85.6%
Emissions	SO$_2$ < 230 g/GJ	Limestone consumption	1.9 t/h
	NO$_x$ < 190 g/GJ	Recirculated fly ash	
	Particles < 15 g/GJ	mass flow rate	3.65 t/h

In addition, besides the listed parameters, the following is important:

(1) Bed height in operation is 1.05 m, and evaporation takes place in the immersed heat exchanger tube banks, removing about 45% of the heat generated.

(2) The remaining heat energy generated is received by water-tubes panel walls in the freeboard and heat transfer surfaces in the convective part of the boiler. The freeboard itself is about 13.5 m high.

(3) Circulation of the water-steam mixture in the boiler is natural, but circulation through horizontal pipes of the exchanger immersed into the fluidized bed is provided by a special circulating pump. The water enters the steam drum from the economizer through the rear and the roof water-tube panel walls, while vertical water-tube walls of the furnace and cooling of the distribution plate are included in the natural circulation water system. Saturated steam from the steam drum passes into the primary and the secondary (irradiated) superheater.

(4) The fluidized bed which has a relatively large area (about 61 m^2) is not divided, but the air plenum chamber is, and it provides 11 independently regulated primary air supplies. A wide power range variation in the range of about 3:1 is provided by slumping the fluidized beds in some parts of the bed. At two points, secondary air is introduced above the bed surface.

(5) Coal, with a size range of 1.5 to 2.5 mm after grinding, is pneumatically fed underbed, in the vicinity of the distribution plate by means of 12 separate pipelines at 36 points evenly distributed in the cross section. A mixture of coal and limestone is fed together simultaneously.

(6) Start-up of the boiler consists of heating of inert bed material by combustion of natural gas in the bed. The gas is introduced and evenly distributed in the bed by means of 20 horizontal perforated pipes.

(7) Separation of fly ash from the combustion products is first carried out by cyclones and then by bag-filters. Trapped particles are returned to the furnace via 8 separate pipelines by means of pneumatic conveying for combustion of the unburnt char particles.

From the published results [24, 25], the 90 MW$_{th}$ AKZO-AFBC boiler operated very successfully and by mid 1990 had achieved more than 10,000 h of operation. The boiler achieved the required design parameters: combustion efficiency of 99%, boiler efficiency of 93.2%, a power increase rate of 4.5%/min, a power reduction rate of 9.0%/min, availability of 87.4%, SO$_2$ emission < 400 mg/kg, NO$_x$ emission < 270 mg/kg, and CO < 110 mg/m^3 at 7% O$_2$ concentration in the flue gases.

After this review of international industrial boiler designs, constructions of three smaller boilers of Yugoslav manufacturers will be discussed.

Figure 5.20 gives a diagram of a boiler for generation of saturated steam, pressure at 18 bar, and a steam capacity 20 t/h, manufactured by Djuro Djaković Co. from Slavonski Brod. The boiler started operation in 1989, and was used to burn coal, wood waste and corncobs [26].

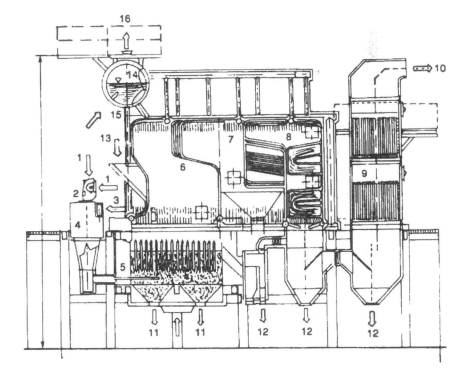

Figure 5.20. *Fluidized bed combustion steam boiler, 20 t/h, designed and constructed by Djuro Djaković Co., Sl. Brod (Croatia) [26]:*
1 – air inlet for start-up burner, 2 – start-up burner, 3 – primary air, 4 – start-up chamber, 5 – distribution plate, 6 – water-tube panel walls, 7 – first convective pass, 8 – second convective pass, 9 – feed water heaters, 10 – toward chimney, 11 – bed material discharge, 12 – fly ash discharge, 13 – coal inlet, 14 – steam drum, 15 – water downcomer, 16 – steam outlet

Figure 5.21 gives a diagram of a boiler for dry saturated steam generation with a pressure of 13 bar and steam capacity 4 t/h, manufactured by MINEL-Kotlogradnja Co. in Belgrade. The boiler was intended for combustion of lignite as mined, with particle size 0 to 20 mm and a heat capacity 9 MJ/kg and started operation towards the end of 1990 [27].

Figure 5.22 shows a diagram of a boiler for superheated steam generation with steam pressure of 6 bar and steam capacity of 15 t/h manufactured by MINEL Kotlogradnja Co. from Belgrade [28]. The boiler was previewed for combustion of wood waste, bark and sawdust from wood-pulp works, with a moisture content of up to 55%.

Both of these two boilers were constructed on the basis of technology developed at the Institute for Thermal Engineering and Energy, VINČA Institute of Nu-

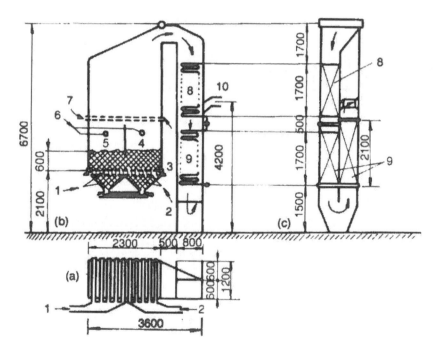

Figure 5.21. *Fluidized bed combustion saturated steam boiler, 4 t/h, designed and constructed by MINEL-Kotlogradnja Co., Belgrade:*
(a) horizontal cross section of the bed, (b) vertical cross section of the boiler, (c) vertical cross section of the convective pass, 1 – primary air inlet, 2 – primary air inlet from the start-up chamber in the starting part of the bed, 3 – sparge tube distribution plate, 4 – starting part of the bed, 5 – second part of the bed, 6 – openings for coal feeding, 7 – secondary air, 8 – evaporator, 9 – feed water heater, 10 – toward chimney

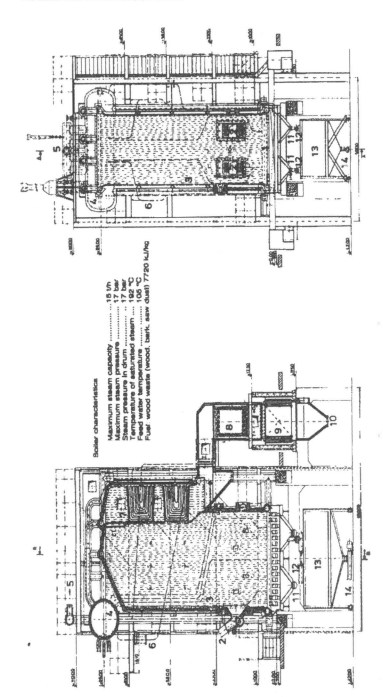

Boiler characteristics

Maximum steam capacity .... 15 t/h
Maximum steam pressure .... 17 bar
Steam pressure in drum .... 17 bar
Temperature of saturated steam .... 192 °C
Feed water temperature .... 105 °C
Fuel wood waste (wood, bark, saw dust) 7720 kJ/kg

Figure 5.22. *Fluidized bed combustion steam boiler for wood waste burning, 15 t/h, designed and constructed by MINEL-Kotlogradnja Co., Belgrade:*
1 – sparge tube distribution plate, 2 – coal spreader, 3 – water-tube panel walls, 4 – steam drum, 5 – steam outlet, 6 – water downcomer, 7 – convective pass, 8 – feed water heater, 9 – air heater, 10 - toward chimney, 11 – bed material discharge, 12 – screw transport, 13 – used bed material hopper, 14 – toward particle separator and bed material recirculation system

clear Sciences, Belgrade, using data on fuel suitability for combustion in a fluidized bed, acquired by means of the unique methodology developed by the Institute, as described in Chapter 6.

5.1.3. Bubbling fluidized bed combustion boilers for electricity production

(a) Purpose and application

Boilers for electric energy production operate under specific conditions: a large unit power is required along with high boiler efficiency, safe operation and high availability. That is why utility companies worldwide are extremely conservative and tend to accept only well demonstrated technologies. Nowadays, only pulverized coal combustion boilers are widely used in electric power plants. In modern electric power plants with PCC boilers, in high-developed countries, burners for reduced NO_x emission and desulphurization modules (usually wet scrubbers), for flue gas emission control are used. Pulverized coal combustion technology has developed over several decades and as such has reached technical and construction perfection, which in turn makes it very difficult to achieve a breakthrough in this market for new power generation technologies.

Modern power generating boilers are expected to fulfill a series of requirements:

– flexibility with respect to type and quality of fuel,
– effective environment protection,
– wide range of load following with no liquid fuel support, and
– fast load variation.

The characteristics of fluidized bed combustion that allow this technology to be competitive with pulverized coal combustion boilers are as follows:

– the possibility of simultaneous or alternate combustion of fuels of different qualities,
– the absence of ash sintering, deposit formation and fouling of heat transfer surfaces,
– a high combustion efficiency,
– an ability to successful satisfy all of the increasingly rigorous emission standards (low SO_2 and NO_x and particulate emission),
– the possibility of burning low-rank and low-quality fuels,
– reduced investment costs (no special devices for desulphurization and control of NO_x emission), and
– dry solid waste and spent bed materials.

Despite the fact that fluidized bed combustion fully satisfies current needs for electric power generating boilers, utility companies have still not fully accepted this type of combustion. As was noted in the first part of this book, boilers with bubbling fluidized bed combustion have reached the phase of commercial exploitation and they have become competitive with pulverized coal burning boilers and boilers burning liquid and gaseous fuels. For electric power generation, fluidized bed boilers, both bubbling and circulating, were in 1990s being accepted as demonstrated technology. Long-term and complete demonstration programs testing of full scale bubling fluidized bed boilers are described below [29, 30]:

- TVA (Tennessee Valley Authority), EPRI (Electric Power Research Institute), Commonwealth of Kentucky and Combustion Engineering constructed a boiler of 160 MW$_e$, using a bubbling fluidized bed, at the Shawnee power plant, Paducah, Kentucky (U.S.A.). The boiler started operation in 1988, and the testing program lasted until 1992 [31-33] and the boiler has been in operation for the last 20 years,
- NSP (Northern States Power) reconstructed an old pulverized coal combustion boiler in the Black Dog power plant, Minneapolis, Minnesota (U.S.A.). The boiler's power was 130 MW$_e$, and it started operation in 1985, and its testing program was completed in 1990 [34-36], and
- EPDC (Electric Power Development Co. Ltd.) carried out a program to verify and demonstrate fluidized bed combustion technology by constructing and testing a series of bubbling FBC boilers of various capacities. The largest boiler in the program is the Takehara of 350 MW$_e$, the construction of which began in 1991. Start-up occured in 1994, and testing lasted until 1995, when the boiler was put into commercial operation [37-39]. The program started in 1979 by construction of a pilot, steam-generated boiler with steam capacity 20 t/h. Testing of this boiler lasted from 1982 until 1983. Based on these tests, construction of a demonstration Wakamatsu 50 MW$_e$ boiler began in 1985, with steam capacity 156 t/h. Wakamatsu boiler started operation in 1987, after 8 months of testing which lasted until after 1992.

Demonstration programs showed that while bubbling fluidized bed combustion boilers can satisfy the requirements of large power systems with respect to reliability and availability there were issues, especially around feeding very large beds, and load following. As a result it is now certain that for larger utility applications only circulating fluidized beds are likely to be used in the future.

Besides these demonstration programs for bubbling fluidized bed combustion boilers, demonstration programs for CFBC boilers also took place. More details on these boilers, their characteristics and testing programs, can be found elsewhere [8, 40, 41]. There is however no doubt that for smaller applications, especially burning low quality and waste fuels, bubbling fluidized beds have a secure future. Despite this caveat, for completion sake and because there are a number of large bubbling FBC used for utility applications, the issues for this type of boiler are discussed below.

The most likely applications of bubbling fluidized bed combustion boilers in electric power generation are as noted above for the following situations:

— as new boilers for combustion of low-quality coals,
— in the retrofitting or reconstruction, revitalization and upgrading of old coal burning boilers, if the quality of the available coal changes, if the boiler needs reconstruction in order to achieve greater combustion efficiency and satisfy new or existing emission standards, and
— in reconstruction of liquid fuel burning boilers, if changing to cheaper, domestic fuel is intended, but without affecting environment emissions and the level of automatization.

(b) Basic parameters and requirements

Along with the requirements discussed above, which must be satisfied by electric power generating boilers, the most important requirements are the following: high unit power — over 100 MW_e and high steam parameters, up to 190 bar and 580 °C. These demanding conditions are necessary to achieving high thermodynamic efficiency for the steam cycle and efficient operation of steam turbines. These requirements and parameters necessary for power generating boilers and the problems and complexity involved in achieving them, can be better understood by comparing them with the requirements for industrial boilers:

— combustion efficiency and fuel costs strongly affect the price of power generation, therefore, it is necessary to achieve combustion efficiency of 99% for large boilers used for power generation, while for industrial boilers combustion efficiencies of 95-98% can be satisfactory,
— boiler efficiency of 85-92%, and
— SO_2, NO_x and particulate emission standards are significantly higher due to their large capacity and large absolute quanties of pollutants such a boiler can potentially emit into the environment.

Given the present emission standards in many developed countries, the following emissions are now standard:

— SO_2 < 150-200 mg/m^3, and 90% of sulphur retention for coals with high sulphur content and 70% of sulphur retention for coals with low sulphur content and lignite,
— NO_x < 150 mg/m^3, and
— particulates < 50 g/m^3.

Based on the data available for bubbling FBC boilers, it can be concluded that these boilers have achieved the following parameters:

- demonstration boilers with capacities of 125-160 MW$_e$ and a 350 MW$_e$ demonstration boiler already in existence in Japan,
- steam parameters 130 bar, 538 °C,
- combustion efficiency 98-99%,
- boiler efficiency 86-88%,
- SO$_2$ emission < 250 ppm,
- NO$_x$ emission < 250 ppm,
- particulate emission < 50 mg/m^3,
- load change capability 3:1 up to 5:1, and
- rate of load change 2-4%/min.

(c) Basic design problems

Nonetheless, designing FBC boilers for electric power generating poses serious problems and important issues include:

- practical maximum specific load of bed (furnace) cross section area is 1 to 2 MW$_{th}$/m^2, and
- poor lateral mixing of fuel particles makes uniform distribution of fuel in the bed difficult.

In view of the fact that a 200 MW$_e$ boiler has thermal power of approximately 500-600 MW$_{th}$, it can easily be seen that the fluidized bed cross section area, depending on the fuel type, must be 300-400 m^2. It is not simple to achieve uniform air distribution and uniform fluidization, and solve the problems of thermal expansion of the air distribution plate in such a large area. Uniform distribution of fuel is also a complex problem given the fact that in underbed feeding, one feeding point must serve each 1-2 m^2 of the bed cross section area. Uniform fuel supply to 150-300 feeding points is a complex technological and construction problem, albeit not insuperable. Demonstration programs, however, show that these problems can be resolved in a satisfactory manner.

The other serious problem is the achievement of combustion efficiency of 99%, which necessarily requires introduction of fly ash recycling. Recycling solves another serious problem – limestone consumption can be reduced to a molar ratio 2:1, which is generally considered more acceptable, given that the quantity of wastes from such a boiler firing high sulphur fuel can be extremely high (for some high sulphur fuels limestone feed may assume values of 50-60% of the fuel feed).

Another issue is that the quantity of "fly ash" is much larger (due to limestone consumption) than in conventional pulverized coal combustion boilers with wet scrubbers, and electric resistance of the ash is higher, thus reducing efficiency of electrostatic precipitators.

Reduction of NO$_x$ emission requires introduction of secondary air.

(d) Basic design solutions for large power bubbling FBC boilers

Combustion Engineering's 160 MW$_e$ TVA boiler is presented in Fig. 5.23 and Table 5.3. Design of this boiler began in 1985, and it was constructed in February 1986. A ten-year testing period was originally planned. The boiler started operation in mid 1988. After 18 months of testing and necessary modifications, in 1990, a two-year period of tests to verify the design parameters and detailed investigations began and a final report was issued in September 1992. The unit has since then run commercially, burning mostly coal, but also successfully co-firing petroleum coke for a short period.

Coal and limestone are introduced in a mixture by means of 12 independent pneumatic conveying systems into 12 cells of the fluidized bed. Each of these 12 pneumatic lines is divided into 10 feeding lines for each cell. This means that there are 120 points for underbed feeding. Coal is dried and ground to a size range of 0-10 mm. Start-up of the boiler is achieved by introducing combustion products

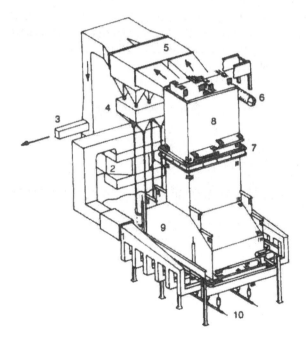

Figure 5.23.
Utility fluidized bed combustion steam boiler, 160 MW$_e$, designed and constructed by Combustion Engineering (Reproduced by kind permission of the ASME from [43]):
1 – distribution of the primary air, 2 – air heater, 3 – flue gases toward coal drying, 4 – hoppers for fly ash recirculation, 5 – flue gas cleaning, 6 – steam drum, 7 – compensator of dilatation, 8 – convective heat exchanger, 9 – freeboard, 10 – bed material discharge system

Table 5.3.   *Basic parameters of the TVA 160 MW$_e$ boiler*

Power	160 MW$_e$	Recirculation ratio	2.5-2.9
Steam production	500 t/h	Combustion efficiency	97%
Steam production	127.5 bar	Bed temperature	200 °C
Steam temperature	540 °C	Bed height	0.66 m
Boiler efficiency	88%	Furnace height	31.5 m
Desulphurization degree	90%	Fluidization velocity	2.5 m/s
SO$_2$ emission	< 200 ppm, < 260 mg/J	Bed cross section area	234 m^2
NO$_x$ emission	< 200 ppm	One feeding point per	1.86 m^2
Coal:		Number of feeding points	120
– moisture content	< 16.5%	Start-up time	8 h 45 min.
– particle size	0-6.3 mm	Range of load change	60-160 Mw$_e$
– calorific value	24-25 MJ/kg		
– sulphur content	4-4.5%		
– Ca/S	2.3		

from two start-up chambers burning liquid fuel, placed on two sides of the boiler. When the entire bed (all 12 cells) reach 260 °C, special start-up chambers start to operate to heat two central cells up to 510 °C, whereupon feeding of coal begins. To increase combustion efficiency there is a system for recycling fly ash. Unburned particles are injected at three points in each of the 12 bed cells. The area of the furnace cross section at the level of the bed surface is very large (234 m^2). At the level of 5.5 m above the bed surface, cross section of the freeboard narrows to the area of 83 m^2.

The boiler has a natural circulation water-steam system, but evaporating surfaces immersed in the fluidized bed are under forced circulation [42-44].

The Black Dog NSP boiler is shown in Fig. 5.24 and Table 5.4. The boiler is actually a reconstruction of an old boiler constructed in 1954. Both boilers have been built by Foster Wheeler. Dismantling the old boiler began in 1984, and mounting of the new equipment in the beginning of 1985. The boiler started operation in the beginning of 1986, and in the middle of that year its commercial operation began. A detailed investigation lasting several years was then carried out. The capacity of the unit was increased from 100 MW$_e$ to 130 MW$_e$ by the reconstruction.

Table 5.4.   *Basic parameters of the 130 MW$_e$ Black Dog boiler*

Power	130 MW$_e$	Area of the main	
Steam production	470 t/h	bed cell	7.3 × 12.8 m^2
Steam pressure	125.4 bar	Two side cells	2 × 7.3 × 5.2 m^2
Steam temperature	540 °C	Total bed area	170 m^2
Bed temperature	840 °C	Range of load change	20-100%
Fluidization velocity	2-3 m/s	Ca/S molar ratio	2.5
Recirculation ratio	1.5	Combustion efficiency	98-99%
Excess air	1.22	SO$_2$ emission	< 150 ppm
Coal:			(< 900 ppm for
– calorific value	8500-15000 Btu/lb		5.7% S)
– moisture content	23.9%	NO$_x$ emission	< 350 ppm
– particle size range	0-20 mm		
– sulphur content	0.3-5.7%		

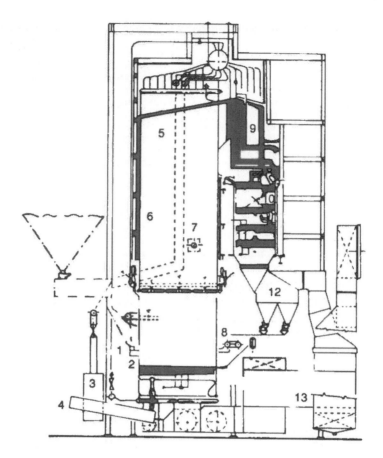

Figure 5.24.
Retrofitted utility fluidized bed combustion boiler Black Dog, 125 MW$_e$, designed and constructed by Foster Wheeler (Reproduced by kind permission of the ASME form [34]):
1 – coal spreader, 2 – immersed heat exchanger, 3 – circulating pump, 4 water cooled screw transport, 5 – water-tube panel furnace walls, 6 irradiated superheater, 7 – liquid fuel burner in the freeboard, 8 – secondary air, 9 – feed-water heater, 10 – primary superheater, 11 – feed-water heater, 12 – mechanical particle separator, 13 – air heater

The fluidized bed is divided into three parts with independent air supply – the central part (about 55% of the area) with two cells, and two side beds (altogether 45% of the area). Central cells are separated from each other only by independent air supply. Side cells are separated from the central ones by a water-tube wall placed above the distribution plate. Coal is fed by spreaders, which are placed at the

wider side of the furnace. In each of the four cells, the region close to the wall with
coal spreaders is separated as a starting cell. There are 12 coal spreaders altogether:
6 for two main cells of the bed and 3 for each of the two side cells. By slumping the
bed in the individual cells a power change over the limits 20-100% is possible. At
low rates, the spreaders feed coal only at the start-up part of the bed (up to the distance of 2.1 m). At higher rates, coal can be spread all the way to the opposite furnace wall at the distance of 7.3 m. The boiler has a natural circulation in the
water-steam system [34, 35, 45].

A diagram of Wakamatsu 50 MW$_e$ boiler is presented in Fig. 5.25 and Table
5.5. The 50 MW$_e$ Wakamatsu boiler was constructed in EPDC research center on

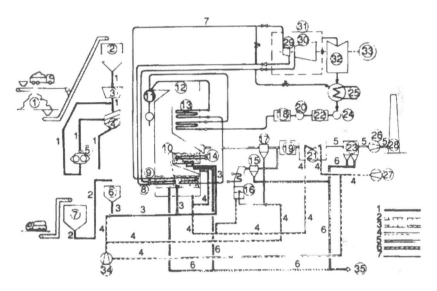

Figure 5.25. *Demonstration fluidized bed combustion boiler Wakamatsu, 50 MW$_e$, Japan:
1 – coal disposal, 2 – coal hopper, 3 – coal drying, 4 – vibrating grid, 5 –
crusher, 6 – intermediate coal hopper, 7 – limestone hopper, 8 – reheater, 9 –
superheater, 10 – underbed coal feeding points, 11 – steam drum, 12 – boiler,
13 – convective heat transfer surfaces, 14 – evaporator, 15 – cyclone, 16 – cell
for burning unburned char particles, 17 – multicyclone, 18 – high pressure
water heater, 19 – catalatic DeNO$_x$ reactor, 20 – feed-water pump, 21 – air
heater, 22 – low pressure water heater, 23 – beg filters, 24 – pump for condensate, 25 – condenser, 26 – flue gases draught fan, 27 – high pressure fan, 28 –
chimney, 29 – high pressure steam turbine, 30 – medium pressure steam turbine, 31 – steam turbine for super high temperatures, 32 – low pressure steam
turbine, 33 – electrical generator, 34 – main fan, 35 – ash discharge:
Lines: 1 – coal feeding system, 2 – limestone feeding system, 3 – transport
lines for coal-limestone mixture, 4 – air ducts, 5 – flue gases ducts, 6 – ash removal system, 7 – water and steam circulation system*

Table 5.5.   *Basic parameters of the 50 MW$_e$ Wakamatsu boiler*

Power	50 MW$_e$	Rate of load change	2%/min.
Steam capacity	156 t/h	Coal:	
Steam pressure	125 bar	– particle size range	0-10 mm
Steam temperature	593 °C (second	– fine particles <0.1 mm	20%
	degree 652 °C)	– calorific value	21-28 MJ/kg
Boiler efficiency	87.72%	Main bed:	
SO$_2$ emission	< 100 ppm	– bed area	99 m^2
NO$_x$ emission	< 200 ppm	– fluidization velocity	1.5 m/s
Particle emission	< 0.01 g/Nm^3	– number of feeding points	86
Desulphurization degree	> 90%	Burn-up cell:	
Ca/S ratio	5	– bed area	17 m^2
Main bed temperature	770-860 °C	– fluidization velocity	1.3 m/s
Burn-up cell bed		– number of points for feeding	
temperature	960-1000 °C	unburned particles	10
Combustion efficiency	98-99%	– furnace height	28 m
Minimum power	40%		

the location of a former thermal electric power plant, which ceased operation. Specific to its design are two fluidized beds placed one above the other in the boiler furnace in order to reduce the cross section area of the furnace, as well as a special cell with fluidized bed (furnace) for burning out unburned char particles. Coal is fed underbed, which is why it is first dried, and then mixed with limestone.

In the upper main boiler bed, a horizontal evaporator is situated, and in the lower part, the secondary, the third and the final superheater are located. Since the evaporators are horizontal, the boiler has partially forced circulation. A part of the evaporator is placed into the char burn-up cell. In each bed only separated cells are started up, where there are no heat transfer surfaces [37-39, 46-49].

5.1.4. Choice of boiler concept – problems and the choice of basic parameters

The choice of whether to employ a bubbling fluidized bed combustion boiler, its construction, basic dimensions, and hydrodynamic and thermodynamic parameters, as well as the choice and construction of auxiliary systems depend primarily on the following:

- boiler power,
- the fuel that will be used,
- the requirements for:
    - combustion efficiency,
    - range of load change,
    - rate of load change,
    - SO$_2$ and NO$_x$ emissions, and
    - level of the boiler automatic operation control.

The choice of different concepts, hydrodynamic and thermodynamic parameters affects mostly the dimensions and the shape of the furnace, and the paraeters that define combustion conditions. For reliable operation of an FBC boiler, the most important thing is to provide uniform and good fluidization, good fuel mixing and ensure its uniform distribution in the bed. The other problems relate to the following issues:
- boiler power (dimensions of the bed cross-section area),
- fluidized bed height,
- freeboard height,
- inert material particle size,
- fluidization velocity,
- combustion temperature (fluidized bed temperature),
- combustion efficiency,
- excess air,
- dimensions of heat transfer surfaces immersed into the bed,
- location for fuel feeding,
- fly ash recirculation,
- draining of the inert bed material,
- boiler start-up, and
- load change.

The parameters and processes that are directly connected with the furnace design will be discussed in this section. The remaining problems (the type of fuel feed system, recirculation of fly ash, draining of the inert bed material, boiler start-up and the manner in which power will be changed), which are linked more to the auxiliary systems, will be discussed to more detail in the following section.

Boiler unit power (fluidized bed cross-section size). Large capacity FBC boilers can be achieved in two ways: by increasing the cross sectional area of the bed or by increasing the maximum specific heat generation per unit of cross-section area of the bed or the furnace.

It was already noted that the cross-sectional heat release rate of the bubbling FBC boilers is 0.5-2 MW_{th}/m^2. This datum refers to the normally used inert material with particle size 0.5-1 mm and fluidization velocity 1-2 m/s. The actual value of the cross-sectional heat release rate depends on fuel quality, excess air and the heat that can be removed by immersed heat exchangers from the bed. For very high-quality coal combustion, with low moisture and volatile matter content, one may achieve an upper limit in the range of 2 MW_{th}/m^2, since the majority of heat is generated in the fluidized bed, where the advantages of fluidized bed combustion become most prominent. Increase of the cross sectional heat release rate for a given fuel can be achieved only by increasing fluidization velocity, which causes increase of inert material particle size. Some boiler constructors use inert material of 1-2 mm and more, but in that case a high pressure primary air fan is necessary, which is also a problem. Erosion of internals and heat transfer surfaces in this environment may also become a significant problem.

If the required boiler power exceeds the reasonable heat release rate, it can be achieved only by an increase of fluidized bed cross section area. For electric power generating boilers of 500-600 MW_{th}, technological issues in building a fluidized bed having an area of 150-200 m^2 are the greatest problem. Despite that, companies such as Combustion Engineering, in the case of the TVA 160 MW_e FBC boiler, did choose a single bed with an area of about 235 m^2. This solution earned the nickname "RANCH" among some engineers in the U.S.A., which indicates the degree of challenge in building such a large bubbling FBC.

The other possibility is to divide the fluidized bed into several stages, one above the other, along the furnace height. Although there have been several examples of such approaches to resolving the problem of increasing the power of bubbling FBC boilers (for example 30 MW_e demonstration boiler in Riversville, U.S.A. [44], 50 MW_e Wakamatsu, Fig. 5.25, Worsmer boilers, Great Britain, and Stal-Laval, Sweden), this type of combustion system never became very popular. In fact, one of the most important reasons for development of circulating fluidized bed boilers was the problem of large bed areas in large power bubbling FBC boilers and currently it can probably fairly be said that for units with a capacity of greater than 25-50 MW_e, any new FBC boiler will almost certainly be of the circulating fluidized bed type.

Fuel mixing in fluidized bed. Mixing of fuel and its uniform distribution in a fluidized bed is one of the greatest problems in furnace construction. Bad mixing leads to formation of local hot spots in the bed, ash melting and agglomeration of the material in the bed, uneven fluidization and, finally, to bed defluidization and interruption of boiler operation. Bad fuel mixing can also contribute to reduction of combustion efficiency. This disadvantage of combustion in a bubbling fluidized bed is especially noticeable for large capacity boilers and in the combustion of waste fuels, refuse derived fuels and industrial waste.

Section 2.3.6 of Chapter 2 dealt in detail with mixing of solid particles in a fluidized bed. It was stressed that mixing in the axial direction is very good, but the low intensity of mixing of fuel in a lateral direction presents a problem. Some practical aspects of the problem will be discussed below.

Good mixing of fuel can be achieved in several ways:

- design of the air distribution plate to achieve a uniform and intense fluidization (this will be discussed in some detail in the following section),
- by a correct choice of the manner of fuel feeding, depending on fuel characteristics,
- by a correct choice of the positions of fuel feeding points (see Sections 2.3.6, Figs. 2.32 and 2.33),
- by a large number of correctly placed feeding points (see 160 MW_e TVA boiler), which will be discussed in the following section, and
- by design solutions which enable directed, organized movement of inert material in the bed, which can be achieved in lower capacity boilers.

Effective directed circulation of inert bed material can be achieved in several ways (see Fig. 2.38 in Section 2.3.6). Such mixing increases the residence time of the fuel particles in the bed and it may be necessary for combustion of low-quality and waste fuels.

The schematic view of a Fluidfire (Great Britain) boiler that used directed circulation of inert bed material is presented in Fig. 5.10 [2, 4, 12, 50]. The following advantages of its construction are stated to be:

- increase of residence time of coarse fuel particles in the bed and reduction of elutriation of smaller particles,
- collecting of ash and oversized material in the middle of the bed where it is more easy to drain them, and
- in cases of low power, heat transfer to water-tube furnace walls are sufficient for heat removal from the bed, so it is unecessary to introduce additional immersed heat transfer surfaces.

A similar, highly successful design was developed by a Japanese firm EBARA Corporation [51] for combustion of various types of waste, even entire car tires (Fig. 5.26). A British firm, Deborah, also used a similar type of mixing arrangement, see Fig. 2.38 [2, 50].

Fluidized bed height. In practice, the height of fluidized bed in industrial boilers ranges from 0.5 to 1.2 m. The height of the fluidized bed depends on many

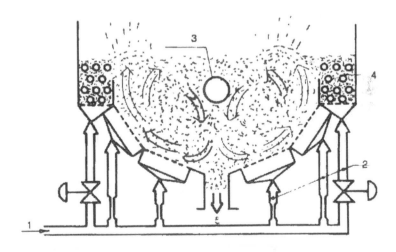

Figure 5.26.
Fluidized bed furnace with internal bed material circulation, designed by EBARA Co. (Japan) (Reproduced by permission of Routledge, inc., part of the Taylor & Fransis Group. Copyright 1986 from [108]):
1 – primary air inlet, 2 – air distribution plate, 3 – opening for coal feeding,
4 – immersed tube exchanger in auxiliary bed, 5 – bed material discharge

factors: desired combustion efficiency, dimensions of immersed heat exchanger (if any), the type of load following necessary, the type of fuel and others. The height of the bed is often dependant and limited by the available pressure head of the primary air fan. In principle, high-quality coals require communition, feeding in the vicinity of the air distribution plate, a large area of the heat exchanger immersed into the bed and a large bed height. Geologically younger coals (with particle size 25-50 mm), can be fed directly on to the bed surface despite their high volatile content, and can be burned efficiently even with a bed height of only 500 mm. However, combustion of "small sized" biomass, which also have a high volatile content, requires deeper beds.

A larger bed height ensures combustion of fine particles and volatiles in the bed and increased combustion efficiency, and permits the use of a larger immersed heat exchanger, bed height variation for the sake of load variation, better utilization of limestone and a higher degree of desulphurization. However, deeper beds also mean a more complex furance construction, a primary air fan with a larger pressure head (over 2000 kPa), and a higher self power or parasitic power consumption for the boiler itself. Very successful boilers were designed with bed heights up to 500 mm, primarily for combustion of geologically younger coals which can be efficiently burnt in shallow beds despite difficulties with volatile matter combustion.

Freeboard height. In order to achieve a high combustion efficiency, a sufficient height above the bed is necessary. It was once believed that a large furnace height was not required for fluidized bed combustion. Considerably lower furnace height, that is a smaller furnace volume, was stated to be a major advantage of this type of combustion. Although this characteristic of FBC boiler still remains an advantage compared to conventional boilers, the furnace height of bubbling fluidized bed combustion boilers is nonetheless significant. For instance, the 90 MW_{th} AKZO boiler had a furnace which was about 13.5 m high, while the furnace height of the 160 MW_e TVA FBC boiler is 31.5 m, and that of the 130 MW_e Black Dog boiler is 28.5 m, and that of the 50 MW_e Wakamatsu boiler was 28 m high. Heat release rate per furnace volume is still higher in FBC boilers than in conventional boilers, and ranges from about 1.2 to 1.7 MW_{th}/m^3. High furnaces (i. e. the space above the bed) provides sufficient time for combustion of fine char particles (important for burning high-quality coals) and for volatile matter combustion (which is important for burning lignite and biomass).

Freeboard height depends on the fluidization velocity and velocity selected for above the bed. This volume must be sufficiently high to provide for any necessary heat transfer to furnace walls, injection of secondary air and the gas mixing necessary for complete volatile mater combustion.

Inert material particle size. The initial inert bed material in FBC boilers is typically just silica sand, although in some systems operators may startup with "saved" bed material. During operation, and depending on the fuel type and characteristics and behavior of ash, and any requirements for making "sand makeup," a 'certain steady state mixture of ash and sand components is likely to be achieved. If limestone is used then the bed is likely to be comprised of primarily limestone derived products. The average particle size (and size distribution) of the inert bed material is a very important design parameter. It is important that the boiler maintain the design specifications for average particle size, to ensure reliable and safe operation of the boiler. A minimum particle size of inert material is maintained "automatically" by the elutriation of small particles depending on the fluidization velocity. However, the maximum particle size must be controlled during boiler operation, by means of a system for draining bed material, and if appropriate by sieving and/or grinding this material and using it as "bed makeup." Typically, the sizing for inert material in a modern bubbling FBC boilers is between 0.5 and 1.0 mm, although in some designs a slightly larger size range of 1 to 2 mm may be used.

Ash behavior in the bed. Ash resulting from the combustion of coal can remain (and accumulate) in the bed or be carried out by gaseous combustion products. The ratio of fly ash and the ash remaining in the bed depends on coal characteristics and in particular its mineral characteristics, as well as the the fluidization velocity. Ash behavior in the bed decisively affects the design and the capacity of auxiliary systems: the system for recirculation of unburned particles, the system for flue gas cleaning, and the system for removal of the solid material from the bed.

The sintering temperature of ash affects the choice of nominal and maximum bed temperature. A low sintering temperature increases the danger of ash sintering and bed material agglomeration, which can result in defluidization, interruption of boiler operation and under some circumstances substantial damage to the boiler. Section 5.6 of this chapter will deal with bed material agglomeration and ash sintering in more detail.

Fluidization velocity. In operating bubbling FBC boilers, the fluidization velocity ranges from 1 to 2.5 m/s, and in some cases even higher velocities of 2 to 4 m/s are employed.

The choice of fluidization velocity depends on the size and type (density) of inert bed material, the desired combustion efficiency and planned range of boiler load change to be achieved by changing the fluidization velocity. The fluidization velocity must be above 4-6 times the minimum fluidization velocity of the inert bed material at nominal operation bed temperature to ensure vigerous fluidization.

Higher fluidization velocity implies more large bubbles and better and more intense mixing in the bed, but also higher elutriation of fine ash and char particles. While the combustion in the bed will be improved, more unburned particles will be elutriated from the bed. Typically, higher fluidization velocity implies lower combustion efficiency, and will likely require the use of unburned flyash recirculation and higher capacity for devices used for particle separation from flue gases.

The fluidization velocity must be sufficiently high to ensure that the fluidization intensity is sufficient to enable good mixing of material in the fluidized bed during load control by reducing the primary air flow rate and corresponding reduction of the fuel flow rate. At minimum boiler power (i. e. minimum flow rate of the primary air), the fluidization velocity should not be below 3 v_{mf}.

Combustion temperature. One of the most important characteristics and advantages of fluidized bed combustion is that combustion is highly efficient, at temperatures much lower than in conventional boilers. In operating boilers, combustion temperature is chosen and maintained within narrow limits between 800 and 900 °C. Towards the higher value, temperature is limited by ash softening and sintering, and towards the lower value, by combustion efficiency. Fortunately, this is also the temperature range in which the reaction rate between SO_2 and CaO is the greatest and for fuels with significant sulphur contents this is nearly always the range chosen. For FBC furnaces which use significantly higher temperatures, sulphur capture is likely not to be a consideration, but in cases of units operating much above 900 °C, NO_x emissions may then start to become a problem.

It is possible to maintain operating temperature below 800 °C, but seldom below 750 °C, and then only for biomass fuels, or for short periods of time for other fuels, to ensure the widest possible boiler power range.

Combustion efficiency. Although combustion efficiency depends significantly on the type of fuel, this is not an a priori determined parameter. Depending on the boiler purpose, ratio of investment and operation expenses, a boiler designer may choose any given combustion efficiency, by a suitable choice of boiler design, and operating parameters. The combustion efficiency will be discussed further in Section 5.3, but typically it ranges from 95% to over 99%.

Excess air. In bubbling fluidization, a portion of the air passes through the bed in bubbles and may not actively participate in the combustion process in the bed. To the extent intense mass (oxygen) transfer between bubbles and emulsion phase is provided, it is possible to ensure adequate combustion in the bubbling fluidized bed with less excess air. Despite this deficiency, industrial bubbling fluidized bed boilers operate with an excess air of 1.2 to 1.3, which is close to the values typical for conventional pulverized coal combustion boilers.

The choice of excess air is also a matter of achieving an optimum in terms of performance. A lower excess air implies a risk of incomplete combustion, but also lower energy losses with the flue gases. Higher excess air provides higher combustion efficiency, but also higher exit losses. In practice, by combining different possibilities for improving combustion efficiency, the designer may choose lower excess air and ensure overall boiler efficiency within the range from 85 to 92%.

Size of heat transfer surfaces immersed in the fluidized bed. Maintenance of appropriate combustion temperature in the bubbling fluidized bed within the range from 800 to 900 °C, often requires heat removal directly from the bed. The quantity of heat which must be directly removed from the bed, depends mostly on the type of fuel. With high-quality coals, with low moisture and volatile matter contents, it is necessary to remove up to 50% of the total heat generated by the fuel from the bed (with excess air of about 1.25). Combustion of low-quality fuels, with high moisture and volatiles contents, can be accomplished without the use of immersed heat transfer surfaces. In some cases, it may even be necessary to cover some or all of the furnace walls area with firebricks in order to prevent excess cooling of the bed, and loss of heat through the furnaces water-tube walls (e.g. for burning very wet biomass or biomass derived fuels for instance).

Generally, excess heat can be removed from the bed in two ways: by use of high excess air or by means of heat transfer surfaces immersed into the bed. High excess air use is typical for hot air or hot gas generation, described in Section 5.1.1. In FBC boilers, excess air is normally chosen to be as close as possible to the stoichiometric value, and the combustion temperature is maintained by removing heat using immersed heat transfer surfaces.

In calculating the quantity of heat which must be removed from the bed, or for calculating the area needed for heat removal, the area of water-tube furnace walls in contact with fluidized bed at operating bed height, must be known, and also the area of any heat exchange surface in the form of tube bundles immersed into the bed. Appropriate correlation for the heat transfer coefficients should also be used in such calculations (see Section 3.5.2 of Chapter 3).

The water-tube wall area which is in contact with the fluidized bed is to some extent predetermined by the given (and limited) furnace dimensions – size and shape of the bed cross-section and the bed height. The size of the immersed exchanger must then be determined depending on amount of heat that must be removed. In any such design the bed height must be sufficient to accommodate the proposed heat exchanger to be immersed into the bed.

This procedure presented here follows reference [2], although there are various, albeit essentially similar, approaches [52]. The determined heat balance is the basis of all engineering calculation procedures to determine the basic dimensions for FBC boiler furnaces, and must be included in the type of mathematical models described in Section 4.7.

First, the necessary fuel flow rate is determined for the required boiler power and the given fuel:

$$\dot{m}_F = \frac{Q_o}{\eta_o H_c} \tag{5.1}$$

Then, based on assumed excess air, the necessary air quantity is determined:

$$\dot{m}_v = \dot{m}_F \, \lambda \tag{5.2}$$

Boiler efficiency (which includes combustion efficiency) and excess air are the parameters the designer chooses based on experience from previously constructed boilers.

The proportion of the heat energy generated by coal combustion in the bed, depends on the type and characteristics of fuel, as well as the furnace and boiler construction. This parameter is very important in FBC boiler design and has been the subject of detailed and comprehensive investigations (see Chapter 6). When this "so-called" degree of inbed combustion is either assumed or known, the quantity of heat released in the bed by combustion is given by:

$$Q_b = \eta_b Q_o \tag{5.3}$$

The quantity of heat to be taken away by combustion products from the bed is given by the following:

$$Q_g = c_g (\dot{m}_F + \dot{m}_v)(T_b - T_{vo}) \tag{5.4}$$

If the quantity of heat needed to heat the fuel, air and limestone up to bed temperature is taken into account, as well as the amount of heat necessary for the calcination process, then the quantity of heat to be removed by the inbed heat exchanger and by water-tube walls, is:

$$Q_l = Q_b - Q_g - Q_F - Q_k \tag{5.5}$$

The surface necessary for heat transfer can be calculated by an expression having the following form:

$$Q_l = \alpha A_s (T_b - T_s) + \varepsilon_s \sigma_c A_s (T_b^4 - T_s^4) \tag{5.6}$$

Based on the characteristics of the bed material, the average bed particle size and the assumed fluidization velocity, the fluidized bed cross-section area may be calculated:

$$A_b = \frac{\dot{m}_v}{v_f \rho_g} \tag{5.7}$$

With the previously assumed bed height H_b, it is then possible to calculate the area of the furnace water-tube walls required to be in contact with the fluidized bed. Following this, and using the correlations given in Section 3.5.2, and relation similar to (5.6), it is possible to calculate an amount of heat Q_w, transfered to the furnace walls.
If:

$$Q_w > Q_l$$

it is not necessary to have inmersed heat exchanger and it may be necessary to cover part of water-tube walls with firebricks, to insulate the furnace.
If:

$$Q_w < Q_l$$

it is necessary to incorporate additional heat transfer surface into the fluidized bed in order to remove the heat surplus $Q_l - Q_w$.

If there is no possibility of incorporating the necessary surfaces into the bed, for a prescribed bed height, a greater bed height should be used, excess air increased, or other measures taken to increase the amount of heat removed from the bed.

The amount of heat which is removed from the bed by combustion products Q_K, determines the size of remaining exchange surfaces – in the freeboard and in the convective section of the boiler. For calculation of the necessary heat to be extracted from the furnace by water-tube walls in the freeboard, expressions such as those given in Section 3.6 should be used. For irradiated superheaters in the furnace and heat transfer surfaces in the convective part of the boiler, the customary engineering methods used in designing conventional boiler calculations are recommended [53, 54].

The engineering heat balance calculation for bubbling FBC furnace shown here, indicate that this type of calculation is based on a series of assumptions and empirical data, and does not take into account actual processes that occur in the bed. Furthermore, these methods of calculation rely on the methods and empirical data developed and verified for conventional boiler design, which have not been fully tested for application under conditions appropriate for FBC boilers.

The reliability of these engineering balance calculations has to be verified by comprehensive experimental investigations of combustion of various coals in laboratory scale furnaces, pilot plants and demonstration boilers, and by supplying data on the behavior of a large number of furnaces over long periods of operation. Initially, with any such combustion technology, there is insufficient operational data and a significant amount of time is needed to gather and interpret such data, which is why so many countries initially involved themselves in demonstration programs. The character and the results of these investigations, which were aimed at revealing the influences of coal characteristics on the design and operation of FBC boilers are described in detail in Chapter 6.

5.2.   The purpose and description of auxiliary systems in FBC boilers

Auxiliary systems appropriate to bubbling FBC boilers have been mentioned and listed in several places in this book, together with discussions on their basic purpose. In this section these systems are described in further detail, and issues relating to their design and application are discussed. As well, efforts have been made to discuss some of the most important parameters in choosing such a system, and practical operational experience is presented on these systems. In practice, information on auxiliary systems and their behaviour during operation are infrequently presented in the literature. Many data and design solutions are considered to be proprietary secrets. Further, such data as is available, tends to be scattered in the papers presented at international conferences (for example [19-21]), or papers describing operational experiences on FBC boilers. Further, as the technology matures it becomes more uncommon to discuss such issues. Therefore results obtained by investigations on auxiliary systems in pilot and demonstration plants [31-33, 42, 44] are of special interest. Most of the material presented in this book is based on data from papers presented at the three important international conferences and on material found in references [2, 12, 50], and especially in the book [55], which was an important source of data and recommendations for the engineers who design some of the most important fluidized boilers in the U.S.A..

Air distribution plate. The main task of the distribution plate is to provide uniform distribution of air for combustion and uniform fluidization over the whole cross section of the bed. At the same time, the distributitor plate must be designed to fulfill the following tasks:

– to prevent inert bed material from passing (weeping) into the air plenum chamber under the distributor plate,
– to support the weight of the static bed and to accommodate thermal expansion while maintaining a gas-tight seal, both in the nominal operational regime and during start-up, and
– to enable removal of any surplus inert bed material or oversized material.

The correct design of the distribution plate is also important in order to decrease erosion of metal surfaces in contact with the fluidized bed. Among the numerous shapes and designs of air distributors for achieving fluidization in chemical and other reactors, hoppers and pneumatic conveying systems, only two basic designs of distribution plates (Fig. 5.27) are used in FBC boilers:

– air distributors in the form of a water cooled base plate with built in nozzles ("bubble caps"), of different shapes (Fig. 5.27a and b), and
– air distributors in the form of a grid made of sparge pipes with drilled holes in the underside or fitted with nozzles ("bubble caps"), (Fig. 5.27c and d).

An interesting construction of the air distributor in the form of grid made of sparge tubes with rectangular cross section, and longitudinal slits was developed by CER, Čačak (Yugoslavia) for their FBC furnaces and FBC boilers built in cooperation with MINEL boiler Co., Belgrade, Fig. 5.27e. Different shapes of "bubble caps" are shown in Fig. 5.27.

Protection of the air distributor from high bed temperatures, can be achieved in several ways:

– by means of a stagnant layer of inert bed material between the "bubble caps" (Fig. 5.27a),
– by means of cooled tubes built into the distribution plate (Fig. 5.27b), and
– by means of air flow through the plate (Fig. 5.27d and e).

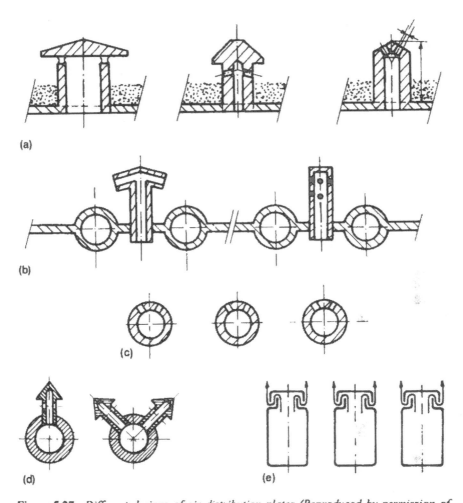

Figure 5.27. *Different designs of air distribution plates (Reproduced by permission of Routledge, inc., part of The Taylor & Francis Group. Copyright 1986, from [108]):*
(a) bubble caps of different shapes, (b) water cooled distribution plate with bubble caps, (c) sparge distribution plate with circular holes, (d) sparge distribution plate with bubble caps, (e) sparge distribution plate with longitudinal slots.

Thermal protection and compensation for thermal expansion of the air distributor are two much more difficult problems during the start-up period, when the combustion products of liquid or gas fuel are passing through at temperatures of up to 600 °C. For this reason the air distributor is often built of stainless steel.

An air distributor in the shape of a plate with "bubble caps" is more expensive to produce, but it has the following advantages:

- more uniform fluidization can be achieved ("bubble caps" are arranged at a uniform pitch of 75-100 mm), and
- thermal protection of the plate is easier to accomplish (height of the "bubble caps" being 50-100 mm).

It is more difficult to achieve uniform fluidization by means of sparge tube air distributors, but this concept has the following advantages:

- it is cheaper and easier to construct, and
- it is suitable for combustion of low-rank, high ash coals "as received," because it is easier to remove larger amounts of excess bed material from the bed, in the form of large ash particles, stones and/or agglomerated bed material, which tend to segregate and accumulate on the distributor.

The basic recommendations for the design and calculation of the air distributor can be summarized in the form of two ratios [2,55]:

(a) the ratio between the pressure drop across the distributor and the fluidized bed should be chosen to be about 0.1 over the nominal boiler operating regime (following the recommendations given by Kunii and Levenspiel from 1969 [56]). Higher values can also be found in the literature, including an extremely high recommended value of 0.3 [57]. When choosing the distributor pressure drop one should note that the air velocity changes with the change of the boiler power according to the square of air velocity, while the bed pressure drop remains constant. In order to provide good fluidization even under minimum regimes for boiler operation, it is better to choose the pressure drop ratio to be 0.1, and

(b) according to Kunii and Levespiel: "the open area of the distributor should be in the range 0.5 to 2% of the total bed cross section."

Detailed recommendations for distributor design can be found in the book [55], and in other books devoted to fluidization [56-58, 60-63], and in references on pneumatic conveying, but one should bear in mind that in FBC boilers the distribution plate operates under rather different conditions than appropriate for much of that literature.

Coal feeding system. Coal feeding in bubbling FBC boilers is one of the basic problems that has to be solved in order to achieve efficient combustion. Poor mixing of inert bed particles and fuel in horizontal direction is a characteristic of the bubbling fluidization, which was described in detail in Section 2.3.6. Problems that arise from this phenomenon are also mentioned in the same section as well as in Section 5.1.4. Among these problems, the most important ones are: short char residence time and residence time of volatiles in the bed and therefore, their incomplete combustion, appearance of hot spots in the bed with temperature higher than the designed ones, and the resulting danger of ash sintering and agglomeration of inert material. Many of these problems can be solved or at least reduced by proper coal feeding.

The task of any FBC coal feeding system is to distribute the coal uniformly over the bed surface and throughout the bed volume, to ensure good combustion conditions. At the same time, combustion products should be prevented from penetrating into the coal hopper, or the furnace's cold air entrance.

Fuel feeding in the fluidized bed furnace is usually arranged in two ways:

- by spreading fuel onto the free bed surface, and
- by injection fuel underbed.

The choice of one of these two procedures for feeding fuel into the bed is primarily influenced by the following fuel properties:

- geologically younger coals, primarily lignites, with high volatile content, or more reactive coals, are usually fed to the furnace as supplied (or as mined), without previous separation, washing and drying. Coal are fed with particle size between 0-25 mm (sometimes even up to 50 mm), most frequently overbed, and
- higher quality coals, which must be crushed to the size < 5 mm before combustion, because of their low reactivity, and fuels with high volatile content, with small particle size (for example biomass, sawdust), are usually injected underbed, near the air distributor.

Industrial boilers and small boilers have a wider range of potential solutions for designs for coal feeding, because of their smaller cross-sectional area. For large units, possible solutions are limited to systems, which can distribute fuel over very large area.

Three possible systems for **overbed FBC coal feeding** are shown in Fig. 5.28.

Gravity feeding (Fig. 5.28a) is achieved by supplying fuel by a screw feeder or a rotary valve (which also serves as a sealing device) to a chute, along which the coal passes through the action of gravity and falls onto the bed surface. The basic advantage of this type of feeding is its simplicity of construction and its high reliablity. Its potential deficiencies include the possiblity of devolatilisation, ignition and combustion of volatiles in the chute, and possible formation of a fuel rich regions and lack of oxygen in the region where fuel falls onto the bed surface. For this reason, it is important to ensure that fuel falls in a region of the bed where there is a strong downward movement of the particles into the interior of the bed (see Section 2.3.6, Figs. 2.32 and 2.33).

Following the recommendations given in [55] one such feeding point is needed for every 3-4 m^2 of the bed surface.

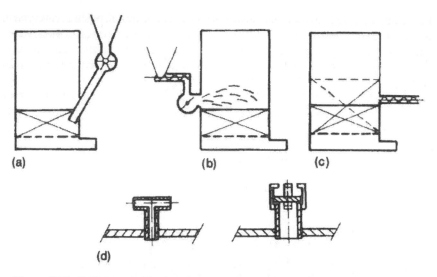

Figure 5.28. *Different coal feeding systems for fluidized bed combustion furnaces (Reproduced by permission of Routledge, inc., part of The Taylor & Francis Croup. Copyright 1986, from [108]):*
(a) gravity feeding, (b) overbed coal spreading, (c) overbed screw feeding, (d) different designs of the underbed pneumatic coal feeding devices

Fuel can be fed directly overbed by means of **screw feeders** (Fig. 5.28b) in a very similar way, and with the same advantages and limitations. Fuel can be injected inbed by means of a variable pitch screw feeder. This type of fuel feeding can also serve 3-4 m^2 of the bed surface, which is why these methods of feeding are applicable only to relatively small boilers. For example, the boiler in Fig. 5.21 has two screw feeders placed on one side of the boiler, for a total furnace cross section area of about 3 m^2.

For boilers of medium and high capacity, **systems for fuel spreading** onto the fluidized bed surface are used (Fig. 5.28c). These feeding devices were developed for conventional spreader stoker boilers with grate combustion, and have been successfully applied in FBC boilers. The basic advantage of this type of feeding is the possibility of ensuring a relatively uniform distribution of coal over a much larger bed surface area. According to the data given in [55] these devices can distribute coal over a surface of 1.5×7 m^2, and in [8, 11] it is noted that one feeding point is needed for 5-15 m^2. The Black Dog 125 MW_e boiler has 12 spreaders, i. e. one for each 14 m^2 [8, 34]. The advantage of this type of feeding system is that it draws on the long experience in designing and operating conventional boilers.

A general advantage of all overbed feeding systems is the possibility of feeding moist fuels "as received," with a particle size range between 0-50 mm, although it is recommended that percentage of particles smaller than 0.5 mm should not be higher than 10% for this type of system [55]. The deficiencies of this type of feeding are its limits in terms of the need to specify a particular particle size range and the likelihood of elutriation of small particles and their subsequent combustion above the bed. A further general deficiency of this type of system is the requirement for recycling unburned char particle, given the high elutriation losses that are likely to be experienced. System for draining, separation, crushing and returning suitable sized material into the furnace are also probably inevitable, since together with coal, large sizes of ash, stones and tramp material will enter the furnace. Futhermore, the limestone utilization for SO_2 capture is likely to decrease due to combustion of significant amount of the fuel above the bed and release of SO_2 in a region where the solid loadings are low.

Inbed coal injection. This is exceptionally accomplished by means of a screw feeder, however, normally it is achieved solely by means of pneumatic conveying, with nozzles of different shapes (see Fig. 5.28d) placed in the vicinity of the air distributor, but above the level of the "bubble caps" for air injection. In smaller boilers, pipelines for conveying coal can enter the furnace through sidewalls. In larger boilers, nozzles for coal injection are arranged uniformly between the "bubble caps" for air injection.

The basic advantages of this type of coal feeding are: (1) large residence time of the coal (char) particles in the bed and higher combustion efficiency, (2) the improved utilization of limestone, and (3) the fact that ash and stones are crushed to the size which does not negatively affect fluidization, making it easier to ensure removal of excess inert bed material.

The basic deficiencies are: (1) complexity of the feeding and coal pneumatic conveying system, (2) erosion of the pipelines for pneumatic conveying, and (3) the need for precrushing and drying coal.

For this type of coal feeding, according to [55], one injection point is needed per each 1.5 m² of the bed, and according to [11], per each 1-1.25 m², making these systems very complex. Even ensuring a proper particle flow distribution for a large number of pipelines is especially difficult. For example, the AKZO boiler of 90 MW$_{th}$ has two coal crushers, each having 6 pipelines for coal mixture conveying, and each supplying 3 feeding points (totally 36 coal feeding points), while the TVA-160 MW$_e$ boiler has 120 coal feeding points.

From the recommendations given in [55] coal should be crushed to a size of 5-6 mm, and dried to ensure moisture content not higher than 6-8%, prior to pneumatic transport and feeding. Nonetheless, despite its deficiencies, pneumatic inbed feeding of coal is the only way of feeding low reactive coals (anthracite, coke and bituminous coal), coal dust and fine biomass.

Limestone feeding systems. Handling of limestone and its introduction into the furnace do not pose major problems. For this reason these systems will not be described here. When feeding coal overbed, limestone feeding is usually ensured by means of a separate feed system, but there are cases where it is premixed with the fuel and introduced together with the fuel.

Inert bed material removal systems. The reliable operation of a FBC boiler demands control of its bed height and ensuring that a specified mean size of inert bed material is maintained. During continuous operation, the initial size distribution of the inert bed material changes, due to the following processes: (1) elutriation of small particles from the bed, (2) constant addition of limestone (if used), (3) accumulation in the bed of ash particles, and (4) the accumulation of tramp material fed with the coal, which will tend to accumulate in the bed and ultimately can defluidize the bed. Also if local overheating occurs, ash sintering and agglomeration of inert bed material are possible during boiler operation. In general, the particles mean size and bed height increase during operation, although there are some systems in which makeup bed material must be continuously introduced in the bed. Due to these processes the composition, size and the amount of inert bed material change during long term boiler operation.

The task of the system for draining inert material from the bed is to remove any surplus material. Drained material must also be cooled (usually to 150 °C), oversized particles separated and/or crushed. Particles of correct size may be returned to the furnace, stored for later use or simply disposed of.

Some boiler designs have overflow openings at the nominal level of the bed surface, however, most frequently material is drained from the bottom of the bed. If a distribution plate with "bubble caps" is used, material is removed by gravity using downflow tubes. Openings for spent inert bed material may be uniformly distributed across the distribution plate, with one opening for each 1.5 m^2, according to [50]. However, the Foster Wheeler boiler shown in Figs. 5.12 to 5.18 has only two openings, and the AKZO 90 MW_{th} has only one such opening. The number of openings depends mostly on the properties and the type of coal feeding system, so that it is difficult to regard any such choice as generally valid. During draining of the inert bed material in the fluidized state, the recommended diameter of the downflow tube is about 150 mm.

If the air distributor is in the form of a sparge tube grid, draining of coarse particles from the bed becomes easier, but one should remember that the material under the grid is not in the fluidized state. For this reason, an adequate slope at the bottom and the size of the opening should ensure the free drainage of the material in the packed bed under the action of gravity.

Material drained from the bed under high temperature (800-900 °C) must be cooled usually to 150 °C before discharge. In the case of continuous drainage of a large amount of material, this must be cooled after leaving the bed, in order to decrease heat losses and increase the boiler efficiency. Cooling is achieved by means of cooled screw conveyors, or by special heat exchangers. In boilers with tube distributors, some cooling can be arranged by directing the openings for air injection sideways or downwards, or by special air injection, which passes through the material deposited under the distributor.

If necessary, the inert material can be sieved and crushed using conventional mechanical devices (rotational sieve, vibration sieve and crusher), and then mechanically or pneumatically returned to the furnace. Diagram for systems for bed material drainage can be seen in Fig. 5.12 and Figs. 5.17-5.22.

Flue-gas cleaning systems. FBC boilers, as well as other conventional boilers, must observe environmental regulations for the emission of solid particulates. In developed countries particle emission less than 50 mg/m^3 are normally required. Systems for flue-gas cleaning from solid particles in FBC boilers apply conventional devices: cyclones, multicyclones, bag filters and electrostatic precipitators, and they do not differ much from systems in conventional boilers.

Besides elimination of particulates, solid collections systems can have another important task in FBC boilers – to collect unburned char particles that may be recycled to achieve burn out. For that reason, particle separation is usually done in two steps. First, particles are separated and returned to the furnace by mechanical (inertial) dust collectors or multicyclones at the exit from the second or the third boiler gas pass. Final flue gas cleaning is most frequently done by bag-filters and in larger boilers by electrostatic precipitators. To achieve extremely high combustion efficiency, particles from bag filters (or electrostatic precipitators) can also be partialy returned to the furnacel.

Compared to conventional pulverized coal combustion boilers flue gas cleaning system have to cope with the following: (1) higher concentration of particles in flue gases, due to the presence of the products from limestone use (lime and anhydrite), (2) smaller particles, (3) the need to separate particles for reinjection into the furnace, and (4) different particle characteristics (and higher electrical resistance), which can cause problems in operation of filters and electrostatic precipitators. The problems of solid particle emission control in FBC boilers will be discussed in more detail in Chapter 7.

System for recirculation of unburned particles. Bubbling fluidized bed boilers cannot achieve high combustion efficiency, especially when low-reactive coals are burned, if unburned char particles are not reinjected to achieve burn out. The need for recirculation and an optimum particle reinjection mass flow rate depend on the properties of the coal (reactivity, volatile content), the manner of feeding, particle size distribution of the coal (i. e. amount of particles smaller than 0.5 mm) and the desired combustion efficiency.

Recirculation of unburned particles can increase combustion efficiency up to 10% [55, 64] and under less favourable conditions by up to 2-5%. The intensity of recirculation is expressed by the ratio of mass flow rate of recirculated particles and mass flow rate of coal feed. This parameter is called recirculation (recycling) ratio or degree of recirculation. In industrial boilers the recirculation ratio will typically adopt values of about 2-3. Higher recirculation ratios do not contribute much to the increase of combustion efficiency, but they do significantly increase the price of the boiler and operation expenses. These moderate recirculation ratios are usually good enough to ensure combustion efficiency of 98-99% [55].

The return of particles collected in multicyclones, or even better, before final cooling of combustion products is more favourable. Particles that are returned into the furnace are larger, it is easier to ensure their transportation and they enter the furnace at temperatures of 350-450 °C, which is better for their final combustion. Particles separated in final dust collectors (particle size 10 μm, and cooled to about 150 °C) are more difficult to transport and they readily elutriate from the bed.

Particles are returned into the bed by pneumatic conveying, and are reinjected inbed.

Boiler start-up systems. A system for FBC boiler start-up should provide safe heating of inert material in the bed, from ambient to the temperature for coal ignition, and good fluidization when coal feeding starts.

The rate of and the start-up procedure itself, as well as the start-up temperature (the temperature when coal feeding begins) depend on various factors, of which probably the most important are the fuel properties. However, there are other factors, such as the cost of the system, the need to decrease liquid or gas fuel consumption during start-up, and the purpose of the boiler. The choice of start-up temperature will be described in detail in the next chapter.

Boiler start-up is a very sensitive process, which must be well planned by the boiler designer. Although operation of FBC boiler is automatic, start-up of the boiler is usually done by manual control, especially for smaller boilers.

More reactive, younger coals – lignites, permit start-up of a boiler from very low bed temperatures. For example, the boilers shown in Figs. 5.21 and 5.22, and the furnaces shown in Figs. 5.1 and 5.2 are designed in such a way that when burning lignites, coal introduction can begin when bed temperatures reaches 350-450 °C. During start-up from such low bed temperatures there are two critical moments. Up to the temperature of 500-600 °C, there is no combustion of volatiles, so that a larger amount of char accumulates in the bed than needed for normal boiler operation. When the ignition temperature of volatiles is achieved, a sudden increase of bed temperature may occur, as well as bed overheating and even ash sintering and agglomeration of inert material. In order to prevent such sudden temperature rises, the start-up of the boiler is begun with a smaller coal flow rate than needed for steady state operation, and coal feeding may be periodically interrupted, and then restarted, when bed temperature drops are noticed.

For less reactive coals, the volatiles will ignite during the start-up (at temperatures of 500-600 °C), however, they tend to burn in the freeboard. Heat generated by combustion of the volatiles does not therefore influence bed heating and bed temperature rise. At the same time, coal char in the bed will tend to accumulate in the bed due to the low bed temperature, which ensures that it tends to have a relatively low combustion rate. Again, for such conditions, there is too much fuel in the bed, and the bed temperature may rise suddenly, causing the permissible temperature to be exceeded, and local sintering or even bed defluidization to occur. For this reason, the start-up process must be carefully controlled during the start-up period from 600 °C to 800-900 °C. Coal feeding can be continuous, but with lower flow rates than in the steady state, with gradual approach to the bed temperature and gradual increase of coal flow rate to the steady state values.

For the above reasons, a start-up temperature higher than 500 °C is recommended in order to avoid at least one start-up critical point [2, 50]. Practically all industrial FBC boilers have a start-up temperature between 500-600 °C. When assessing this choice one should bear in mind that older coals, brown and hard coal do not ignite below 500-600 °C, and that anthracite or petroleum coke may demand a minimum temperature of 700 °C for a safe and sufficiently quick start-up.

A simple rule can be formulated: the higher the start-up temperature, the safer, less dangerous and faster the FBC boiler start-up, but this comes with great technical and construction difficulties and the investment costs for the boiler are higher.

Three procedures have so far been applied for heating a bed of inert material in a FBC boiler:

- gas or liquid fuel burners placed above the bed surface and directed towards the bed surface to heat inert material at bed surface. Such systems are typically only used for smaller units and have two major deficiencies: (1) much of the heat released does not go into bed heating, and (2) overheating of the bed surface may also occur, and if fuels with low ash melting point are used, sintering may occur on the bed surface. The advantage of this start-up procedure is that the distribution plate is not thermally loaded,
- the bed is heated by direct combustion of gas and an air mixture which is introduced into the bed in the vicinity of the distribution plate. When the bed is cold, the gas ignites and burns on the surface of the fixed bed, and then the flame front gradually descends into the bed allowing the fluidization of the part of the bed above the flame front. This start-up system is suitable for larger boilers (see for example 90 MW$_{th}$ boiler in Fig. 5.19). Such systems however come with a risk of explosion problems, and the standards and regulations appropriate to the design and operation of such gas burning systems are still somewhat problematic, and
- the bed can be heated by the products of gas or liquid fuel combustion introduced through the distribution plate. This is the most frequently applied procedure for furnace and FBC boilers start-up, and it can be applied for boilers of all capacities, both small and large. Liquid or gas fuel burns in a special start chamber, or directly in the air plenum chamber under the air distribution plate. Combustion products having temperatures of 600-800 °C, enter the plenum chamber, pass through the plate and gradually heat-up the bed. The advantage of this start-up procedure is that most of the heat is used for bed heating itself. The deficiency of this approach lies in the need for thermal isolation of the pipelines and the air plenum chamber, and cooling of the distributor, or its construction with an expensive material such as stainless steel. Thermal expansion of the plate can also be an important design problem.

A large amount of heat is necessary to heating the inert material during start-up of FBC boilers due to the great thermal capacity of the bed and built-in heat exchanger surfaces. This is a significant problem in large capacity boilers, which necessarily contain a huge amount of "inert material." For 1 m² of air distributor surface, at a mean fixed bed height of 0.5 m, there are typically about 700 kg of inert material. To heat this mass from 20-500 °C, the quantity of heat necessary is about 335 MJ, and a FBC boiler of capacity of 20 MW$_{th}$ has about 7-10 t of inert material in the bed.

The heat absorbed during limestone calcination should also be added to the amount of heat necessary. If the heat taken away by the heat exchanger were added to this, a very high power start-up chamber would be needed, as well as a large consumption of fuel during the start-up period. Ideally, the start-up chamber capacity should not be more than 10% of the boiler capacity.

The following construction and design solution can reduce the amount of heat and power needed by the start-up chamber for reasonably fast start-up of FBC boilers:

– division of the fluidized bed (furnace) into several sections. Only one, the smaller, section is heated by combustion products and the fuel is injected in it first. The remaining, larger section of the furnace does not operate during start-up. Heating of the remaining part of the furnace is performed by periodical fluidization and mixing of bed material with heated material from the start-up section. Fuel feeding into this section of the furnace begins when adequate temperature is achieved,

– heat exchanger surfaces are typically not built into the start-up section of the furnace, and even any furnace water-tube walls are coated with firebricks in this section, and

– start-up of the boiler begins using a decreased amount of inert material in the bed, so that the heat transfer surface remains above or outside this section of the bed.

In all the FBC boiler designs, except boilers of very low power (for example a few MW$_{th}$.) separation of a smaller start-up section of the furnace is used. Division of the furnace can be physical, but in any case, air injection is separated for each section of the furnace. The boiler shown in Fig. 5.12, is separated into four sections, two of them being start-up sections. The AKZO 90 MW$_{th}$ boiler (Fig. 5.19) is separated into 11 cells in order to attain better load following, but even when gas combustion in the bed is used, only 3/5 of the furnace operates at start-up [23, 24]. Boilers shown in Figs. 5.21 and 5.22 are started with only half of the furnace. A hot water demonstration boiler of 9.7 MW$_{th}$, built in the Institute of Thermal Engineering and Energy of the VINČA Institute of Nuclear Sciences (Belgrade) by reconstructing an old boiler burning liquid fuel, has a special design solution and starts up with only 1/4 of the furnace employed [6, 66].

The minimum fluidization velocity of the inert bed material at ambient temperature is several times higher than the minimum fluidization velocity under operating conditions at 800-900 °C. If it were necessary that all the inert material be fluidized at the beginning of the start-up period, the capacity of the primary air fan would have to be a several times larger. At the same time, due to high velocities in the air distributor openings, the air distributor pressure drop would be excessively high during start-up. Fortunately, fans of such large capacities and power are not needed during steady state boiler operation.

Boiler start-up begins with small air flow rates, and the inert material is not fluidized at the beginning of the lightup process. During start-up, air flow rate should be chosen so as to ensure that the inert material is fluidized when one approaches the start-up temperature, i. e. when coal feeding begins, in order to provide proper mixing of fuel.

Usually the flow rate is chosen to provide fluidization velocity of about $2 v_{mf}$ at the start-up temperature of the furnace. The degree of fluidization increases with the increase of bed temperature with constant air mass flow rate, so that a decrease of flow rate is possible until the nominal value is reached.

Typically, a few hours (2-4 h) are needed for bubbling FBC boilers to start-up from the cold state, dependent on the fuel type and start-up procedure. However, the high heat capacity of the inert material enables the bed to remain heated for up to a day or so after shut down. Start-up from this heated state is possible without using the start-up system or with only minimal support with combustion products from the start-up chamber burner.

System for water and steam circulation. System for water and steam circulation, as well as arrangement of heat transfer surfaces in the FBC boiler do not differ from the circulation in the conventional boilers of corresponding power. This is evident from the boiler diagrams shown in figures of Sections 5.1.2 and 5.1.3. The only particularity is that a high amount of heat (up to 50% of the boiler thermal power) must be removed from the fluidized bed (except when burning very moist fuels such a bark or a range of biomass fuels). To remove this amount of heat, the heat exchanger surface immersed in the bed can be much smaller than would be necessary in conventional boilers for the same amount of heat, because the heat transfer coefficients in fluidized bed are high. For this reason, the total surface for heat transfer, and thus the boilers size, is significantly smaller than in conventional boilers with the same power output.

Heat transfer surfaces in the bed are most frequently horizontal or positioned with a slight inclination of 10-20°.

Natural as well as forced circulation is used. In the case of natural circulation, a special circulating pump is introduced into the circulation scheme (for example in the boiler in Fig. 5.19) in order to enable circulation through horizontal heat transfer tubes in the bed. This pump is indispensable if evaporation surface is involved, but it is not needed if a steam superheater is placed in the bed.

5.3.   Efficiency of solid fuel
combustion in the FBC boilers

Comparison of FBC boiler performance with other boiler types should be made based on the following parameters: combustion efficiency, overall boiler efficiency, flexibility with respect to the ability to burn different fuels and emissions of harmful gases and particulates in the environment. The ability to efficiently burn different and unusual low-quality fuels has been discussed several times in this and previous chapters. In this section therefore, only data concerning combustion efficiency and overall boiler efficiency will be provided and discussed. An entire chapter (Chapter 7) is devoted to the emission of harmful gases and particulates.

Combustion efficiency is defined for fluidized bed combustion boilers in the same way as for conventional boilers. Efficiency of combustion is primarily affected by the heat losses due to elutriation of unburned char particles in fly ash and to a much lesser extent by chemically incomplete combustion expressed in terms of unburned hydrocarbons and carbon monoxide. Fluidized bed combustion boilers suffer another loss of combustible solids, namely unburned solids in any excess bed material that must be removed either intermittently or continuously from the furnace. This loss is rarely measured separately and data on efficiency often does not include this loss, as bed char loadings are normally not high. However, for some fuels, such as very unreactive high ash coals, it may be important and should be included in efficiency calculations.

Fluidized bed boiler efficiency depends mainly on the temperature of the flue gases, combustion efficiency, and of course, consumption of electric power needed by the boiler to running its auxiliary systems (fans, pumps, feeders etc.), the so called parastic power requirement for a boiler. According to the data in references [40, 68, 69], the self power consumption of FBC boilers per 1 kWh of generated energy is similar to that for pulverized coal combustion boilers. Also, it may be possible to adopt lower flue gas exit temperature when designing FBC boilers due to significantly smaller SO_2 concentration in flue gases. That is why it is often possible to achieve very high degrees of efficiency in FBC boilers (up to 90%) which are comparable with those seen with pulverized coal combustion boilers, and normally significantly higher than efficiency in grate fired boilers.

As noted in the introduction, the combustion efficiency of boilers with a bubbling, or stationary fluidized bed, is typically up to levels of about 90% without recirculation of fly ash, and up to levels of about 98% with particle recirculation [4, 11, 14], depending on the type of fuel and boiler design. Data available in the open literature also indicate that even higher efficiencies can sometimes be achieved, see for example references [19-21].

As an illustration, of the potential efficiency of combustion in FBC boilers, we will cite data obtained during the operation of some of the biggest bubbling fluidized bed boilers ever built (Table 5.6).

Table 5.6   *Combustion and boiler efficiency data for some very large bubbling FBC boilers*

Boiler	Combustion efficiency [%]	Overall boiler efficiency [%]
TVA, 160 MW$_e$ [42]	96.00-98.00	87.46
Black Dog, 130 MW$_e$ [36]	98.70-99.04	88.60
Wakamatsu, 50 MW$_e$ [46, 48, 70]	98.00-99.50	88.61-91.32
AKZO, 90 MW$_{th}$ [70]	98.85-99.15	92.50-93.24

Numerous factors influence combustion efficiency in boilers with fluidized bed combustion. Practically all of the parameters discussed in Section 5.1.4 influence combustion efficiency. These parameters can be classified into three basic groups: fuel properties, parameters determining operating conditions and boiler design. Fuel properties influencing combustion efficiency the most are: volatile content, char reactivity, fuel particle size distribution. Basic parameters, which define the combustion regime, are bed temperature, fluidization velocity and excess air. Design solutions that influence the efficiency of combustion are primarily: the degree of fly ash recirculation, ratio of secondary and primary air flow rates and intensity of mixing in the freeboard, place and manner of fuel feeding, bed height and freeboard height.

Data on combustion efficiency is either obtained on industrial boilers or from pilot and experimental furnaces and boilers [19-21]. Data obtained by measurements from industrial plants are the maximum efficiency of the plant measured under steady state operating conditions. For industrial scale boilers, it is usually impossible to investigate the influence of each parameter on combustion efficiency. Instead, the influence of such parameters is determined by detailed experimental programs carried out in experimental furnaces or pilot boilers. The majority of such plants typically have cross sections of the order of 300×300 mm up to perhaps 1×1 m or higher, the thermal power from such units may range from less than 1 to 20 MW$_{th}$. The best known plants of this type are the TVA 20 MW$_{th}$ in the U.S.A., 20 t/h Wakamatsu in Japan, TNO 4 MW$_{th}$ in the Netherlands and the 16 MW$_{th}$ at Chalmers University in Göteborg. For these units a range of detailed studies have been carried out on the effect of the various parameters on combustion efficiency. When analyzing these data, one should bear in mind that in practice it is not really possible to change only one parameter and keep the others constant and there are always changes in other process parameters if one looks at the data closely enough. Combustion efficiency data obtained by such investigations not surprisingly shows that for these smaller plants they are usually somewhat lower than the values seen with large boilers. The basic reason is probably the fact that industrial boilers usually operate with recirculation of fly ash and higher freeboard heights, which usually allows sufficient time for more complete combustion of char particles. It

should also be borne in mind that for some of these fuels (e. g., anthracite, or petroleum coke), combustion may be quite difficult in more conventional boilers, especially when those fuels are employed in smaller systems such as stokers and so it is
important to take into account the fuel when comparing boiler performance .

5.3.1. Influence of fuel properties

Coals with high percentage of volatiles (lignite and subbituminous coals) as
a rule have lower fixed carbon content and the possibility of elutriation of small unburned char particles is lower. At the same time, these type of coals produce highly
reactive char, which facilitates char combustion possible either in fluidized bed itself or the freeboard. During combustion of such coals combustion inefficency, due
to to unburned char particles escaping the boiler is very small [1,71-73].

Influence of the C_{fix}/VM_0. The influence of coal properties on combustion
efficiency is expressed most frequently by means of the dependence of combustion
efficiency on the so called fuel ratio, i. e. ratio of fixed carbon to volatile mater content, C_{fix}/VM_0. Detailed investigations of combustion efficiency for different coal
types using this correlation were performed on the Wakamatsu pilot boiler in Japan.

Results from tests with seven different coals having volatile contents ranging from 23.9 to 40.8% [74] are shown in Fig. 5.29. Investigations were carried out
under nominally identical conditions (bed temperature 850 °C and excess air 1.2)
in two plants – the pilot-boiler with the steam capacity of 20 t/h, and the demonstration Wakamatsu boiler of 50 MW_e. Combustion efficiency for high volatile coal
was over 99% in Wakamatsu 50 MW_e boiler which has a special burn out cell for
combustion of unburned char particles (which is in effect equivalent to
recirculation). When burning coal with low volatile content this boiler achieved a
combustion efficiency higher than 98.5%. In the pilot plant without recirculation,
the combustion efficiency for this coal was below 90%. The results of long term
comparative investigations of combustion efficiency for 20 coal types using the experimental furnace 300 × 300 mm at the Institute for Thermal Engineering and Energy of the VINČA Institute are also described in detail in Chapter 6. These
investigations also show a strong dependence of combustion efficiency in a
fluidized bed on the ratio C_{fix}/VM_0. Combustion efficiency (without recirculation)
for lignites was higher than 99% and for anthracite and coke with small particle size
it took much lower values of only 65-75% (Fig. 6.9). The influence of volatile content on combustion efficiency can be seen in Fig. 6.11 in the next chapter, where results are compared for four different coal types burned in plants of different sizes.

Influence of coal (char) reactivity. Fuels which have low char combustion
rate (low reactive coals), such as anthracite, coke and high rank coals burn in FBC
boilers with very low efficiency independent of their volatile content.

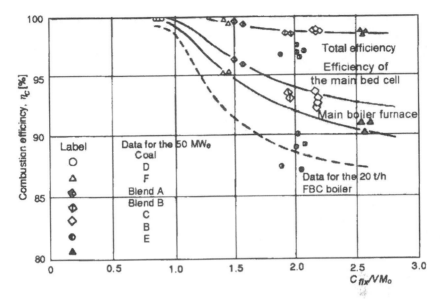

Figure 5.29. *Combustion efficiency as a function of the C_{fix}/VM_o ratio for the different coals [74]*

The basic reason for this is that the residence time of char particles smaller than 1 mm is insufficient to allow for complete combustion. Losses due to incomplete combustion of particles, which are elutriated from the furnace, can be very high. When burning these coals it is essential to recirculate the fly ash particles caught in cyclones or bag filters in order to increase efficiency. Combustion temperature is the dominant parameter influencing the combustion rate of low reactive coals, and therefore combustion efficiency, so in these cases a higher bed temperature must be employed, although this may adversely affect sulphur capture or NO_x emissions.

Results from investigations performed with the mixture of coal and anthracite in different ratios [48] can be used to illustrate the influence of reactivity on combustion efficiency. In an investigation with an experimental boiler with a cross section of 300×300 mm^2 and height of 4.7 m under identical conditions, the combustion efficiency seen with coal firing of 90% changed to about 60% for a 100% anthracite (Fig. 5.30).

A qualitative presentation of combustion efficiency achieved in Ahlstrom plants when burning different coals and the influence of recirculation and bed temperature when burning coals of different reactivity are shown in Figs. 5.31 and 5.32 respectively.

Influence of volatile matter content. Very high fuel volatile matter content can have an unfavourable influence on combustion efficiency. Volatile matter can escape

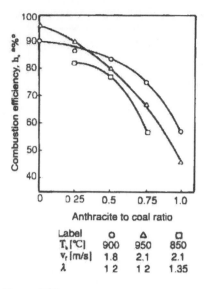

Label	O	△	□
T, [°C]	900	950	850
v_f [m/s]	1.8	2.1	2.1
λ	1 2	1 2	1.35

Figure 5.30.
Influence of anthracite content in the anthracite/coal blend on the combustion efficiency (Reproduced by kind permission of the ASME from [48])

from the furnace mostly as unburned, as CO or unburned hydrocarbons, because of poor mixing with oxygen in the bed or above it. When burning this type of fuel secondary air should be introduced above the bed in order to provide adequate oxygen for combustion and to increase the mixing intensity in the furnace. The illustration of the possible negative influence of volatile matter content on combustion efficiency is shown in Fig. 5.33 [49], which shows the results of a study on a mixture of coal (about 37% of volatile matter) and fuel waste (waste with about 57% of volatile matter) at different ratios. In Fig. 5.33, it can be seen that the decrease of total combustion efficiency is caused by incomplete combustion of volatiles, and evidently, favourable conditions for their combustion were not achieved in this plant. This study was carried out using an experimental furnace with cross section 300 × 300 mm and insufficient height, without introducing secondary air.

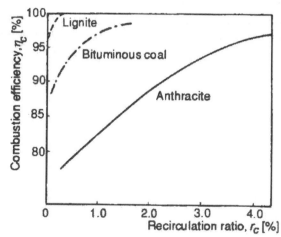

Figure 5.31
Influence of the recirculation ratio on combustion efficiency for coals with different reactivity (Reproduced by kind permission of the ASME from [75])

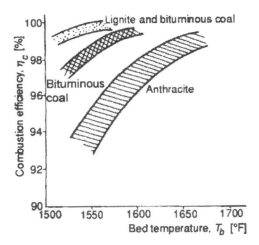

Figure 5.32.
Influence of the bed temperature on the combustion efficiency for coals with different reactivity (Reproduced by kind permission of the ASME from [75])

Influence of fuel particle size distribution. It is very complicated to study the effect of fuel particle size and to separate it from the effects of other parameters as for example fluidization velocity, fuel reactivity, fragmentation and attrition of fuel particles, etc. Further, the very act of feeding will change the particle size. It is also important to distinguish between the problems associated with the content of fine particles (fuel particles smaller than 1 mm), which will tend to be elutriated from the bed, and the real influence of mean fuel particle diameter burning in the

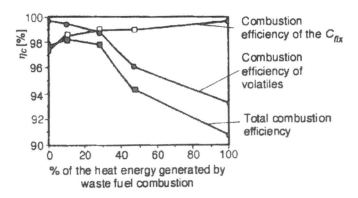

Figure 5.33. *Reduction of the combustion efficiency due to the incomplete volatiles combustion (Reproduced by kind permission of the ASME from [49])*

bed. Combustion efficiency for fuels with different particle sizes is also strongly influenced by the manner of feeding. When fed inbed, particles have a higher probability of complete burnout, but the mean fuel particle size for such type of feeding
must be smaller, and the yield of fine particles will therefore be greater. When fed
overbed, the mean size of the particles is larger (up to 50 mm), but most of the fine
particles smaller than 1 mm are unlikely to reach the bed surface and experience
complete burnout in the furnace, instead they will tend to be elutriated.

When studying the influence of mean particle size on combustion efficiency
one should have in mind two competing influences. With the decrease of particle
size, burnout time decreases but the possibility of elutriation also increases and the
particle residence time in the bed will decrease. Experimental data in the literature
are contradictory because of the simultaneous influence of these and other factors,
and they show that both decreases and increases of combustion efficiency are possible when particle size decreases [55]. Theoretical studies show that there exists a
critical diameter of fuel particle for which particle burnout time reaches a minimum. The dependence of combustion efficiency on the fuel particle size when
feeding Ohio No. 6 coal under bed is shown in Fig. 5.34. This study was performed
in an experimental boiler with a cross-section 6 × 6 ft in the U.S.A. [55].

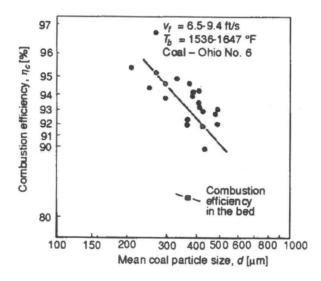

Figure 5.34.
*Influence of fuel particle size on combustion efficiency during
over bed feeding (Reproduced by kind permission of MIT,
Cambridge, from [55])*

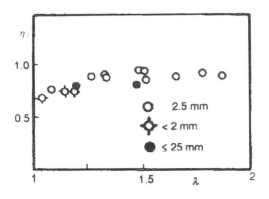

Figure 5.35.
Influence of particle size on combustion efficiency for high reactivity coal [72]

When feeding coal overbed, the greatest influence on combustion efficiency can come from particles smaller than 1 mm which will most probably not reach the bed surface. In this case, the mean diameter of the particles, which reach the bed may not have any major influence on combustion efficiency since the losses can be predominantly from the elutriated char particles. However, since feeding overbed is most frequently used when burning highly reactive coals (lignite), small particles can still burn out before leaving the furnace depending on freeboard hight. Results of an investigation on a small experimental furnace having a diameter of 147 mm, presented in Fig. 5.35, show that the influence of particles smaller than 1 mm on combustion efficiency of lignite is not significant [72].

5.3.2. Influence of combustion regime parameters

Influence of fluidization velocity. The general conclusion of numerous experimental investigations [19-21, 55, 77-79] is that combustion efficiency decreases with the increase of fluidization velocity. The basic reason for this negative influence is the increase of probability that small char particles will be elutriated unburned from the bed and from the furnace. The magnitude of this influence depends on other factors as well, for example: fuel reactivity, the size of fuel particles, and velocity range in which investigation is done. Also, if the fluidization velocity is too low this may itself lead to a decrease of combustion efficiency, because of lower mixing intensity in the bed. In practice, such a low fluidization velocity is not adopted for many other reasons, first of all, poor fluidization might itself cause defluidization, hotspots and sintering, and secondly because the output of the furnace is determined by the amount of air being fed per unit area so that very low fluidizing velocites cannot be associated with efficient boiler performance for a given boiler size. As an example, the influence of fluidization velocity during combustion of two types of Indian coal with high ash

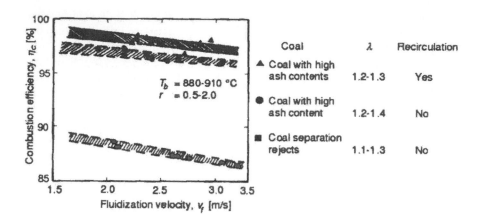

Figure 5.36. *Influence of fluidization velocity on combustion efficiency (Reproduced by kind permission of the ASME from [77])*

content in the experimental furnace with 1 x 1 m² cross section and the freeboard height of 8 m at ORNL (U.S.A.) is shown in Fig. 5.36 [77].

    Influence of excess air. Increase of oxygen concentration, i.e. more oxygen available for combustion, decreases chars burnout time and increases the chances for combustion of volatiles. For this reason, excluding elutriation losses, an increase of combustion efficiency should be expected with the increase of excess air in the region close to the stoichiometric ratio. However, under real operating conditions, with usual values of excess air in FBC boilers (of about 1.1 to 1.2), further increases of excess air do not signifcantly contribute to an increase of combustion efficiency, as can be seen from Fig. 5.37 [55], which shows that an increase of excess air over 1.2, does not in practice cause a decrease of heat losses and hence does not contribute to combustion efficiency.

    This results for two reasons. First, char combustion takes place in the emulsion phase in which the amount of air is limited by the magnitude of minimum fluidization velocity. Any increase airflow will tend to bypass the bed in the form of bubbles, thus taking no part in the combustion process, although assuming that it is not connected with a simultaneous increase of fluidization velocity, an increase of excess air does not influence particle elutriation from the furnace. Second, higher gas flowrates are associated with greater heat losses in the flue gases.

    Influence of bed temperature. Combustion temperature (bed temperature) influences the rate of oxidation chemical reaction exponentially, causing the

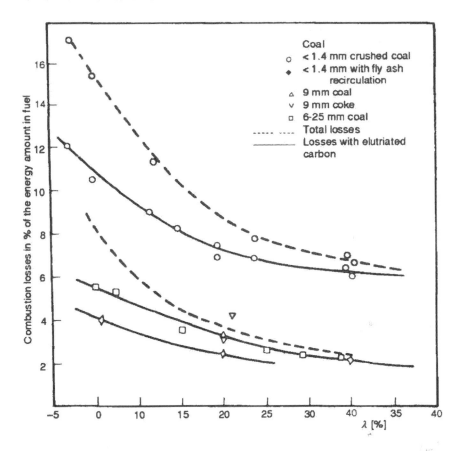

Figure 5.37. *Influence of excess air on combustion losses in fluidized bed combustion (Reproduced by kind permission of MIT, Cambridge, from [55])*

char burnout time to decreasing significantly, which contributes to an increase of combustion efficiency. All of the available experimental data show an increase of combustion efficiency with the increase of bed temperature [19-21, 55, 73, 76, 78-80]. The influence of temperature is higher for the combustion of low-reactive coals, when char combustion is controlled by kinetics of the chemical reaction (see Fig. 5.32). In combustion of reactive coals (lignite), with coarse particles, the control process is diffusion of oxygen towards the fuel particle and the combustion temperature has a much smaller influence. When considering the influence of bed temperature on combustion efficiency under real operating conditions, it is important to note two facts: (a) the influence of temperature is higher at lower tempera-

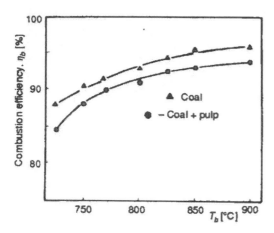

Figure 5.38.
Influence of bed temperature on combustion efficiency (Reproduced by kind permission of the ASME from [73])

tures (700-800 °C) which are normally not adopted as operating temperatures (except for biomass), and (b) the choice of the bed temperature where sulphur capture is a feature will be determined by the optimum temperature for sulphur capture with limestone (about 850 °C). In general, it is unlikely that a bubbling fluidized bed will operate much above 850 °C. In consequence, increasing combustion efficiency by choosing a higher bed temperature when designing FBC boilers, is a fairly limited option and in any case the influence of temperature on burn-out time is smaller in the temperature region between 850 and 900 °C. Examples of the influence of temperature on combustion efficiency are shown in Fig. 5.38 [73], based on experimental results from the combustion of brown coal (volatile content about 30%) and a mixture of coal and waste pulp from paper production in industrial boilers having steam capacities between 10 and 22.5 t/h.

5.3.3. Influence of furnace design

The influence of recirculation of unburned particles. Highest losses during combustion in fluidized bed arise from the elutriation of unburned particles from the bed and the furnace itself. For that reason when burning low reactive coals or coals with high percentage of particles smaller than 1 mm, it is not possible to achieve acceptable combustion efficiency, or high boiler efficiency, without fly ash recycle of the unburned solids collected in cyclones or in the bagfilters. Introducing recirculation increases the complexity of the plant and increases overall boiler costs, but the gain in increased combustion efficiency often exceeds the increased cost of using this option. Based on investigations in pilot plants and demonstration boilers, it has been established that the degree of recirculation (the ratio between mass flow rate of recirculated particles and mass flow rate of coal) when burning unreactive coals has the highest influence on combustion efficiency (see Fig. 5.31) in the range 0 to 2.0 for this ratio [69, 74-76, 78, 79, 81-84].

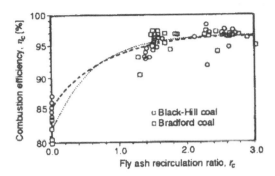

Figure 5.39.
Influence of fly ash recirculation ratio on combustion efficiency (Reproduced by kind permission of the ASME from [81])

Further increase of recirculation degree do not significantly influence the combustion efficiency. As an illustration of the influence of the degree of recirculation one can look at the results of detailed combustion efficiency measurements for two types of brown coal in an experimental furnace with a cross section $1 \times 1 \ ft^2$, Fig. 5.39 [81].

The influence of fluidized bed height. The increase of fluidized bed height causes an increase of the bed residence time of fuel particles and the probability of combustion of volatile matter burn-out, while char concentration in the bed decreases, all of which, generally have a favourable influence on combustion efficiency. The influence of fluidized bed height itself on combustion efficiency has rarely been investigated in detail [55, 84]. This is because, first of all, bed height is determined by the available fan pressure head, the space needed for the placement of immersed heat transfer surfaces, the necessary bed height for heat transfer to water tube walls, etc. In general it is better to choose the greatest possible bed height compatible with those limits.

The influence of freeboard height. It is always recommended to choose furnace height as great as possible, if there is no important reason to do otherwise. A high furnace enables char particles and volatiles to increase their residence time in the furnace, and achieve better burnout, thus greater height means higher combustion efficiency [55].

The influence of secondary air. In contemporary FBC boiler designs, injection of secondary air above the bed surface is nearly always used. In the bed, combustion is expected to take place under stoichiometric or even substoichimetric conditions. Injection of secondary air allows the following: (a) achievement of staged combustion in order to decrease NO_x emissons, and (b) a supply of a high enough amount of air and intense mixing in the freeboard, to achieve more efficient combustion of volatiles. Furthermore, in the case of burning high volatile fuels (for example biomass), secondary air is indispensable to achieve acceptable combustion efficiencies and low CO emissions. Experimental data [76, 85] shows that injection of secondary air and staged combustion does not normally have a negative

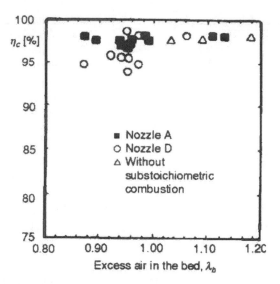

Figure 5.40.
Influence of secondary air on combustion efficiency
(Reproduced by kind permission of the ASME from [85])

influence on combustion efficiency. However, a poor design of nozzles for injecting secondary air can produce undesirable results. The combustion efficiency for
different ratios between primary and secondary air flow rates is shown in Fig. 5.40
[85]. Investigations were carried out using a 4 MW_{th} TNO experimental boiler,
with substoichiometic combustion in the bed.

The influence of fuel feeding. In order to achieve high values of combustion
efficiency, the location for fuel feeding must be adapted to the properties of fuel and
its particle size distribution. Low reactive fuels (anthracite, low volatile bituminous
coals and most of the brown coals) should be fed inbed, especially if they contain
significant fines. The number of feeding points should be chosen in such a way as to
avoid the production of local substoichiometric conditions, which could cause local overheating in the bed and lower combustion efficiency. Highly reactive fuels
(lignite) with coarse particles, can be fed overbed and high combustion efficiencies
can be achieved. There appear to be no detailed studies in the open literature giving
a direct comparison of combustion efficiency when feeding the fuel overbed surface and underbed, while maintaining all the other conditions constant. Based on
comparative investigation of a large number of different fuels at the Institute of
Thermal Engineering and Energy Research [73, 76, 80, 86, 87], in a PhD thesis
[88], the influence of feeding was analyzed. An illustration of this influence is
given in Fig. 5.41 together with a comparison with the results of calculations using
the model developed in [88].

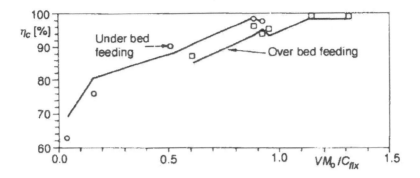

Figure 5.41. *Influence of coal feeding on combustion efficiency [88]*

5.4. Load control in FBC boilers

Regular operation of a bubbling fluidized bed combustion boiler demands keeping the fluidization velocity and bed temperature within very narrow limits. Fluidization velocity cannot be decreased too much, in order not to threaten good fluidization and fuel mixing in the bed. The bed temperature must be limited to a very narrow interval between 800-900 °C, if desulphurization is required. If a fuel with low sulphur content is burned, a minimum bed temperature of about 700-750 °C is necessary for good combustion even with biomass, while the maximum temperature cannot exceed 900-950 °C because of problems with ash softening and possible bed sintering and increased NO_x emissions.

These demands limit the possibility of load changing in bubbling bed fluidized boilers. In order to achieve the same turn-down range as in conventional boilers, it is necessary to take special design measures, which considerably complicate boiler design. In Section 1.8 it was noted that this major deficiency of first generation FBC boilers was one of the main incentives to develop circulating fluidized bed boilers.

A decrease of coal and air fed to the bed allows some decrease in the amount of energy generated by the boiler, but if the amount of heat to be removed from the bed by immersed heat transfer surfaces and water-tube furnace walls is not decreased, the temperature of the bed must fall. In the case where bed temperature decrease is allowed over the interval mentioned above, a continuous decrease of boiler power would be possible, dependent on the type of heat exchanger in the bed:

– by about 23%, if hot water tubes are in the bed,
– by about 27% for evaporator tubes in the bed, and
– by about 50% for superheater tubes in the bed.

Rembering that there are usually different heat transfer surfaces in the bed, one cannot expect variation of boiler output over 25%, simply by changing the bed temperature.

Further decrease of boiler output is possible – see eq. (5.6) – either by changing the heat transfer coefficient or by changing the size of the exchanger surface in contact with the fluidized bed.

Since the heat transfer coefficient in the fluidized bed is practically constant within the limits of possible changes in fluidization velocity, a greater turn-down range for a bubbling fluidized bed boiler can only be produced by changing the size of the heat transfer surface immersed in the bed. The most common way to achieve the wanted turndown range is division of the furnace (physically – by a wall, or by subdivision of the air supply only) into several compartments. By decreasing the fluidization in certain compartments (or by turning parts of the furnace off), transfer of heat on the immersed surfaces in that cell practically stops, i. e. the heat transfer surface in the bed is decreased. In the zones where fluidization is terminated, the fuel supply should also be stopped. This method of load change has been commonly applied in practice, as can be seen in the figures and sections 5.1.2 and 5.1.3, which show diagrams of commercial FBC boilers in operation.

An alternative approach to load control is by varying the area of immersed heat transfer surface by decreasing the bed height, and by arrangement of the exchanger tubes in such a way, so as to ensure that some part of the surface remains above the bed surface. The height of the bed can be decreased either by decreasing fluidization velocity or by decreasing the amount of inert material in the bed. This procedure for load change in FBC boiler is rarely applied in bubbling atmospheric FBC boilers but has been used in pressurized FBC boilers, which have inherently deep beds.

By applying these procedures, bubbling FBC boilers can achieve turndown range up to 4:1, and in special cases even more. The procedures for load control for bubbling FBC boilers are presented in a systematic manner in Table 5.7 along with a brief description of their defficiencies from [89].

Besides the defficiencies noted above, load control by slumping the bed in certain compartments of the furnace can permit a serious problem to occur, namely slow char burning in the non-operating sector of the bed, which may cause local overheating and ash sintering. At the same time, char and coal ejected from the operating sector may accumulate on the surface of the slumped bed. It is also not possible to prevent some air from bypassing through the slumped bed, either because of streaming from the operating bed region or because of failures in sealing and the phenomenon of natural air circulation. Periodic fluidization of such slumped bed regions is recommended for cooling, mixing and equalizing the temperature [66-68]. When constructing the subdivision of furnace into compartments and building in heat transfer surfaces into the bed, care should be taken to avoid a change of the ratio of the heat transferred to other heat exchangers, while changing heat transfer to specific immersed surfaces.

The load decrease produced by bed slumping is practically instantaneous, but the time to increase or ramp load again is limited by the large heat capacity of inert bed material. If the non-operational sector does not cool down below the start-up temperature, however, ramping may be very rapid.

Table 5.7. *Different approaches for load control in the bubbling fluidized bed combustion boilers*

Method			Principle	Defects
Bed temperature variation			Heat transfer coefficient in the bed is constant. Temperature difference reduced.	Decrease of bed temperature is limited by combustion and desulphurization kinetics.
Bed slumping in one sector			Heat transfer in one of the bed sectors decreases due to bed slumping (Fig. 5.12).	Penetration of air into bed sector which is not fluidized causes sintering. Transfer of material from the active to the inactive bed sector hinders restart.
Fluidization velocity decrease			Surface for heat transfer decreases due to decreased bed height (Fig. 5.11).	Erosion of tubes in splash zone. Change of bed height is insufficient for wide turn-down range.
Inert material decrease in the bed			Surface for heat transfer is decreased due to decreased bed height.	Erosion of tubes in splash zone. Handling material with high temperature is difficult.
Additional heat transfer surfaces			Heat transfer coefficient changes by independent fluidization of separated bed section (Fig. 5.26).	Heat transfer coefficient is either insensitive or too sensitive to velocity variation. Application limited to small units.

When changing power by decreasing bed height, a rate of change of up to 5% of maximum power can be achieved in one minute. However, load control remains a serious problem for designers and builders of bubbling fluidized bed combustion boilers, and the failure to resolve these problems is one of the main reasons, along with the problems of evenly feeding fuel to very large beds, which is why bubbling FBC is likely to be restricted to size ranges below 25-50 MW_e in the future.

5.5. Erosion of heat transfer surfaces immersed into the fluidized bed

Fluidized bed combustion technology is a relatively young technology. The oldest boiler units have been in operation for only ten to twenty years, furthermore FBC technology has been rapidly changing during this period as various ideas and subsystems have been explored (e. g. the use of carbon burnout cells, or the construction of different fluidized bed boiler concepts such as units operating with multiple beds, etc.). Moreover, the range of fuels that have been burned in FBC boilers is also extremely large. For this reason there are not enough data about operating problems for this type of boiler [90]. However, it has been known from the very beginning of the development of this technology, that there would be a serious operating problem – erosion, or metal wastage from internals in contact with the fluidized material in the furnace. Experience with the first industrial FBC boilers confirmed these predictions, although there is a fair variation in the results.

The negative effects of erosion of metal have been noticed on furnace walls of FBC boilers and on the heat exchangers immersed in the fluidized bed, during long-term operation. Erosion directly influences the operating life of the furnace, operating conditions and need for repairs and maintenance. Potential effects of leakage from worn out tubes in the furnaces of FBC boilers may be severe. Tube bundles comprising any heat exchanger in FBC boilers are usually tightly packed, and any steam leaks through cracks caused by erosion can cause channelling of solid flows, thus speeding up the erosion on neighbouring surfaces and leading to a "chain wearing" of adjacent tubes in the "heat exchange bundle." Even if this phenomenon is detected from for example, inbed thermocouple readings, the thermal heat inertia of the many tons of solids in an industrial FBC system prevent any instantaneous responses to prevent this process. There is also a risk of bed agglomeration in the furnace and defluidization. In the case of limestone derived beds, if they become flooded with water during the cool down process, then the bed materials may require major efforts to remove the mass of cement like material that will form once liquid water is present. For this reason, special attention should be paid to the processes of erosion and design of any heat exchangers in the bed.

Originally corrosion and not erosion was regarded as a potentially important problem in fluidized bed boilers with heat exchangers. However, over a period of time the issue of erorsion or more propely metal wastage has become the major issue. A review of data concerning metal wastage in FBC boilers based on the relatively scarce data available in the open literature will be given here. The results of this type of work is normally proprietary, and it is difficult to obtain reliable data. Data in this section are based on a detailed analysis performed in reference [90], on papers presented at conferences [19-21], and internal company reports [71, 91-107]. One can distinguish between at least three different types of "wear" of metal surfaces [92, 93]:

– abrasion – or losses of material from the surface as a result of direct contact between particles of inert material and their impact pressure and movement across metal surfaces,
– erosion – or loss of material resulting from collision of particles with metal surface, and
– erosion/corrosion – or loss of material caused by common effect of erosion/abrasion and corrosion.

"Erosion" most frequently means the sum of abrasion and erosion effects, while erosion/corrosion is described by a single term "corrosion," and for the sake of simplicity it is often better just to use the term metal wastage, where there is uncertainty about the causes of metal loss.

Some general conclusions on erosion were obtained from experimental studies [91]:

- rate of material wear (mass of eroded material in a unit of time and per unit of mass of particles colliding with the surface) is independent of mass flux of particles,
- in the majority of cases, it is possible to reach a situation, in which the rate of wear is constant,
- for particles greater than 20 μm, the rate of wear does not depend on particle diameter,
- experimental data obtained by investigating erosion on experimental, pilot and industrial plants show that hardness of the bed material investigated does not necessarily influence its wear rate,
- relation between hardness of particles and erosion has not been established, and
- thermal treatment of metals with the aim of increasing their hardness has not been generally successful.

Up to now, investigations of metal erosion in fluidized bed have pointed out the following as the most important causes of this phenomenon [91]:

- effect of "slow" but relatively big particles,
- effect of particles which were "pushed into" the surface by other particles,
- effect of "fast" particles, resulting from bubbles,
- effect of "fast" particles as the result of moving of fast bubbles along vertical or steep surfaces. When changing the inclination of a surface, these particles may collide with the curves and elbows on tubes,
- erosion caused by particles colliding with the surface when the bubbles burst on the bed surface,
- erosion in the region around the bed surface (splash zone), as the result of collision of fast particles ejected when bubbles explode on the bed surface,
- erosion resulting from jets near the openings for inbed coal feed, and near the openings for recirculation of particles, as well as other jets, for example of secondary air, jets of primary air from air distributor "bubble caps," and or jets designed to move oversized material into a bed drain system,
- erosion resulting from directed, circulating movement of particles in the bed - streaming down the furnace walls, or rising in the centre of the furnace, circulating flows in larger furnaces designed to ehanced bed mixing, etc., and
- wear resulting from geometrical irregularities. For example, vertical tube welded on a horizontal support, or a horizontal tube (measuring probe) may cause local wear.

Besides erosion, care should be taken to reduce material fatigue and tube cracking at places where they are "fixed" to the furnace wall, which can arise from oscillation due to bed movement such as the generally oscillating behavior of the fluidized bed which are caused by bubble phenomena. Unfortunately, detailed investigations and data on these phenomena are extremely limited and this subject will not be addressed further here.

Among all the of the causes discussed here, erosion due to the effect of local jets is perhaps the most important and most problematic for the designers and fabricators of FBC furnace. Other causes of erosion tend to be slower and hence are more long term issues. The effect of jets, however, is local and may cause rapid and masssive metal wastage, in a manner that is difficult to predict.

The intense movement of fluidized bed particles and their collisions with the metal surfaces immersed in the bed, or for that matter with furnace walls, are ample reasons to believe that the erosion phenomena are inevitable. However, it is evident that there are FBC furnaces where erosion was not shown to be a major problem [71], and so it is clear that it is possible to produce engineering solutions which minimize such problems.

Reliable data on erosion of metal surfaces in contact with fluidized bed may be obtained only in controlled, steady state conditions from the long-term operation of the unit under constant conditions. Unfortunately, such investigations are both expensive and time consuming, and for this reason there are very few such studies are available in the open literature. Even when such data on metal wear in fluidized bed are provided, important data about the operating conditions and specific details (thickness of the tubes, type of erosion, size and type of inert material particles, distance between the tube bundle from the distribution plate, etc.), are rarely published because of commercial considerations. Despite the fact that such systematized data are missing from Table 5.8 [91], this table gives a good insight into metal wear rate data available from data taken from various pilot and industrial boilers.

When analyzing the erosion data given in Table 5.8, one should remember that metal wastage rates of up to 35 nm/h are normally considered to be acceptable. The minimum acceptable lifetime of tubes is about 50,000 h. For a tube 6 mm thick, with maximum acceptable loss of 4 mm, a loss of 80 nm/h would be acceptable. It can be seen from the table, that wear of material is different in different plants, and that it changes by more than two orders of magnitude, and up to few thousand of nanometres per hour. Besides these results, data from furnaces where there is "no wear" of material appears in the literature [71, 91]. Babcock Power Co. in United Kingdom investigating a boiler furnace in Renfrew, Scotland, over several years, where there was no significant loss of material in the furnacer. Results from a boiler operated by Wormser (Texas, U.S.A.) also indicated the effective absence of erosion after 7000 hours of operation. Similar results were seen for the boiler in Shamokin (U.S.A.) after 10000 h of operation. In this boiler, the heat exchange tubes are vertical, and fluidization velocity is about 1 m/s. A boiler in East Strouds-

Table 5.8. *Results of investigating the loss of material on the exchanger surfaces of the furnace*

Type of exchanger surface	Operation time [h]	Fluidization velocity [m/s]	Wear rate [nm/h]	Remarks
Georgetown University, Washington DC, U.S.A. [107]				
Water tube wall	1000	2.2	15-127	After first tests, tubes were protected by Al$_2$O$_3$ coating. This layer eroded in about
Exchanger in the bed	5850	2.5	755	1000 h, i. e. it did not effectively decrease the erosion rate
People's Republic of China (cited by Georgetown University) [107]				
Exchanger in the bed	1200	2.70	833	– Bottom of the lowest bundle in the bed
	5000	2.74	max 460	– Min. rate 100 nm/s
Exchanger in the bed	–	3.00	1300	
Side water wall	–	3.00	100	
Stork TNO, the Netherlands (steam temp. 200 °C) [91]				
I tube-bundle	1800	1-3	max 2278	– Mean value 778 nm/h for the inbed
	600	1-3	4000	exchanger. Distance from distribution
	2400	1-3	1292	plate to bundle bottom 365 mm. Wear increases towards the bundle top.
II tube-bundle Nitrated tubes in	532	1-3	850	– Mean maximum for tube 430 nm/h
the bed	532	1-3	max 320	– Distance from distribution plate to tube-bundle bottom 198 mm
III tube-bundle	342	1-2	max 690	
Babcock & Wilcox, Alliance, Ohio, U.S.A. (6 × 6 inch AFBC) [97]				
	2414	2-2.5	420	– Maximum erosion in the bed centre, above the recirculation opening Measurement is not completely reliable
	7300		100.000	– Intense erosion above the opening for ash recirculation
Northen States Power, Minnesota, U.S.A. [96]				
Evaporator in the bed	200	4	254	
Central Soya, Marin, Ohio, U.S.A. [96]				
Exchanger in bed	11000	2.4-3.7	454	– Approximate values. Initial thickness of
Water tube wall	14000	2.4-366	366	tube wall was unknown
TVA/EPRI Padukah, Kentucky, U.S.A. (20 MW) [103, 107]				
B&W exchanger	5558	2.5	112	– Probably much more
Water tube wall	5558	2.5	250	– Distance from
Evaporator in bed	5558	2.5	96	distribution plate to tube
Superheater in bed	5558	2.5	630	bundle 580 mm
Corrosion of superheater	2823	2.5	max 1500	
C-E evaporator	3900	2.5	max 1150	– Bottom of tube bundle
Great Lakes, U.S.A. [91]				
Evaporator in bed	5300	2.1	127	– On vertical sections
Evaporator in bed	4722	2.1	max 365	– 119 nm/h mean value
Vertical evaporator sections in bed	4722	2.1	742	– Two tubes are worn by erosion; rate calculated based on initial thickness – It is possible to achieve rates of 25 nm/h by correct design.
Battelle Columbus, Ohio, U.S.A. [95]				
Water-cooled tubes	1100	2.5	5600	
Chalmers University, Sweden [98, 99]				
Evaporator	1210	2.5	1000	– Distance between distribution plate and tube bundle is 600 mm

Table 5.8. *Continued*

Type of exchanger surface	Operation time [h]	Fluidization velocity [m/s]	Wear rate [nm/h]	Remarks
Superheater	1210	2.5	600	– At the bottom of tube bundle
Vertical tube	1210	2.5	1200	– Wear highest on the level 0.5-0.9 m
Cebu, Philipines [91]				
	2500	2.8	1800	– Distance between distribution plate and tube bundle is 200 mm Damaged tubes were replaced by a bundle lowered to 40 mm. Erosion decreased
Allones, France [91]				
Inclined tubes	1000	2.2	4000	– Distance between distribution plate and tube bundle is 400 mm at one end and 1000
Membrane wall	1700	2.2	600	mm at the other end. Tubes were destroyed
Wakamatsu, Japan [97]				
Evaporator	3200	–	330	– Maximum wear on tube
Superheater	1700	–	100	curve on the bundle top
Volkingen, Germany [91]				
Exchanger in the bed	3000	1.3	330	Distance from distribution plate 900 mm. On replacement, exchanger was lowered to 600 mm. Erosion decreased
Studsvik, Sweeden [91]				
	4000	2.2	1000	
Gibson Wells, Great Britain [71]				
Unit 1				
bare tubes	5100	2.3	max 750	mean value 60 nm/h
with ribs	11900	3.2	max 240	mean value 33 nm/h
ribs	5000	3.2	max 120	mean value 80 nm/h
Unit 2				
bare tubes	3200	2.5	max 700	mean value 170 nm/h
with ribs	3000	3.1	max 114	mean value 360 nm/h
General Electric LTHTFF, Malta, New York, U.S.A. (small unit) [91]				
Water-cooled spiral				
1. spiral	1050	0.7-0.9	1974	– Intense erosion down the vertical line was moderated by built-in ring screens
2. spiral	1195	0.7-0.9	1195	
3. spiral	104	0.7-0.9	1270	
Nova Scotia Power Co. Point Tupper, Canada [101]				
	10000	2.4	max 300	– Total wear including oxidation and sulphation. Erosion is predominant

burg (U.S.A.) after 3500 h of operation showed no signs of erosion, either. The fluidization velocity was 1-1.75 m/s. The fact that there are boilers, for which there are no significant erosion problems indicates that this is not necessarily an intractable problem.

Based on the data presented in Table 5.8 it is not possible to precisely determine the relationship between erosion phenomenon and boiler operating conditions. However, the following parameters may be regarded as critical:

– fluidization velocity; a choice of lower fluidization velocity is recommended. Results from cold models show that there exists a velocity threshold above which the intensity of erosion may suddenly increase. Some manufacturers recommend velocities lower than 1.8 m/s,

- bed material; many authors think that hardness of particles influences the intensity of erosion. Results of investigating erosion in the 20 MW TVA/EPRI boiler (U.S.A.) [107], showed an influence of limestone on intensity of erosion,
- size of bed particles; investigations show an increase of erosion intensity with the increase of particle size. This influence is most probably connected with increased fluidization velocity and change of bubbling regime [94, 99],
- tube material; current experience suggests erosion intensity doesn't depend on hardness or strength of the material. But there are data that suggest that stainless steels are less liable to erosion [100, 105, 106],
- temperature of metal; data from industrial plants and research results clearly show that erosion decreases with increasing metal temperature. As a rule, wear of steam superheaters is lower than the erosion of evaporation surfaces in the bed [103, 104]. It is recommended to place exchangers with higher tube temperature in the regions where more intense erosion is expected,
- shielding/cladding; many studies have been done on the effects of cladding or shielding on heat exchange surfaces and metal internals that are expected to be subject to erosion. Systematic investigations at Chalmers University in Sweden, on three shielding materials (Metco 101, Metco 2185, and Metcoloy 2) have shown that they have very good resistance to erosion [98, 99]. Metcoloy 2 and Metco 2185 are alloys based on chromium, and Metco 101 is a Al_2O_3 material with addition of TiO_2. Further the oxide layer formed on the tube surfaces can have a shielding role as well.
- ribbing and studding on the tubes; experience from many manufacturers of FBC equipment show that it is possible to decrease erosion considerably by welding ribs, or studs, or small cylinders or spheres along the tubes (on the lower side of the tube). A greater wall thickness is also recommended in order to allow for "a reserve" of material for wear and to permit an increase temperature of the tube surface,
- distances between the bundle of tubes and the distribution plate; greater distance enables free growth of bubbles may lead to intense wear of exchanger tubes in the lower rows of any tube bundle. Distance of 200 to 600 mm are recommended to minimize erosion,
- geometry of tube bundle; corridor or tightly packed arrangement of tubes in the heat exchanger are recommended to avoid the growth of bubbles, which are seen as a cause of erosion,
- rigid heat exchanger; oscillation of the exchanger under the influence of collision of particles and bubbles all contribute to material fatigue and to the increase of erosion intensity. Therefore any heat exchangers should be "rigid" to avoid these deleterious effects of oscillations, and
- influence of fuel type; there are indications that chlorine can contribute to metal wear [102].

According to industrial experience, erosion intensity of up to 2000 nm/h should be expected, if protection measures are not taken. Testing for erosion and tube wear should be incorporated in regular boiler maintenance.

5.6.   Ash sintering during combustion in fluidized bed

Proper design and choice of operating parameters for FBC boilers can prevent ash sintering during normal operation. Combustion temperature, which are typically between 800-900 °C, must always be lower than ash sintering temperature for the coal in question, but operational experience has shown that it is even possible to burn lignites in a FBC without major problems, despite their potentially high alkali metal content. While literature data on ash sintering is not particularly common [109-113], it should be noted that both manufactures and boiler operators may be fairly hesitant to publish details of their problems in the open literature and although there are "users" groups in various countries, such as the U.S.A. based Council of Industrial Boiler Owners (CIBO) [114], many of their concerns may not be presented in the formal open literature. It can be hoped that with the recent involement of CIBO in joint organization of the ASME FBC conferences, more such practical data will be found in the open literature.

In order to create operating conditions in the furnaces of FBC boilers, in which ash sintering does not occur, one must prevent hotspots or local superheating of the bed above nominal operating temperature. Local superheating is less likely if:

— there is uniform and intense fluidization,
— there is good mixing and favourable conditions for combustion of volatile matter,
— there is sufficient oxygen for combustion of char in the whole volume of the bed, and
— there is good mixing of volatile mater above the bed to prevent high temperatures in the freeboard.

Sintering in a FBC boilers may occur [110]:

— during start-up of the boiler and load change, i. e. during transient regimes,
— during operation with a part of furnace, i. e. operation with reduced power, and
— due to incompatibility of ash and the chosen inert bed material.

Sintering of ash during boiler start-up occurs if start-up is from a low starting temperature. For younger and more reactive coals and biomass, the startup temperature can be chosen to be in the interval from 350-400 °C. Under such conditions the char combustion rate in the bed is low, combustion is incomplete, and volatile matter tends not to ignite during the initial start-up period. Hence, char particles accumulate in the bed and the char loadings will be higher than needed for operation in the steady state regime. When the bed temperature rises above 600 °C, volatiles are ignited, the combustion of char accelerates "exponentially" and an "explosive" increase of bed temperature may occur (even with temperature spikes over 1000 °C) causing "instantaneous" sintering, defluidization, and interruption of boiler operation. These events can be avoided by correct choice of start-up procedure [110].

Sintering of ash during the operation of one section of the bed. Reducing the power of FBC boilers is often managed by dividing the furnace into compartments. If the boiler operates under low load, only a section of the bed in which combustion takes place is fluidized. Under such conditions, sintering may occur in two ways: on the surface [110] of the bed section, which is not operating, or in its interior [110, 111].

Sintering on the surface of the non-operating section of the bed results from transfer of "red-hot" burning char particles from the operating to the non-operating section in the furnace. These particles accumulate on the surface of the non-operating section of the bed and burn under substoichiometric conditions, using air from the active bed area or air filtering through the slumped bed because the system is not properly sealed. Under such conditions, inert material around the particles in the non-operating section of the bed surface may be overheated and form the nucleus for the formation of large sintered masses which can cause defluidization. This phenomenon can be particularly important when burning younger coals, lignites [110], which have lower densities and tend to burn near the surface of the bed or actually float on the bed surface. This phenomenon can be prevented by building in partitions walls between the bed sections.

Sintering in the interior of the non-operating bed this results from insufficient sealing of the non-operating bed section, and inevitable gas bypassing from the operating fluidized bed section [67, 111]. When slumping the bed, a certain amount of hot char and coal particles remains in the non-operating bed section. Local superheating and sintering occurs when these particles burn in the bed interior. Sintering of this type can be avoided by complete separation of bed sections, by good sealing among other methods [110, 111].

Sintering caused by incompatibility of ash and inert bed material appears during combustion of fuel with high percentage of alkaline metal compounds in the ash. Some types of lignite have 5-10% of Na_2O and K_2O in their ash, and some biomasses up to 30% of these compounds in their ash [113]. During operation of the boiler, these compounds can accumulate in the fluidized bed by forming eutectic compounds with the sand (SiO_2), whose melting temperature is as low as 720-750 °C. These mixed oxides – potassium-sodium-silicon oxide, can cause sintering and defluidization. One interesting solution is to chose a suitable inert bed material, e. g. Fe_2O_3, as it was done in the case of burning corncobs [113] or perhaps by the use of additives which can chemically combine with the alkaline components such as bauxite or kaoline [115]. The only caveat in such cases is that amounts of such materials to be used may in some cases become very large, especially where fuel ash is high (and so must be removed from the bed) or the fuel ash contains high amounts of tramp material.

5.7.   Niche markets for bubbling FBC
(written by Dr. E. J. Anthony)

In recent years there has been more attention paid to distributive power production, and hence increasing interest in technologies suitable for small units. While admitting the indisputable advantages of CFBC boilers, when compared with bubbling fluidized bed combustion (BFBC) boilers for utility applications, as noted above, in distributive power production smaller units are necessary, so interest in bubbling fluidized bed boilers and incinerators is again recovering. In consequence, BFBC boilers currently have and in the future will likely continue to have a significant role in distributive energy production. Also, in countries in which the economic situation is more difficult than that typical of most western European nations, and where the structure of primary energy resources are limited, there is and will remain a need to use local fuels [116]. This means that coal, biomass and wastes will be in many cases the fuels of choice. In these situations, BFBC boilers and hot-gas generators can have a significant role in heat and power production in the industrial and agricultural sector as well as for district heating.

Bubbling FBC are cheaper than CFBC for smaller units (<25 MW$_e$) [117], which means that they are ideally suited for a wider range of waste fuels, which are not present in large quantities, but either pose a disposal problem, and/or represent an opportunity fuel, which can be used at a given location. In this context biomass is a particularly good example of a fuel, which is typically not available in very large quantities in any given locality, and so lends itself to use with bubbling FBC technology. The same is true for coals from small mines, which may represent an available cheap local fuel, which must be burned in small boilers and hence is suitable for combustion in small industrial scale FBC boilers.

Coal from small local mines

Many developing countries are poor in energy resources, and are faced with the necessity of importing large amounts of expensive fuels such as oil and natural gas. To satisfy their growing energy needs such developing countries also need to be able to use all of their available domestic energy resources in an efficient and environmentally acceptable manner. One of the most commonly available resources is coal produced from small mines. The low production capacity of such mines cannot justify large transportation costs or expensive pretreatment of coal – drying, washing, and pulverization required for large conventional boilers. In such cases the "small sized coal" (less than 15 mm) is simply disposed of by landfill or some type of surface deposition, since it cannot be used in conventional boilers – grate boilers or pulverized coal combustion boilers. Given the combustion performance of bubbling fluidized bed technology, such low quality coals can be easily burned in an "as mined" state while satisfying modern emission standards. Equally, if there is a need to burn a variety of fuels from several small mines, or substitute with another fuel or co-fire or fire with fuels such as blended coal, biomass and wastes, then bubbling fluidized bed combustion boilers and hot-gas generators can easily cope with such a situation.

Biomass fuels

The term biomass can cover an enormous range of fuels, from good quality wood chips, to bark, hogfuel, straw, grasses, etc., and fuels comprised of various commercial and industrial products, e.g. municipal solid wastes (MSW), refuse derived fuels (RDF) and construction wastes. Even natural biomass fuels may have widely different properties and, in particular, concentrations of alkali metals (i. e., Na and K) can vary from less than 1 to 6-7%, as in the case of straw. It is, therefore, not surprising that the combustion performance of different biomass fuels can vary substantially. There is, however, an increasing desire to use these fuels to achieve a "CO_2 neutral" strategy (one in which biomass is grown at the same rate as it is converted into energy), and also to avoid landfilling. Also worth noting is that biomass fuels, unlike coal or petroleum coke, produce negligible amounts of N_2O, and hence the use of these fuels in FBC boilers is not compromised by the production of this greenhouse gas [118, 119].

A number of general comments can be made about biomass fuels. First they are usually bulky, low-density fuels, so in the absence of some external constraint (e. g., a carbon tax), it is usually not practical to transport them much over 150 to 300 kilometres [120]. This constraint sets an effective upper limit on boiler size, independent of choice of boiler technology, unless co-firing of another fuel is practiced, and the largest unit burning solely biomass is currently the 160 MW_{th} Orebro CFBC boiler in Sweden [119, 121].

For a commercial operation there are various issues that must be resolved in order to successfully burn biomass fuels. For a unit of any size it is essential to ensure sufficient availability of those fuels for continuous operation, and the failure to do this has in some cases forced owners to co-fire with coal, for example. Second, it is necessary where fuels are purchased, to develop appropriate strategies to characterize these highly variable fuels (e. g., a system of weigh stations, and moisture determination per truck load of biomass will be necessary to ensure both fuel quality and price per load). The other issue to be borne in mind is that these fuels are often fibrous and difficult to feed, and unless the feed system has been properly designed, it may be problematic to use widely different sources of biomass fuels in a single boiler.

Finally, in discussing the use of bubbling FBC for this type of application, it is worth reiterating that there has been a considerable reduction of the number of companies offering FBC technology on the market. This is at least partially a result of an extremely difficult energy market in Europe and North America over the last 10 years or so, and at this time there is no indication that this situation is likely to change and the reduction in the number of companies offering such technology is likely to continue. However, sometimes the new parent company may offer the original technology. To exemplify this situation the CANMET Energy Technology Centre carried out a feasibility study in 2001, for a bark burning FBC boiler producing 30,000 kg/h of steam, and identified only 12 companies offering such technology internationally.

Pulping and deinking sludges

These fuels are increasingly burned in North America using FBC technology. There appear to be at least 6 new units at various stages of development in Canada, for example. Generally, the emission results have been extremely good, and it is evident that this fuel type can be burned with extremely low emissions of SO_2, and negligible polyaromatic hydrocarbons, dioxins and furans [122].

As these fuels contain high moisture, and may often not have sufficient heating value to support combustion by themselves without employing strategies like combustion air preheat, they are often co-fired with a premium fuel such as natural gas or coal. There is, however, one important consideration when co-firing coal, namely that NO_x emissions are not a simple linear function of the relative amount of coal being fed. While the amount of fuel nitrogen in coals is significantly higher than that for most biomass fuels, the actual percentage conversion of fuel nitrogen is relatively low. This is because coal use results in high char loading in the bed that, in turn, ensures that the majority of the fuel NO_x is reduced. By contrast, biomass fuels have low fixed carbon contents, and also usually have very low fuel nitrogen content and the corresponding char concentrations in the bed are therefore low due both to the low fixed carbon content of the fuel and the very reactive nature of biomass chars. This, unfortunately, ensures a much higher conversion of fuel nitrogen to NO_x. Therefore, in situations where a small amount of coal is being fed (roughly in the 20-30% weight range), it is possible to be in a regime where the coal contributes significantly to the overall fuel nitrogen, while not providing enough char to the bed to ensure good NO_x reduction, resulting in disproportionately high NO_x emissions [118].

Municipal solid wastes

FBC technology has been widely used for the combustion of MSW and RDF [123-125]. There is resurgent interest in this technology in Europe, due to the fact that the high calorific value of MSW has been increasing, making it more difficult to burn in grate systems. Key factors determining plant success include: ensuring that fuel pre-processing systems are correctly engineered; and evaluating overall plant economics for the "locality" where the plant is situated, especially if the project depends on revenues from fuel separation and sales of side streams such as metals, glass, etc. Generally, the siting process for such plants is contentious, expensive and lengthy, as advocacy groups will become involved in the permitting. An interesting method of resolving public perception problems for the permitting process was employed by BCH Energy, in connection with a 2×30 MW$_{th}$ plant in Fayetteville, North Carolina. This project employed BFBC technology from Kvaerner Pulping AB [126]. In this case the plant was deliberately sited in an industrial location, thus avoiding the strong reaction that "greenfield" projects normally engender. Unfortunately, the plant is now closed down due to what appears to be a failure to properly develop the plant's economic plan.

In terms of emissions it appears that modern plants can reduce their heavy metal output below European and North American standards. For example, the Madrid FBC MSW incinerator (three 30 MW$_{th}$ Ebara twin revolving bed units) has reported emissions considerably lower than the standards: CO of 6 mg/Nm^3 (limit 80); SO$_2$ of 3 mg/Nm^3 (limit 240); and HF < 0.02 mg/Nm^3 (limit 1.6) [127]. In the case of dioxins and furans (PCDD/F) the situation is a little more complicated. Most plants can achieve dioxins/furans emissions in the range of several ng/Nm^3 expressed as a TEQ. However, to meet the German standards of < 0.1 ng/Nm^3, it is necessary to use flue gas treatment, such as activated carbon injection in the baghouse, and perhaps some type of lime scrubber. The use of these backend technologies is also effective in further reducing heavy metal emissions. A lime scrubber is also often necessary to minimize HCl emissions, which are not effectively captured by lime sorbents at bed temperatures. While such subsystems clearly perform well, they do add significantly to the cost of the overall plant, and tend to reduce the competitive advantage of FBC over grate technologies. The Madrid plant uses calcium hydroxide and activated carbon injection to reduce PCDD/F emissions to < 0.1ng/Nm^3 [127].

Hazardous and special wastes

Special and hazardous wastes would not normally be thought of as fuels for FBC boilers. However, there is an opportunity to improve the project economics of such boilers if permission can be obtained to co-fire these wastes. Providing the amounts of these wastes are relatively small (<10-20% by weight of the fuel feed), it is unlikely that they (e. g., PCBs, contaminated soils, etc.) will affect the overall boiler performance, and the author is aware, for example, of a 20 MW$_e$ coal-fired CFBC boiler being used to successfully remediate contaminated soil.

That FBC can burn wastes more effectively than conventional technologies at lower temperatures has also been acknowledged by the EPA for MSW combustion [128]. Equally, those studies done on organic emissions (both PAHs and PCDD/F) from coal burning in FBCs have suggested extremely low emissions [129-131]. Given that some coals contain significant amounts of Cl (up to 1%) and relatively high volatiles, why can FBC boilers, which are generally recognized to be low-temperature systems, perform so well?

Bulewicz, Anthony and co-workers have proposed an answer to that question, as follows. First, they suggest the reason a very high temperature (980 °C) [132] is generally required for incineration of special wastes is that most combustors, unlike FBC units, are not isothermal. In consequence, a high temperature is necessary to ensure that bypassing of unburned organics, which give rise to PAHs or PCDD/F emissions, can be avoided. Second, they argue that FBC operates in the same way as high temperature combustion systems, because the conversion of CO and volatiles is controlled by superequilibrium concentrations of free radicals [133-135], which means that despite its low temperatures FBC can be regarded as a continuum of flaming combustion.

High-sulphur pitch

Pitch is another high heating value fuel, which ought to be very promising for FBC systems. While these fuels have been examined at the pilot-scale [136, 137], until recently there were no industrial examples of this class of fuel being successfully burned in a FBC boiler. One major project to destroy 700,000 tons of coal tar pitch deposit (in a number of open lagoons) was terminated in 1995 because of the failure of the feed system chosen for the project, although FBC trials on the combustion of the pitch indicated successful results [138, 139]. Recently, however, an FBC boiler supplied by CSIR Mattek, based in Pretoria, South Africa, has been successfully deployed for this purpose. Contractor for the project was IMS Process Plant (a division of IMS Projects) based in Johannesburg, and the client was Sasol Chemical Industries, based in Sasolburg, South Africa. The following is a description of the project, provided by CSIR Mattek [140].

The primary purpose of this FBC incinerator is to destroy 2400 kg/h of a high sulphur pitch (containing up to 10% sulphur), while capturing in excess of 85% of the sulphur by limestone addition to the bed. Additionally, a stream of 2000 kg/h of a dilute phenolic effluent is incinerated using the energy liberated from the combustion of the pitch.

The FBC designed by CSIR to incinerate these streams has a rating of about 26 MW_{th} and a plan area of 20 m^2. The pitch (preheated to 80 °C to enable it to be pumped) is injected through 12 nozzles to ensure a high degree of in-bed combustion and even bed temperature. The phenolic effluent is injected in the freeboard area. Limestone is supplied to the bed through two air assisted feeders. Coal, which is used during start-up to get the bed to 900 °C, is also supplied through these feeders.

The off-gases are used to generate 21 t/h of steam in a waste heat boiler (therefore also resulting in a net reduction of CO_2 emissions from the plant) before being cleaned of particulates in a bag filter. Most of the commissioning activities related to limestone selection and pitch feeding. Several limestones were tried before an optimal limestone was found (essentially the best trade-off of efficiency and cost), originating from Northern Province, in the northern part of South Africa. With this limestone a sulphur capture of about 88% was achieved. The pitch injectors were modified to increase in-bed combustion, so that the bed and freeboard could be controlled at a temperature conducive to sulphur capture. The plant has been running successfully since January 1997, and has met all design requirements in terms of incineration capability and sulphur emissions reduction.

Nomenclature

A_b	cross sectional area of the fluidized bed, [m²]
A_s	outside surface area of the heat exchanger immersed in the bed, [m²]
C_{fix}	fixed carbon content in coal, as measured by proximate analysis, [%]
C_g	specific heat of gas (or combustion products), [J/kgK]
d	mean fuel particle diameter, [mm] or [m]
H_c	calorific value of the fuel, [J/kg]
$\dot{m}_f$	fuel mass flow rate, [kg/s]
$\dot{m}_v$	air mass flow rate, [kg/s]
Q_b	heat energy generated in the bed, [W]
Q_f	amount of heat necessary to heat fuel from ambient to bed temperature, [W]
Q_l	amount of heat removed by immersed heat exchanger, [W]
Q_k	amount of heat necessary to heat limestone from ambient to bed temperature, [W]
Q_o	thermal power of the boiler, [W]
Q_w	amount of heat removed by furnace walls, [W]
r_c	recirculation ratio (recycle ratio)
T_b	bed temperature, [K] or [°C]
T_s	immersed heat exchanger surface temperature, [K] or [°C]
T_{vo}	inlet air temperature, [K] or [°C]
v_f	fluidization velocity, [m/s]
VM_o	volatile matter content as measured by proximate analysis, [%]

Greek symbols

α	bed-to-surface heat transfer coefficient, [W/m²K]
ε_s	emissivity of the immersed heat transfer surface
η_b	degree of combustion in the fluidized bed, ratio of the heat released in the bed to the total heat released by fuel combustion
η_c	combustion efficiency
η_o	boiler efficiency
λ	excess air
λ_b	excess air in the bed, during stage combustion and secondary air injection over the bed surface
ρ_g	gas density (density of the combustion products), [kg/m³]
σ_c	Stefan-Boltzman constant, [W/m²K⁴]

References

[1]   S Oka, B Arsić, D Dakić, M Urošević. Experience and problems of the lignite
      fluidized bed combustion. Proceeding of 3rd European Coal Utilisation Conference,
      Amsterdam, 1983, Vol. 2, pp. 87-97

[2]   J Highly, WG Kaye. Fluidized bed industrial boilers and furnace. In: JH Howard, ed.
      Fluidized Bed – Combustion and Application. London: Applied Science Publishers,
      1983, pp. 77-169.

[3]   Department of Industry – Fluidized Bed Combustion Boilers for Industrial Uses,
      CEGB, 1982.

[4]   A National Coal Board, Report Fluidized Bed Combustion of Coal, 1985.

[5]   S Oka. Fluidized bed combustion, a new technology for combustion of coal and
      other solid fuels (in Serbian). In: Energy and Development. Belgrade: Society
      Nikola Tesla, 1986, pp. 147–156.

[6]   S Oka, B Arsić, D Dakić. Development of the FBC furnaces and boilers (in Serbian).
      Primenjena nauka 1:25-31, 1985.

[7]   Fluidized Bed Devices – Product Guide. Modern Power Systems, Nov. 1990.

[8]   S Oka, B Grubor. Circulating fluidized bed combustion boilers. State-of-the-art
      (Monograph in Serbian). Report of the Institute of Nuclear Sciences Boris Kidrič,
      Vinča, Belgrade, IBK-ITE-645, 1987.

[9]   PF Fennelly. Fluidized bed combustion. American Scientist 3:254-261, 1984.

[10] MA Conway. Has fluidised combustion kept its promise? Modern Power Sistems
      Dec/Jan:19-22, 1984/85.

[11] Sh Ehrlich. Fluidised combustion: Is it achieving its promise? (Keynote address).
      Proceedings of 3rd International FBC Conference Fluidised Combustion: Is It
      Achieving Its Promise?, London, 1984, Vol. 2, KA/1/1-29.

[12] J Makansi, B Schwieger. Fluidised bed boilers. Power 8:s-1-s-16, 1982.

[13] Fluidised bed devices. Part A. Equipment offered. Modern Power Systems
      Dec/Jan:67-77, 1984/85.

[14] D Wiegan. Technical and economical status of FBC in West-Germany. Presented at
      International Conference on Coal Combustion, Copenhagen, 1986.

[15] U Renz. State of the art in fluidid bed coal combustion in Germany. Yugo-
      slav-German Coloquium Low Pollution and Efficient Combustion of Low Grade
      Coals, Sarajevo (Yugoslavia), 1986.

[16] J Jacobs, B Becker. Fluidized bed combustion furnaces. Survey. Report given at the
      Meeting of an ECE Working Group, Turku (Finland), 1989.

[17] G Leithner. Einfluss unterschiedlicher WSF - Systeme auf Auslegung, Konstruktion
      und Betriebsweise der Dampferzeuger. VGB Kraftwerktechnik 69, Heft 7, 1989.

[18] RT Hokvilainen. Application of fluidized bed combustion. UN Economic Commis-
      sion for Europe, Report ENV/WP. 1/R, 77/Rev. 1, 1986.

[19] Proceedings of 9th International Conference on FBC, Boston, 1987, Vol. 1, Vol. 2.

[20] Proceedings of 10th International Conference on FBC, San Francisco, 1989, Vol. 1,
      Vol. 2.

[21] Proceedings of 11th International Conference on FBC, Montreal, 1991, Vol. 1, Vol. 2,
      Vol. 3.

[22] JL Klei. Experience with AFBC boiler at Shell Europort. Proceedings of 3rd European Coal Utilisation Conference, Amsterdam, 1983. Rotterdam: Industrial Presentation Group, Vol. 2, pp. 57-71.

[23] F Verhoeff. The design of a large industrial fluidised boiler. M Radovanović, ed. Fluidized bed Combustion. New York: Hemisphere Publ. Co.,1986, pp. 279-301

[24] F Verhoeff. AKZO-90-MW$_{th}$ SFBC boiler in Holland: Noteworthy results of a two-year demonstration program. Proceedings of 11th International Conference on FBC, Montreal, 1991, Vol. 1, pp. 251-261.

[25] G Brem. Mathematical modeling of coal conversion processes, PhD dissertation, University of Twente, Enshede (the Netherlands), 1990.

[26] F Kolobarić, M Beronja, S Jelinić. Development of FBC Boilers in Djuro Djaković Co. (in Croatian). Termotekhnika 3-4:271-284, 1990.

[27] B Grubor. M Ristić, D Dakić, S Kostić. Industrial FBC Boiler for Saturated Steam Production with Capacity 2×4 t/h (in Serbian). Termotekhnika 3-4:285–292, 1990.

[28] B Grubor, S Oka, B Arsić, J Jovanović. Design and concept of the FBC Boiler for Burning Wood Waste (in Serbian). Termotekhnika 3-4:261–270, 1990.

[29] AM Manaker. Status of utility fluidized bed commercial development in the United States. Presented at ASME/IEEE Power Generation Conference, Milwaukee, 1985.

[30] MM Delong, KJ Heinschel, et al. Overview of the AFBC demonstration projects. Proceedings of 9th International Conference on FBC, Boston, 1987, Vol. 1, pp. 132-139

[31] MD High. Overview of TVA's current activity in FBC. Proceedings of 9th International Conference on FBC, Boston, 1987, Vol. 1, pp. 6-10

[32] AM Manaker, PB West. TVA orders 160 MWe demonstration AFBC power station. Modern power station. Modern Power Systems Dec/Jan:59-65, 1984/85.

[33] JW Bass, JL Golden, BM Long, RL Lumpkin, AM Manaker. Overview of the utility development of AFBC technology TVA. Proceedings of 9th International Conference on FBC, Boston, 1987, Vol. 1, pp. 146-152

[34] RE Tollet, M Friedman, D Parham, WJ Larva. Start-up activites at the black dog AFBC conversion. Proceedings of IX International Conference on FBC, Boston, 1987, Vol. 1, pp. 153-160.

[35] GM Goblirsch, TL Weisbecker, SM Rosendahl. AFBC retrofit at Black Dog. A project review. Proceedings of 9th International Conference on FBC, Boston, U.S.A., 1987, Vol. 1, pp. 185-190.

[36] WJ Larva, SM Rosendhal, JF Swedberg, DP Thimsen. AFBC retrofit at Black Dog – Testing update. Proceedings of 10th International Conference on FBC, San Francisco, 1989, Vol. 2, pp. 729-738

[37] The coal mining research centre, Japan, Annual Report – Summary 1986, R&D on the Coal Utilisation Technology. Presented at 13th IEA-AFBC Technical Meeting, Tokyo, 1987.

[38] EPDC, Wakamatsu 50 MW$_e$ FBC demonstration plant – Outline and operation, Presented at 13th IEA-AFBC Technical Meeting, Tokyo, 1987.

[39] T Taniguchi. Development of Takehara 350 MW$_e$ atmospheric FBC demonstration plant. Presented at 15th IEA-AFBC Technical Meeting, Amsterdam, 1988.

[40] S Oka. Circulating fluidized bed boilers – State-of-the-art and experience in exploitation (in Serbian). Proceedings of Symposium JUGEL Development of Electricity

Production in Yugoslavia from 1991 till 2000, Ohrid (Yugoslavia). Belgrade: JUGEL, Vol. 2, 1990, pp. 593-600.

[41]   S Oka. Research and development of the boilers for burning Yugoslav lignites (in Serbian). Proceedings of Symposium JUGEL Development of Electricity Production in Yugoslavia from 1991 till 2000, Ohrid (Yugoslavia). Belgrade: JUGEL, 1990, Vol. 2, pp. 557-566.

[42]   R Carson, J Wheldon, J Castelman, P Hansen. TVA-160 MWe AFBC demonstration plant process performance. Proceedings of 11th International Conference on FBC, Montreal, 1991, Vol. 1, pp. 391-401.

[43]   EA Kopetz, WB O'Brien. Start-up and operating experience of the 160 MWe AFBC demonstration plant at TVA Shwanee fossil plant. Proceedings of 10th International Conference on FBC, San Francisco, 1989, Vol. 1, pp. 709-716.

[44]   JW Bass, JL Golden, BM Long, RL Lumpkin, AM Manaker. Overview of the utility development of AFBC technology Thennessee Valley Authority (TVA). Proceedings of 9th International Conference on FBC, Boston, 1987, Vol. 1, pp. 146-152.

[45]   D Thimsen, J Stallings. EPRI perspective on the NSP company Black Dog unit No. 2 AFBC retrofit experience to date. Proceedings of 11th International Conference on FBC, Montreal, 1991, Vol. 2, pp. 811-816.

[46]   S Ikeda. Wakamatsu 50 MWe AFBC combustion test results and EPDC development schedule of FBC. Presented at 21st IEA-AFBC Technical Meeting, Belgrade, 1990.

[47]   Present status of the FBC boilers in Japan. Presented at 16th IEA-AFBC Technical Meeting, Padukah (U.S.A.), 1989.

[48]   JE Son, CK Yi, YS Park, JH Choi, PS Ji. Combustion of antracite-bitumious coal blend in a fluidized bed. Proceedings of 11th International Conference on FBC, Montreal, 1991, Vol. 3, pp. 1415-1420.

[49]   CR McGowin, EM Petrill, MA Perna, DR Rowley. Fluidized bed combustion testing of coal/refuse-derived fuel mixtures. Proceedings of 10th International Conference on FBC, San Francisco, 1989, Vol. 1, pp. 7-15.

[50]   M Valk. Fluidized bed combustors. In: M Radovanović, ed. Fluidized Bed Combustion. New York: Hemisphere Publ. Co., 1986, pp. 7-35.

[51]   EBARA internally circulating FBC boiler. Report EBARA Corporation, 1989. Presented at 21st IEA-AFBC Technical Meeting, Belgrade, 1990.

[52]   B Repić, D Dakić. Computer program for calculation thermal and mass balance in FBC boilers (in Serbian). Report of the Institute for Nuclear Sciences Boris Kidrič, Vinča, Belgrade, IBK-ITE-617, 1987.

[53]   R Dolezal. Large Boiler Furnaces. New York, London, Amsterdam: Elsevier Publ. Co., 1967.

[54]   HV Kuznjecov, VV Mitor, IE Dubovski, ES Karasina, eds. Thermal Calculation Method for Boiler Design – Standard Method (in Russian). Moscow: Ehnergiya, 1973.

[55]   SE Tung, GC Williams, eds. Atmospheric fluidized-bed combustion. A technical source book (Final report). MIT, Cambridge, and US Department of Energy, 1987, DOE/MC/14536-2544 (DE88001042).

[56]   D Kunii, O Lewenspiel. Fluidization Engineering. New York: R. E. Krieger Publ. Co., 1977.

[57] FA Zenz. Regimes of Fluidized Behevior. In: JF Davidson, D Harrison, eds. Fluidization, Ist ed. London: Academic Press, 1971.

[58] JF Davidson, D Harrison, eds. Fluidization, 2nd ed. London: Academic Press, 1985.

[59] VA Boroduglia. Small FBC boilers in U.S.S.R.. Presented at the 21st IEA-AFBC Technical Meeting, Belgrade, 1990.

[60] AP Cheremisinoff, PN Cheremisinoff. Hydrodynamics of Gas Solids Fluidization, ed. London: Gulf. Publ. Co., 1984.

[61] NP Muhlenov, BS Sazhin, VF Frolov. eds. Calculation of Apparatuses with Bubbling Fluidized Beds (in Russian). Leningrad: Khimia, 1986.

[62] OM Todes, OB Citovich. Apparatuses with Bubbling Fluidized Beds (in Russian). Leningrad: Khimia, 1981.

[63] FA Zenz, DF Othmer. Fluidization and Fluid-Particle Systems. New York: Reinhold, 1960.

[64] M Radovanović. Combustion in fluidized beds. New York: Hemisphere Publ. Co., 1986, pp. 128-183.

[65] D Dakić, S Oka. Results of the testing in operation of the FBC hot-air generator with 4.5. MW$_t$ in power, mounted in Karadjordjevo (in Serbian). Report of the Institute of Nuclear Sciences Boris Kidrič, Vinča, Belgrade, IBK-ITE-434, 1983.

[66] B Grubor, D Dakić, S Oka. Analisys of operation of the retrofited FBC boiler with thermal power 9.3 MW$_{th}$ (in Serbian). Report of the Institute of Nuclear Sciences Boris Kidrič, Vinča, Belgrade, IBK-ITE-639, 1987.

[67] B Grubor, D Dakić, S Oka, M Ilić. Report about the testing of the FBC boiler in OTEKS-Ohrid (in Serbian). Report of the Institute of Nuclear Sciences Boris Kidrič, Vinča, Belgrade, IBK-ITE-811,1989.

[68] WW Hoskins, RJ Keeth, S Tavoulareas. Technical and economic comparison of circulating AFBC v. s. pulverised coal plants. Proceedings of 10th International Conference on FBC, San Francisco, 1989, Vol. 2, pp. 175-180.

[69] RE Allen, AJ Karalis, BJ Manaker, JH Chin, CR Rozzuto. Comparison of a year 2000 atmospheric circulating fluidized bed and conventional coal-fired power plant. Technical features and costs. Proceedings of 10th International Conference on FBC, San Francisco, 1989, Vol. 2, pp. 511-518.

[70] EPDC-report, Test results of 150 t/h bubblig type FBC for electric power utility. Presented at 22nd IEA-AFBC Technical Meeting, Montreal, 1991.

[71] SA Brain, EA Rogers. Experience of erosion of metal surfaces. In: U. K. Fluid Bed Boilers, Report British Coal Corporation, Coal Research Establishment, Cheltenham (G.B.).

[72] S Oka, B Grubor, B Arsić, D Dakić. The methodology for investigation of fuel suitability for FBC and the results of comparative study of different coals. Proceedings of 4th International FBC Conference, London, 1988, pp. 1/8/1-19.

[73] JS Cho, SK Lee. Operating experience in fluidized bed boilers with deep/shallow bed in Korea. Proceedings of 9th International Conference on FBC, Boston, 1987, Vol. 1, pp. 424-429.

[74] T Yoshioka, S Ikeda. Wakamatsu 50 MW$_e$ atmospheric fluidized bed combustion tests results & future test plan. Presented of 20th IEA-AFBC Technical Meetnig, Lisbon, 1990.

[75] JT Tang, F Engstrom. Technical assessment on the Ahlstrom pyroflow circulating and conventional bubbling fluidized bed combustion systems. Proceedings of 9th International Conference on FBC, Boston, Vol. 1, 1987, pp. 38-53.

[76] M Valk, EA Bramer, HHJ Tossaint. Optimal staged combustion conditions in a fluidized bed for simultaneous low NO_x and SO_2 emission levels. Proceedings of 10th International Conference on FBC, San Francisco, 1989, Vol. 2, pp. 995-1001.

[77] RP Krishnan, EJ Anthony, M Rajavel, S Srinivasan, AJ Rao, S Rajaram. Performance testing with high ash Indian coals and coal washery rejects in an AFBC pilot plant. Proceedings of 10th International Conference on FBC, San Francisco, 1989, Vol. 2, pp. 1245-1250.

[78] J Lan, et al. The experimental investigations of fine ash recycle in an AFB burning lean coal. Proceedings of 9th International Conference on FBC, Boston, 1987, Vol. 2, pp. 1096-1104.

[79] BJ Zobeck, MD Maun, DR Hajicek, RJ Kadrmas. Western U.S. coal performance in a pilot-scale fluidized bed combustor. Proceedings of 9th International Conference on FBC, Boston, 1987, Vol. 1, pp. 330–337.

[80] EJ Anthony, HA Becker, RK Code, JR Stephanson. Pilot-scale trials on AFB combustion of a petroleum coke and a coal-water slurry. Proceedings of 10th International Conference on FBC, San Francisco, 1989, Vol. 2, pp. 653-660.

[81] MK Senary, J Rirkey. Limestone characterization for AFBC applications. Proceedings of 10th International Conference on FBC, San Francisco, 1989, Vol. 1, pp. 341-350.

[82] AM Manaker, J Fishbangler, HH Vronn. Project overview for the 160 MWe AFBC demonstration plant at TVA Shawnee fossil plant regenaration. Proceedings of 11th International Conference on FBC, Montreal, 1991, Vol. 1, pp. 507-514.

[83] F Verhoeff, PHG van Heek. Two-year operating experience with the AKZO 90 MWth coal-fired AFBC boiler in Holland. Proceedings of 10th International Conference on FBC, San Francisco, 1989, Vol. 2, pp. 289-296.

[84] UHC Bijvoet, JW Wormgoor, HHJ Tossinaint. The characterization of coal and staged combustion in the TNO 4 MWth AFBC research facility. Proceedings of 10th International Conference on FBC, San Francisco, 1989, Vol. 2, pp. 667-673.

[85] JW Wormgoor, UHC Bijvoet, BJ Gerrits, MLG van Gasselt. Enchanced environmental and economical performance of atmospheric fluidised bed boilers. Proceedings of 11th International Conference on FBC, Montreal, 1991, Vol. 2, pp. 665-676.

[86] B Grubor, D Dakić, S Oka. Investigation of the combustion of coke dust in fluidized bed and design and concept of the FBC boiler (in Serbian). Report of the Institute of Nuclear Sciences Boris Kidrič, Vinča, Belgrade, IBK-ITE-581, 1986.

[87] D Dakić, M Ilić, S Damljanović, B Arsić. Investigation of the suitability of coal Bogovina for FBC (in Serbian). Report of the Institute of Nuclear Sciences Boris Kidrič, Vinča, Belgrade, IBK-ITE-819, 1991.

[88] B Grubor. Investigation and mathematical modeling of processes in during coal combustion in fluidized bed (in Serbian). PhD dissertation, Mechanical Engineering Faculty, University of Belgrade, 1992.

[89] JD Acierno, G Garver, B Fisher. Design concepts for industrial coal-fired fluidized-bed steam generations. Proceedings of 8th International Conference on FBC, Houston, 1985, Vol. 1, pp. 406-415

[90]   M Trifunović. Erosion and corrosion in FBC boilers – Review of the world's experience (in Serbian). Report of the Institute of Nuclear Sciences Boris Kidrič, Vinča, Belgrade, IBK-ITE-665, 1987.

[91]   J Stringer. Current information on metal wastage in fluidized bed combustors. Proceedings of 9th International Conference on FBC, Boston, 1987, Vol. 2, pp. 658-696.

[92]   SA Jansson. Tube wastage mechanisms in fluidized bed combustion systems. Proceedings of 8th International Conference on FBC, Houston, 1985, Vol. 2, pp. 750-759

[93]   JW Byam Jr, MM Madsen, AH Nazemi. Analysis of metal loss from heat exchangers in fluidized bed combustors. Proceedings of 8th International Conference on FBC, Houston, 1985, Vol. 2, pp. 807-821.

[94]   JM Parkinson, et al. Cold metal studies of PFBC tube errosion. Proceedings of 8th International Conference on FBC, Houston, 1985, Vol. 2, pp. 730-738.

[95]   HH Krausse, et al. Erosion-corrosion effects on boiler tube metals in a multisolids fluidized bed coal combustor. Journal of Engineering for Power 1:1-8, 1979.

[96]   HH Stringer, F Ellis, W Stockdale. In-bed tube erosion in atmospheric fluidized bed combustors. Proceedings of 8th International Conference on FBC, Houston, 1985, Vol. 2, pp. 739-749

[97]   KK Babcock-Hitachi. Corrosion/erosion behaviours in 20 t/h pilot plant. Presented at 12th IEA-AFBC Technical Meeting, Vienna,1986.

[98]   F Johnsson, B Leckner. Material loss from the heat exchange surfaces of the Chalmers 16 MWth fluidized bed boiler. Presented at 12th IEA-AFBC Technical Meeting, Vienna, 1986.

[99]   B Leckner, F Johnsson, S Andersson. Erosion in fluidized beds-influence of bubbles. EPRI Fluidized-Bed Meterials Workshop, Port Hawkesbury (Canada), 1985.

[100] AJ Mincher, et al. Materials studies at the IEA Grimethorpe PFBC experimental facility. Proceedings of 8th International Conference on FBC, Houston, 1985, Vol. 2, pp. 760-771.

[101] DW Briggs, NH Andrews. The Canadian FBC materials TESE program. Proceedings of 8th International Conference on FBC, Houston, 1985, Vol. 2, pp. 794-806.

[102] DR Hajicek. et al. Corrosion/erosion resulting from the fluidized bed combustion of low rank coals. Proceedings of 9th International Conference on FBC, Boston, 1987, Vol. 2, pp. 663-671.

[103] VK Sethi, et al. Materials and component performance in the TVA 20 MWth AFBC pilot plant. Proceedings of 9th International Conference on FBC, Boston, 1987, Vol. 2, pp. 629-636.

[104] JW Larson, et al. Summary, components-materials. Proceedings of 8th International Conference on FBC, Houston, 1985, Vol. 2, pp. 711-713.

[105] AV Levy, Yong-Fa Man, N Jee. The erosion-corrosion of heat exchanger tube in FBC. Proceedings of 9th International Conference on FBC, Boston, 1987, Vol. 2, pp. 637-655

[106] JS Lin, DA Stevenson, J Stringer. The role of carbonization in the in-bed corrosion of alloys at FBC. Proceedings of 9th International Conference on FBC, Boston, 1987, Vol. 2, pp. 656-662.

[107]   RQ Vincent, JM Poston, BF Smith. Erosion experience of the TVA 20 MWth AFBC boiler. Proceedings of 9th International Conference on FBC, Boston, 1987, Vol. 2, pp. 672-684.

[108]   M Radovanović, ed. Fluidized Bed Combustion. New York: Hemisphere Publ. Co., 1986.

[109]   M Hupa, BJ Skrifvars, A Moilanen. Measuring the sintering tendency of ash by laboratory method. Journal of the Institute of Energy 452:131-137, 1989.

[110]   S Oka. Some remarks on ash sintering in fluidized bed. Presented at 24th IEA-FBC Technical Meetting, Firenca (Italy), 1991.

[111]   S Andersson. Bed expansion and slumping of fluidized beds. PhD dissertation, Chalmers University, Göteborg (Sweden), 1990.

[112]   BJ Skrifvars, M Hupa, M Hiltunen. Sintering of ash during FBC. Presented at Engineering Foundation Conference on Inorganic Transformations and Ash Deposition During Combustion. Palm Coast (U.S.A.), 1991.

[113]   V Pavasović, M Ilić, V Nedović, S Oka. Inert materials for fluidized bed combustion of biomass. Presented at 19th IEA-AFBC Technical Meeting, Göteborg (Sweden), 1989.

[114]   Council of Industrial Boiler Owners, http://www.cibo.org/

[115]   BM Steenari. Chemical propetries of FBC ashes. PhD dissertation. Chalmers University, Göteborg (Sweden), 1998.

[116]   S Oka. Is the future of bubbling fluidized bed combustion technology in distributive power generation? Thermal Science 2:33-48, 2001.

[117]   D McCann. Design review of biomass bubbling fluidized bed boilers. Proceedings of 14th FBC Conference, Vancouver, 1997, Vol. 1, p. 29-37.

[118]   B Leckner, M Karlsson. Gaseous emissions from circulating fluidized bed combustion of wood. Biomass and Bioenergy Vol. 4, 5:379-389, 1993.

[119]   B Leckner, M Karlsson, M Mjornell, U Hagman. Emissions from a 165 MW$_{th}$ circulating fluidized bed. Journal of the Institute of Energy 464:122-130, 1992.

[120]   D Granatstein. FBC of solid waste fuels for electrical generation. Canadian Electrical Association Report, CEA No. 9334 G 1017, 1994.

[121]   B Skoglund. Six years of experience with Sweden's largest CFB boiler. Proceedings of 14th International Conference on FBC, Vancouver, 1997, Vol. 1, pp. 47-56.

[122]   EJ Anthony. FBC of alternate solid fuels, status, successes and problems of the technology. N Chigier, ed. Progress in Energy and Combustion Science, 1995, 21, pp. 239-268.

[123]   RC Howe, RJ Divilio. FBC of Alternate Fuels. EPRI Report TR-1000547s, 1993.

[124]   R Legros. Energy from waste – Review in the field of FBC of municipal solid wastes. Report to CANMET under DSS contract 28SS.23440-1-9070, 1993.

[125]   NM Patel, P Wheeler. FBC of municipal solid wastes: A status report for the International Energy Agency. Task XI: MSW Conversion to Energy, Activity: MSW and RDF. ETSU, Harwell (G.B.), 1994.

[126]   M Lundberg, U Hagman, B-Å Andersson. Environmental performance of the Kvaerner BFB boiler for MSW combustion – Analysis of gaseous emissions and solid residues. FDS Preto, ed. Proceedings of 14th International Conference on FBC, ASME, Vancouver, 1997, pp. 7-13

[127]   A Maillo. Private communication. Urbaser, s. a., Madrid, 1997.

[128]   LP Nelson. Municipal waste combustion assessment: Fluidized bed combustion. U.S. EPA contract 68-03-3365, EPA-600/8-89-061, 1989.

[129]   DA Orr, OW Hargrove, T Boyd, W Chow. FBC air toxics and the clean air act. Proceedings of Conference Application of Fluidized-Bed Combustion for Power Generation. EPRI T-101816, 1993

[130]   D Cianciarelli. Characterization of semi-volatile organic emissions from the Chatham 20 MW circulating fluidized bed demonstration unit. Environment Canada Report, File 4030-7-15, 1989.

[131]   R Mortazavi. Characterization of semi-volatile organic compounds (SVOCs) and volatile organic compounds (VOCs) from the Point Aconi coal-fired power plant. Environment Canada Report PMD/96-7, 1996.

[132]   Seminar Publications: Operational Parameters for Hazardous Waste Combustion Devices. EPA Report EPA/625/R-93/008, 1993.

[133]   D Liang, EJ Anthony, BK Loewen, DJ Yates. Halogen capture by limestone during FBC. EJ Anthony, ed. Proceedings of International Conference on Fluidized Bed Combustion, ASME, Montreal, 1991, pp. 917-921.

[134]   EJ Anthony, EM Bulewicz, FDS Preto. The combustion of halogenated wastes in FBC Systems. TF Wukasch, ed. Proceedings of 49th Industrial Waste Conference, West Lafayette (U.S.A.): Lewis Publishers, 1994, pp. 673-680.

[135]   EJ Anthony, EM Bulewicz, E Janicka, S Kandefer. Chemical links between different pollutant emissions from a small bubbling FBC. Fuel 7:713-728, 1998.

[136]   C Brereton, JR Grace, CJ Lim, J Zhu, R Legros, JR Muir, J Zhao, RC Senior, A Luckos, N Inumaru, J Zhang, I Hwang. Environmental aspects, control and scale-up of circulating fluidized bed combustion for applications in Western Canada. Final Report to Energy Mines and Resources under contract 55SSS 23440-8-9243, 1991.

[137]   EJ Anthony, DY Lu, JQ Zhang. Combustion of heavy liquid fuels in a bubbling fluidized bed. Journal of Energy Recources Technology, Vol. 124, 1:40-45, 2002.

[138]   GW Boraston. Revolving fluidized bed technology for the treatment of hazardous waste materials. Proceedings of CANMET Conference on Energy and the Environment, Toronto, CANMET, 1991.

[139]   http://www.serl-ns.com/textversion/aua.html

[140]   B North. Private communication. CSIR Mattek, South Africa, 1997.

6.

INVESTIGATION OF COAL SUITABILITY FOR FLUIDIZED BED COMBUSTION

6.1. Effects of fuel characteristics on the design and FBC boiler concept and its operational behavior

When comparing conventional boilers, the major advantage of FBC boilers (besides low SO_2 and NO_x emissions) is the possibility of burning fuels with a very heterogeneous character, either simultaneously (i. e., co-firing), or periodically as cheaper fuels become available. Currently, bubbling FBC boilers hold this advantage in relation to conventional boilers, but in particular circulating FBC boilers (see Chapter 1) have an advantage in this respect. A conventional boiler designed for a certain type of coal cannot easily burn another type of coal. In comparison with conventional boilers, a well-designed FBC boiler can to some extent be considered to be independent of the fuel type.

However, despite this high fuel flexibility, when designing the FBC boiler and its auxiliary systems, it is still necessary to know in advance, and keep in mind, what spectrum of fuels the boiler will likely be used for. For proper FBC boiler design it is far more important to have the auxiliary systems designed so that they can operate with various fuels. For instance, in combustion of "as-received" low-grade fuels, the equipment and systems for feeding fuel and draining bed materials must be of far greater capacity than when burning high-grade low-ash fuels.

In Sections 5.1.4 and 5.2, a list was provided of the numerous fundamental dilemmas and unknown quantities that a boiler designer should ideally know in order to make a boiler suitable for a specific fuel. The design must also fulfill the purpose of the boiler. Decision-making concerning these issues is indispensable as the calculation depends mostly on fuel characteristics and its behavior during combustion in the bubbling fluidized bed.

The significance and severity of the problems that boiler designers are faced with can best be illustrated by a list of different fuels whose combustion was successfully demonstrated in experimental and pilot facilities in the U.S.A. [1–4]. The list is as follows:

– coal separation and washing waste,	– electric cable insulation,
– peat,	– heavy oils,
– sawdust,	– chemical industry waste waters,
– bark,	– classified industrial waste,
– coke gasification remnants,	– wooden railroad sleepers,
– various pulps from process industries,	– wooden plates,
– brown coals,	– oil shale,
– wood processing waste,	– anthracite culm,
– petroleum coke,	– lignites,
– paper waste,	– battery cases (lead removed), and
– plastics,	– photographic films (silver removed).
– car tires,	

This list also includes the successful combustion of 19 different fuels tested in experimental furnaces at the Laboratory for Thermal Engineering and Energy at VINČA Institute of Nuclear Sciences, Belgrade, Yugoslavia (hereinafter: ITE-IBK). Of these 19 fuels, 16 were coals, ranging from lignite to anthracite, and three were different types of biomass: corncobs, tree bark and sawdust.

The standard ultimate and proximate analyses of coals and other types of fuel are insufficient to provide answers to all the questions an FBC boiler designer is expected to resolve. The data necessary for the calculation and design of FBC boilers can be obtained only experimentally, by measurements and observation of coal behavior in a fluidized bed under real operating conditions.

Comprehensive and long-term research programs have been carried out in many countries. The main aims of such programs were to investigate the suitability of different fuels for FBC combustion. The behavior of its organic and mineral components during combustion can reveal the effects of fuel characteristics on boiler concept and design.

Investigation and fuel testing in large-scale boilers in industry are, however, very expensive and cannot answer all the questions that arise during boiler design. For those reasons, fuel suitability studies have been carried out in a number of the technologically advanced countries (U.S.A., Japan, the Netherlands, Great Britain) which rely primarily on experiments in small- and medium-size experimental furnaces. In the U.S.A., a long-term project titled "Characterization of Fuel" was run by EPRI (Electric Power Research Institute), and carried out by Babcock & Wilcox [1, 5-8]. Within the EPRI program, investigations were conducted in facilities with 3 in, 6 in, and 1×1 ft^2 cross section furnaces. A similar program was carried out in

Japan [9, 10]. By contrast to the programs in other countries, those in the U.S.A. and Japan also involved investigations on large-scale pilot and demonstration facilities. Here, the aim was to check if the results and conclusions obtained in small-scale furnaces could be used when designing large industrial boilers (scale-up). The EPRI program included investigations on the following large facilities: a pilot boiler with a furnace cross section 6 x 6 ft^2, built in 1977 in Alliance (B&W), a pilot boiler with capacity 20 t/h, built in the thermal Shawnee Power Plant, Paducah, and several minor 25-50 t/h demonstration boilers. Two such facilities used in this program were the 160 MW$_e$ TVA and 130 MW$_e$ Black Dog demonstration boilers. In addition to checking the scale-up issues, an important aim of these investigation with these large-scale facilities, was an examination of the operation of auxiliary systems.

A similar program was started in Japan in 1977, with the construction of a series of pilot plants and demonstration boilers of capacities of 20 t/h and 160 t/h [9, 10]. This program continued with the construction and investigation of PFBC boilers.

Such comprehensive, long-term and expensive programs cannot be carried out in all countries, mainly due to a lack of financial resources.

One of the few systematic research programs, carried out with the aim of determining the effects of fuel characteristics on the behavior of fluidized beds and on FBC boiler design, was run by the ITE-IBK. The program started in 1976 and tested a very large number of different fuels. An original methodology was developed to investigate the choice of solid fuel suitability for fluidized beds (in this book it is called the ITE-IBK methodology). The basic approach adopted in ITE-IBK methodology was that reliable data on fuel behavior in fluidized bed, and the data necessary for boiler design, could be obtained by investigations in small-scale experimental furnaces. This approach is nowadays predominant in other countries too, as can be seen from a published analysis on different approaches to this problem [11].

6.2.   ITE-IBK methodology for investigation of solid fuel suitability for combustion in fluidized beds

The material presented in this chapter is based on results obtained by the author and his associates from a long-term investigation of solid fuel combustion in fluidized beds. The program started in 1976 and still continues, with the only difference being that investigations are now directed towards processes in circulating fluidized beds [12, 13]. During this period, nineteen different solid fuels were investigated (ranging from lignite to anthracite and coke, plus three types of biomass). Results of these investigations were published in more than 40 publications, many of which are presented in this book [14-31]. Some of the results were published in internal reports of the ITE-IBK and they are quoted in [14]. An analysis of results for all 19 fuels and a description of the ITE-IBK methodology are presented in references [14, 22, 23, 32-34]. In this book, conclusions drawn from these investigations are substantiated, supplemented and illustrated by results obtained by other authors, primarily from large-scale facilities.

6.2.1. Principles of the ITE-IBK methodology

The ITE-IBK methodology for investigation of solid fuel suitability for combustion in fluidized beds is based on the following:

– knowledge of a series of empirical data is necessary for calculation and design of a boiler,
– standard investigations of coal characteristics are insufficient to obtain these data,
– effects of fuel characteristics on its behavior in combustion and on boiler design can only be determined experimentally,
– in order to fully exploit the possibility of a FBC boiler (or furnace) to burn very different fuels, it is necessary to know the distinctive behavior of each fuel and adjust the design of the boiler (furnace) to them, and
– nowadays it is still impossible to design a furnace of a FBC boiler using mathematical modelling based solely on the data obtained by proximate and ultimate coal analysis. Even if we are now closer to this objective, it was certainly not the case at the end of the 1970's.

From the very beginning of methodology development, the following objectives were set:

– to apply standard methods of fuel investigation to the largest possible degree,
– the methodology must provide all the data necessary for choice of boiler concept, calculation of the main boiler parameters and its dimensions, and the design of the boiler itself and its auxiliary systems,
– experimental investigations ought to be carried out in small-scale, cheap experimental furnaces at conditions as close to the real ones as possible, in order to be able to use the results with more certainty in design and calculation for industrial-size furnaces and boilers, and
– experiments are expected to offer data on the choice of optimum operation parameters of the nominal regime, and also for the parameters necessary for boiler start-up.

In order to obtain useful results for boiler designers, it is necessary to answer the following general questions:

– Is the fuel whose combustion is expected in the FBC boiler appropriate for combustion in a fluidized bed?
– What are the optimum conditions of combustion?
– How should the choice of the FBC boiler concept be made?

and
- What is the range of different fuels that can burn in any given FBC boiler, and how can that range be made as wide as possible within given economic boundaries?

Answers to these general questions can be obtained if the designer has answers to the following specific questions and issues:

- What is:
    - the ash sintering temperature in the bed,
    - the optimum combustion temperature in the stationary regime,
    - the possible temperature change range,
    - the fluidization velocity,
    - the permissible range of fluidization velocity change,
    - the start-up temperature,
    - the necessary bed height,
    - the necessary height of the freeboard,
    - the optimum excess air,
    - the char burning rate, and
    - the freeboard temperature in adiabatic conditions?
- How does ash behave during combustion?
- What is the quantity of heat generated in the bed, and above the bed?
- What are the CO, SO_2, and NO_x emissions?
- How to design a flue gas cleaning system for particle removal, and whether the following are necessary:
    - the system for fuel feeding above or below the bed,
    - the system for recirculation of unburned particles,
    - the system for continuous or periodical draining of material from the bed,
    - the system for limestone introduction, and
    - the system for secondary air injection.

Planning of the methods and experiments (methodology) which will give answers to these questions requires preliminary consideration of three groups of problems:

(1) What physical and chemical characteristics of fuel (coal) affect its behavior in fluidized bed combustion the most, and how do they affect the processes, the concept and the design of the boiler furnace and its auxiliary systems?

(2) What physical and chemical processes should be investigated in FBC furnaces and what quantities should be measured in order to establish differences in the behavior of different fuels and to obtain data for furnace, boiler and auxiliary system calculation, and choice of their concept and design?

and

> (3) Under what conditions should experimental investigations be run in or-
> der to obtain real answers to these questions and in order to be able to use
> the results in large-scale furnace and boiler design?

**Physical and chemical characteristics of solid fuel influencing its behavior
in fluidized bed combustion.** In many parts of this book, especially in Chapters 4 and
5, much has been said about the influence of fuel characteristics on processes before
and during ignition and combustion. In the text that follows, the characteristics of solid
fuel and a discussion on their influence on the boiler concept and design will be systematized. These parameters are: particle size distribution, moisture content, volatiles content, ash content and sulphur content, chemical and petrographic composition of ash,
ash sintering temperature, ignition temperature and burning rate.

The size distribution of coal is one of its most important characteristics. It
determines the way coal is fed (above or below bed), as well as the number, distribution and position of fuel feed points.

The quantity of particles below 1mm in size can also greatly affect particle
elutriation, amount of unburned particles in the fly ash, the freeboard temperature,
combustion efficiency, and the need to recirculate unburned particles.

Coal particle size also helps determine the degree of elutriation of its ash
components, and any tramp material, or the amount of ash reporting to the bed and
the choice of bed height. In particular, defluidization of large particles on the bottom of the bed helps determine the capacity and operation of the drainage system
for bed material, and this depends on the coal particle size.

The moisture content affects fuel behavior in the conveying and feeding
systems, and what is especially significant, the size of the heat transfer surfaces, if
any, that are immersed in the bed.

The content of volatiles affects the ratio of heat quantity generated in the bed
and in the freeboard, and the freeboard temperature, e. g., the distribution of heat
transfer surfaces in the furnace and choice of bed height. Coal feeding and distribution of feed points depend on the quantity of volatiles. In addition, the start-up temperature, combustion efficiency, the need to introduce secondary air and the
freeboard height also depend on the volatile content.

The moisture content, volatile matter content and ash characteristics affect
particle fragmentation in the fluidized bed and, therefore, indirectly affect the char
inventory in the bed, the combustion efficiency (particle elutriation) and the response rate of the boiler to load change.

The sulphur content determines whether it is necessary to introduce limestone into the bed, and the capacity of any limestone transport and feeding system
in order to achieve the prescribed SO_2 emissions.

The physical and chemical characteristics of the ash affect the choice of bed
operation temperature (so as to avoid sintering), and the ash behavior in the bed

(elutriation or remaining). An extremely high percentage of sodium and potassium compounds in the ash may affect the choice and type of inert material. Equally, the quantity, content and behavior of ash during combustion in a fluidized bed affect the capacity and design of the system for draining material from the bed and the capacity of the particle separation device from flue gases. The lime content in the ash also determines whether it is possible to use the ash to capture SO_2 and hence reduce the amount of limestone used for desulphurization.

Coal ignition temperature affects the choice of start-up temperature, and the manner of bed heating at start-up, the size of start-up equipment and the start-up procedure.

The coal (char) burning rate affects the choice of fuel particle size, the coal feeding system and the start-up temperature. Char inventory in the bed in the stationary regime and response rate of the furnace to load change also depend on the burning rate.

Systematization of these and a proper understanding of the characteristics of the fuel that affect the FBC boiler concept and design are presented in Table 6.4 at the end of this chapter.

Parameters and processes characteristic for testing of fuel behavior in fluidized bed combustion. Testing of fuel behavior during combustion in fluidized beds requires an experimental determination of the physical and chemical fuel characteristics and investigation of a series of processes in conditions as close as possible to the real ones. The following have to be examined: the behavior of fuel in the handling and feeding system; the influence of the location and the manner of fuel feeding; combustion in the stationary regime and start-up period; particle fragmentation in a fluidized bed; particle elutriation from the bed; distribution of heat generation in the furnace and ash behavior during long stationary furnace operation.

In order to follow and analyze these processes, it is necessary to determine and measure the following parameters, classified into three groups according to type:

(a) Parameters characteristic for fuel:
 – the physical and chemical characteristics of fuel,
 – the physical and chemical characteristics of ash,
 – the ash sintering temperature,
 – the ignition (start-up) temperature,
 – the burning rate, and
 – the fuel particle fragmentation characteristics.
(b) Parameters characteristic for the combustion process:
 – the quantity of heat generated in the bed,
 – the quantity of heat generated above the bed,
 – the optimum combustion temperature, and
 – the optimum excess air.

(c) Parameters referring to the furnace in general:
- the quantity and size distribution of ash particles elutriated from the bed,
- the quantity and size distribution of ash particles remaining in the bed,
- the percentage of unburned particles in fly ash, and
- the concentration of SO_2, NO_x, and CO at the furnace exit.

Experimental conditions. In experimental investigations of physical and chemical processes there is always an effort to create conditions similar to those in real processes. If such similarity is fully achieved, it is possible to apply the results of the investigations to all similar phenomena, regardless of the specific conditions in which they take place. In engineering, the most important thing is to be able to apply results of investigations carried out in experimental devices of small dimensions for the calculation of processes in full-scale facilities, i. e., to make the so-called "scale-up" possible. It is well known that two physical phenomena are similar if values of the so-called similarity criteria are the same for both phenomena. That means there is geometric, hydrodynamic and thermodynamic similarity, and here that must also include the "similarity of chemical reactions" [35, 36]. In modelling processes for solid fuel combustion it is impossible to achieve equality of all criteria resulting from geometric, hydrodynamic and thermodynamic similarity. In consequence, the achievement of so-called partial similarity is necessary [37]. Unfortunately, achievement of partial similarity depends significantly on personal assessment of the experimenter on which processes and parameters are essential for a proper investigation of the phenomena under study. This means that achieved similarity depends both on the nature of the investigated phenomenon and the skills of the investigator as well.

In planning experiments on fluidized bed combustion and fuel suitability, it is suggested here that it is necessary to achieve the following partial conditions of similarity.

Geometric similarity: provision of fuel particles and inert bed material particles similar or identical to those in the real furnace, given the unavoidability of deviation from large-scale geometric similarity or equality of furnace dimensions, that is, of the space where combustion takes place.

Hydrodynamic similarity: achievement of hydrodynamic similarity of fuel and inert material particles, which is provided by identical fluidization velocity. In order to achieve similarity of conditions for chemical reactions, it was decided to provide hydrodynamic similarity for the bubbles. This was accomplished by adoption of the real bed height appropriate to a full-scale furnace and ensuring that there was a sufficiently large furnace cross section thus preventing any limitation of growth of bubbles due to wall effects. Complete similarity of particle mixing in the bed and similarity of the particle flows and circulation in the bed unfortunately cannot be achieved. In this respect there are large differences between experimental and real, industrial furnaces. However, there are similar types of differences between large-scale industrial furnaces as well, so one experimental study could never satisfy the requirement of providing such data for all furnace sizes and designs.

Thermodynamic similarity is provided by identical combustion temperature, by using the same fuel and the same excess air. Adiabatic furnace walls are adopted, but with heat removal from the furnace depending on fuel type and desired excess air. However, in this approach similarity of geometry and distribution of heat transfer surfaces in the bed are achieved.

6.2.2. Description of the ITE-IBK methodology

Testing of solid fuel suitability for combustion in a fluidized bed using the ITE-IBK methodology is carried out in three phases, in a series of experimental furnaces suitable for investigation of different processes. When it is impossible to arrive at reliable quantitative data due to the same testing methodology used, a comparative analysis of results for different fuels is possible, in addition, there is a determination of qualitative effects of fuel characteristics [22, 32, 33].

Phase one of testing includes the following:

- the determination of fuel size distribution, and especially the percentage of fuel particles below 1 mm,
- a standard proximate fuel analysis,
- a standard ultimate fuel analysis,
- a chemical analysis of ash,
- a "float or sink" analysis of ash, and
- the determination of ash sintering temperature.

All of the analyses listed in the first phase of ITE-IBK methodology are carried out using standard procedures, while characteristics of the fuel are determined in the form in which it will be used in a real furnace or boiler. The determination of ash sintering temperature is specific for the first phase. In addition to the standard procedure used to determine characteristic ash softening temperatures (beginning of sintering, softening point, hemisphere point, and melting point), the ash sintering temperature is determined directly in a small laboratory furnace 150 mm in diameter, under conditions close to the real ones. For this furnace, the desired bed temperature is obtained by gas combustion [24]. Figure 6.1 gives a diagram of this furnace.

The process of sintering is investigated as follows. A portion of the fuel or prepared laboratory ash from this is introduced into a fluidized bed of inert material previously heated to the desired temperature by gas combustion. By visual obser-

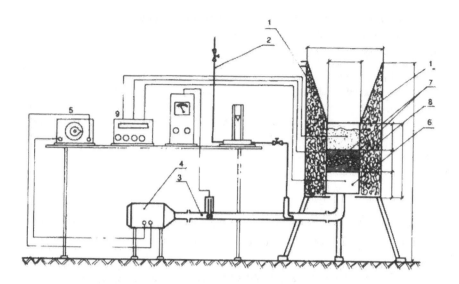

Figure 6.1. *Laboratory furnace for determination ash sintering temperature during fluidized bed combustion:*
1 – thermal isolation of the furnace, 2 – inlet and flow rate measurement of the gas burning in the fluidized bed, 3 – inlet and flow rate measurements of the fluidizing air, 4 – fan, 5 – variable transformer, 6 – air plenum chamber beneath the distribution plate, 7 – distribution plate made from sintered metal, 8 – gravel between distribution plates, 9 – thermocouple for measurement of the gas temperatures beneath distribution plate, temperature of the distribution plate and fluidized bed temperature

vation it is possible to determine whether the ash is melted and sintered and the inert material agglomerated. The appearance of agglomeration can also be observed by following the bed pressure drop. If sintering does not occur, the procedure is repeated at a higher temperature, until the bed temperatures for ash sintering and inert material agglomeration are determined. This test then allows a maximum bed temperature to be recommended, which then must not be exceeded in real furnaces.

From a standard, so called, "float or sink" analysis, the density of coal, ash and waste are determined. These data then allow an estimate of the quantity of material that will settle to the furnace bottom during combustion. Evidently, this material needs to be drained from the furnace by a bed material drainage system.

Phase two testing includes the following:

– investigation of the process of primary coal fragmentation during devolatilization and determination of the mean critical particle fragmentation diameter,

- the determination of start-up temperature, and
- the determination of burning rate.

The manner in which the fragmentation process is tested is described in detail in references [24, 38, 39]. Figure 6.2 presents a diagram of an experimental furnace 78 mm in cross section diameter, used in these tests. Desired bed temperature is achieved by means of electric heaters. Testing of particle fragmentation takes place in inert atmosphere, by fluidizing inert material with nitrogen, and by interrupting the process by sudden draining of inert material from the bed and its rapid cooling. Coal fragmentation is determined by shooting fuel particles before and after devolatilization. Results of the particle fragmentation investigation are presented in detail in Section 4.3 of Chapter 4.

Determination of start-up temperature and burning rate are described in references [17, 19, 26-29, 40, 41] and will be presented in detail in Section 6.2.4.

Phase three. Fuel combustion is tested in stationary operating conditions. Testing is performed in an experimental furnace and a pilot facility, in conditions close to the real ones, with the fuel in the form it will be burned in a real furnace. Similarity requirements described in this section are fully achieved in this experi-

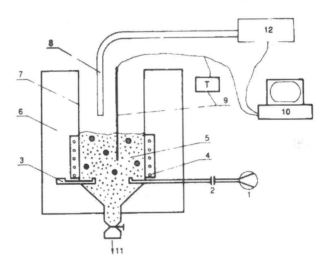

Figure 6.2. *Experimental furnace for investigation of primary fragmentation of coal particles:*
1 – fan for fluidizing air, or inlet of fluidizing nitrogen, 2 – measuring orifice, 3 – distribution plate with bubble caps, 4 – electrical heater, 5 – fluidized bed, 6 – thermal isolation, 7 – ceramic tube – furnace wall, 8 – probe for gas sampling, 9 – thermocouple with measuring instrument, 10 – personal computer for data acquisition, 11 – opening for bed material draining into the bed material cooler, 12 – gas analyzer

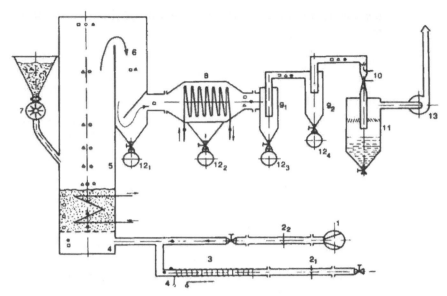

Figure 6.3. *Large experimental furnace for investigation of coal combustion in fluidized bed in real operating conditions:*
1 – main fan, 2₁, 2₂ – air flow meters, 3 – air heater, 4 – air plenum chamber beneath the distribution plate, 5 – furnace, 6 – flue gas outlet from the furnace, 7 – coal feeder, 8 – heat exchanger for flue gas cooling, 9₁, 9₂ – cyclones, 10 – ejector, 11 – wet scrubber, 12₁-12₄ – separators of solid particles, 13 – flue gas drought fan, • – thermocouple for air temperature and flue gas temperature measurements, ○ – gas temperature measurements using aspiration pyrometers, □ – pressure taps, △ – measurements of the composition of combustion products

mental furnace. The power of the experimental furnace is 200 kW$_{th}$ when operating with an immersed heat exchanger. Cross section of the furnace is 300 × 300 mm^2 and height 4 m. A diagram of the furnace is presented in Fig. 6.3.

The furnace enables combustion testing during coal feeding on to the bed or under the bed surface, but it also enables introducing limestone in a mixture with fuel or separately. All tests were carried out with silica sand with particle size 0.3-1.2 mm, as bed material. After elutriation of fine particles, mean diameter of inert particles ranged between 0.6 and 0.8 mm. Depending on fuel type the other testing parameters were chosen in the following ranges:

- combustion temperature 700-900 °C,
- excess air 1.1-1.4 (even up to 2 in combustion conditions of hot air generators),
- fluidization number v_f/v_{mf} = 3-6, and
- fixed bed height 300-500 mm.

Testing was usually performed in the following three characteristic stationary regimes:

- maximum combustion temperature regime,
- minimum combustion temperature regime, and
- minimum fluidization velocity regime.

Stationary operation is maintained for 2-4 h, depending on the fuel type and ash quantity. All quantities necessary for determining heat and mass balance (both gaseous and solid materials) were measured.

Apart from determining the quantities necessary for heat and mass balance, the following parameters were either measured or determined:

- the fluidized bed temperature,
- the temperature distribution along freeboard height,
- the quantity of heat removed from the bed,
- the quantity of heat generated above the bed,
- the change of combustion product composition – CO, CO_2, O_2, and SO_2 along the freeboard height,
- the quantity and size distribution of ash remaining in the bed,
- the quantity and size distribution of fly ash,
- losses due to unburned particles in fly ash,
- losses due to incomplete combustion, and
- O_2, CO, CO_2, SO_2, and NO_x concentrations at the furnace exit.

6.2.3. Characteristics of the investigated fuels

Data on tested fuels obtained by proximate analysis are given in Table 6.1. The same table includes the range of particle sizes of tested fuels and some results that will be considered later: sintering temperature, start-up temperature and ash percentage elutriated from the bed as fly ash.

Table 6.2 presents results of an ultimate analysis for the tested coals and biomass. Some results for the proximate analysis were repeated in order to make it easier to follow changes linked to fuel characteristics. Table 6.2 also includes measured values of combustion efficiency. Here, only the losses due to unburned solid particles in the fly ash were taken into consideration.

Tables 6.1 and 6.2 show that a broad spectrum of different fuels was tested. Moisture contents of the tested fuels ranged from 2 to 58%, ash content from 9 to 37% (about 2% in biomass), volatile content ranged from 9 to 40% (68% in biomass). While the calorific value in the "as-received state" varied from 8.5 to 28 MJ/kg, or when calculated on a dry ash free basis from 18 to 35 MJ/kg.

Table 6.1. *Proximate analysis of the coals tested using ITE-IBK methodology*

Fuel	Particle size d [mm]	Moisture content V_w [%]	Ash P [%]	Volatile matter content VM_o [%]	Content of C_{fix} [%]	Sulphur S_n [%]	H_c [MJ/kg]	Sintering temperature T_m [°C]	Start-up temperature T_{rf} [°C]	Particles with size <1 mm [%]	Fly ash U [%]
No. 1	15-30	14.66	11.73	32.84	0.99	0.98	18.46	1035+	350	≈0	95
No. 2	0-30 5-15	58.34	9.09	19.13	14.31	0.53	8.521	1040+	350	6.86	50-68
No. 3	0-5	47.20	12.18	24.15	15.30	1.2	11.889	930	600	27.76	95-90
No. 4	0-30	34.14	16.43	27.18	21.36	1.79	12.526	885	500	2-10	20-30 90 (d < 2 mm)
No. 5	0-5	2.42	33.27	8.81	54.48	0.67	20.932	1025	750-800	62	≈0 (d > 2 mm)
No. 6	0-35	32.51	19.55	23.40	20.36	2.12	12.623	1060+	500	11	35-65 90 (d < 2)
No. 7	0-2	15.15	13.22	2.70	67.37	1.43	23.472	1025+	750	45.91	≈0 (d > 2 mm)
No. 8	0-5	41.49	0.75	47.94	9.82	–	11.243	765+	400-450	–	≈100
No. 9	0-10	57.96	1.81	31.79	8.45	–	8.403	780+	400-450	–	≈100
No. 10	0-30 5-15	8.04	2.41	69.06	20.49	–	16.841	855	400	–	≈100
No. 11	0-30	17.88	16.71	30.32	33.81	1.40	19.253	900	600	0-1.5	60-70
No. 12	0-25	15.76	15.76	40.39	27.36	10.33	28.911	1000	600	1.92	95
No. 13	0-5	13.32	33.03	25.91	25.50	2.91	15.540	895+	400-500	57	30
No. 14	0-25	11.32	18.05	37.52	31.51	2.91	21.273	910	–	–	–
No. 15	0-2	10.00	37.11	24.38	25.98	2.67	14.487	880	600	7.5	75
No. 16	0-15	42.43	16.42	23.08	17.63	0.51	9.79	1030+	400	6.32	90
No. 17	0-15	21.94	20.66	27.27	28.94	1.69	15.464	855+	–	18.25	25
No. 18	0-3	24.13	17.08	23.57	23.49	2.23	13.650	920+	500	60	100
No. 19	0-10	20.57	26.45	20.21	21.00	5.02	14.500	920	500-550	24	60-70

Remark: All values are based on coal as received
+ In reducing conditions

Table 6.2. *Ultimate analysis of the coals tested using ITE-IBK methodology*

Fuel	Hydrogen [H] [%]	Oxygen [O] [%]	Carbon [C] [%]	Fixed carbon C_{fix} [%]	Calorific value H_{daf} [MJ/kg]	Volatile matter content VM_{daf}	C_{fix}/VM_{daf}	Combustion efficiency η_c^* [%]
No. 1	5.50	26.97	65.27	44.61	25.763	54.82	0.81	97-99
No. 2	5.89	28.15	63.92	42.50	25.305	56.80	0.75	86-87
No. 3	5.77	26.87	61.93	37.67	29.264	59.45	0.63	97-98
No. 4	5.44	25.20	64.62	43.22	25.341	54.99	0.79	99
No. 5	4.31	4.25	88.27	85.27	32.551	13.69	6.23	80
No. 6	6.24	20.71	66.67	44.39	27.525	51.03	0.87	98-99
No. 7	2.03	2.66	93.33	94.04	32.765	3.77	24.97	63-66
No. 8	6.19	4.91	48.24	17.01	19.464	82.99	0.20	99
No. 9	6.63	41.55	50.48	20.99	20.886	79.07	0.27	99
No. 10	5.90	45.88	47.57	22.39	18.316	75.10	0.30	99
No. 11	5.04	27.43	62.73	51.70	29.437	46.36	1.12	97-99
No. 12	6.08	5.98	74.16	33.27	35.151	49.11	0.68	98-99
No. 13	6.51	18.43	69.41	47.53	28.967	48.30	0.98	97
No. 14	5.20	15.35	74.52	44.63	30.12	53.12	0.84	–
No. 15	5.10	19.57	69.78	49.11	27.391	46.10	1.07	90-94
No. 16	5.32	30.05	61.26	42.84	23.795	56.10	0.76	98.50
No. 17	5.56	23.63	67.55	50.42	26.94	47.50	1.06	98.50
No. 18	5.13	26.06	62.88	48.14	27.975	48.30	1.00	92.1-97.4
No. 19	2.78	24.50	64.99	47.65	31.214	45.85	1.04	97-98.5

Remark: All values are based on daf basis

* Only carbon losses in fly ash are taken into account

The particle size of the coals tested was 0-30 mm and the majority of the coals tested were in an "as-received" state so that the percentage of particles smaller than 1 mm was extremely high in some cases.

The scope of quality and characteristic properties of tested coals can be seen in Figs. 4.7 and 4.8 (Chapter 4) and Fig. 6.4, plotted based on data given in Tables 6.1, 6.2 and 6.3.

In Fig. 4.8, the characteristics of coals given in Table 6.2, are shown in a diagram that illustrates the dependence of calorific value Q_{daf}, on the fixed carbon content in fuel C_{fix}.

Figures 4.7 and 6.4 show the characteristics of tested coals in two other ways: by demonstrating the connection between oxygen and carbon content in the coal, and as a dependence of calorific value on the total carbon content in fuel. Although most of the tested coals are geologically young coals (lignites and brown coals), a broad spectrum of fuels was covered – up to and including anthracite. Coal No.12 is a high-rank coal, but with exceptionally high content of volatile matter which deviates from the general behavior shown in preceding figures. The tested types of biomass (Nos. 8-10) also do not fit into the general behavior shown in Figs. 4.7, 4.8, and 6.4 that refer to coals.

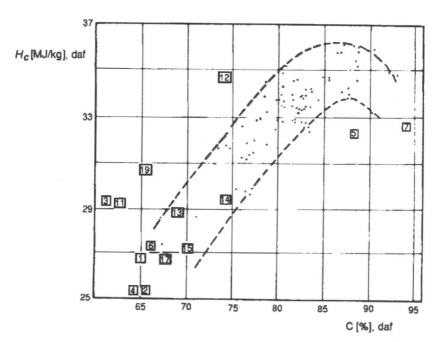

Figure 6.4.   *Calorific value of the coals investigated by ITE-IBK methodology as a function of total carbon content of the coal*

Table 6.3. Chemical analysis of ash for the coals tested using ITE-IBK methodology

Fuel	SiO_2 [%]	Al_2O_3 [%]	Fe_2O_3 [%]	CaO [%]	MgO [%]	Na_2O [%]	K_2O [%]	TiO_2 [%]	P_2O_5 [%]	SO_3 [%]	K [%]
No. 1	96.83	21.84	11.18	11.92	3.0?	0.15	0.72	0.82	0.11	3.02	69.49
No. 2	53.85	21.40	9.50	7.47	2.8?	0.22	0.84	1.03	0.08	2.33	76.28
No. 3	46.20	20.24	7.97	7.00	1.5	0.99	1.44	-	-	7.22	66.44
No. 4	-	-	-	-	-	-	-	-	-	-	-
No. 5	50.00	27.97	7.16	6.90	1.4	0.43	1.0	1.40	0.06	3.60	79.37
No. 6	32.87	17.12	7.45	20.17	3.0	2.25	1.67	0.74	0.10	13.85	61.58
No. 7	42.02	17.83	22.18	5.50	1.2	0.64	2.63	0.94	0.20	6.37	60.79
No. 8	6.45	0.66	2.05	73.05	0.5	0.17	4.50	0.02	1.90	1.93	7.13
No. 9	5.88	0.20	0.33	80.72	0.2?	1.00	2.57	0.02	1.30	0.95	6.10
No. 10	32.74	3.05	0.70	2.03	6.83	1.38	29.24	0.02	4.58	2.12	35.81
No. 11	42.48	16.28	25.27	5.45	1.8	2.16	1.40	1.60	0.48	3.02	60.36
No. 12	6.86	7.00	4.58	39.63	5.42	1.37	0.45	0.50	0.05	34.06	14.36
No. 13	49.55	19.80	15.30	3.06	3.1?	1.74	0.53	1.38	0.12	1.72	70.73
No. 14	-	-	-	-	-	-	-	-	-	-	-
No. 15	34.18	13.95	6.48	24.28	1.6	0.20	1.20	0.86	0.10	16.60	48.99
No. 16	49.98	13.65	7.28	14.02	2.2	0.27	1.68	1.06	0.16	9.58	64.69
No. 17	60.77	21.32	11.22	1.52	1.0?	0.22	1.12	1.77	0.08	0.83	83.86
No. 18	49.03	16.97	8.50	13.62	2.47	0.14	1.03	0.52	0.08	7.43	66.52
No. 19	17.64	6.53	2.29	53.03	1.5?	0.11	0.38	0.30	0.08	17.70	24.47

$K = SiO_2 + AlO_3 + TiO_2$

Tested fuels are marked in the figures with numbers given in Tables 6.1 and 6.2. The other points represent different fuels from mines of many countries around the world [42].

Table 6.3 presents data on the chemical composition of ash from the fuels tested. At the end of the table, in the last column, the value of the so-called acid number is given: $K = SiO_2 + Al_2O_3 + TiO_2$. Ash chemical composition is necessary for the analysis of ash behavior during combustion in fluidized beds.

6.2.4. Determination of the start-up temperature

In order to plan the FBC boiler start-up and analyze the transient behavior of a boiler, it is necessary to know three quantities: the ignition delay of fuel particles, the start-up temperature and the burning rate. The significance of the latter two quantities for the design and selection of the boiler start-up system is discussed in Section 5.2.

Of these three quantities, burning rate was given the greatest attention, as is evident from the comprehensive data given in Section 4.6.5. The burning rate was determined most frequently at temperatures ranging from 700-900 °C. Experimental conditions were similar to the nominal stationary boiler regimes. While the burning rate at lower temperatures is interesting for the boiler start-up period, it was determined only in references [17, 19, 25-29, 40, 41, 43].

The ignition delay was tested experimentally by Siemens [44]. It was shown that the ignition delay at temperatures lower than 600 °C is considerably longer than the time necessary for a fuel particle ($d = 8.6$ mm) to reach the bed temperature.

The start-up temperature and its experimental determination will be discussed below in some detail.

Start-up temperatures should not be confused with coal particle ignition temperature, although the definitions and even the nature of these two temperatures are similar.

The coal ignition temperature was carefully investigated for small particles, 50-500 µm, i. e., the particles employed in pulverized coal combustion boilers or in investigations devoted to studying the problems of self-ignition and coal dust explosions. In the literature ignition temperature is defined in different ways:

(1) as a temperature at which ignition of two to three fuel particles takes place,

(2) as a temperature at which spontaneous ignition and flame spread occur, and

(3) as a particle temperature at which heat generation during particle combustion is equal to heat losses of the particle.

The first two definitions in fact satisfy the requirement stated in the third one: for a single particle in the first case, and for a cluster of particles in the second case.

There is no need in this book to go deeply into the problems of determining ignition temperature or to present detailed results of experiments carried out for coal particles of such small dimensions, and a detailed review of such experiments and their results is available from Essenhigh [45] and Harker [46]. Here, we will only cite qualitative conclusions from these investigations. Ignition temperature following definition (1) is lower than the ignition temperature following definition (3). For 50-500 μm particles, ignition temperature using definition (1) ranges from 400 °C for lignite to 600 °C for high-rank coals, and using definition (2) from 600-730 °C.

For both definitions, the ignition temperature depends primarily on the coal rank, that is, on its volatiles content. Ignition temperatures of lignites with high volatiles content are near the lower limit of these ranges, and those of high-rank pit coals and anthracite in the higher range. While the ignition temperature rises with an increase of particle size, it does not depend strongly on ash and moisture content.

. Here, we have not attempted to explain how the volatiles content is associated with ignition temperature given that lignite ignites at temperatures below the ignition temperatures of volatile matter itself, namely at 500-600 °C. It appears that highly reactive coals, i. e., coals with higher burning rates, ignite at lower temperatures.

If we analyze ignition condition (3) qualitatively, the effect of the various quantities on ignition temperature is quite reasonable:

-   a higher burning rate contributes to faster heat generation during particle combustion, so an equilibrium between generated and removed heat is established at lower temperatures,
-   a larger quantity of volatile matter, produced during devolatilization, forms a highly porous char particle, which also tends to result in higher heat generation, and
-   with increases of particle size, the specific area of the particle external surface per unit of particle volume is reduced. In such small particles (100-300 μm), the burning rate is proportional to diameter squared, so the quantity of generated heat will be smaller, if other conditions remain the same.

Coal ignition temperature in fluidized beds has not been investigated much. The definition of ignition temperature (3) is valid for fluidized bed combustion too, but the results obtained earlier by investigating ignition of fine particles in air [45, 46] cannot be used, due to the very different heat transfer between such particles and their environment.

In considering the heat energy balance for such particles Siemens [44] analyzed qualitatively the process of coal ignition in the fluidized bed and reached the conclusion that the ignition process takes place in the so-called subcritical regime. In this regime particle temperature gradually increases to a stationary value. However, specific values of ignition temperature are not given in this reference.

A detailed investigation of ignition conditions in fluidized beds for five different coals (from lignite to anthracite) was also carried out by Read and Minchener [43]. Their objective was to obtain data necessary to plan the start-up of industrial boilers, and hence they determined start-up temperatures.

In order to define start-up temperature, definition (3) can be used, but adapted to conditions and dimensions of the furnace. Start-up temperature is defined as temperature at which heat generated due to coal combustion is equal to the quantity of heat removed from the bed (by combustion products, through furnace walls or by immersed heat transfer surfaces).

As part of an effort to experimentally determine activation energy for a number of tested coals, the coal ignition temperature in a fluidized bed was determined in reference [43] based on the heat balance for an experimental furnace by applying definition (3). For tested coals in question, with mean particle sizes of 0.5-1.0 mm, the ignition temperature determined for lignite (49% volatiles on dry, ash free basis) was about 330 °C, and for anthracite (6.6% of volatiles) it had a value of about 500 °C.

Scarce data about coal ignition temperature in fluidized beds, and the need to determine start-up temperatures for FBC boiler design, were the motivations for the determination of start-up temperature as part of standard ITE-IBK methodology. For this purpose, a special methodology described in detail in references [17, 19, 26, 28, 29, 40, 41, 47] was developed. Start-up temperatures and burning rate were determined for particle sizes 5, 10, 15, and 25 mm.

However, it was concluded that for provision of a reliable furnace start-up, the start-up temperature should be high enough to enable a sufficiently rapid increase of bed temperature up to the stationary state and hence, instead of condition (3), a different criterion was used for start-up temperature definition.

The start-up temperature was determined based on experimental results for the burning rate determined according to methodology described in Section 4.6.5. Testing was performed in two experimental furnaces: a smaller one of 78 mm in diameter (Fig. 6.2), and a larger one with a cross section diameter 147 mm and continuous coal feed (Fig. 6.5) and a thermal power capacity of 20 kW_{th}.

The following criterion was adopted for the determination of start-up temperature: "start-up temperature should be chosen as the temperature at which the coal burning rate is sufficiently high to provide a quick rise of bed temperature to the stationary state."

In this definition of start-up temperature there are two undetermined criteria:

— what burning rate is sufficiently high?
and
— what rise of bed temperature is considered sufficiently quick?

A series of investigations in experimental furnaces together with "checking" in an industrial furnace were carried out. From this work it was concluded that

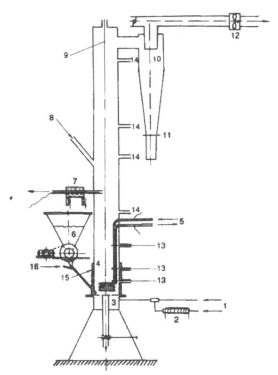

Figure 6.5.
Experimental furnace for determination of the combustion rate and start-up temperature:
1 – inlet of fluidizing air, 2 – air heater, 3 – plenum chamber beneath the distribution plate, 4 – furnace, 5 – water inlet in immersed heat exchanger in the bed, 6 – feeder for coal feeding under the bed surface, 7 – aspiration pyrometer, 8 – opening for coal feeding on the bed surface, 9 – moving thermocouple, 10 – cyclone, 11 – device for flying ash collection, 12 – flue gas drought fan, 13 – thermocouple for measuring bed temperature, 14 – openings for aspiration pyrometers, 15 – electric heater, 16 – auxiliary air for coal feeding under the bed surface

a sufficiently quick furnace start-up (with no immersed heat transfer surfaces and with insulated walls), can be achieved during the combustion of dried Kolubara lignite (No. 1 in Tables 6.1, 6.2, and 6.3) at a bed temperature of 350 °C . According to Fig. 4.53, a burning rate $2 \cdot 10^3$ kg/m²s (calculated for the external surface of 5 mm particle size) corresponds to these conditions, i. e., $1.57 \cdot 10^{-6}$ kg/s.

By determining the burning rate for each coal at different temperatures, it was possible to determine the temperature at which burning rate is $1.57 \cdot 10^{-6}$ kg/s for 5 mm particles. The temperature chosen in this manner was adopted as the start-up temperature of the respective coal. Typical values obtained for start-up temperatures are given in Table 6.1. The change of burning rate with temperature for different coals can be seen in Fig. 4.53, and the start-up temperatures determined are shown in the hatched section. Start-up temperature was determined for 19 tested coals and varied within the range of 350 to 750 °C.

It should be noted that in all experiments, the combustion of coals tested and the rise of bed temperature were also registered at bed temperatures below the start-up temperature determined according to the criteria listed above. This means that start-up temperatures obtained according to these criteria are higher than the coal ignition temperatures determined according to the criterion applied in reference [43]: that the quantity of generated heat was equal to heat removed from the bed. An analysis of results given in this paper leads to the conclusion that in fact the same criterion was used as in ITE-IBK methodology. The burning rate determined in reference [43] for different coals at the ignition temperature has the same value. Ignition temperatures obtained in [43] and [46] (330-500 °C), are lower than start-up temperatures determined by the criterion of the ITE-IBK methodology, probably due to the fact that start-up temperature determined in this way ensures the desired increase of bed temperature, but not the balance between generated and removed heat.

Particle sizes, which in [43] were 0.5-1.0 mm, and in experiments [19, 28, 41] about 5 mm, may also have contributed to the somewhat lower values of ignition temperature.

Experiments following the ITE-IBK methodology [19, 28, 41] show the same qualitative dependence on coal age, that is on volatile content, as determined for small particles [45, 46] and in combustion in fluidized beds [43].

Figure 6.6 presents values of start-up temperature depending on the volatiles content obtained by the ITE-IBK methodology. Despite significant scatter of data that might be the result of other factors and errors in determining burning rate and start-up temperature, the general dependence mentioned above can be observed. The start-up temperature drops with increasing volatiles content. The ignition temperatures determined in [43] and [46] have also been included in the

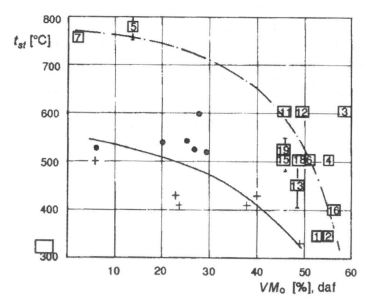

Figure 6.6. *Start-up temperature in dependence on volatile content in coal: Numbers – S. Oka et al., • – Harker and Mellor [46], + – Read and Minchener [43]*

diagram. The same general dependence is observed, but evidently the influence of volatile matter content is small if $VM_{o,daf}$ is below 30%. Lower values of ignition temperature obtained in these references are the result of the following two causes: (1) particle size is smaller (0.1-0.3 mm in [46] and 0.5-1.0 mm in [43]), and (2) values of coal ignition temperature are given in [43] and [46], while the ITE-IBK methodology enables determination of start-up temperature which ensures a stable increase of bed temperature. The start-up temperature according to the ITE-IBK methodology for particles over 5 mm is higher than the one presented in Fig. 6.6.

The analysis of application of the start-up temperature determined according to the ITE-IBK methodology poses two questions:

– for what particle size should start-up temperature be determined?

and

– can results obtained in a small-scale experimental furnace be used in industrial-size FBC boiler furnaces?

Experiments aimed at determination of start-up temperature should be carried out with particle sizes which will be used in practice. However, due to difficulties in interpretation of burning rate definition and data, it is logical to carry out experiments with particles whose size is equal to the mean equivalent diameter of the coal that is used. For this reason, for this investigation a particle size of 5 mm was chosen, as the mean equivalent diameter of coal fed above the bed surface (particles in the range 0-25 mm). Smaller coal particles contribute to faster start-up, and larger particles cause an increase of char inventory in the bed, which should be taken into account in planning the start-up.

The possibility of using a start-up temperature that is determined in non-stationary experiments for industrial-size boilers can be considered by analyzing eq. (4.103) [19, 40, 41].

Equation (4.103) is obtained based on a heat balance for fluidized bed as the control volume, in which experiments are made by introducing a single batch of coal particles into a previously heated fluidized bed:

$$M_b c_b \frac{dt_b}{d\tau} = Q_{go} + Q_b + Q_E + Q_g \qquad (6.1)$$

Since the steady state is achieved before the coal batch is introduced ($t_b^* =$ const.), the following relationship applies:

$$Q_{go} + Q_E = Q_g^* \qquad (6.2)$$

that is, the quantity of heat carried away by air the temperature of which is t_b^* is equal to the quantity of heat that is brought in from out side. In this case, it is the sum of the quantity of heat carried in by the air flow and the heat generated by heaters placed around the furnace walls.

If eqs. (6.1) and (6.2) are combined, the following is obtained:

$$M_b c_b \frac{dt_b}{dt} = Q_b + Q_g^{\bullet} - Q_g \tag{6.3}$$

After additional simplification, one may write:

$$Q_b = -H_c \frac{dm_c}{dt}$$

$$Q_g = \dot{m}_v c_g \Big|_0^{t_b} t_b$$

$$Q_g^{\bullet} = \dot{m}_v c_g \Big|_0^{t_b^{\bullet}} t_b^{\bullet}$$

$$c_g \Big|_0^{t_b} = c_g \Big|_0^{t_b^{\bullet}} = c_g \tag{6.4}$$

where

$$\frac{dt_b}{dt} > 0, \quad \text{and} \quad \frac{dm_c}{dt} < 0$$

By introducing relationships (6.4) into eq. (6.3), eq. (4.103) is finally obtained in the following form:

$$\frac{d}{dt}(t_b - t_b^{\bullet}) + \frac{\dot{m}_v c_g}{M_b c_b}(t_b - t_b^{\bullet}) = -\frac{H_c m_{\infty}}{M_b c_b} \frac{d}{dt}\left(\frac{m_c}{m_{\infty}}\right) \tag{6.5}$$

which expresses the fact that, at any time, the quantity of generated heat due to coal combustion is used to raise bed temperature and to heat air from inlet temperature to bed temperature.

Equation (6.5) includes two characteristic quantities for the observed process, but for the start-up process as well:

— characteristic time:

$$\tau^{\bullet} = \frac{M_b c_b}{\dot{m}_v c_g} \tag{6.6}$$

and
— characteristic temperature rise:

$$\Delta t_b^* = (t_b - t_b^*) = \frac{H_c \, m_{co}}{M_b \, c_b} \tag{6.7}$$

The characteristic time $\tau*$, is the ratio between the heat capacity of inert bed material and the heat capacity of air flow (combustion products). This ratio expresses the characteristic cooling rate of inert bed material after heat generation is finished. In a time period lasting $\tau*$, bed material cools by about 2/3 of it initial temperature.

The characteristic temperature rise $(t_b - t_b^*)$ is the maximum bed temperature rise that would be achieved by combustion of the total coal quantity m_{co}, if no heat were conveyed away from the bed by combustion products.

If the experimental furnace and a large-scale industrial furnace have the same two characteristic quantities, their rise of temperature will have the same rate. In this case, the start-up temperature determined by this methodology applies to large-scale furnaces as well.

To achieve these conditions, the following is necessary:

- the same fuel,
- the same inert material,
- the same fluidization velocity,
- the same bed height, and
- the same fuel content in total bed mass.

In the case of continuous fuel feeding with an inert material bed which has been previously heated, based on the heat balance, the following equation similar to eq. (6.5) can be obtained:

$$\frac{d}{d\tau}(t_b - t_h^*) + \left[\frac{\dot{m}_v c_g}{M_b c_b} + \frac{\alpha A_S}{M_b c_b}\right](t_b - t_h^*) = -\frac{H_c}{M_b c_b}\frac{dm_c}{d\tau} \tag{6.8}$$

The quantity of heat removed by heat transfer surfaces in the bed is included in the equation, and mc is the transient carbon loading in the fluidized bed, while:

$$\frac{dm_c}{d\tau} \approx \dot{m}_{l} \tag{6.9}$$

From eq. (6.8), it is evident that cooling surfaces immersed in the bed demand a higher heat generation at start-up temperature in order to compensate for the increased heat losses from the bed, i. e., higher start-up temperature. These facts should be considered in choosing start-up temperature and start-up for real FBC furnaces or boilers based on the start-up temperature as determined by the ITE-IBK methodology.

Equation (6.8), according to its physical meaning, is identical to the equation from which ignition temperature [43-45] is defined. This justifies the statement that start-up temperature and ignition temperature are actually physically the same quantities.

6.2.5. *Effects of fuel characteristics on fuel behavior during combustion in fluidized beds*

This section presents fundamental results from the testing of all 19 fuels whose characteristics are described in Tables 6.1, 6.2, and 6.3. Testing was done by burning these fuels in three stationary operating regimes in an experimental furnace with 300×300 mm^2 square cross section (Fig. 6.3). Furnace dimensions and experimental conditions were chosen to achieve the similarity conditions given in Section 6.2.1. All investigations were carried out in the same way, according to the concept of phase three of the ITE-IBK methodology described in Section 6.2.2.

Fuels were burned in the form in which they would be used in real furnaces. Depending on the kind of fuels, they were either fed above or below bed. In certain cases, both types of feeding were investigated. The results of these investigations are presented in references [20, 22, 23, 32-34, 48], and results for each coal separately appear in internal reports that are cited in the bibliography of reference [14]. Table 6.4 represents a systematization of the data obtained in these investigations.

Effects of moisture content. Table 6.1 gives the fuel moisture content, which varied between 3 and 58%. Moisture content did not affect fuel ability to reach and maintain a stable steady state operational regime, or for that matter significantly influence the combustion efficiency. The moisture content affected primarily the operation of the fuel feeding system. Operating experience in industrial FBC boilers supports the view that fuel moisture creates significant problems in the transport and feeding systems, especially when burning "as-received" coal [1, 2-4, 23].

Effects of size distribution. The size distribution of coal affects combustion in several ways. Lignites and brown coals with a maximum particle size between 30 and 40 mm were burned in an experimental furnace without difficulties in feeding above bed. Highly reactive coals (lignites) can burn even with particle size up to 50 mm. Coals with low reactivity (anthracite, coke, bituminous coals with low volatile content) are suitable for combustion in a fluidized bed only when crushed to a size of 4-5 mm. If particles of these coals are larger, ignition and reaching a steady state operating regime and maintenance of stable combustion are very difficult.

The size distribution primarily affects combustion efficiency and heat generation distribution. Particles below 1 mm are elutriated from the bed, but if the coal is highly reactive, combustion of these particles will take place above the bed. In this case, freeboard temperature will be much higher than bed temperature, and there will be few unburned particles in the fly ash. Combustion efficiency will, therefore, still be high despite the high percentage of fine particles.

Figure 6.7 shows the freeboard-to-bed temperature difference during combustion of a highly reactive coal (No. 1), but with different size distributions. Measurements were made in a 20 kW$_{th}$ (Fig. 6.5) and a 200 kW$_{th}$ (Fig. 6.3) furnace, and in an industrial 4 MW$_{th}$ (Fig. 5.2) furnace. For the combustion of coals without fine particles (below 1 mm) and with above bed feeding, the temperature difference was 20 °C. For the same size distribution, but with fine particles that mostly burn above the bed, freeboard temperature could be 150 °C higher than the bed temperature. For the combustion of low reactivity coals, the temperature difference was only 20 °C.

When burning highly reactive coals (lignites), particles with a size below 1 mm do not greatly affect combustion efficiency. In the combustion of dried and washed lignite (No. 1) with no fine particles and a size distribution 5-25 mm, the combustion efficiency was about 5% greater than for the same coal as mined (No. 2) with a particle size distribution 0-50 mm and with 6-7% of particles smaller than 1 mm. The combustion efficiency of the lignite as mined was about 87% at a bed temperature of 800 °C.

During combustion of coke (No. 7) and anthracite (No. 5), losses due to unburned particles in fly ash were 33%, and 20% respectively, although feeding was below bed (see Table 6.2).

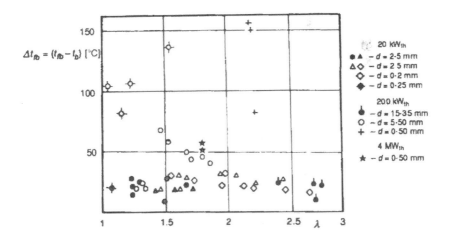

Figure 6.7. *Influence of coal particle size on freeboard temperature for high reactivity coal during coal feeding above bed [48]*

Effects of volatile matter content. Coals with high volatile content normally have lower ignition and start-up temperatures (see Section 6.2.4 and Fig. 6.6), greater burning rate (Fig. 4.53), and lower losses in terms of unburned particles in fly ash (Table 6.1, last column). However, they also have more problems with volatiles combustion. Measured difference between freeboard temperature and bed temperature when burning these fuels reached significant levels during both furnace start-up and in steady state operation at temperature of 800-850 °C.

During start-up, the freeboard-to-bed temperature difference depended greatly on the combustion temperature in the fluidized bed, and this must be taken into account in the planning of start-up and in the choice of start-up temperature. Figure 6.8 shows this temperature difference during steady state combustion at different bed temperatures, when burning dried lignite with particle size distribution 5-25 mm (No. 1). Measurements were made in experimental 20 kW$_{th}$ (Fig. 6.5) and 200 kW$_{th}$ (Fig. 6.3) furnaces and in an industrial 4 MW$_{th}$ (Fig. 5.2) furnace. At lower temperatures (<600 °C), when conditions for ignition of volatile matter are still not met, the measured temperature difference is relatively small 20-60 °C.

Favorable conditions for combustion of volatile matter are obviously experienced first in the freeboard. Due to combustion of volatiles in the freeboard, at bed temperatures above 600 to 650 °C, the temperature difference can reach values of up to 150 °C. With a further rise of bed temperature, an increasing amount of volatiles burns in the bed. During normal operation in furnaces with this coal (800-850 °C), the freeboard-to-bed temperature difference is reduced to 20-50 °C.

During biomass combustion (30-70% of volatile matter, No. 8, 9, and 10), temperatures above the bed may be 200-300 °C [21] higher than bed temperature, depending on particle size and the manner of feeding. Due to poor mixing of fuel in the bed and low biomass density, biomass tends to "float" on the bed surface or burns near the surface.

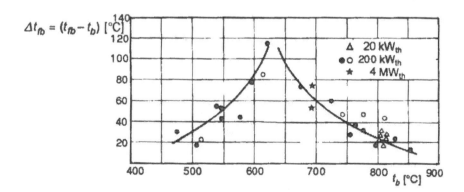

Figure 6.8. *Difference between freeboard temperature and bed temperature during the start-up period of FBC furnace [20]*

The preceding section deals in detail with volatile matter effects on start-up temperature. Results show that start-up temperature is somewhere between 350 °C (for lignites and biomass) and 750 °C for coke and anthracite.

Coals with higher volatile contents are more reactive. Figure 4.53 indicates that at a bed temperature of 700 °C and for particle size 5 mm, the burning rate for anthracite (No. 5) is $(1-2)\cdot 10^{-3}$ kg/m^{2}s. For the same conditions, the burning rate for lignite (No. 1) is $(14-16)\cdot 10^{-3}$ kg/m^{2}s.

According to results on the primary fragmentation process, coals with a larger PRN number (ratio of volatile and analytical moisture content) have a smaller critical fragmentation diameter, meaning they are inclined to fragmentation (see Section 4.3 and [8, 24, 39]).

Effects of physical and chemical characteristics of ash. The behavior of the mineral part of the fuel during fluidized bed combustion is not determined only by the physical and chemical characteristics of the ash, but also by the nature of the coal preparation, in particular, the resulting particle size distribution and coal particle fragmentation. This is the reason why the results obtained by testing these 19 fuel types show considerable scatter. In spite of this scatter it is evident that ash behavior in the bed is affected by chemical composition.

Coals burned in large pieces, especially if they contain a high percentage of stones (coals used as mined), leave a higher percentage of ash (stones) in the bed regardless of their chemical ash composition (compare coals No. 1 and No. 2, Table 6.1 and 6.3). Coals with a higher percentage of fine particles (<1 mm) and with a tendency towards fragmentation, produce a higher percentage of fly ash that is elutriated.

When speaking only of the behavior of the so-called internal ash, its behavior depends mostly on chemical composition. For this reason, a "float or sink" analysis was introduced into the methodology, since it offers data on the quantity of stones in coal.

The majority of tested lignites obviously burn according to the so-called shrinking core model, and the ash at the burned layer of particles separates due to mechanical effects of inert material particles. Testing of these 19 fuels shows that fly ash particle size is between 300 and 400 μm [48], and that it is elutriated.

Coals with low reactivity (anthracite, bituminous and brown coals) most frequently burn following the reacting core model and they leave behind a particle ash skeleton whose size is practically unchanged from the parent particle. These coals leave the most ash in the bed during testing.

Possibility of generalization of results and prediction of fuel behavior during combustion in fluidized beds. On the basis of the comparative testing of all 19 fuels, using the same methodology and under the same conditions, it can be concluded that in general their behavior depends strongly on their physical and chemical characteristics, despite the fact that coal is a highly complex material. Ex-

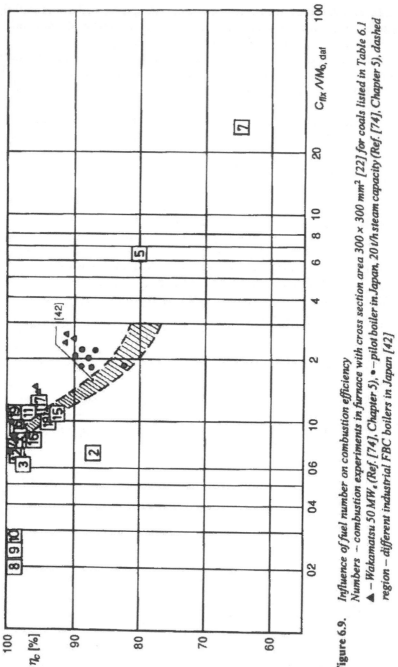

Figure 6.9. Influence of fuel number on combustion efficiency
Numbers – combustion experiments in furnace with cross section area 300 × 300 mm² [22] for coals listed in Table 6.1
▲ – Wakamatsu 50 MWₑ (Ref. [74], Chapter 5), ● – pilot boiler in Japan, 20 t/h steam capacity (Ref. [74], Chapter 5), dashed
region – different industrial FBC boilers in Japan [42]

perimental investigations from other authors (see Chapter 5, references [1, 5-7]) and operational experience gained from industrial facilities support this view.

Establishing the specific influence of individual physical and chemical characteristics of fuels requires even more detailed research, but the results of investigation of different coals by applying the ITE-IBK methodology offer sufficient grounds for generalization of these results and for an attempt to present them in a unique way.

Figure 6.9 shows the combustion efficiency of all 19 fuels during combustion in an experimental 300×300 mm^2 furnace (Fig. 6.3) using the fuel ratio: $F = (C_{fix}/VM_o)_{daf}$. Evidently, fuels with a higher fuel ratio show a lower combustion efficiency. An increase of combustion efficiency may be achieved only by recirculation of unburned particles. Also, it is evident that lignites, biomass, and geologically older coals with high volatile contents (for example No. 15), all have high combustion efficiency.

Figure 6.9 also gives data on combustion efficiency from Japanese pilot-scale research (shown as a hatched belt) [42], and from reference [74] in Chapter 5. Deviations from the general tendency given in Fig. 6.9 are the consequence of different particle size distributions for the tested coals (see for example, coal No. 1, which was tested after separation, washing and drying, but note that No. 2 is the same coal but tested "as mined").

The effects of chemical composition on ash behavior during coal combustion in a fluidized bed are shown in Fig. 6.10. The ordinate gives the percentage of total ash that remains in the bed, and the abscissa gives the acid number. Scattering of the measured results would certainly have been lower, had the "stones" been sep-

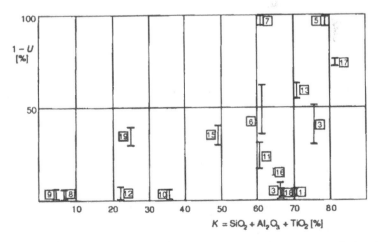

Figure 6.10. *Influence of ash chemical composition on ash behavior in a fluidized bed*

arated from coal before testing, in which case only the internal ash elutriated from the bed would be determined. Variations in the size distribution of tested coals must also contribute to deviations from the general rule.

Based on phase one of the ITE-IBK methodology and the general rules indicated by diagrams in Figs. 6.9 and 6.10, it is possible to predict the most significant aspects of coal behavior during FBC and to plan the combustion and design concept of a future FBC boiler.

The results and experience acquired in testing these 19 fuels, as well as exhaustive data provided in the literature, allow a systematization of the knowledge about coal behavior in fluidized beds, as shown in Table 6.4 [14] for use by engineers/designers.

Table 6.4.   *Data on coal which influence operation parameters, concept and design of a FBC boiler*

No.	Name	Operation parameter, process, system, equipment or characteristic/concept of a boiler that depends on the given quantity (the manner of determination)
1		GENERAL CHARACTERISTICS, PREPARATION, ORIGIN AND COAL RANK
		Based on these data, general concept of the boiler and auxiliary systems can be given.
		Data usualy obtained from coal suppliers
1.1	**Coal as mined**	
1.1.1	Open mine	Coal with high moisture content and a large amount of stones. Broad particle size distribution. Young coal, usually lignite. Low ignition temperature and high volatile content. If the boiler is not near the mine, preparation of fuel is desirable – separation, drying, washing. The boiler will have several auxiliary systems. A system for draining material from the bed and a system for recirculation of unburned particles are needed. Low start-up temperature. Most probably with no heat exchangers in the bed and with refractory-lined furnace walls.
1.1.2	Pit mining	Characteristics similar to above, but low moisture content. Usually high-rank coals with low volatiles content. Coal with low reactivity, high start-up temperature. A large start-up chamber or a separate start-up section of the bed are also needed. Heat exchangers in the bed probably necessary. Difficult start-up. Auxiliary systems as above.
1.2	Washed coal	No fine particles (<1 mm). Increased combustion efficiency. No recirculation system for young coals.
1.3	Separated coal	Small quantity of ash and stones. Feeding on the bed for young coals. Probably with no systems for draining material from the bed.
1.4	Washed and separated coal	Coal with no ballast material or fine particles. Suitable for combustion in FBC boilers with few auxiliary systems. Concept depends on coal rank.

1.5	Dried, separated washed	Coal with no stones, fine particles, usually reactive. Suitable for combustion in FBC boilers of simple concept, without many auxiliary systems.
1.6	Waste slurry, coal-water mixture	Very fine particles and usually with low reactivity. Must be fed under bed. Compulsory recirculation system and bed draining system. After drying, feeding is possible by pneumatic conveying or with high-pressure pumps in liquid state.
1.7	Young coals (lignites)	High moisture and volatile content. Reactive coals, low start-up temperature. Suitable for combustion in FBC boilers with few auxiliary systems. High combustion efficiency. With no heat transfer surfaces in the bed.
1.8	Brown coals	Significant ash content, average moisture and volatiles content. Moderately reactive coals. Complex boilers with all auxiliary systems – recirculation, draining from the bed. Feeding depends on particle size distribution. Heat transfer surfaces in the bed. Medium start-up temperature.
1.9	High-rank, bituminous, anthracite, coke	Considerable quantity of ash, low reactivity. High start-up temperature. Feeding only under the bed surface with particle size about 2 mm. Necessity for recirculation and draining systems, as well as heat transfer surfaces in the bed.

2   PHYSICAL AND CHEMICAL CHARACTERISTICS OF COAL AND ASH

2.1   Particle size distribution of supplied coal
Determined by standard sieving analysis

2.1.1	Size distribution of coal as supplied	Affects the choice of storage, design of daily consumption bunkers, transportation from coal disposal to boiler, feeding system, manner and position of coal feeding into the boiler and number of feeding points per m^2 of furnace cross section. Necessary datum for choice of boiler concept and consideration of possible preparation of coal.
2.1.2	High content of pieces >50 mm	Necessary grid for separation of large pieces at the coal disposal. Necessary coal crushing.
2.1.3	Considerable percentage of particles <2 mm	Possible accumulation of ash in the bed – draining system, separation and inert bed material return system necessary.
2.1.4	Considerable percentage of <1 mm particles	Low combustion efficiency – unburned particle recirculation system necessary. Particle separation system – cyclones and filters – greatly dependent on particle load. Compulsory feeding under the bed surface, greater bed height, introduction of secondary air.

2.2   Particle size distribution of coal at boiler entrance – after transport and feeding systems

Has a decisive influence on boiler concept, combustion efficiency, elutriation of particles from the bed and furnace, heat generation distribution, freeboard temperature, introduction of secondary air, quantity of char in the bed, manner of feeding, auxiliary systems and others according to analysis in 2.1. It can significantly change boiler concept conceived on the basis of an analysis of supplied coal. (Determined on the basis of a sample obtained after long-term operation of transport and feeding systems in pilot or demonstration boilers or prototypes of these devices).

2.3	Proximate coal and ash analyses as supplied (Determined in a standard manner. Ash sintering temperature is determined in a fluidized bed)	
2.3.1	Moisture content	Affects the choice of coal transportation and feeding; the quantity of heat removed from the bed; size of heat exchangers immersed in the bed; stationary operating stability; choice of flue gas exit temperature (dew point); necessary measures for increased boiler efficiency.
2.3.2	Volatiles content	Affects uniformity of the temperature field in the fluidized bed and above it; heat generation distribution in furnace; manner, place and number of coal feeding points; bed height and freeboard height; introduction of secondary air; freeboard temperature; boiler start-up temperature; CO emission; particle fragmentation; bed and freeboard cross section dimensions; choice of fluidization velocity.
2.3.3	Ash content	Affects capacity, design and characteristics of bed draining system and flue gas cleaning system.
2.3.4	Ash sintering temperature	Affects combustion temperature choice, range of bed temperature change; sintering and choice of inert material. It speaks about risk/danger of ash and sand sintering in the bed; affects choice of start-up type (start-up prechamber, liquid fuel burners above the bed or direct gas combustion in the bed). (It is determined in a standard procedure and in real conditions in an experimental furnace with fluidized bed combustion.)
2.3.5	Chemical composition of ash	Affects ash behavior in fluidized bed; ash particle disintegration; ash remaining or elutriation from the bed; high percentage of K_2O and Na_2O in ash increases the risk of sintering and it might affect the choice of bed material; high percentages of SiO_2 and Al_2O_3 cause an increased percentage of ash which remains in the bed and capacity and characteristics of draining and separation systems of bed material.
2.3.6	External ash, stones	Causes an increased percentage of ash which remains in the bed. It is determined by "float or sink" method.
2.3.7	Density of ash and stones	Affects (together with particle size) distribution (segregation) of ash along bed height; ash can float on the surface of the bed or sink to its bottom; number and position of bed material draining openings; ash and stones should be considered separately.
2.3.8	Calorific value	Quantity necessary for calculation of heat balance in the boiler calculation. Affects the capacity of coal transportation and feeding systems; excess air; general boiler concept; size of heat transfer surfaces in the bed; range of different fuels which can burn in the boiler.
2.4	Ultimate analysis of coal (C, H, O, N, S) (Determined in a standard manner) Necessary datum for calculation of boiler heat and material balance, excess air and composition of combustion products; affects NO_x and SO_2 emission.	
2.4.1	Sulphur content in coal	Affects capacity of limestone feeding system to the furnace, necessary limestone flow rate. (Ca/S molar ratio).
2.4.2	Nitrogen content in coal	Affects NO_x emission, need for introducing staged combustion and secondary air.

3	CHARACTERISTIC QUANTITIES DETERMINED BY COMBUSTION IN EXPERIMENTAL FURNACES	
3.1	Maximum particle size that will not fragment during devolatilization	Affects char quantity in the bed; combustion efficiency; quantity of particles elutriated from the bed; furnace start-up time and start-up temperature; start-up chamber size. Determined in laboratory furnace in planned experiments.
3.2	Char burning rate	Affects choice of start-up temperature, start-up time period, need for coal crushing, feeding manner and place, quantity of char in the bed, rate of boiler load change. Determined in experimental furnace in planned experiments.
3.3	Coal ignition temperature	Affects choice of start-up temperature, start-up chamber size, boiler start-up time period, possible usage of auxiliary fuel for start-up, division of furnace into start-up and operating sections. Determined in experimental furnace in planned experiments.
3.4	Ratio of ash that remains in the bed and is elutriated from the bed	Affects the need for possible preliminary preparation of coal; affects incorporation, capacity and design of draining system, need for drained bed material cooling, system for ash separation and return of bed material of suitable dimensions; affects the system for flue gas cleaning, design and dimensions of gas tract and convective heat transfer boiler surfaces. Determined during long-term combustion of coal as supplied in real conditions in experimental furnaces and pilot facilities or during testing coal combustion in demonstration and industrial boilers.
3.5	Fuel number. C_{fix}/VM_o ratio	Based on this ratio, combustion efficiency can be assessed (heat losses due to the unburned particles), affects general boiler concept (furnace cross section, secondary air, fly ash recirculation system). Determined based on proximate analysis.
3.6	Heat generation distribution in the bed and above it	Affects dimensions of heat transfer surfaces immersed, freeboard temperature, need to introduce secondary air, ratio between primary and secondary air in the bed. Determined experimentally based on heat balance during stationary combustion regime of coal as supplied in real conditions in experimental furnaces or pilot facilities.

6.3. Justification of the application of laboratory furnace investigation results in designing industrial boilers

The transfer of laboratory investigation results to large-scale facilities, i. e., scale-up, is one of the major problems in all spheres of science and technology. It is especially prominent in the development of new technology, when there is insufficient experience available from industrial facilities under real operating conditions.

The scale-up problem is especially important for investigating processes (such as solid fuel combustion) when the fulfillment of all similarity conditions is impossible. For that reason, any highly developed and economically powerful countries will be forced to carry out expensive and long-term investigations of pilot and demonstration plants operating under realistic conditions.

Therefore, it is justifiable to ask the following question: can results obtained by the ITE-IBK methodology or similar methodologies [5-8, 11] be used in calculation, design, choice of concept and construction of large-scale boilers and furnaces? We will give a few facts that justify the application of this methodology for coal testing:

(1) The methodology offers all data necessary for the calculation and design of boilers, and resolves all problems discussed in Section 5.1.4,

(2) Results obtained by investigation of combustion of coal and its behavior in a fluidized bed, according to analysis of results presented in Section 6.1, are both logical and physically justified,

(3) Results obtained by this methodology were used in calculations and design of more than one hundred 1-10 MW_{th} industrial furnaces for hot-gas and hot-air generation, all of which operated for periods of years and found use in industries and agriculture,

(4) Results obtained by this methodology were used in the choice of concept, design and calculation of boilers shown in Figs. 5.21 and 5.22, and

(5) Similar methodologies are either being used (at least in part) or proposed in other countries [5-8, 11].

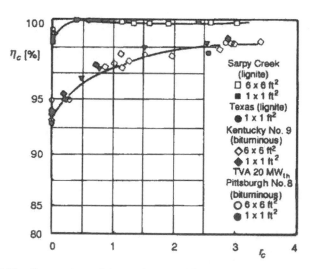

Figure 6.11. *Comparison of the combustion efficiency obtained by burning the same coal in furnaces of different sizes during under-bed coal feeding (Reproduced by kind permission of the ASME from [5])*

Comparisons made with the U.S.A. [1, 5] and Japan [9, 10] show that the results obtained in small experimental furnaces (300×300 mm^2 or 1×1 ft^2, which are the smallest cross sections recommended) agree well with results obtained in large pilot, demonstration or industrial furnaces.

The results of these comparisons demonstrate the utility of this methodology for coal investigation. Fig. 6.11 shows the results for determining the combustion efficiency of four American coals (Texas lignite, Sarpy Creek brown coal, and two bituminous coals, Kentucky No. 9 and Pittsburgh No. 8). Measurements were made in three facilities of different size: 1×1 ft^2 experimental furnace, 6×6 ft^2 pilot plant and the TVA 20 MW$_{th}$ demonstration boiler discussed in Section 6.1, the cross section of which is 18×12 ft^2. The dimensions for the cross sections of these facilities differ by a factor of 216, that is by two orders of magnitude.

As can be seen, the measured combustion efficiency did not depend on size of the furnace. Somewhat greater differences, but still within expected limits for measurement error, were obtained for combustion without fly ash recirculation ($r_c = 0$), probably due to the fact that freeboard height in the 6×6 ft^2 furnace is greater than in the 1×1 ft^2 furnace.

Nomenclature

A_S	outside surface area of the heat exchanger immersed in the bed, [m^2]
c_b	specific heat of particle of the inert bed material, [J/kgK]
c_g	specific heat of gas at constant pressure, [J/kgK]
$c_g\big|_0^t$	mean specific heat of gas in temperature range 0 to t °C, [J/kgK]
C_{fix}	fixed carbon content in coal, [%]
d	mean fuel particle diameter, [mm] or [m]
daf	"dry, ash free basis"
H_c	calorific value of the fuel, [J/kg] or [MJ/kg]
K	acid number of the ash, $(SiO_2 + Al_2O_3 + TiO_2)$, [%]
m_c	total mass of char particles in the bed (bed inventory), at the moment τ, [kg]
m_{co}	initial mass of carbon in the bed, [kg]
$\dot{m}_F$	fuel mass flow rate, [kg/s]
$\dot{m}_v$	air mass flow rate, [kg/s]
M_b	mass of inert material in the fluidized bed, [kg]
P	ash content in the coal, on "as received basis," [%]
Q_b	heat generated in the bed by fuel combustion, [W]
Q_E	amount of heat transferred to the bed from electrical heater, [W]
Q_g	amount of heat taken away from the bed by flue gases, [W]
Q_{gv}	amount of heat brought to the bed by inlet fluidizing air, [W]
Q_g^*	amount of heat taken away from the bed by fluidizing air at bed temperature t_b^*, [W]
r_c	recirculation ratio (recycle ratio)
S_u	total sulphur content in coal on "as received" basis, [%]
t_b^*	bed temperature, [°C]
t_b^*	constant bed temperature at the beginning of coal "batch" combustion experiments, [°C]

Δt_b^* characteristic maximum temperature rise during "batch" combustion experiments, [°C]
t_{fb} freeboard temperature, [°C]
Δt_{fb} freeboard-to-bed temperature difference, [°C]
T_{sn} ash sintering temperature, [°C]
T_{st} start-up temperature, [°C]
U amount of the ash elutriated from the fluidized bed, from the total ash content in coal, [%]
V_u total moisture content in coal, on "as received" basis, [%]
VM_o volatile matter content in coal, as measured by proximate analysis, on "as received" basis, [%]

Greek symbols

α heat transfer coefficient for heat exchanger immersed in the fluidized bed, [W/m²K]
η_c combustion efficiency
λ excess air
τ time, [s]
τ^* characteristic bed cooling time, [s]

References

[1]   SE Tung, GC Williams, eds. Atmospheric fluidized-bed combustion; A technical source book (Final report). MIT, Cambridge, and U.S. Department of Energy, 1987, DOE/MC/14536–2544 (DE88001042)

[2]   Proceedings of 9th International Conference on FBC, Boston, 1987, Vol. 1, Vol. 2.

[3]   Proceedings of 10th International Conference on FBC, San Francisco, 1989, Vol. 1, Vol. 2.

[4]   Proceedings of 11th International Conference on FBC, Montreal, 1991, Vol. 1, Vol. 2, Vol. 3.

[5]   RR Chandran, JN Duqum. A performance code for AFBC scale-up. Part I: In-bed combustion. Proceedings of 9th International Conference on FBC, Boston, 1987, Vol. 1, pp. 300-306.

[6]   RR Chandran, JN Duqum, MA Perna, HC Jafari, DR Rowey. A new method for AFBC fuels characterization. Proceedings of 9th International Conference on FBC, Boston, 1987, Vol. 1, pp. 292-299.

[7]   RR Chandran, et al. Ranking fuels for utility-scale AFBC application. Proceedings of 10th International Conference on FBC, San Francisco, 1989, Vol. 1, pp. 313-322.

[8]   CS Daw, DR Rawley, et al. Characterization of fuels for AFBC. Proceedings of 11th International Conference on FBC, Montreal, 1991, Vol. 1, pp. 157-166.

[9]   K Furuya. EPCD's fluidized bed combustion RD&D: A progress report on Wakamatsu 50 MW$_e$ demonstration test and the world's largest FBC retrofit project. Proceedings of 10th International Conference on FBC, San Francisco, 1989, Vol. 1, pp. 811-827.

[10]   Y Nakabayashi. Demonstration test program of the 50 MW$_e$ AFBC boiler in Japan. Proceedings of 9th International Conference on FBC, Boston, 1987, Vol. 1, pp. 177-184.

[11]   A Delebare, S Brunello, T Queva. Characterization of coals by pilot scale tests. Proceedings of 11th International Conference on FBC, Montreal, 1991, Vol. 1, pp. 151-156.

[12]   S Oka. Research & development program: Development of technology and devices for efficient combustion of low quality solid fuels (in Serbian). Proceedings of 8th Yugoslav Symposium of Thermal Engineers YUTERM'90, Neum (Yugoslavia), 1990, pp. 443-449.

[13]   S Oka. Research and development of the boilers for burning Yugoslav lignites (in Serbian). Symposium JUGEL Development of Electricity Production in Yugoslavia from 1991 till 2000, Ohrid (Yugoslavia), 1990, Vol. 2, pp. 557-566.

[14]   S Oka, D Dakić, B Arsić, B Grubor. Coal behavior data base necessary for choice of concept, calculation and design of the FBC boilers (in Serbian). Report of the Institute of Nuclear Sciences Boris Kidrič, Vinča, Belgrade, IBK-ITE-816, 1989.

[15]   S Oka. Fluidized bed combustion, a new technology for combustion of coal and other solid fuels (in Serbian). In: Energy and Developmented. Belgrade: Society Nikola Tesla, 1986, pp. 147-156.

[16]   S Oka, B Arsić, D Dakić. Development of the FBC furnaces and boilers (in Serbian). Primenjena nauka 1:25, 1985.

[17]   S Oka. Kinetics of coal combustion in fluidized beds. Proceedings of International Seminar Heat and Mass Transfer in Fluidized Bed, Dubrovnik (Yugoslavia), 1984, New York: Hemisphere Publ. Co. 1986, pp. 371-383.

[18]   M Ilić, B Arsić, S Oka. Devolatilization of coal particles at different temperatures (in Serbian). Proceedings of 8th Yugoslav Symposium of Thermal Engineers, YUTERM'90, Neum (Yugoslavia), 1990, pp. 443-449.

[19]   S Oka. Investigation of coal particles combustion kinetics in fluidized bed (in Serbian). Report of the Institute of Nuclear Sciences Boris Kidrič, Vinča, Belgrade, IBK-ITE-391, 1983.

[20]   S Oka, B Arsić, D Dakić, M Urošević. Experience and problems of the lignite fluidized bed combustion. Proceedings of 3rd European Coal Utilization Conference, Amsterdam, 1983, Vol. 2, pp. 87-97.

[21]   S Oka, B Arsić. Biomass combustion in fluidized bed (in Russian). In: SS Kutateladze, S Oka, eds. Transport Processes in High Temperature and Chemically Reacting Flows. Novosibirsk: Siberian Branch of the U.S.S.R. Academy of Sciences, 1982, pp. 40-49

[22]   S Oka, B Grubor, B Arsić, D Dakić. The methodology for investigation of fuel suitability for FBC and the results of comparative study of different coals. Proceedings of 4th International FBC Conference, London, 1988, pp. 1/8/1-19.

[23]   S Oka, D Dakić, B Grubor, M Trifunović, B Arsić. Comparative analysis of fluidized bed combustion of different coals (in Serbian). Proceedings of 2nd Yugoslav Congress of Chemical Engineering, Dubrovnik (Yugoslavia), 1987, Vol. II, pp. 355-358.

[24] D Dakić. Influence of solid fuel characteristics and thermodynamic parameters on combustion in fluidized bed (in Serbian). MSc thesis, Mechanical Engineering Faculty, University of Belgrade, 1990.

[25] S Oka, M Ilić, I Milosavljević. Coal particle combustion kinetics in fluidized bed at different temperatures (in Serbian). Proceedings of 8th Yugoslav Symposium of Thermal Engineers YUTERM'90, Neum (Yugoslavia), 1990, pp. 409-416.

[26] M Ilić, S Oka, I Milosavljević, B Arsić. Influence of volatile content on coal particle reactivity (in Serbian). Report of the Institute of Nuclear Sciences Boris Kidrič, Vinča, Belgrade, IBK-ITE-800, 1990.

[27] S Oka, M Radovanović. Combustion kinetics of different high volatile coals in a fluidized bed (in Serbian). Proceedings of 2nd Yugoslav Congress for Chemical Engineering, Dubrovnik (Yugoslavia), 1987, Vol. II, pp. 358-361.

[28] I Milosavljević, M Ilić, S Oka. Experimental investigation of coal combustion kinetics in a fluidized bed (in Serbian), Termotekhnika 3-4:151-165, 1990.

[29] M Ilić, I Milosavljević, S Oka. Gas concentration measurements during transient combustion processes in a fluidized bed (in Serbian). Termotekhnika 3-4:165-176, 1990.

[30] B Grubor, M Ristić, D Dakić, S Kostić. Industrial saturated steam FBC boiler with 2×4 t/h steam capacity (in Serbian). Termotekhnika 3-4:285-292, 1990.

[31] B Grubor, S Oka, B Arsić, J Jovanović. Concept and design of the FBC steam boiler for burning wood waste (in Serbian). Termotekhnika 3-4:261-270, 1990.

[32] S Oka, B Grubor, B Arsić, D Dakić. Methodology and results of investigation of solid fuel suitability for combustion in a bubbling fluidized bed (in Russian). Proceedings of Heat and Mass Transfer Problems in Combustion and Gasification of Solid Fuels. Minsk: A.V. Luikov Heat and Mass Transfer Institute, Academy of Sciences B.S.S.R., 1988, Vol. I, pp. 32-51.

[33] S Oka, B Grubor, B Arsić, D Dakić. Methodology and investigation of suitability of solid fuels for FBC (in Serbian). Termotekhnika 3-4:221-237, 1990.

[34] S Oka. Some comments on the influence of coal characteristics on behavior of coal in FBC furnaces. Presented at 17th IEA-AFBC Technical Meeting, Amsterdam, 1998.

[35] LI Sedov. Similarity and Scaling in Mechanics (in Russian). 5th ed. Moscow: Nauka, 1965.

[36] SS Kutateladze. Similarity Analysis in Thermophysics (in Russian). Novosibirsk: Siberian Branch of the U.S.S.R. Academy of Sciences, 1982.

[37] DB Spalding. Some Fundamentals of Combustion, Oxford: Pergamon Press, 1956.

[38] D Dakić, G van der Honing, M Valk. Fragmentation and swelling of various coals during devolatilization in a fluidized bed. Fuel 7:911-916, 1989.

[39] D Dakić, G van der Honing, B Grubor. Mathematical simulation of coal fragmentation during devolatilization in a fluidized bed. Presented at the 20th IEA-AFBC Technical Meeting, Göteborg (Sweden), 1989.

[40] S Oka. Coal combustion kinetics in a fluidized bed (in Serbian). Termotekhnika 3-4:99-123, 1983.

[41] AB van Engelen, G van der Honing. Char characterization by combustion experiments in a small scale fluidized bed interpreted by an analytical single porous particle conversion model. Presented at the IEA-AFBC Math. Modeling Meeting, Lisbon, 1990.

[42] H Terada. Combustibility of high volatile coal in AFBC. Presented at the 14th IEA-AFBC Technical Meeting, Boston, 1987.

[43] DC Read, AJ Minchener. The low temperature combustion characteristics of various coals in a fluidized bed combustion. Proceedings of 3rd International FBC Conference Fluidized Combustion: Is It Achieving Its Promise? London, 1984, Vol. 1, DISC/26/224-231.

[44] RV Siemons. The mechanism of char ignition in fluidized bed combustors. Combustion and Flame 70:191-206, 1987.

[45] RH Essenhigh. Ignition of coal particles. In: MA Elliot, ed. Chemistry of Coal Utilization. New York. 1981, 2nd Supplement volume, pp. 1285-1297.

[46] JH Harker, NS Mellor. Ignition temperatures of coal. Journal of the Institute of Energy 440:154-159, 1986.

[47] S Oka. Preliminary analysis of solid fuel combustion kinetics in a fluidized bed (in Serbian). Report of the Institute of Nuclear Sciences Boris Kidrič, Vinča, Belgrade, IBK-ITE-328, 1982.

[48] S Oka, B Arsić, D Dakić. Some characteristics of fluidized bed combustion of crushed lignite, with and without fines. Proceedings of 3rd International FBC Conference Fluidized Combustion: Is It Achieving Its Promise? London, 1984, Vol. 1, pp. 189-194

[49] J Makansi, B Schwieger. Fluidized bed boilers, Power 8:s1-s16, 1982.

7.

HARMFUL MATTER EMISSION
FROM FBC BOILERS

7.1.   Introduction

In the introductory chapter of this book, the requirements that must be met by modern coal combustion technologies were discussed in some detail. Without going into details of how to rank these requirements, three issues may be distinguished – combustion efficiency, flexibility with respect to fuel quality and low emission of toxic pollutants into the atmosphere. These are three major criteria in the development and evaluation of a new combustion technology and the evaluation of boiler design, all of which directly depend on the nature of the combustion process. Combustion efficiency in FBC boilers and their ability to burn various fuels were considered in Chapters 5 and 6. The final evaluation of fluidized bed combustion technology and the success of its development hinge on its ability to reduce environmental air pollution to the lowest possible level. It was stated in Chapter 1 that in this respect, fluidized bed combustion has significant advantages over other combustion technologies. In this chapter this will be discussed in some detail. Specific data obtained from measurements in pilot facilities and industrial boilers will be presented to demonstrate the potential of this technology. The purpose of this chapter is to show the characteristics and behavior of FBC boilers with respect to harmful air emissions. Also, the influence of various parameters on emission level will be considered. In light of the significance of resolving problems of harmful air emissions from FBC boilers, the references accompanying this chapter will include a much more comprehensive list of papers than it would be possible to discuss in detail in this book. Space does not permit one to provide detailed insight into the physical and chemical processes of formation of harmful gases, but interested readers will find in the additional references detailed explanations of the numerous problems and processes and directions on their further investigation.

7.1.1. Combustion of coal and formation of harmful matter

A series of gaseous combustion products and a solid incombustible remnant are created by coal combustion. Due to the complexity of the chemical composition of coal, gaseous and solid combustion products are varied and complex.

Solid combustion products such as fly ash are separated in devices for flue gas cleaning (cyclones, bag filters and electrostatic precipitators). The portion of ash which remains in the fluidized bed is removed from the furnace by a system for removing inert bed material, and it may also serve to replace some of the initial inert material in the furnace.

The problems of deposition, storage and use of solid products of combustion will not be considered here. While solid combustion products also pollute the environment – during transportation and storage in open-air depots, this is also considered to be a secondary issue. Although it should be noted that the possibility of exploitation of solid combustion products from FBC boilers is under serious investigation. Interested readers may find papers dealing with this topic in the Proceedings of the Conferences on FBC, particularly in more recent conferences (see Chapter 5, references [19-21]).

A discussion of the emission of harmful gases produced during fluidized bed combustion is important for two reasons: (1) because FBC boilers are known for their very low emission of such gases, and (2) because low emissions from FBC boilers are the result of the combustion process itself, which is the main subject of this book.

The main flue gas products of coal and hydrocarbon fuels combustion are: CO_2, CO, and H_2O. Strictly speaking all three products affect the atmosphere either in the vicinity of a thermal power plant or, in the case of CO_2, globally, and can be denoted as pollutants. Nevertheless, until recently carbon monoxide, while recognized as pollutant, was primarily considered as a sign of poor combustion efficiency; this situation is now changing and in the case of carbon dioxide there are increasing concerns about its production but this subject is outside of the scope of this book.

High concentration of industries and thermal power plants in highly developed countries, and the direct effects of numerous harmful compounds discharged into the atmosphere on the surrounding vegetation, animals and the health of people, have forced experts in various professions during the fifties to start seriously dealing with the issue of air pollution caused by gaseous combustion products.

Even today, new health impacts of the gases formed by combustion are being recognized as health issues, and in that connection one can think of the increasing concern about Hg. A list of 40 or more harmful compounds that can be detected in combustion products can now be found in the literature. At the same time, the methods and instruments for the detection and measurement of the steadily increasing number of components that are now considered as pollutants in combustion products are constantly being developed.

The effect of increases of water in the local environment due to combustion and water vapor from cooling towers and for that matter the thermal pollution of the environment from thermal power plants is a separate scientific topic which, while it has significant environmental implications, will not be discussed here. Instead, we will first discuss the effects of emission of SO_2 and NO_x compounds.

Sulphur dioxide and nitrogen oxides are gaseous compounds that were first noticed in terms of their harmful effect on the environment. In the past several decades, a series of technologies for the reduction of SO_2 and NO_x emission has been developed, as well as technologies for their elimination from combustion products. One of these technologies is fluidized bed combustion.

By emission of SO_2, it is customary to imply the total concentration of SO_2 and SO_3 (typically SO_3 forms only 3-4% of the total emission of sulphur oxides), and NO_x compounds include NO and NO_2 (with NO being typically about 95% of the NO_x produced from coal combustion). It should be noted that NO_x does not include N_2O which is considered separately.

Along with resolving the problem of SO_2 and NO_x emissions, both the ambition and possibilities of elimination of the other harmful compounds grew. References show that the number of papers devoted to SO_2 has now been substantially reduced, while the number of those which deals with reduction of NO_x remains large. This mean that the problem of NO_x control is not yet resolved in a satisfactory, efficient and economical manner. Interest, and therefore an increase in the number of papers which deal with the formation and emission of N_2O has increased greatly, but this interest also appears to be peaking.

The characteristics of bubbling FBC boilers will be considered in this chapter with respect to solid particulates, CO, SO_2, NO_x and N_2O emissions, because the concentrations of these compounds from combustion are the highest and the technologies to deal with them and the general understanding is relatively well established, making it possible to provide a clear discussion on this subject.

Various other compounds, such as the compounds of chlorine and fluorine, hydrocarbons of different composition – polycyclic organic matter and the emission of heavy metals and their compounds (mercury, lead, arsenic, nickel, chrome, cadmium, zinc and cobalt) will not be discussed here. Similarly, the problem of CO_2 emission as the main and unavoidable combustion product of all fossil fuels and also the gas that contributes the most to the "greenhouse" effect, is not discussed here. At the current level of development of combustion technologies, reduction of CO_2 emission is primarily accomplished by efficient combustion, with increased thermodynamic efficiency and energy saving in general. Fluidized bed combustion technology can contribute significantly here, with the development of pressurized fluidized bed combustion boilers and gasifiers and the possibilities offered due to combined gas-steam cycles and the so-called "topping cycle."

7.1.2. Regulations on air protection

The steady increase of air pollution in highly developed industrial countries encouraged the development of environmental regulations. In consequence, such legislation and recommendations have had an increasing effect on the development of coal combustion technologies. At first, legal regulations were limited to regulate the so-called emission limits. Emissions were measured by the quantity of a pollutant per cubic metre of air, over a given time period – 10 min., 15 min., 30 min., 1 h or 24 h.

Soon it was recognized that such emission limits cannot by themselves resolve the problem of pollution, since many gaseous pollutants are capable of being carried over enormous distances. Thus, whereas stack sizes steadily increased with sizes up to 200 m, and did much to improve the local air quality, the same cannot be said for the overall levels of air pollution. Equally, local regulation did not effectively stimulate development of modern, low-polluting combustion technologies and technologies for flue gas cleaning.

The beginning of the seventies saw the introduction of legal regulations on permissible SO_2 and NO_x emissions from thermal power plants. Emission standards are the permissible concentration of harmful materials at the stack exit expressed in ppm, mg/m^3 or mg/MWh, that is mg/MJ of generated energy. The best known regulations on emission are the American Clean Air Act, the German TA-LUFT, and regulations of the countries of the European Community and OECD countries [1-3].

A brief consideration of regulations on emissions is useful here, in order to determine the measures and criteria for evaluation of environmental characteristics of fluidized bed combustion. It is also possible to make a comparison with the situation concerning pollution and emission of thermal power plants in particular countries (in this book pollution in former Yugoslavia is considered in some detail). At the same time, such comparisons allow us to gain an insight into the significance of fluidized bed combustion within general world efforts to reduce pollution. Possible fields of application of FBC boilers both worldwide and in specific countries can be considered too.

Regulations on environment protection, especially on emissions during combustion, are least of all just legal issues. In every individual country, and worldwide, these regulations reflect primarily technological possibilities and achievements, as well as the economic and technological development of a particular country, and its social and political circumstances. Most frequently these regulations are a compromise between wishes and possibilities, and they are least influential when there is a huge gap between the two.

Following the development of technologies and regulations over a long period of time clearly shows a significant interaction and mutual influence between these two elements of modern society. The effect of the American Clean Air Act is a typical example, since a long- term research and development program called the Clean Coal Technology Program was initiated in response to it in the U.S.A., and whose third phase was completed in 1995. This R&D program, significantly financed from state funds, accelerated development and demonstration of numerous new technologies (all with a joint title – Clean Coal Technologies), of which it can be argued combustion in fluidized bed was the most successful of the resulting developments.

Development of new combustion technologies or technologies for flue gas cleaning, on the other hand, enables constant increase in the severity of emission regulations.

Table 7.1 taken from [3] presents a comparison of emission regulations on particulates, SO_2 and NO_x in different countries. Regulations differ for various fuels and facilities of different power, which is an indirect consequence of the possibilities of specific technologies. A comparison of data from Table 7.1 shows that regulations in different countries vary significantly, and that they are more rigid for high-quality fuels and large facilities. It can also be seen that regulations in certain countries are more rigid for fluidized bed combustion, which speaks in favor of the significance and the potential of this technology.

The severity of environmental regulations is constantly increasing as technologies develop. In 1984 in Germany, a special regulation was adopted for large facilities. In 1990 amendments to the Clear Air Act were adopted in the U.S.A. coming into effect in the year 2000 [2].

Contemporary regulations in the most developed countries set the following limits:

Particles	< 50 mg/m^3,
SO_2	< 400 mg/m^3,
NO_x	< 200 mg/m^3, and
CO	< 40 mg/m^3.

American amendments to the Clean Air Act were even more rigid-reduction of total SO_2 emission by the year 2000 by $10 \cdot 10^6$ t/year and NO_x emission by $2.5 \cdot 10^6$ t/year [2, 3]. By contrast, Yugoslavia had no uniform regulations on air protection.

Guidelines from the League for Clean Air of Yugoslavia were an attempt to introduce some order in this area. They were primarily intended to establish the

Table 7.1.   *Emission standards for particles SO₂ and NOₓ*

Country	Fuel	Type of harmful matter [μg/m³]				
		Particles	SO₂		NOₓ	
			150-300 MW	>300 MW	>150 MW	>500 MW
Austria	Coal	50	200	200	300	200-250
	Liquid fuel	50	350	200	200	150
	Gas	10	350	200	200	150
			100-300 MW	>300 MW	<300 MW	>300 MW
Belgium	Coal	50	1200	400	800	650
	Liquid fuel	50	1700	400	450	–
	Gas	10/4	100	100	350	–
Greece	Coal	100(N) 150(O)	None	None	None	None
	Liquid fuel	50				
	Gas	50				
Italy	Coal	50(N)	400(N)	1200(O)	650/1200	
Luxembourg	Coal	75	none		none	
Germany	Coal	50(N) 80-125(O)	400 + 85%		800/1000	200
	Liquid fuel	50	400 + 85%	–	450/700	150
	Gas	5	400		450/700	100
Portugal	Coal		None, It will be after 1991		800	
	Liquid fuel	115 (>50 MW)				
	Gas					
France	Coal	95/625*	None, only local regulation			
	Liquid fuel	125/200*				
		>300 MW	<300 MW	>300 MW		
Switzerland	Coal	50	2000	400	200	
	Liquid fuel	50	1700	400	150	–
	Gas	–	–	–	150	
Spain	Coal	150	2400-antr.			
	Liquid fuel	120 >200 MW	5000-coal 5500-lignite	–	None	–
			<300 MW	>300 MW		–
The Netherlands	Coal	50(N)	700	400 + 85%	500/400	
	Liquid fuel	–	1700	400 + 85%	300	–
	Gas	20(N)	1700	400 + 85%	200	
			<500 MW	>500 MW	–	–
European Community	Coal	100/50	2000	400	650	
	Liquid fuel	50	1700	400	450	–
	Gas	5	1700	400	650	
U.S.A.	Coal	35	–	1270 + 90%	750	–
	Liquid fuel	40		935 + 90%	520	
Japan	Coal	100	133–267	–	400	–
	Liquid fuel	50	133–267		260	
Canada	Coal	116	700	–	614	–
	Liquid fuel	116	700		350	
	Gas	116	700		287	

(O) old plant, (N) new plant; * after/before 1976

base for adoption of a law taking into account TA-LUFT and the regulations of the European Community [1, 3]. However, these Guidelines did not include limits on the overall emission of harmful matter, but only emissions.

Emission norms existed only in Bosnia and Herzegovina and Slovenia following the German regulations of 1988 (Table 7.2) [3, 11]. In 1990, Macedonia adopted the same regulations, while in other republics such regulations were not adopted at all.

Table 7.2. *Limiting values for emission from solid fuel combustion in Slovenia 1988*

	1-50 MW	50-300 MW
Particles [mg/m^3]	50	50
CO [mg/m^3]	250	250
SO$_2$ [mg/m^3]	2000	400
NO$_x$ [mg/m^3]	500	400

7.1.3. Air pollution in Yugoslavia

There are no reliable data on air pollution in Yugoslavia. Overall emissions are measured locally, in the most polluted places and in the vicinity of thermal power plants, and emissions are typically only measured on an occasional basis.

An approximate picture of the general situation concerning air pollution in the vicinity of thermal power plants in Yugoslavia can be obtained on the basis of incomplete and unsystematically collected data given in papers [4-13].

At the moment, no thermal power plant in Yugoslavia has devices for desulphurization of flue gases or elimination of NO$_x$. There are also no systems for taking primary measures for SO$_2$ and NO$_x$ control (two-stage combustion, burners with reduced NO$_x$ emission, introducing CaCO$_3$ or ammonia in the furnace), or for introduction of sorbents into flue gas ducts. The only modern devices for flue gas cleaning in thermal power plants are electrostatic precipitators. Other small thermal power generation facilities have cyclones only or bag filters at best.

In modern thermal power plants (600 MW$_e$, TENT B near Belgrade), particulate emissions in the stack are very low, about 30 mg/m^3 and meet modern European standards. The efficiency of electrostatic precipitators reaches 99.9% [8, 13]. However, particulate emissions in older facilities can be as high as 1500 mg/m^3 [11, 14, 16,].

Given the various changes that have occurred in the region in recent years, it is probably most helpful to look at data for a period of relative stability, and we will for these purposes choose the period of 1980 to 1990. According to the data of the

Federal Hydro-Meteorological Office [12, 13], the total annual SO_2 emission in Yugoslavia in 1990 was $0.74 \cdot 10^6$ t, ranking Yugoslavia 7th in Europe. Austria, by contrast produced only $0.047 \cdot 10^6$ t/year and had the lowest annual SO_2 emission in Europe. Although these values are comparatively low, the fact that SO_2 emission increased from 1980 to 1990 by 13.8% and is still tending to rise, must cause concern. Apart from Yugoslavia, SO_2 emissions increased during that period only in the former Democratic Republic of Germany, Greece, and Poland. All other European countries in that ten-year period saw marked reductions of SO_2 emissions ranging from 20-75%.

When burning lignite from the Kolubara basin, SO_2 emission in combustion products varies from 1000-3000 mg/m^3 [13]. During burning of coal in TE Trbovlje (now Slovenia), SO_2 concentration could rise up to 13,600 mg/m^3 [11] before 1990. After 1990, this power plant applied wet scrubbers, reducing SO_2 emission to European standards.

With its $0.42 \cdot 10^6$ t/year of NO_x compounds, Yugoslavia ranked sixth in Europe, with a slight rise in the past decade. Comparatively low emission of NO_x, although no measures were taken for their control, results from the composition of Yugoslav lignites. The greatest overall NO_x emitters were vehicles, with thermal power plants responsible for only 25% of the overall NO_x production [12, 13]. Austria, by contrast, produced only $0.2 \cdot 10^6$ t/year at the end of that period, and has also seen a 12.5% reduction since 1985.

NO_x emission from thermal power plants that burn lignite is comparatively low: they are 400-650 mg/m^3 [15] in TE Nikola Tesla, and 600-620 mg/m^3 in TE Trbovlje [11]. According to data in [13], NO_x emission in thermal power plants in the Kolubara basin ranges from 200 to 800 mg/m^3, and in boilers with liquid slag drainage, over 2000 mg/m^3.

CO concentration in flue gases of thermal power plants in Yugoslavia has not been systematically measured [13].

A comparison of the data being discussed here with the limiting permitted values for emissions given in Table 7.1 shows that the situation for air protection in Yugoslavia is disturbing. Per unit of generated energy, Yugoslavia had twice as high an emission of SO_2 as did Western European countries [12]. For NO_x emission, the situation is much better, although the emission values from thermal power plants are also 2-3 times higher than those permitted in highly developed industrial countries (Table 7.1).

7.1.4. The role of FBC boilers in reduction of air pollution

By comparison with the permitted values of SO_2 and NO_x emission in different countries, and anticipating the rise in power generation in less developed countries, it can be concluded that there is much to be done to reduce the emission of harmful compounds when burning coal. Developed countries are constantly refining their regulations and making them more stringent. This tendency can be seen in the U.S.A. with their amendments of the Clean Air Act, in Germany in terms of regulations for large furnaces, or in the Netherlands where the construction of coal-fired furnaces has been banned.

At the same time, new technologies are developing in many countries to: (1) eliminate SO_2 and NO_x from combustion products, (2) reduce their formation during combustion by injecting additives or by optimizing the combustion process and (3) implement combined gas-steam cycles. Many of these technologies have been in use for a long period of time, and some of them are still in the phase of demonstration on the industrial level. However, the question here is what are the possibilities, and the role and the place of fluidized bed combustion among these technologies for SO_2 and NO_x emission control in the near to medium term, say the next 10-15 years.

A comparative analysis [17-19] of FBC boilers, both bubbling and circulating, with pulverized coal combustion boilers showed that there are no significant differences in investment costs and power generation costs, if the comparison is made under the same conditions, i. e., for the same SO_2 and NO_x emission at the stack exit. Grate combustion boilers by contrast are much more expensive [20].

Figure 7.1 [2] gives some possibilities of different coal combustion technologies.

Fluidized bed combustion technology, according to values presented in Fig. 7.1 and a detailed analysis of its possibilities listed in reference [2], has significant advantages in relation to other technologies, if the intention is to comply with the demanding requirements of the Clean Air Act Amendments. These advantages are:

— the technology is already demonstrated and commercially available,
— it has lower SO_2 and NO_x and solid particulate emission at the present stage of its development,
— this technology is still being developed and further significant improvements can be expected,
— it is convenient for the realization of combined cycles, and
— it is convenient for introduction into already existing facilities, i. e., retrofit applications.

According to the analysis in [2], in the next period from 2000 or 2010, there will be major success in reducing air pollution with the introduction of FBC boilers.

The present analyses refer to the best performance of fluidized bed combustion technology achieved in boilers with circulating fluidized beds. As data in the next section will show, the possibilities of improvement with bubbling FBC boilers are somewhat smaller because SO_2 and NO_x emission are somewhat higher. How-

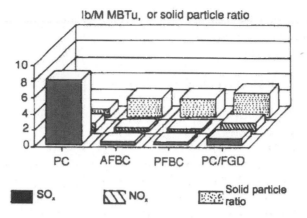

Figure 7.1.
Comparison of the NO_x, SO_2 and particle emissions from differ-
ent types of boilers (Reproduced by kind permission of the
ASME from [2]):
PC – pulverized coal combustion boilers without desulph-
urization of flue gases, AFBC – atmospheric fluidized bed
combustion boilers, PFBC – pressurized fluidized bed com-
bustion boilers, PC/FGD – pulverized coal combustion boilers
with wet scrubbers for desulphurization of flue gases. Solid
particle ratio-ratio of the particle emission for given technol-
ogy and particle emission of pulverized coal combustion
boiler, for PC=1.

ever, in view of the fact that this kind of boiler covers the range of low unit power, it
is probably more reasonable that they be compared with grate combustion boilers
with which they compare favorably [20, 21]. Two more facts should be noted: bubbling FBC boilers are still being improved [21]; further, only a limited number of
coals, enriched or of high quality, can be burned in grate combustion boilers.
Low-quality coals, coal wastes or industrial or municipal waste can all be successfully burned in bubbling FBC boilers.

7.2. Characteristics of the first generation industrial and demonstration FBC boilers in operation – SO2, NOx, CO, and particle emission

The characteristics of bubbling FBC boilers can best be presented if the following types of data are analyzed:

(1) requirements posed to boiler designers, and

(2) data obtained by site measurements at industrial or demonstration facilities during real long-term exploitation.

7.2.1. Design requirements for the first generation FBC boilers

Design requirements for several large industrial and demonstration facilities are listed in Table 7.3. In considering these data, one must bear in mind that design requirements, specific solutions and data measured in an industrial facility depend on specific conditions, such as type and quality of coal, dimensions of the boiler, location, load and exploitation conditions. In the case of emissions, design requirements are most commonly set for SO_2 and NO_x. Low emission of CO is ensured by the requirement for high combustion efficiency, so that in most places there has been no explicit CO requirements. This is similarly the case with particulate emission requirements for conventional facilities.

Table 7.3.   *Design requirements for bubbling FBC boilers*

Boiler	Power	SO_2 emission	NO_x emission	CO emission	Particle emission	Ref.
AKZO The Netherlands in operation	90 MW_{th}	230 mg/MJ 600 mg/m³	190 mg/MJ 500 mg/m³	<86 mg/MJ <225 mg/m³	15 mg/MJ	22 26
Wakamatsu Japan in operation	50 MW_e	100 mg/m³ 90% reduction of SO_2	200 mg/m³	--	10 mg/m³	23 24
TVA-160, U.S.A. demonstration boiler in operation	160 MW_e	90% reduction of SO_2	260 mg/MJ	--	--	36
Takehara Japan in design	350 MW_e	100 mg/m³	200 mg/m³	--	--	23

Due to the size of the facilities and their character as demonstration units, the requirements systematized in Table 7.3 are actually at the top end of the performance for modern bubbling FBC boilers. Smaller FBC boilers, which are described in numerous papers (see Chapter 5, references [19-21]), are not designed to meet such strict requirements. Space limitations simply do not allow the listing of requirements set in construction of the numerous small boilers built in Europe. However, if necessary, even the stringent requirements listed for large boilers can be achieved and this will be discussed in the following section.

7.2.2. Emission measured during operation of several characteristic first generation FBC boilers

There are many data on measured gas and particle emissions of industrial bubbling FBC boilers (see Chapter 5, references [19-21]). Here, we will give comparative data for the largest boilers units constructed that were already described in Chapter 5. The main reason for the choice of these data is reliability, because there have been long-term systematic measurements made on all of these boilers. Further, since these units are also demonstration boilers, combustion of many different types of coal has been tested, making the data more general in character.

The data systematized in Table 7.4 were taken from several papers published in the period 1987-1991 [22-37]. Detailed data on operating regimes of the boilers in which these measurements were made are given in the cited papers.

Table 7.4. *Emission measured in large power bubbling FBC boilers in operation*

Boiler	Power	SO_2 emission	NO_x emission	CO emission	Particle emission	Ref.
AKZO The Netherlands Industrial boiler	90 MW_{th} <2.5% S in coal	100-230	50-180	10-60	–	28
		At 100% load				
		206.6	183.4	46.0	0.53	
		At <50% load				
		150	71	103.9	1.55	
		(all in mg/MJ)				
Values averaged over 2 years		356	254	115	≈10	22, 31
		(all in mg/m³ at 7% O_2)				
Wakamatsu Japan Pilot boiler in operation	20 MW_e	20-150	254	–	–	27
		80-98% of desulphurization				
		(all in ppm at 6% O_2)				
Wakamatsu Japan Demonstration boiler	50 MWe 160 t/h <0.75% S in coal	3-45	167-210	–	1-3 mg/m³	30, 23, 24
		(all in ppm)				
TVA-160 U.S.A. Demonstration boiler	160 t/h 3.5% S in coal	50-300	85-150	300-550 ppm	–	33
		(in mg/MJ, 90-92% desulphurization)				
Values averaged over 4 months		92% desulphurization degree	<110 mg/MJ	300 ppm	–	34
Black Dog U.S.A. Industrial boiler in operation	130 MW_e 0.8 S in coal	at 20-100% load				29
		<250	<200	<500	–	
		(in mg/MJ)				
	with coals 0.8-5.7% S	16-875	272-367	80-280		
		(in ppm at 3.5% O_2, 81-95% desulphurization)				
Hasket Station U.S.A. Industrial boiler in operation	80 MW_e 0.36-2.5% S	mean values in November 1988				32
		<250	<140	–	<24	
		(all in mg/MJ)				

For the sake of comparison with Table 7.4, we will cite data on emissions from some small and medium size industrial boilers. In [38], the following data are cited for boilers of the Japanese firm Babcock Hitachi, with steam capacity 8-57 t/h:

SO_2 emission	50-150 ppm (at % O_2),
NO_x emission	60-200 ppm, and
desulphurization degree	70-90%.

7.2.3. Comparison of emissions from bubbling and circulating FBC boilers

The issues for SO_2 and NO_x emission control of bubbling FBC boilers can be realistically evaluated by making comparisons with circulating fluidized bed combustion boilers. As an improved version of the same technology, it is logical to expect that circulating FBC boilers will achieve lower values of SO_2 and NO_x emissions. However, various operating conditions, various types of coal, and different boiler concept and design, make direct comparison difficult. Thus, attempts on comparison and analysis are often reduced to general statements.

That is why it is significant to cite results of systematic comparisons made in reference [40]. Comparative investigations were performed for a bubbling fluidized bed combustion boiler (16 MW_{th}, Chalmers University, Göteborg) and a circulating fluidized bed combustion boiler (40 MW_{th}, Nykoping). The same coal was used (2% S) with every effort made to realize similar combustion conditions (excess air, combustion temperature, the same limestone and Ca/S molar ratio).

Figure 7.2 presents the changes in the degree of desulphurization for these

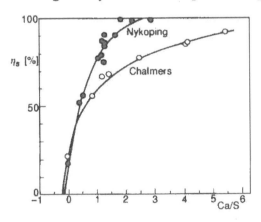

two boilers, with Ca/S ratio from 0 to 5.5. As can be seen, the bubbling fluidized bed combustion boiler requires a considerably higher Ca/S molar ratio (that is the quantity of limestone) to achieve the same degree of desulphurization. For this boiler, 90% desulphurization is achieved only with a Ca/S = 5.5. In the analysis of these data, one should bear in mind that the FBC boiler at Chalmers University operates without fly ash (and limestone) recirculation and that causes a significant increase in limestone use.

Figure 7.2.
Comparison of the degree of desulphurization for BFBC and CFBC boilers at optimum bed temperature (Reproduced by kind permission of the ASME from [40])

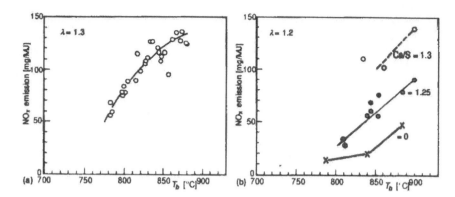

Figure 7.3.   *Effect of bed temperature on NO_x emission (Reproduced by kind permission of the ASME from [40]):*
(a) BFBC demonstration boiler at Chalmers University, (b) CFBC boiler – Nykoping

Figure 7.3a and 7.3b give comparisons of NO_x emission values for both boilers. It was observed that the increase of Ca/S ratio increases the NO_x emission in a boiler with circulating fluidized bed. In spite of this fact, by the choice of opti-

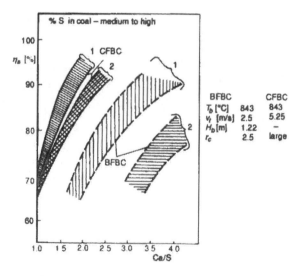

Figure 7.4.
Desulphurization degree as a function of limestone reactivity for BFBC and CFBC boilers designed by Ahlstrom company (Reproduced by kind permission of the ASME from [25]):
1 – limestone of high reactivity, 2 – limestone of medium reactivity

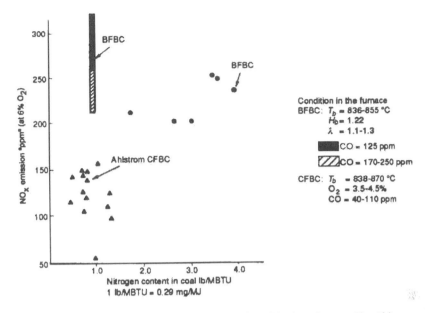

Figure 7.5. *NO_x emission from BFBC and CFBC boilers designed by Ahlstrom company (Reproduced by kind permission of the ASME from [25])*

mal regime parameters, it is possible to achieve lower values of NO_x emission in CFBC boilers even with equal values of SO_2 emission.

Having collected a large number of data from measurements in industrial bubbling FBC boilers and comparing them with data for its own CFBC boilers, Ahlstrom offers two diagrams. Figure 7.4 gives a comparison of desulphurization degree for different Ca/S ratio and limestones of different reactivity. Figure 7.5 gives a comparison of NO_x emission values for different boilers and various nitrogen contents in coal [25].

7.3. Carbon-monoxide emission in bubbling fluidized bed combustion

Efforts to achieve the highest possible combustion efficiency ensures CO emission reduction, that is, reduction of losses due to incomplete combustion. CO concentration in flue gases depends on combustion temperature, excess air, quantity of unburned fine char particles, volatile content in coal, and perhaps most of all, on the intensity of mixing in the freeboard.

The most detailed investigation of the influence of combustion conditions on CO emission was carried out at a 4 MW_{th} pilot facility in Apeldoorn, the Netherlands [37, 41].

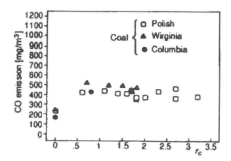

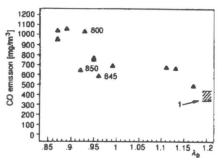

Figure 7.6.
CO emission as a function of recirculation ratio (Reproduced by kind permission of the ASME from [41]):
$v_f = 2$ m/s, $T_b = 850$ °C, $H_b = 1.05$ m

Figure 7.7.
CO emission for staged combustion (Reproduced by kind permission of the ASME from [41]):
1 – dashed region-experiments without secondary air, $T_b = 825$-859 °C. In all experiments excess air was $\lambda = 1.2$

     Figure 7.6 shows that fly ash recirculation, that is, recirculation of unburned char particles, contributes to CO concentration increase in combustion products. The increase of CO concentration is the result of an increased hold-up of char particles in the freeboard, which causes combustion conditions to deteriorate.

     Introduction of staged combustion, i. e., substoichiometric condition in the bed and introduction of secondary air above the bed, which has a favorable effect on NO_x emission control, contributes to CO emission increase (Fig. 7.7). This effect was observed also in paper [42] and it is shown in Fig. 7.8. In this case, one should bear in mind that introduction of secondary air means at the same time re-

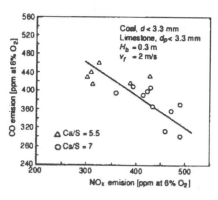

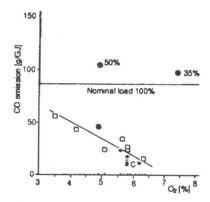

Figure 7.8.
Influence of CO emission on NO_x formation from the measurements of Fernandez (Reproduced by kind permission of the ASME from [42])

Figure 7.9.
Influence of excess air on CO formation from measurements of Verhoeff on AKZO FBC 90 MW$_{th}$ boiler

duction of temperature in the freeboard, as well as more intense mixing, which are opposite effects.

In investigations presented in [37] it was observed that the position and form of nozzles for injection of secondary air significantly affect CO emission. Introduction of secondary air in the vicinity of the bed surface tends to reduce CO concentration. Increase of excess air contributes to better combustion conditions and CO emission reduction, which is clearly seen from measurements in the 90 MW_{th} AKZO boiler [22], (Fig. 7.9).

7.4. Sulphur-dioxide emission in bubbling fluidized bed combustion

Elimination of SO_2 from combustion products during the combustion process by adding limestone (or rarely dolomite) into the bed is one of the major advantages of FBC boilers. This harmful gaseous product of combustion is then converted into an inert solid material, which is also found in nature. Significant savings in investment costs occur because there are no special devices for flue gas desulphurization, and this makes FBC boilers competitive in the market. This advantage has been the driving force for the many comprehensive investigations that have been going on to achieve optimum conditions for sulphur capture under FBC conditions. A significant effort in collecting and systematizing the data on the desulphurization process during combustion in fluidized bed was first made in a book published by the Massachusetts Institute of Technology (MIT) and U.S.A. Department of Energy (DOE) [43, 44].

The degree of desulphurization depends on two factors: the operational parameters and the characteristics of the limestone. Operating parameters that affect desulphurization include combustion conditions, the quantity of limestone used (more properly the Ca/S molar ratio), the boiler concept and design, and the bed temperature. In the sections that follow, the effects of these parameters will be considered on the basis of investigations in experimental furnaces and pilot plants.

7.4.1. Physical and chemical processes controlling rate and degree of limestone sulphation during fluidized bed combustion

During combustion of coal, SO_2 is released or formed during devolatilization and during char combustion. SO_2 formation occurs either on the surface of coal particle or its vicinity. Capture of SO_2 takes place on the surface or in the interior of the limestone (or dolomite) particle pores, which has already gone through the calcination process.

In order to enable SO_2 to react with CaO created from the calcined limestone (or dolomite), a series of physical and chemical processes must take place.

Proper consideration of these processes enables a more complete understanding and explanation of experimental data and measures necessary for efficient SO_2 retention in fluidized bed combustion boilers.

There are two forms of resistance in a series which control reaction of SO_2 and CaO, resistance (i. e., processes) linked to the limestone (dolomite) particle and gas diffusion resistance (processes). Diffusion has to enable the SO_2 created in the vicinity of the coal particle to achieve contact with the particle of calcined limestone. These two resistances are different in nature and depend on different parameters, so it is appropriate to consider them separately. Experimental investigations are also usually carried out separately for these two groups of processes.

Limestone and dolomite are natural mineral stones which are very widespread in the crust of the Earth. The formula of pure limestone is $CaCO_3$. Dolomite is a mixture of calcium carbonate ($CaCO_3$) and magnesium carbonate ($MgCO_3$). Ideally dolomite has a molar ratio of Ca:Mg of 1:1. Natural limestone and dolomite most frequently have various admixtures ("impurities"). A particle of limestone is a set of individual crystals of calcite or dolomite that are connected in a matrix together with other admixtures (Al_2O_3, SiO_2, Fe_2O_3 and others). Table 7.5 lists the usual contents of impurities according to [43], and separately for Yugoslav limestones according to [45, 46].

Table 7.5.   *Impurities in limestone and dolomite*

Component	According [43]	For 8 Yugoslav limestones [45, 46] [%]
SiO_2	0.40-4.4	0.200-3.18
Al_2O_3	0.04-0.6	0.030-0.72
Fe_2O_3	0.02-0.3	0.014-0.40
H_2O	0.00-1.1	—
TiO_2	Not measured	0.010-0.02
MnO	"	0.010-0.12
Na_2O	"	0.010-0.06
K_2O	"	0.010-1.20
P_2O_5	"	0.010-0.08
SrO	"	0.010-0.10
S	"	0.010-0.07
$CaCO_3$	"	90.34-97.89

Calcination begins with heating of limestone. Calcite crystals ($CaCO_3$) disintegrate into very small lime crystals (CaO). In the case of heating dolomite, CaO crystals are found along with crystals of MgO. During heating, the linked water and organic admixtures are released. Large pores are thus created and the released CO_2 diffuses out through the porous particle structure. Simultaneously with the process of calcination, sintering of small CaO crystals to larger ones takes place. CaO crystals have smaller molar volume than the parent limestone, and thus new pores are created. The main result of the calcination process, apart from creation of CaO, is an increase of porosity and formation of particle porous structure, which dominates the subsequent sulphation process.

The calcination process is endothermic [43]:

$$MgCO_3 = MgO + CO_2 \qquad -113 \text{ kJ/mol} \qquad (7.1)$$
$$CaMg(CO_3)_2 = CaCO_3 MgO + CO_2 \qquad -128 \text{ kJ/mol} \qquad (7.2)$$
$$CaCO_3 = CaO + CO_2 \qquad -183 \text{ kJ/mol} \qquad (7.3)$$

The calcination process depends on temperature and partial pressure of CO_2 in the vicinity of the particle. The usual conditions during combustion in fluidized beds are $T = 750$-$900\,°C$, and concentrations of CO_2 (13-15%). Considering data on thermodynamic equilibrium, if the overall partial pressure of CO_2 is 1 bar, the decomposition temperatures are the following:

for $MgCO_3$ 402 °C,
for $CaMg(CO_3)_2$ 466 °C, and
for $CaCO_3$ 889 °C.

This means that favorable thermodynamic conditions always exist for decomposition of magnesium carbonate, and dolomites under atmospheric conditions [43] and hence one might always expect dolomitic stones to have a porous structure, even though the Mg component cannot react with SO_2 at typical FBC conditions. However, there is a somewhat narrower range of temperatures for which favorable thermodynamic conditions exist for calcination of $CaCO_3$, and at partial pressures of CO_2 in the 0.12-0.15 bar range during combustion, decomposition of $CaCO_3$ will always occur above about 790 °C. This temperature is below the lower limit of the usual fluidized bed combustion operating temperature range with sulphur capture.

The process of total calcination can last several minutes. According to measurements in fluidized bed conditions, the calcination process is much faster than the sulphation process [43, 45, 47]. Figure 7.10 shows CO_2 and SO_2 concentration change during the limestone sulphation process in an experimental reactor with a fluidized bed [45]. According to these measurements, the calcination process for particles in the 0.4 to 0.5 mm size range is practically completed after 50 s. The

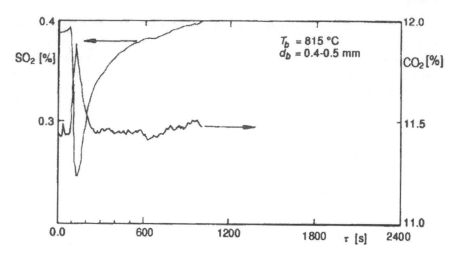

Figure 7.10. *Change of CO₂ and SO₂ concentrations during calcination and sulphation of limestone No. 7, Table 7.6 from the measurements of Arsić [45]*

sulphation process under the same conditions lasts 1200 s. That is why calcination time and rate are often disregarded in modelling the sulphation process, and an instantaneous transformation of $CaCO_3$ into CaO is assumed [48].

Sulphur dioxide penetrates through the porous structure of the particle that has endured calcination, and comes into contact with the surface of the CaO crystal, where a heterogeneous chemical reaction (sulphation), takes place. The resulting $CaSO_4$ has a higher molar volume than the CaO crystal, and this causes reduction of particle porosity. As the sulphation process proceeds, porosity diminishes, and so does the particle surface available for the reaction. Due to final blockage of pores and creation of a layer of $CaSO_4$ on the particle external surface, penetration of SO_2 to unreacted CaO crystals is finally prevented, and the sulphation process effectively stops, although, as the following sections will show, 40-70% of CaO remains unused.

The heterogeneous chemical reaction of sulphation is exothermic:

$$CaO + SO_2 + \tfrac{1}{2}O_2 = CaSO_4 + 486 \text{ kJ/mol} \tag{7.4}$$

Significant attention has been devoted to investigation of reducing conditions (increased CO content) on decomposing of the already created $CaSO_4$ [49, 50]. The following reactions are possible under reducing conditions:

$$CaSO_4 + CO \rightarrow CaO + SO_2 + CO_2 \tag{7.5}$$

or

$$CaSO_4 + 4CO \rightarrow CaS + 4CO_2 \tag{7.6}$$

Whether these reactions will take place, and which of them is more probable, depends on the temperature and concentration of SO_2 and CO [49, 50]. The effects of reducing conditions may in part be the cause of the differences in the sulphur capture behavior of bubbling and circulating fluidized bed boilers [49].

In the presence of oxygen, the following reactions are also possible:

$$CaS + 2O_2 = CaSO_4 \tag{7.7}$$

and

$$CaS + \tfrac{1}{2}O_2 = CaO + SO_2 \tag{7.8}$$

We will discuss further only the basic sulphation processes presented in eq. (7.4).

Phase diagrams of thermodynamic equilibrium [43] show that, apart from the fundamental reaction (7.4), other reactions are also possible. These reactions lead to incomplete or intermediate products such as CaS, $CaSO_3$ or decomposition of $CaSO_4$ into CaS and O_2. The probability and extent of these reactions depend on temperature and the partial pressure of SO_2, O_2, and CO. In conditions typical of atmospheric fluidized bed combustion (overall pressure 1 bar, 50-3000 ppm of SO_2 at 3% O_2), a thermodynamically stable reaction product ($CaSO_4$) exists in the temperature range 800-1000 °C and outside this range, $CaSO_4$ is not thermodynamically stable. This is in part the reason for the existence of an optimum temperature at which the maximum degree of sulphation is achieved. This temperature maximum for sulphur retention has been seen in experimental and in industrial plants.

The sulphation rate depends on a series of connected processes (resistances). Processes controlling sulphation rate are mass transfer to the surface of the particle, SO_2 diffusion through a porous particle structure, rate of heterogeneous chemical reactions (7.4) and (7.6), and the process of pore blocking (that is, sulphation degree). Numerous investigations and analyses of these processes show that the sulphation rate and sulphation degree of a limestone particle are influenced by temperature, limestone structure, limestone particle porosity after calcination, limestone particle size, CaO content and impurities, and gas velocity around the particle.

The Ca/S molar ratio that should be used in bubbling FBC boilers is typically in the range 2-5 in order to obtain an acceptable SO_2 emission. Transfer of SO_2 from the place where it is formed or released to the limestone particle surface depends on particle mixing (coal, inert material, limestone) and gas mixing (O_2, CO, SO_2) processes in the fluidized bed and the freeboard. Conditions for efficient mixing, flow and appropriate thermodynamic conditions in these two characteristic zones of the fluidized bed boiler furnace are different. It is possible to say in advance that conditions for capturing SO_2 with limestone in the freeboard are relatively unfavorable. Therefore, most of the efforts at optimizing sulphur capture are directed to ensuring that the sulphation process is optimized in the fluidized bed itself.

A systematization of the processes and conditions for two characteristic zones in a FBC furnace is presented in Fig. 7.11 [48]. This figure can help facilitate consideration of the conditions and parameters that affect the sulphur retention process. The following characteristic processes take place in both zones (in and above the bed): release or formation of SO_2 during devolatilization and char combustion; mixing of gas and particles; flow (motion) of gas and particles; fragmentation and attrition of coal and limestone particles; and elutriation of fine coal and limestone particles (see middle column of Fig. 7.11).

Six key characteristics (or conditions) that affect the sulphation process are listed in the left column of Fig. 7.11. These conditions may be significantly different in the bed and the freeboard.

(1) Mean particle size and limestone particle size distribution. In the freeboard, the mean limestone particles size is considerably smaller and close to the size that will be elutriated.

Characteristics of the contact between SO_2 and $CaCO_3$ in FBC furnace	Processes in the FBC furnace	Control parameters
Limestone particle size distribution		Characteristics of coal Coal particle size distribution Feeding point Coal particle mixing
Limestone reactivity Conditions in bed Distribution along the furnace height: – Temperature – O_2 – CO_2 Distribution of the SO_2 formation along the furnace height	Devolatilization Combustion Particle mixing Atrition Elutriation	Limestone characteristics Limestone size Limestone feeding point Limestone mixing
Limestone particle concentration Gas-limestone particles contact time		Recirculation ratio Injection point Distribution of fly ash reinjection points
Limestone particles residence time		Temperature Pressure Fluidization velocity Bed height Excess air Air distribution

Figure 7.11. *Control parameters, processes and characteristics of desulphurization in FBC boilers. According to [48]*

(2) Conditions for achieving maximum limestone reactivity. In the free-board, the temperature is considerably lower with high CO_2 concentration and a small O_2 concentration. Further, the CO concentration is also considerably higher than in the bed.

(3) Release of SO_2. Most of the SO_2 is released in the bed. During coal feeding to the bed surface and combustion of high volatile coals, a considerable part of the SO_2 is released above the bed where conditions for capture by limestone are not so favorable.

(4) The concentration of limestone particles is considerably higher in the bed.

(5) The time available for contact between the gas and limestone particles is considerably longer in the bed. In the freeboard, limestone particles residence time is very short, because the particles are very small and they can be elutriated from the combustion zone.

(6) Residence time and "lifetime" of limestone particles differ in these two zones of the furnace.

Parameters which affect these processes are listed in the right-hand column of Fig. 7.11. It is obvious that good sulphur retention, i. e., a high degree of limestone utilization (small Ca/S ratio), can be achieved by a good boiler design concept, and proper attention to operating conditions. Among the listed parameters, the manner and location of the coal and limestone feeding points, and the degree of fly ash recirculation are clearly among the most important parameters. The influence of these parameters and limestone characteristics will be discussed in the following two sections.

7.4.2. The effects of design and operating parameters on SO_2 emission in fluidized bed combustion

The effect of operating parameters has usually been studied in small experimental furnaces or pilot facilities, as it is much easier to change combustion conditions in such installations, and it is also possible to vary a broad range of parameters. Large demonstration or industrial boilers do not permit easy change of parameters and it is very expensive and time consuming to obtain meaningful data from such facilities. Nevertheless, measurements have also been made in such facilities, and these verify results obtained from laboratory-scale equipment. Quantitative differences are observed, and they are normally ascribed to the consequence of different mixing conditions and different particle residence time in such large facilities.

Effect of combustion temperature. Numerous experiments have shown that an optimum temperature exists for which the efficiency of SO_2 capture is maximized. The optimum temperature ranges from 800 to 900 °C. The most important reason for the choice of bed temperature within this range is the creation of the most favorable conditions for SO_2 retention. Discussions on the effects of temperature on this temperature maximum are presented in some detail in references [40, 42, 51, 52].

Reference [42] gives the results of an investigation in which different optimum temperatures 820, 840, and 860 °C, were obtained for the combustion of lignite from three different basins using the same limestone. Different optimum temperatures were also obtained from measurements in industrial boilers at Chalmers University and Nykoping [40], (Fig. 7.12). The authors could not explain those measured differences, but comprehensive measurements in different furnaces with different types of coal and limestone of various origins showed that below 780 °C and above 960 °C the sulphation efficiency is significantly reduced.

Despite the existence of the optimum temperature, which varies from unit to unit, a combustion temperature can be chosen within a broad range, because such sulphation efficiency changes in the vicinity of the optimum temperature tend not to be dramatic.

Effect of Ca/S ratio. Almost all experimental investigations aimed at SO_2 emission reduction in fluidized bed combustion have explored the effect of varying the Ca/S molar ratio. Investigations have been carried out both in industrial and

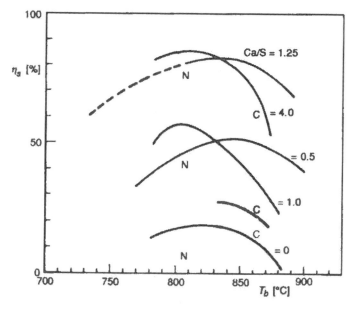

Figure 7.12. *Effect of bed temperature on the degree of desulphurization, from the measurements in CFBC Nykoping boiler (N), and BFBC boiler in Chalmers (C) (Reproduced by kind permision of the ASME from [40])*

demonstration facilities (see references in Section 7.2) and in experimental furnaces and pilot plants [40-43, 48-58]. Experiments were carried out with different types of limestone and coal, and in different facilities, varying the Ca/S molar ratio from 1 to 7. The general conclusion is that the degree of desulphurization increases with the Ca/S ratio, but that the actual efficiency achieved depends on numerous factors. All the parameters mentioned in Section 7.4.1, and presented in Fig. 7.11, influence the degree of desulphurization that can be achieved with a given Ca/S ratio.

The effect of Ca/S ratio on SO_2 emission can be seen in Figs. 7.2, 7.4, and 7.5. As an illustration of the effect of Ca/S ratio we will cite results of the comparative study carried out in four different facilities [51]. In Fig. 7.13 values of desulphurization efficiency obtained during combustion of the same coal and using the same limestone in different facilities are given. The overall conclusion from this comprehensive investigation is that it is not possible to determine a general relationship for the degree of desulphurization on the Ca/S ratio. This investigation was carried out employing six facilities, with nine types of fuels (from lignite to coke) and with four different limestones.

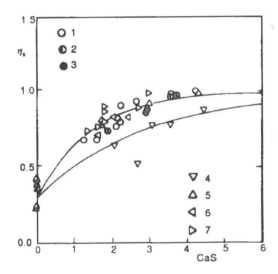

Figure 7.13.
Effect of Ca/S ratio on the degree of desulphurization when burning Minto coal in 4 different boilers (Reproduced by kind permission of the ASME from [51])
Boiler, limestone particle size, fluidization velocity:
1 – Queen's Univ., BFBC, 0.49-0.78 mm, 1.6-1.9 m/s, 2 – Queen's Univ., BFBC, 0.49-0.78 mm, 2.0-2.3 m/s, 3 – Queen's Univ., BFBC, 0.49-0.78 mm, 2.7-2.8 m/s, 4 – Queen's Univ., BFBC, 0.49-0.78 mm, 1.72–1.98 m/s, 5 –experimental boiler MARK1, BFBC, 1.1 mm, 6– district heating boiler Summerside, BFBC, 0.85-2.36 mm, 7 – UBC CFBC boiler, 0.21-0.45 mm

The choice of Ca/S ratio is always the result of a compromise: the effort to reduce SO_2 emission, cost savings and elimination of problems that appear with the use of a large quantity of limestone. The effect of limestone reactivity will be discussed in the sections that follow.

In commercial facilities the molar ratio Ca/S is typically between 2-4. In each specific case, in the design and construction of a boiler, it is best to predetermine the reactivity of the available types of limestone, and choose the most efficient one, providing some other factor like limestone transportation costs does not become a limiting factor. For the chosen limestone, the Ca/S ratio that will enable the desired degree of desulphurization, must be determined in an experimental or pilot furnace. While existing mathematical models [43, 59, 60] may enable the determination of the degree of desulphurization for the given Ca/S ratio, they also demand experimental determinations of limestone characteristics (reactivity).

Effect of bed height. Increased residence time of limestone particles in the fluidized bed, and prolonged contact with SO_2 increase the degree of desulphurization. Increasing bed height affects both of the above-mentioned quantities, and that is why bed height may be increased when it is necessary to achieve very low SO_2 emissions.

Effect of fluidization velocity. The effects of fluidization velocity have not been systematically investigated. There are two possible reasons for this: (a) fluidization velocity is predetermined by the chosen mean particle size of inert material in the bed, and (b), fluidization velocity influence depends on a range of other parameters. Typically, the mean inert material particle size range in a bubbling bed is between 0.5 and 2 mm, so that fluidization velocities in existing facilities are chosen within the range from 1-3 m/s. Results of measurements in pilot facilities given in reference [61] do not suggest a major influence of fluidization velocity on sulphur retention.

The influences of fluidization velocity on the sulphation process are various and not necessarily easy to quantify. An increase of fluidization velocity increases the elutriation of small limestone particles. Thus, for a higher fluidization velocity a larger limestone particle mean size should be chosen and this will reduce the specific reaction area associated with a limestone particle. Similarly, higher velocities reduce gas residence time in the fluidized bed and in the freeboard, but increase the intensity of mixing and attrition of the sulphated layer on the limestone particle surface. Due to these opposing effects, it is difficult to anticipate the effect of fluidization velocity in a specific case. Systematic measurements of sulphur retention in a pilot facility, a 4 MW_{th} boiler in the Netherlands [41, 62], showed that a somewhat lower degree of desulphurization was achieved in the bed with large particles, which might be a result of higher fluidization velocity.

During load control by increasing the fluidization velocity, increased emission of SO_2 should normally be expected. Generally, it is possible to claim that higher fluidization velocity most probably means a lower degree of desulphurization. Improvements in sulphur capture can generally be produced by increasing Ca/S ratio or the degree of fly ash recirculation.

Effect of excess air. An increase of excess air should mean an increase of desulphurization efficiency. The probability of the existence of zones with reduction conditions ought to be decreased, in which decomposition of $CaSO_4$ can occur [25, 49, 50]. Results of measurements given in [52] show that an increase of excess air from 1.3 to 1.4 reduces SO_2 emission by about 5-10%.

Effect of the ratio between primary and secondary air. The introduction of two-stage combustion is common in fluidized bed combustion boilers in order to achieve NO_x emission reductions. There are numerous investigations of the effects of two-stage combustion and the ratio of primary and secondary air on SO_2 and NO_x emission [37, 41, 52, 62, 63]. Due to substoichiometric combustion in the bed, introduction of secondary air unfavorably affects desulphurization. Figure 7.14 [37] shows results of an investigation carried out in a 4 MW_{th} TNO boiler, where reduction of primary air share from 1.2 to 0.87, due to the reduction of available oxygen, caused SO_2 retention to be reduced from 90% to 80%. Therefore, adoption of higher total excess air is recommended when employing two-stage combustion.

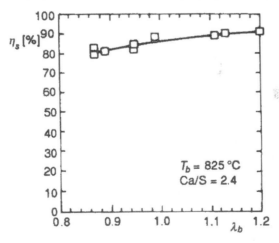

Figure 7.14.
Effect of primary to secondary air ratio on the degree of desulphurization, from the measurements of Wormgoor et al. (Reproduced by kind permission of the ASME from [37])

Effects of coal feed. Figure 7.15 [25] shows a comparison of desulphurization during feeding of the same coal above the bed surface and under the bed surface, based on investigations in a TVA 20 MW$_{th}$ pilot boiler. The effects of fine particle percentage in coal (below 0.6 mm) are also shown.

Significant reduction of sulphur retention are seen with an increase in the fine particle percentage in the coal, and this indicates the cause of reductions in the desulphurization efficiency for overbed feeding. In overbed feeding, a significant amount of SO_2 is released in the freeboard, during devolatilization and combustion of fine coal particles. The residence time of this portion of SO_2 in the furnace is shorter, and the possibility of its "encounter" with limestone particles is less, which leads to a lower sulphur retention. If limestone is fed together with the coal on the bed surface (which is the most frequent case), due to elutriation of fine limestone particles, the possibility of high limestone utilization becomes even smaller.

Effect of fly ash recirculation. The greatest possibilities for SO_2 emission reduction and more efficient use of limestone are offered by introduction of recirculation of fly ash captured by cyclones, bag filters or electrostatic precipitators. With high recirculation ratios (ratio between the mass flow rate of recirculated fly ash particles and coal), bubbling FBC boilers are closer to characteristics of circulating fluidized bed boilers. Bubbling FBC boilers can fulfill modern regulations

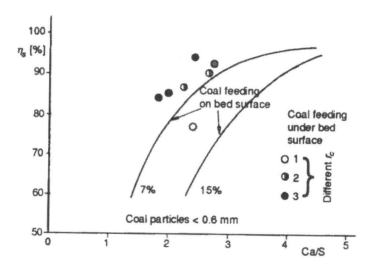

Figure 7.15. *Effect of coal feeding on desulphurization efficiency in bubbling fluidized bed combustion according to Tan and Engstrom (Reproduced by kind permission of the ASME from [25])*

on SO_2 emission only by introducing fly ash recirculation. Systematic investigations of the effects of recirculation [54, 57, 62, 64, 65] established a strong influence on sulphur retention. Application of recirculation enables desulphurization efficiency of up to 99% to be achieved. Figure 7.16 shows the effect of recirculation ratio measured in the TVA 20 t/h pilot boiler [57].

With a recirculation ratio of about 2, a desulphurization level of 95% is achieved, as opposed to 90% without recirculation. Results of measurements in the same facility [64] show an increase of desulphurization efficiency from 75% to 95% when feeding above the bed surface, after introduction of fly ash recirculation. Data obtained in industrial facilities show the same tendency [31]. The customary recirculation ratio range in industrial facilities is 0.5-1.

Effects of characteristics of coal and ash. Investigations of SO_2 emission during combustion of different coals show great differences. There were no systematic investigations of volatile content effects, because it was impossible to maintain all of the other combustion conditions at the same level. The observed differences could, therefore, be the result of effects that were already considered characteristics of coal or characteristics of ash.

Reasons why an increased amount of volatiles may cause an increase of SO_2 emission have been already mentioned. The release and formation of SO_2 and H_2S during devolatilization may be the basic cause. In addition, high volatile coals (most frequently lignites) are typically fed above the bed surface, which can also increase SO_2 emission.

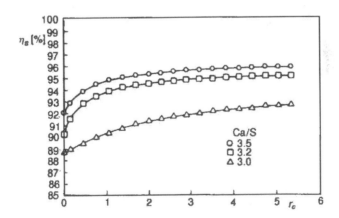

Figure 7.16. *Effect of fly ash recirculation ratio on the degree of desulphurization from the measurements of Saroff (Reproduced by kind permission of the ASME from [57])*

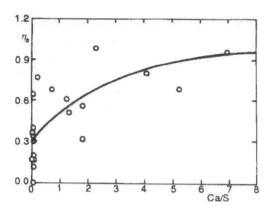

Figure 7.17.
*Effect of Ca/S ratio in coal ash on the desulphuriza-
tion level (Reproduced by kind permission of the
ASME from [51])*

It has been irrefutably demonstrated [51, 66, 67] that calcium contained in coal ash can
capture a considerable amount
of SO_2. Figure 7.17 shows results of measurements of sulphur retention by coal ash
obtained in 6 different facilities
for 23 different types of coal
[51]. The internal Ca/S molar
ratio for different types of coal
can even be above 6 and ash
from certain types of coal can
capture well over 90% of the
sulphur. Despite significant
scatter of results due to different
combustion conditions and
boiler designs, a dependence of
self desulphurization degree on Ca/S ratio in coal is obvious. The data scattering
may also be the result of a catalytic effect of Fe_2O_3 on the reaction between CaO
and SO_2 [66]. Sulphur self-capture in coal ash can significantly contribute to limestone consumption reduction in FBC boilers.

7.4.3. Effects of limestone characteristics

When analyzing the resistances (i. e., processes) that affect limestone
sulphation rate and degree given in section 7.4.1, it is evident that limestone type
and characteristics greatly affect the desulphurization process. The reactivity of
limestone and the size of its particles are perhaps the most important parameters
here. Systematic comparative investigations [25, 58, 65, 68] showed that characteristics and the origin of limestone greatly affect sulphur retention. In order to
achieve the same sulphur retention, widely different quantities of limestone must
be used (different Ca/S ratios). Investigation of 6 types of Kentucky limestones
[65] showed that it was necessary to vary the Ca/S ratio between 1.6 to 3.0, in order
to achieve 90% sulphur retention. Figure 7.18 [58] shows that for 6 different types
of limestone during combustion of the same coal under the same conditions in a pilot facility, the desulphurization levels ranged from 65% to over 90%.

It is obvious that different types of limestone have different capabilities for
SO_2 capture. In designing a boiler and selecting the optimum combustion regime
the choice of limestone can be very significant for reduction of SO_2 emission and
the economical operation of the facility. That is the reason why so much attention

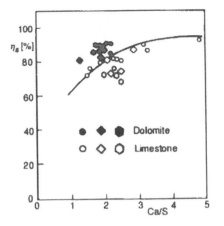

Figure 7.18.
Influence of limestone type on desulpuri-zation levels. All other conditions were the same and the results are taken from the mea-surements of Snell et al. In a pilot plant facil-ity (Reproduced by kind permission of the ASME from [58])

has been devoted to limestone reactiv-ity measurements. Methods of deter-mination of limestone efficiency in capturing SO_2 and the causes of differ-ences between limestone types of dif-ferent origins will be discussed in the following section.

It has been noted that soft lime-stones that tend to crumble and are readily subject to attrition have a higher capability for SO_2 capture. Constant removal of the sulphated layer at the particle surface facilitates SO_2 penetration through the porous structure, and allows it to reach the un-used CaO [68].

High percentages of CaO in lime-stone need not necessarily mean a higher capacity for SO_2 capture. It is possible that Fe_2O_3 may have a cata-lytic effect, and it is possible that other impurities influence the limestone capacity for sulphur capture. In an investigation of 9 different types of limestone in [68], a limestone with 30% MgO achieved the highest sulphur retention.

Selection of limestone particle size is a matter of a well-selected compro-mise: smaller particles have higher specific external area available for reaction, but they are liable to higher losses due to elutriation. Limestone size is normally deter-mined by requirements for an appropriate size close to the size of the bed material – 0.5-2.0 mm.

7.4.4. Efficiency of limestone utilization

Analysis of the effects of operating parameters and limestone characteris-tics shows that sulphur retention depends most on combustion temperature, recirculation degree and limestone type. Other parameters have minor effects, and their choice is affected more by combustion conditions, type of coal or design con-siderations. However, the combustion temperature has to be chosen in the range which corresponds to optimum desulphurization conditions. Fly ash recirculation is introduced most frequently in order to increase combustion efficiency, increas-ing at the same time limestone performance. It is evident that the proper selection of limestone is decisive in achieving low SO_2 emissions.

Selection of the best limestone is a difficult task both for researchers and boiler designers. Differences in desulphurization levels illustrated in Fig. 7.18 show that a poor selection of limestone may have both environmental and economical consequences. Similarly, a change of limestone, after the construction of the boiler, may require major reconstruction of the limestone handling system and the flue gas particle removal system. That is why a large number of limestone types have been investigated in the R&D programs of many countries (in Canada 19 [69] and 12 [70], in the U.S.A. 6 [65] and 9 [68], and in Yugoslavia 8 [71]).

Determination of limestone performance, that is capability of limestone to capture SO_2, can be done in two basic ways:

(a) by fundamental investigation of the calcination and sulphation processes and effects of limestone characteristics and the thermodynamic parameters. Kinetic constants of these processes have to be determined as well as the other data necessary for development of mathematical models. Limestone efficiency is predicted from the result of mathematical modelling, and

(b) by comparative experimental analysis of limestone characteristics and limestone efficiency, without examining the characteristics of the process.

The sulphation level used to compare limestone capabilities to capture SO_2 can be represented as follows

$$X_{CaO} = \frac{\text{Amount of Ca bound as } CaSO_4}{\text{Available amount of CaO in limestone}} \qquad (7.9)$$

The sulphation degree is an important design parameter necessary in determination or valuation of sulphur retention during design of FBC boilers and the selection of limestone.

The maximum values of the sulphation obtained after a sufficiently long reaction time $-(X_{CaO})_{max}$, are used for comparison. For the tested limestone types this quantity falls typically within limits of 0.15 to 0.6.

Reactivity of limestone in the sulphation process. In considering physical and chemical processes in Section 7.4.1, calcination and sulphation were denoted as key processes in capturing SO_2. Although these two processes in reality partially coincide in time, it is believed that they may be studied separately. Investigation can be carried out in modified TGA devices or small chemical reactors with fluidized beds. The measured sample of $CaCO_3$ is exposed to heating and then sulphated with a mixture of gases N_2, CO_2, SO_2 and O_2.

Transformations of limestone particles during calcination are decisive for the sulphation process and capability of limestone to capture SO_2 [42, 70, 72]. Figure 7.19 shows calcination degree (percentage $CaCO_3$ converted to CaO) at differ-

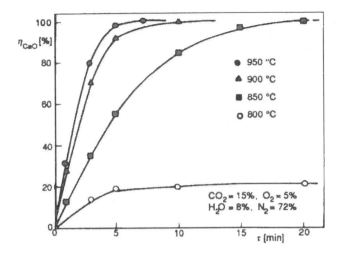

Figure 7.19. *Calcination degree at different temperatures as a function of time, from the measurements of Kim (Reproduced by kind permission of the ASME from [47])*

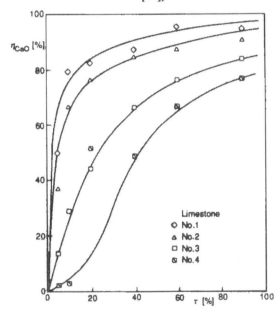

Figure 7.20.
Calcination degree for different limestones as a function of time, from the measurement of Haji-Suleiman (Reproduced by kind permission of the ASME from [72])

ent temperatures [47]. Decomposition of $CaCO_3$ depends on the heating rate experienced by the limestone particles. At lower temperatures, the final conversion of $CaCO_3$ is achieved after 20 minutes. This time is significantly longer than the limestone particle heating time. The calcination rate depends on the type of limestone as well. Figure 7.20 shows that the character of the calcination process for four tested types of limestone [72] is quite different. Limestone No. 1 reached 80% of $CaCO_3$ conversion in less than 10 minutes, while for limestone No. 4, less than 5% conversion occurs over the same time. The time necessary for the calcination process may significantly reduce the quantity of CaO available for SO_2 capture. In unfavorable conditions, sulphation will begin and end before conversion of $CaCO_3$ to CaO is completed. If one knows that the residence time of limestone particle in the bed is 2-3 minutes, the significance of the calcination rate becomes obvious.

According to investigations in [72], "impurities" in the limestone affect calcination time the most. Limestone with about 17% impurities had the shortest calcination time, and a limestone with only 2% of impurities had the longest. Impurities in the limestone affect the creation of pores during the calcination process and retard their blockage during $CaSO_4$ formation, and perhaps can also act as a catalyst during the sulphation process [72].

Increase of particle porosity is the most significant result of the calcination process that affects sulphation. It has not yet been clarified whether the pore volume or specific area is decisive for the sulphation process. According to some authors [47, 69, 70, 72], the pore volume or mean diameter affects the sulphation process decisively. Results of other investigations [58, 71] indicate the effect of specific area of the pores. Further investigations on the creation and effects of the porous structure of limestone particles after calcination are necessary, because both hypotheses are reasonable. Large pores that contribute to the volume of pores the most, enable easy access of SO_2 to the surface of CaO crystals. Pores of a small diameter contribute to the size of specific reaction surface the most. Depending on the conditions and limestone structure, either SO_2 diffusion or reaction kinetics can control the sulphation process and depending on this, the volume of the pores or their area will be decisive.

In any case, limestone particle porosity significantly changes during the calcination process. According to measurements [47, 68, 70, 72], limestone porosity may change by an order of magnitude during calcination. Calcined limestone porosity may even exceed 0.4 cm^3/g.

Porosity change depends on temperature, calcination time and CO_2 concentration, that is on the final conversion degree of $CaCO_3$ into CaO. Figure 7.21 shows that this porosity change is not the same for different pore dimensions. The tested limestone in different experimental conditions [47] suffered volume changes for pores over 250 Å (1 Å = 10^{-1} nm). Porosity increase occurred only in the vicinity of the external particle surface.

If large pores in the vicinity of the external particle surface are created during calcination, SO_2 diffusion towards the particle interior will be easier, and blockage of these pores during sulphation will be more difficult. By contrast, if pores below 200 Å are created, the probability that pores will be filled up during the sulphation process and reduction of the sulphation degree is higher. Further investigations are needed to establish how different types of limestone will behave during the calcination process.

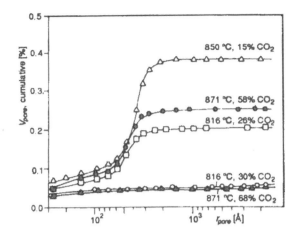

Figure 7.21.
Cumulative pore volume in Fedonia limestone particles, after calcination, from the measurements of Kim (Reproduced by kind permission of the ASME from [47])

Calcination reaction rate depends on particle temperature and size. With temperature below 850 °C, a much higher energy of activation was obtained (about 280 kJ/mol), regardless of particle size. For temperatures higher than 850 °C an activation energy was obtained in the range 30-100 kJ/mol, for the limestone tested with particle sizes from 0.2-0.6 to 2.0-2.4 mm. Temperatures over 850 °C, and smaller particles are favorable for shortening calcination time and for increasing $CaCO_3$ conversion into CaO.

The sulphation process was investigated to detail in references [47, 58, 68-70, 72, 73]. Figure 7.22 [70] shows the course of the sulphation process for different SO_2 concentrations in the gas. The "final" sulphation degree is achieved after a considerably longer time (about 2 h) than is needed for conversion of $CaCO_3$ into CaO. This justifies the assumption that the calcination process can be considered and modelled separately from the sulphation process.

In the initial period of sulphation, the rate of the process is high and depends on SO_2 concentration. In that period the sulphation process is controlled by SO_2 external diffusion towards the particle surface. As the sulphation process proceeds, the rate reduces exponentially, and is no longer affected by SO_2 concentration. During this period, until the final sulphation level is achieved, the reaction rate is controlled by the process of filling up of the pores due to $CaSO_4$ formation, that is by SO_2 diffusion through the pores. The process is finally interrupted, because SO_2 is prevented further access to unreacted CaO in the interior of the particle. A signifi-

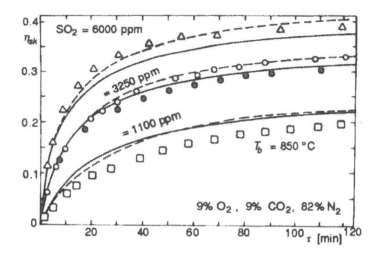

Figure 7.22. *Effect of SO₂ concentration on rate and degree of sulphation, from the measurements of Couturier et al. (Reproduced by kind permission of the ASME from [70])*

cant effect of SO_2 concentration on the final sulphation level is also obtained after a long exposure time. The final value of sulphation is the most significant quantity in practice. The main objective of the majority of experiments is determination of the dependence of this quantity on various parameters, and in particular the effect of limestone characteristics on this parameter.

The dimension of particle pores after calcination has the greatest effect on the final limestone utilization. A good illustration of this dependence is given in Fig. 7.23, which is based on investigations of 12 different types of Canadian limestone.

Depending on particle size, temperature affects the final limestone sulphation efficiency. The highest degree of limestone utilization is obtained over the temperature range 850 to 900 °C. For particles below 1 mm, the temperature is not highly important within this range, but according to investigations in [70], for particles of 2 mm, the final sulphation degree drops drastically at temperature above 850 °C.

If SO_2 diffusion through the porous particle structure is the only controlling mechanism, the final conversion degree in these cases would be lower for larger particles. In the majority of experiments, such dependence has been seen [47, 72, 73]. However, in investigations of some types of limestone [70], an effect of particle size on sulphation levels has not been shown for particles between 0.5-1.0 mm over the temperature range 850-900°C. This means that there is another process that controls the sulphation process. It is assumed [70] that this is the sintering pro-

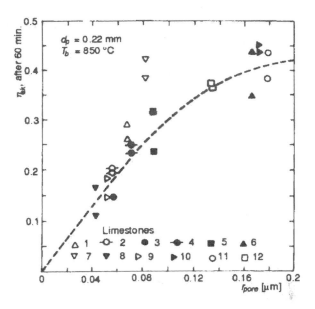

Figure 7.23. *Final degree of limestone utilization as a function of mean pore size, from the measurements of Couturier et al. (Reproduced by kind permission of the ASME from [70])*

cess during calcination that contributes to the creation of large pores that are filled up at a lower rate during $CaSO_4$ formation. These processes require much more extensive investigation.

A detailed presentation of mathematical models of calcination and sulphation processes exceeds the limits of this book. These models are aimed at determining of limestone conversion based on knowledge of limestone characteristics [59-61, 70, 72, 73]. More recent models take into account changes of limestone particle porosity during the sulphation process [60, 70, 73]. Agreement with experimental results is satisfactory, but all such models still require input data obtained by investigation of the relevant type of limestone. Some of these models (see [60]), have been incorporated into general mathematical models that describe all processes in FBC boiler furnaces. Comparison of modelled SO_2 concentration at the exit of the furnaces with measurements in real plants offers relatively good agreement (see Chapter 4, references [104, 105]).

Methods of comparative analysis of limestones. The diversity of limestone characteristics in nature is very high. According to available data, some characteristic limestone properties have a broad range of values. Prior to calcination, porosity is about 0.067-0.34 cm^3/g [71], and after calcination, pore volume is increased to 0.2-0.42 cm^3/g. Specific pore surface is 2.5-36.7 m^2/g and mean pore diameter 40-500 nm [58].

The complexity of calcination and sulphation processes and a large number of factors that affect these processes, make selection of the most appropriate limestone very complex. As noted above, these processes are insufficiently studied, and mathematical models are still in development and cannot be used without experimental input data. To solve the problem, a large number of boiler manufacturers and research institutions need to develop practical, comparative methods for limestone ranking. This kind of method should offer reliable data for the choice and calculation of the limestone quantity necessary to obtain desired sulphur retention in the boiler.

The following methods for ranking limestone are available in the literature: (1) Argon National Laboratory (ANL) developed a method for U.S. Department of Energy, (2) Westinghouse, (3) Canada Centre for Mineral and Energy Technology (CANMET), (4) Tennessee Valley Authority (TVA), (5) German Ministry of Development and Technology, and (6) Babcock and Wilcox. Based on an analysis of these methods [5, 6, 76], in 1986-1987 at the Institute for Thermal Engineering and Energy (IBK-ITE) VINČA Institute, Belgrade, an original methodology of comparative investigation of limestone efficiency was subsequently developed and tested.

The basic aim of these methodologies was ranking limestone according to its capability to capture SO_2, or according to efficiency in reducing SO_2 emission during application in FBC boilers.

The methods can be classified in three groups:

(a) methods based on determination of physical and chemical limestone characteristics,

(b) methods based on determination of sulphation degree in conditions similar to conditions in FBC boilers, and

(c) complex methods which include all limestone changes from the introduction of a limestone to its final conversion in a boiler.

At the base of methods of type (a) is a conviction that limestone characteristics are decisive for the ability of a limestone to capture SO_2. The following quantities are determined: chemical composition, pore volume, specific area and mean diameter of the pores, microstructure of limestone, calcination and sulphation analysis in a TGA device, particle erosion and attrition. Limestone is generally ranked according to porosity after calcination and limestone erosion and attrition intensity.

The group of methods (b) relies on determination of limestone sulphation levels in chemical reactors of small dimensions, at conditions close to those in a real furnace. Investigations are carried out in fluidized bed with the same bed particle sizes as in a real facility, with the same bed temperature and fluidization velocity. Combustion is simulated with test gas, with N_2, O_2, CO_2, and SO_2 concentrations similar to those in gaseous combustion products of a corresponding coal. Some methodologies [5, 6, 45, 77], apart from determining sulphation degree, also provide a determination of most of the characteristics required by the methods a). Final limestone ranking is achieved based on measured final sulphation degree, because it includes the effect of particle erosion, attrition and elutriation.

The most comprehensive, but therefore most expensive investigations are those included by methods from group (c):

- limestone erosion and attrition in systems for limestone transportation and feeding,
- simulation of the influence of cold limestone feeding into the hot fluidized bed at temperatures of 850-900 °C,
- investigation of the calcination process and limestone particle structure change,
- determination of limestone particle erosion and attrition in a fluidized bed, and
- determination of limestone sulphation degree in small chemical reactors.

For the selected highly ranking types of limestone, it is best to determine desulphurization degree in experimental furnaces or pilot facilities, in order to anticipate possible minimal SO_2 emission and limestone consumption (Ca/S ratio) for specific coal in real conditions.

A large number of comparative investigations of limestone were made by these methodologies. Results have been published in the following references – Canadian limestones in references [61, 69, 70], American in [47, 58, 65, 68, 72], Yugoslav in [45, 46, 71, 75, 76], Dutch in [77]. Results of these investigations have already been presented in several figures in this chapter.

We will describe the methodology for comparative analysis of limestone proposed by the Institute for Thermal Engineering and Energy, in some detail. This methodology is a combination of three mentioned methods [5, 6, 45, 75, 76].

Physical and chemical characteristics of limestone are determined by standard methods in the first phase:

- chemical composition (13 compounds),
- moisture content,
- particle density, and
- limestone particle porosity.

In the second phase, the physical and flow parameters necessary for design of the storage bunker and calculation of the removal, transportation and pneumatic systems for limestone feeding into the furnace have to be determined:

- aerated bulk density,
- packed bulk density,
- fixed bed porosity,
- minimum fluidization velocity,
- free fall velocity,

  – parameters of flowability (by the standard R. L. Carr method, using an instrument from the Hosokawa firm):

    – compressibility of the fixed bed,
    – angle of repose,
    – angle of spatula, and
    – index of flowability,

  – parameters of floodability (by the standard R. L. Carr method, using an instrument from the Hosokawa firm):

    – angle of fall,
    – angle of difference,
    – dispersibility, and
    – index of floodability.

Non-standard experimental apparatuses are used in the better part of these investigations.

In the third phase of the investigation, calcination time and degree and the maximum, final sulphation degree are then determined in a small-scale fluidized bed chemical reactor.

In the fourth phase of the investigation, if necessary, the desulphurization efficiency is determined and the necessary limestone consumption (Ca/S ratio) during combustion of the coal under consideration is determined in an experimental furnace with a cross section of 300×300 mm^2.

The key link in this methodology is the determination of limestone sulphation level in the chemical reactor (1) with a fluidized bed, of 39 mm cross section. A diagram of the reactor is given in Fig. 7.24. A 10 g quantity of limestone under test is introduced into a fluidized bed of sand (200 g) with adequate particle size distribution and 70 mm in height. Fluidization takes place with a simulated gas mixture of a predetermined composition – N_2, CO_2, O_2 and SO_2 (6). The composition of the gas mixture is selected according to the composition of coal and combustion conditions under consideration, and maintained by measuring relevant mass flow rates (7, 8, 10). The composition of gas mixture is measured by a gas analyzer (12) and heating of the fluidizing gas mixture and the fluidized bed is achieved by electric heaters (2 and 3), while the gas temperature above the bed is also maintained constant by an electrical heater (4). After introducing a limestone batch, the gas concentrations after desulphurization are measured continuously (12) and the concentrations of O_2, CO_2 and SO_2 are measured throughout the experiment. Typical profiles for the measured SO_2 and CO_2 concentrations are given in Fig. 7.9. Measurement continues until SO_2 concentration returns to the initial value, that is, until the final limestone sulphation degree is reached. Based on the curve of SO_2 concentration change over time, it is possible to determine the total quantity of SO_2 absorbed during the experiment. Then, based on the measured CaO quantity in the

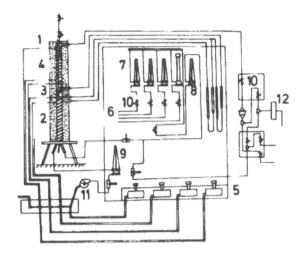

Figure 7.24.
*Experimental installation for determination of sulpha-
tization levels according to ITE-IBK methodology [46, 47]:
1 – fluidized bed reactor, 2 – fluidizing gas preheater, 3 –
fluidized bed heater, 4 – gas reheater above bed surface, 5 –
regulation transformers, 6 – inlet of gas from bottles, 7 –
gas flow meters, 8 – flow meter for gas mixture, 9 – air flow
meter, 10 – regulation valve, 11 – fan, 12 – gas analyzer*

initial limestone, the sulphation degree of limestone X at any moment and at the
end of the experiment can be determined.

An illustration of this method is shown in Fig. 7.25, where the change in
sulphation levels is presented for 8 Yugoslav limestones. Chemical composition
and some physical characteristics of the examined limestones are given in Table
7.6. Based on the curves in Fig. 7.25 it can be seen that, in spite of the very close
chemical composition of the limestone types (except in one case, where the $CaCO_3$
content ranges from 96 to 98%), limestone utilization efficiency varies significantly. Limestone No. 2 has by far the greatest efficiency, with the lowest percentage of $CaCO_3$. These results correspond well with the results of other authors and
the view that impurities in limestone significantly increase limestone utilization efficiency. Limestone No. 2 has the highest total percentage of impurities, and especially considerably more Fe_2O_3 (0.4%). At the same time the pore volume in this
type of limestone is also considerably higher. In limestones Nos. 2, 4, and 5, porosity was determined before calcination. Sulphation degree of these three types of
limestone did not fall in line with the sequence of their porosity. This may be the result of the effect of some other factor or the fact that porosity significantly changed
after calcination.

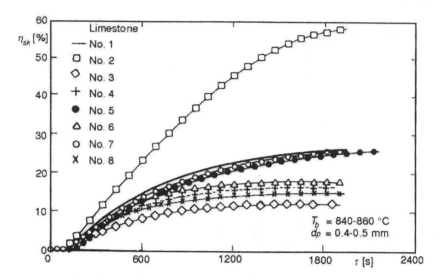

Figure 7.25. *Sulphatization levels for 8 different Yugoslav limestones (Table 7.6) as a function of time, from the measurements of Arsić et al. [71]*

Table 7.6. *Composition of the limestones investigated [71]*

Contents	No. 1	No. 2	No. 3	No. 4	No. 5	No. 6	No. 7	No. 8
SiO_2	0.75	3.18	0.60	1.48	0.66	1.21	0.22	0.20
Al_2O_3	0.20	0.72	0.17	0.44	0.12	0.33	0.03	0.15
TiO_2	0.01	0.02	0.01	0.01	0.01	0.01	0.01	0.01
MnO	0.12	0.08	0.11	0.02	0.02	0.01	0.07	0.07
Fe_2O_3	0.16	0.40	0.12	0.21	0.04	0.014	0.14	0.18
CaO	54.05	50.62	54.75	54.48	54.62	53.81	54.05	54.85
MgO	0.45	1.77	0.40	0.18	0.48	0.53	1.12	0.42
Na_2O	0.01	0.06	0.01	0.05	0.01	0.04	0.01	0.02
K_2O	0.05	0.10	0.04	0.15	0.22	1.20	0.02	0.03
P_2O_5	0.02	0.08	0.01	0.01	0.01	0.01	0.01	0.01
SrO	0.05	0.10	0.04	0.01	0.02	0.02	0.02	0.02
S	0.04	0.07	0.07	0.01	0.01	0.01	0.04	0.05
CO_2	43.45	42.64	43.48	42.81	42.92	42.28	43.75	43.53
$CaCO_3$	96.47	90.34	97.72	97.23	97.48	96.04	96.47	97.89

Figure 7.26 presents the maximum, and final values of the limestone sulphation levels for all 8 types of limestone as a function of temperature. In agreement with other experiments, it can be seen that there is an optimum temperature

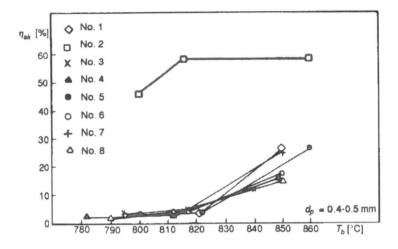

Figure 7.26. *Effect of bed temperature on final degree of limestone utilization for 8 different Yugoslav limestones (Table 7.6), from the measurements of Arsić et al. [71]*

value (about 850 °C). Limestone sulphation efficiency is quite low at temperatures below 820 °C, for all types of limestone.

The effect of limestone particle size is shown in Fig. 7.27. For limestone No. 5 an optimum particle size exists, in this case 0.4-0.5 mm, which is, as discussed above, the result of many conflicting effects.

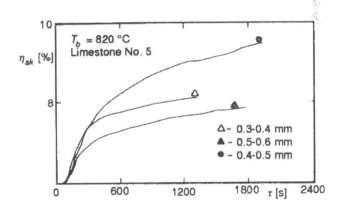

Figure 7.27. *Limestone sulphation degree for different limestone particle size. According to the measurements of B. Arsić et al. [71]*

The main disadvantage of the methods described here for limestone characterization is that they cannot adequately predict limestone sulphation levels without experimental tests being carried out on the limestone under consideration. Prediction of limestone sulphation efficiency by mathematical modelling still has not been developed to the level of an engineering tool.

At the end of this section, certain recommendations can be given based on available knowledge on the calcination process, desulphurization process and investigation of limestone characteristics, which ensure the lowest SO_2 emission possible with minimum limestone consumption:

- The highest possible release of SO_2 in the fluidized bed should occur close to the distribution plate.
- The lowest possible limestone particle size, compatible with the fluidization velocity and inert material particle size should be chosen.
- The most reactive limestone should be selected.
- The most intense mixing should be achieved in the bed and above it.
- Optimum temperature should be chosen.
- The longest possible limestone particle residence time in the furnace should be provided.
- Regions with reducing conditions should be avoided, for two-stage combustion, by increasing total excess air.
- Fly ash recirculation should be introduced.

7.5. Emission of NO_x and N_2O in bubbling fluidized bed combustion

7.5.1. Nitrogen balance during coal combustion in FBC boilers

Natural sources, such as biological denitrification of the soil and waters are believed to amount to about 10 Mt of N_2O, which leaves about 4 Mt derived from anthropogenic activities. One of these activities is combustion and it is further assumed that a portion of the NO emitted by combustion and deposited to the soil, is also probably converted into N_2O and reemitted to the atmosphere via the process of denitrification [82].

Environmental regulations for boilers have so far limited themselves mainly to SO_2, NO_x and particulate emission, and there are no existing regulations limiting N_2O emission. Given that conventional boilers for combustion coal, liquid and gaseous fuel emit less than 5 ppm of N_2O, future application of FBC boilers and their competitiveness may depend on the possibility of N_2O emission control, although it is worth noting that for one important application this is not a problem, as FBC boilers firing biomass emit practically no N_2O.

Numerous investigations of NO_x and N_2O, show that emission of these gas species show an "opposite effect" for many changes in operating parameters. For instance, a bed temperature rise increases NO_x while simultaneously reduces N_2O. In consequence, many studies have been directed to how to achieve reduction of N_2O emissions without simultaneously increasing NO_x emissions. Another complication is that N_2O studies were restricted in the past by measurement difficulties, and reliable and cheap instruments for continuous measurements of N_2O are a relatively new development [83-86].

Typical values for NO_x emissions from industrial scale FBC boilers are given in Section 7.2 and in particular Tables 7.3 and 7.4. However, Table 7.7 also presents, along with NO_x emissions, some values of N_2O emission from industrial-scale FBC boilers.

Table 7.7. *NO$_x$ and N$_2$O emissions from some industrial size FBC boilers*

Boiler type	Boiler capacity	Coal type	Bed temperature range [°C]	NO$_x$ [ppm]	N$_2$O [ppm]	Ref.
BFBC furnace	4 MW$_{th}$	Bituminous low volatile coal	780-880	10-90	125-35	85
				50-180	100-20	
		high volatile		at 7% O$_2$		
CFBC boiler	150 t/h steam	Bituminous high volatile without CaCO$_3$	800-900	50-120	240-100	84
		With CaCO$_3$		110-180	220-80	
				at 7% O$_2$		
CFBC boiler	90 t/h steam	High volatile bituminous coal without CaCO$_3$	800-900	60-160	220-360	84
				at 7% O$_2$		
Nucla CFBC boiler	110 MW$_e$	With CaCO$_3$	865-925	140-250	90-45	83
				at 3% O$_2$		
TVA BFBC boiler	160 MW$_e$	With CaCO$_3$	810-870	220-320	55-35	86
				at 6% O$_2$		
CFBC boilers	Different capacities 4 boilers	Coal and ind. waste	780-850	11-113	153-40	86
				at 6% O$_2$		
BFBC boilers	Different capacities 7 boilers	Coal and ind. waste	780-900	35-243	103-6	86
				at 6% O$_2$		
Conventional pulverized coal combustion boilers Averaged data for 3 units		Different coals	1000-1400	130-460	3-4	86

Table 7.7 shows that there is not a large difference between bubbling and circulating FBC boilers and that generally, CFBC boilers have both slightly lower NO_x and slightly higher N_2O emission than bubbling FBC boilers. This data also indicates that it is possible to achieve very low N_2O emission, 6-30 ppm. The results from a number of comprehensive studies on N_2O formation and destruction reactions will be described in some detail in the following sections of this chapter.

It has been noted that total N_2O and NO_x emissions are approximately constant, although they slightly decrease with increasing bed temperature, as can be seen from data on N_2O and NO_x emissions variations with the bed temperature from a 4 MW_{th} bubbling FBC boiler shown in Fig. 7.28. Here, N_2O concentration decreases from 250 to 100 mg/m^3 as the bed temperature increases for a low volatile bituminous coal. At the same time NO_x concentration rises from 0 to 150 mg/m^3 [85, 87]. Overall resulting total $N_2O + NO_x$ concentration decreases slightly from 250 to 200 mg/m^3 for this coal. During the combustion of coal with high volatile content more NO_x is formed, but less N_2O, and the same behavior is present.

The effect of the coal type on nitrogen oxide and nitrous oxide emissions was examined in a comprehensive Japanese study [86] on an industrial EBARA BFBC boiler with steam capacity 10 t/h. Seven different types of coal were examined and the proximate and ultimate analysis of these coals is given in Table 7.8.

Table 7.8.   *Coals tested in an industrial EBARA BFBC boiler [86]*

Coal	A	B	C	D	E	H	G
Total moisture content [%]	4.6	6.5	4.5	16.5	5.0	4.7	26.4
Heat capacity [MJ/kg]	28.25	27.8	29.4	27.6	25.6	29.5	26.9
Equilibrium moisture content [%]	2.8	5.0	2.5	7.8	1.3	1.1	9.2
Ash content [%]	14.7	13.4	11.2	5.9	24.5	16.1	1.3
Volatile content [%]	25.8	40.2	34.0	33.5	35.6	11.6	43.7
Fixed carbon C_{fix} [%]	56.7	41.4	52.3	52.8	38.6	71.2	45.8
C [%]	85.97	81.97	84.22	81.45	82.67	91.04	74.58
H [%]	4.81	6.08	5.32	4.85	5.87	4.04	5.45
N [%]	1.79	1.61	1.82	1.10	1.33	1.59	1.16
O [%]	6.88	9.73	8.16	12.46	7.79	3.08	18.78
S [%]	0.55	0.62	0.46	0.14	2.34	0.25	0.04

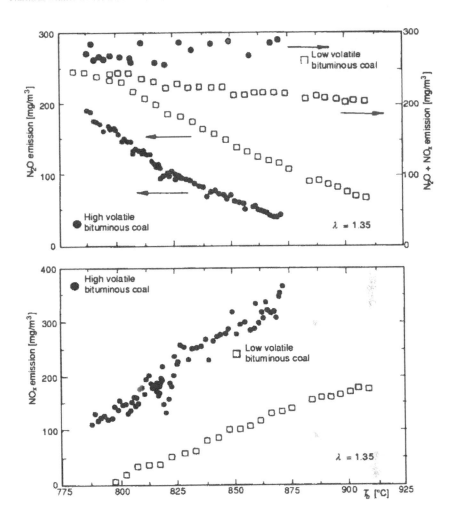

Figure 7.28. *Effect of bed temperature on N_2O and NO_x emissions for two different coals, from the measurements of Braun (Reproduced by kind permission of the ASME from [85])*

The results for NO_x and N_2O emission from this industrial boiler are shown in Figs. 7.29 and 7.30 [86]. Significant effects due to the coal type are observed. Combustion of high volatile coal with a high percentage of nitrogen as a rule results in higher N_2O emissions and lower NO_x emission, while the total $N_2O + NO_x$ emission is approximately unchanged with a change in bed temperature. These investigations also noted significant effects of excess air and bed char inventory on N_2O and NO_x emission, which will be discussed below.

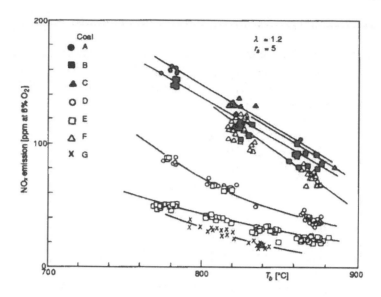

Figure 7.29. *Effect of bed temperature on NO$_x$ emission when burning different coals in several industrial Japanese BFBC boilers from Harada [86]*

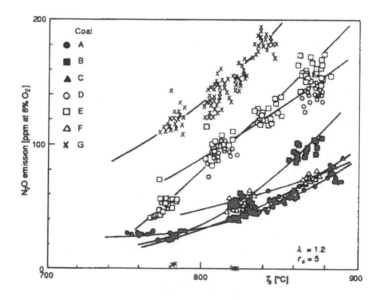

Figure 7.30. *Effect of bed temperature on N$_2$O emission when burning different coals in several industrial Japanese BFBC boilers from Harada [86]*

7.5.2. Mechanisms of N_2O and NO_x formation and destruction

The data on nitrogen oxide and nitrous oxide emission in FBC industrial boilers indicates that N_2O and NO_x concentrations are strongly influenced by the following parameters: the coal type and its characteristics, bed temperature, excess air, freeboard temperature, char hold-up in the bed, the percentage of secondary air, and the type of inert bed material.

Elucidation of the effects of such a large number of parameters is only possible if the mechanisms for N_2O and NO_x formation and destruction are well understood for fluidized bed combustion conditions.

Nitrogen oxides can be formed from molecular nitrogen in the air or from nitrogen in fuel in three ways:

- via a reaction between nitrogen and oxygen in the combustion air, the so-called "thermal" NO_x,
- in a reaction between hydrocarbon radicals and molecular nitrogen, the so-called "fast" or "prompt" NO_x, and finally
- by oxidation of fuel nitrogen compounds, the so-called "fuel" NO_x.

Thermal NO_x is formed by the following reactions:

$$N_2 + O = NO + N \tag{7.10}$$
$$N + O_2 = NO + O \tag{7.11}$$
$$N + OH = NO + H \tag{7.12}$$

Prompt NO_x derives from the following reactions:

$$CH + N_2 = HCN + N \tag{7.13}$$
$$CH_2 + N_2 = HCN + HN \tag{7.14}$$

and from these compounds, NO is formed by reaction with oxygen.

"Thermal" and "prompt" NO formation is low below 1000 °C, and also for low excess air. Thus, although air has about 80% nitrogen content and a typical nitrogen content for coal is only 1-2%, at typical fluidized bed combustion temperatures (800-900 °C), just a few percentage of the nitrogen oxides produced come from atmospheric nitrogen [80].

Mechanisms for N_2O and NO_x formation and destruction in fluidized bed combustion are complex and are still insufficiently well understood. Johnsson [91, 97], Hulgaard [80], and Gustavson [82], indicate that there are over 80 possible reactions for nitrogen oxide and nitrous oxide formation, while over 90 possible reactions with HCN are believed to play a key role in NO_x and N_2O formation from the nitrogen released during devolatilization. Important investigations on NO_x and N_2O formation mechanisms in fluidized bed combustion conditions were carried out in Japan [80, 86, 92-95], Sweden [78, 81, 82, 88, 90, 96], and Denmark [80, 91, 97].

The greatest source of NO_x and N_2O during combustion in fluidized bed is as noted the coal nitrogen. According to experimental results given in [87] about

10% of the fuel nitrogen is oxidizing during combustion to NO_x and N_2O, while data from reference [89] indicates about 17-30% of the fuel nitrogen results in NO_x and N_2O production, depending on the coal type.

Nitrogen in the coal is released either as part of the coal volatiles or it remains in the char. There are therefore two separate paths for nitrogen oxide (NO_x) and nitrous oxide (N_2O) formation: (1) one involves homogeneous oxidation reactions and gaseous compound reduction involving amines (NCO), hydrogen cyanide (HCN) and ammonium (NH_3), released during coal devolatilisation and pyrolysis; (2) while heterogeneous oxidation of nitrogen in char particles, or nitrogen reduction on the surface of char particles are the other sources.

Moritomi [92, 93], Fig. 7.31, has provided a plausible picture for N_2O and NO_x formation, which has been supported by the results obtained by other researchers [79, 81, 82, 88, 90]. During coal devolatilisation in fluidized bed, the NH_3 and HCN released as part of the volatile matter, are the main sources of gaseous NO compounds, although it is not completely clear as to which of the two compounds (NH_3 or HCN) play a more prominent role in NO formation. According to [92 and 98], formation of NO_x from NH_3 is significantly enhanced by the catalytic activity of calcium oxide (CaO) which is present in a fluidized bed where limestone is used for sulphur capture. It is, nevertheless, believed that NO and N_2O are formed over HCN, based on homogenous reactions [82]:

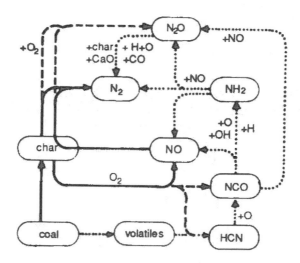

Figure 7.31.
Most probable scheme for N_2O and NO_x formation and destruction during fluidized bed combustion of coal (Reproduced by kind permission of the United Engineering Foundation from [93])

$$HCN + O \leftrightarrow NCO + H \tag{7.15}$$
$$NCO + NO \leftrightarrow N_2O + CO \tag{7.16}$$
$$NCO + OH \leftrightarrow NO + CO + H \tag{7.17}$$
$$NCO + O \leftrightarrow NO + CO \tag{7.18}$$

or directly over NH_3:

$$NH_3 + 5/4\ O_2 = NO + 3/2\ H_2O \tag{7.19}$$

Apart from reaction with NCO, nitrous oxide can also be formed by a heterogeneous reaction of nitrogen from char and oxygen:

$$N\text{-from char} + O_2 \leftrightarrow N_2O \tag{7.20}$$

or in a reaction of NO formed on the surface of the char particle:

$$N\text{-from char} + NO \leftrightarrow N_2O \tag{7.21}$$

There are no indications that N_2O is formed directly during devolatilization [93]. However, molecular nitrogen may be formed during devolatilization by reduction with carbon:

$$C + 2NO \leftrightarrow CO_2 + N_2 \tag{7.22}$$

These reactions explain the role of coal characteristics (volatile content) in NO_x and N_2O formation, and the influence of char concentration in the bed and freeboard, as well as the effect of temperature and the degree of devolatilization.

Possible paths for NO_x and N_2O formation can also involve NH_i compounds, for example:

$$NH_2 + NO \leftrightarrow N_2O + H_2 \tag{7.23}$$
$$NH + NO \leftrightarrow N_2O + H \tag{7.24}$$
$$NH + O \leftrightarrow NO + H \tag{7.25}$$
$$NH + OH \leftrightarrow NO + H_2 \tag{7.26}$$

The role of bed solids, especially char, is highly important in the destruction of NO_x and N_2O and for the formation of molecular nitrogen, as is evident from Fig. 7.31 and eq. (7.22). Detailed investigations on the effect of silica sand, CaO, $CaSO_4$ and ash [81], also indicate a catalytic effect of CaO and fly ash, probably via the following routes:

$$N_2O + \text{particles} \tag{7.27}$$
$$N_2O + H_2 + \text{particles} \tag{7.28}$$
$$N_2O + NH_3 + \text{particles} \tag{7.29}$$
$$N_2 + CO + \text{particles} \tag{7.30}$$

with various possible products.

Destruction of N_2O is possible by the reaction [82, 79]:

$$N_2O + H \leftrightarrow N_2 + OH \tag{7.31}$$

or to a minor extent by:

$$N_2O + OH \leftrightarrow N_2 + HO_2 \tag{7.32}$$
$$N_2O + CO \leftrightarrow N_2 + CO_2 \tag{7.33}$$

Summing up the various paths for NO_x and N_2O formation, it can be concluded that it is possible for them:

(a) to be formed directly from nitrogen in char, and
(b) to be formed indirectly:

– by homogeneous reactions of nitrogen compounds in volatiles (NH_3, HCN, NH_2),
– by homogeneous reactions of gases from char, and
– by heterogeneous reactions (reduction) of gases, and char.

In many of these reactions, CaO particles and ash also have a catalytic effect.

The available values for kinetic constants of over 250 possible reactions are listed in references [80, 82]. It is also concluded that many of these values are not reliable, and that for some reactions reliable data do not exist at all, although this situation is rapidly changing. These data were used for mathematical modeling of N_2O and NO_x formation and destruction during fluidized bed combustion [80-82, 97].

The mechanisms described for NO_x and N_2O formation during combustion in a fluidized bed indicate the importance of the following:

– the oxygen levels (excess air, secondary air),
– gas residence time at combustion temperature (freeboard temperature and height of the furnace),
– coal type and coal characteristics,
– the type of inert bed material (Ca/S ratio, ash composition), and
– char concentration in the furnace (recirculation degree, coal type, coal rank).

A good knowledge of the influence of the parameters listed above, makes it possible to influence N_2O and NO_x emission level in FBC boilers by a suitable selection of operating conditions, furnace design and also by the addition of various nitrogen compounds (NH_3, urea and others) in the freeboard.

7.5.3. Effects of coal characteristics on NO_x and N_2O formation

Data on industrial N_2O and NO_x emissions presented in Figs. 7.28, 7.29 and 7.30, clearly indicate significant differences during combustion of different types of coal. The mechanisms of formation and destruction of nitrogen oxides and nitrous oxides discussed indicate that the volatile matter and char content may have the greatest effect on their emissions. The results of studies on the effects of coal type and characteristics in laboratory conditions are presented in references [88, 89, 92-94] and elsewhere. It is believed that the nitrogen content in the coal affects N_2O and NO_x formation and the level of their emission less than the amount of fixed carbon in the coal and the char hold-up content in the bed and freeboard [92].

The conclusion of all these investigations is that high volatile coals produce considerably lower N_2O concentration and considerably higher NO_x concentration. Vice versa, coals with lower volatile matter content produce low NO_x concentration and higher N_2O concentration. Also, the total quantity of $N_2O + NO_x$ is slightly higher for coals with high volatile content. The effect of coal type can be seen in Fig. 7.28. The main reason for such behavior with different types of coal in fluidized bed combustion is the increased char concentration in the furnace (in the fluidized bed and in the freeboard). This ensures favorable conditions for NO_x reduction and formation of molecular nitrogen. At the same time, a high volatile matter content is the main source of nitrous oxide from homogeneous chemical reactions [89]. Comparative investigation on the combustion of a bituminous coal and its char [88] showed that only 1/3 of the nitrous oxide formed during combustion of the parent coal is produced from char bound nitrogen.

Detailed investigations of N_2O emission during combustion of char formed at different temperatures [93, 94, 89], show that with a rise of devolatilization temperature (i. e. with the reduction of volatile matter content in char), N_2O emission is reduced.

7.5.4. Effects of operating parameters

Effect of bed temperature. There is consensus that during fluidized bed combustion, combustion temperature has the largest effect on N_2O and NO_x emissions (Figs. 7.28-7.30). It can be clearly seen that the concentration of NO_x rises with the bed temperature while the concentration of N_2O drops. The total emission of $N_2O + NO_x$ also falls slightly with the rise of temperature.

The effects of bed temperature can be explained by its effect on the chemical reaction rate, which leads to a change in relative effects for reactions for NO_x and N_2O formation and destruction. The strong effect of bed temperature on NO_x emission is primarily the results of the reduced char content in the bed at higher temperature, which reduces the degree of nitrogen oxide reduction into molecular nitrogen [99]. According to Moritomi [94], the catalytic effects of char and CaO on N_2O destruction reaction is also higher at high temperatures.

There are still problems understanding N_2O and NO_x emission behavior in bubbling fluidized bed boilers. According to [94], during combustion in a bubbling fluidized bed, N_2O and NO_x emission is not affected by the coal volatile content. By contrast, Amand and Anderson [78] noted a significant effect during combustion in a circulating fluidized bed. A similar contradiction exists concerning the effect of bed temperature on N_2O emission. While the majority of investigations [79, 85, 94] showed a reduction of N_2O emission with bed temperature rise, reference [78] suggested an increase of N_2O emission for a 16 MW_{th} boiler.

Effect of freeboard temperature. During combustion of coal with high volatile content, the temperature above the bed, especially in refractory lined furnaces, can be significantly higher than the bed temperature. In bubbling fluidized bed combustion, the temperature of combustion products above the bed significantly changes by the time they leave the furnace exit, while in circulating fluidized bed furnaces temperature is often effectively constant. These differences in temperature profile in the furnace can cause different N_2O and NO_x emission in bubbling and circulating fluidized bed boilers. Changes in NO_x and N_2O concentration along the furnace height are therefore often connected with temperature change above the bed.

Systematic investigations on the effects of freeboard temperature at constant bed temperature were carried out by Braun [87]. Results of these measurements presented in Fig. 7.32 show that when the temperature above the bed rises by 40 °C, the N_2O concentration drops by a factor of two, from 220 mg/m^3 to about 100 mg/m^3. At the same time, the NO_x concentration remains practically unchanged, at about 200 mg/m^3. Similar results were reported in reference [94].

A change of NO_x and N_2O concentrations along the furnace height depend on the temperature profile in freeboard, on the region for coal particle devolatalization [94] and on the catalytic effect of inert material particles. Different conditions in the freeboard are therefore the main reason for the different behaviour of NO_x and N_2O emissions in BFBC and CFBC boilers. An increase of temperature above the bed due to the release and combustion of volatile matter during over-bed coal feeding may result in N_2O reduction.

Measurements made in references [82, 85] show that NO_x emission remains practically unchanged along the bed height, while there is N_2O reduction with a change in height during combustion in a circulating fluidized bed [93].

Effect of excess air. Apart from the temperature effects, the effect of the excess air on NO_x and N_2O formation in fluidized bed combustion has been investigated in detail [62, 80, 87, 90, 93, 94, 100, 101]. In the majority of cases, studies show that an increase of excess air leads to a significant NO_x increase, while the concentration of N_2O remains practically unchanged or increases only slightly. However, there are also measurements that show a simultaneous significant increase of N_2O concentration [78, 83, 88].

Analysis of the effects of excess air is difficult because in most of the experiments excess air change is connected with bed or freeboard temperature change. However, measurements from the Nucla and TVA-160 MW_e [83] showed a significant NO_x and N_2O emission increase with increasing excess air, despite increased bed and freeboard temperatures, which should lead to N_2O emission reduction. By

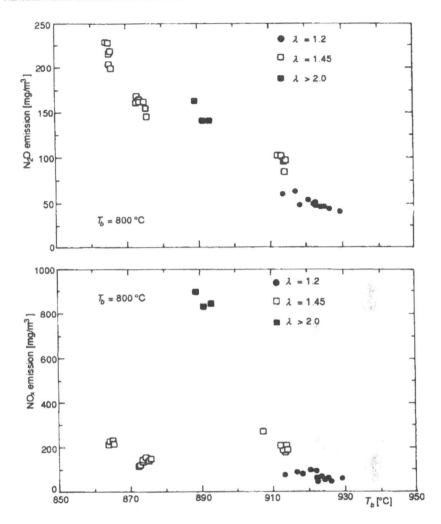

Figure 7.32. *Effect of freeboard temperature on N_2O and NO_x emissions for fluidized bed combustion of high volatile bituminous coal from Braun [87]*

contrast, measurements in a 4 MW_{th} furnace [87] showed a considerable NO_x concentration increase, while there was a practically constant level of N_2O for excess air change from 30 to 270%, with the same bed temperature of about 800-825 °C.

The effects of excess air have been often investigated simultaneously in studies on the effects of substoichiometric combustion [88, 99]. An increased quantity of air simultaneously affects the formation and reduction of nitrogen oxides. A large quantity of "available oxygen" creates favorable conditions for the oxidation of nitrogen compounds released by volatiles (e. g. NH_3). At the same time the char concentration is reduced, which reduces NO_x reduction. All investigations show a significant increase of NO_x with increasing excess air. A catalytic effect of CaO particles as the "inert bed material" on NH_3 oxidation has also been observed. This catalytic effect leads to significant NO_x emission differences from combustion in a silica sand bed [99]. However, the mechanisms by which excess air influences N_2O are still not sufficiently clear.

In the reference given aboves, there are many figures that illustrate NO_x and N_2O change with excess air. In practically all cases, a rise of NO_x and N_2O can be observed with increasing excess air. However, it must be pointed out that in these investigations there was also a change of many other parameters: bed temperature, temperature above the bed, substoichiometric conditions in the bed, the quantity and location of secondary air introduction etc. As an illustration of the effect of excess air, we have therefore selected Braun's results [87], because in his experiments different excess air ratios were achieved at effectively the same bed temperature (Fig. 7.33).

Effect of char hold-up in the bed. The effect of char hold-up in the bed was already considered in the discussion about the effects of the coal type, bed temperature and excess air. The role of heterogeneous reactions which take place on the surface of char particles, and catalytic effects of these particles, has been highlighted by many studies on NO_x and N_2O formation and destruction mechanisms. The effect of char inventory on reduction of NO_x emission in FBC has also been demonstrated.

In this section, we will discuss the effect of char concentration by means of a discussion of the effect of the fly ash recirculation rate. Increasing fly ash recirculation is used in bubbling fluidized bed combustion boilers to improve the combustion efficiency. It also leads to an increase of char concentration in the bed and in the freeboard. Combustion in a circulating fluidized bed can be considered to be a case of extremely high recirculation, which explains in part the lower observed emission of NO_x in these types of boilers. The effect of the degree of recirculation was investigated in references [62, 63, 79, 102]. As expected (Fig. 7.34), emission of nitrogen oxide decreased with increasing fly ash recirculation. Simultaneous measurements of NO_x and N_2O with a changing degree of recirculation was presented only in reference [79]. For both tested coals, NO_x concentration fell, and N_2O concentration rose as the degree of recirculation increased. The possible reason for N_2O concentration increase may be explained by NO reduction on the surface of char particles. Results in reference [102] indicate that after a certain value of the degree of recirculation, there was no further reduction of NO_x emission.

Effect of substoichiometric combustion and secondary air. Combustion under substoichiometric conditions with subsequent addition of secondary air is a well-known method for reduction of nitrogen oxide emission during the combus-

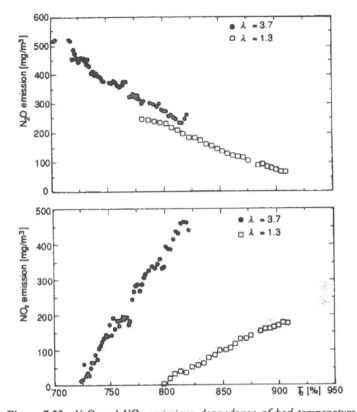

Figure 7.33. *N_2O and NO_x emissions dependance of bed temperature during fluidized bed combustion of low volatile bituminous coal, with two different excess airs from Braun [87]*

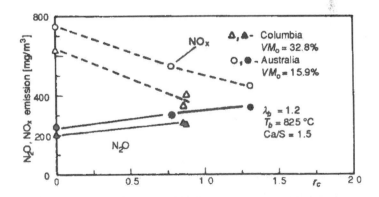

Figure 7.34. *Effect of recirculation ratio on N_2O and NO_x emissions for two different coals from Bramer and Valk [79]*

tion of fossil fuels. Numerous investigations also show that this is effective in reducing NO_x emissions for coal combustion in fluidized beds [62, 63, 79, 88, 99, 102].

All investigations indicate that the division of air into primary and secondary, and substoichiometric combustion in the bed, lead to a reduction of NO_x and N_2O emission for the same total excess air. With substoichiometric combustion, char hold-up significantly increases in the fluidized bed, as well as carbon monoxide concentration, creating in addition to the lack of O_2, favorable conditions for NO_x reduction into molecular nitrogen over char. At the same time, in these conditions, the quantity of nitrogenous compounds released as volatiles increases (for example NH_3) which can later contribute to further reduction of NO_x compounds in the freeboard [99].

The effect of two-stage combustion is more evident during combustion in bubbling fluidized bed. In these boilers the fuel and hence most of the combustion are concentrated in a small part of the furnace, namely the dense fluidized bed, where substoichiometric conditions tend to prevail.

A change of NO_x and N_2O emission with excess air in the bed is shown in Fig. 7.35. Investigations were carried out with a $1\ MW_{th}$ furnace, by keeping all parameters except one constant. When the excess air in the bed was changed, from the standard value of 1.2 to the substoichiometric 0.8, a reduction of NO_x emission from 550 to 375 mg/m^3 was seen. For the same coal, a reduction of N_2O emission from 310 to 230 mg/m^3 was also observed.

Effect of inert bed material type. A catalytic effect of CaO particles on the formation and reduction of NO_x and N_2O has been frequently noted [80, 81, 93]. It was also noted in the earliest investigations [40], that the influence of bed solids on

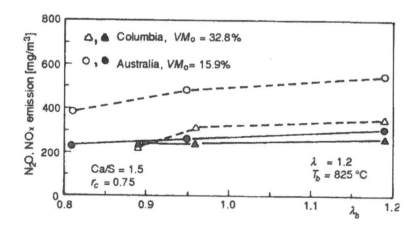

Figure 7.35. *Effect of primary to secondary air ratio on N_2O and NO_x emissions for two different coals from Bramer and Valk [79]*

NO_x differed in bubbling and circulating fluidized beds, although this could not be well explained at the time. Adding limestone to a 16 MW_{th} FBC boiler (within Ca/S ratio range 0-4) did not affect NO_x concentration, while measurements in a 40 MW_{th} CFBC boiler clearly showed an increase of NO_x emission by 2-3 times for a change of Ca/S ratio from 0 to 2.3. The catalytic effect of CaO evidently varies, depending on O_2 concentration. Under substoichiometric condtions, CaO may cause a reduction of NO_x, while in the presence of an excess of oxygen, NO_x formation is increased [40].

In the majority of experiments, a reduction of N_2O emission was noted for limestone addition to a sand bed [93, 81, 103], for both laboratory-scale circulating and bubbling fluidized beds. Figures 7.36 and 7.37 show results for NO_x and N_2O concentrations for the same boiler [40, 55, 103].

In bubbling fluidized bed combustion, results presented in reference [87] show a different behavior (Fig. 7.38). For a significant change of Ca/S, the N_2O remains the same. By contrast, the measurements presented in reference [40], show NO_x increased.

The different behavior of NO_x and N_2O emission in bubbling and circulating fluidized bed boilers indicates the complexity of these phenomena and the effect of other quantities – e. g. the effect of temperature and excess air profiles along the furnace height.

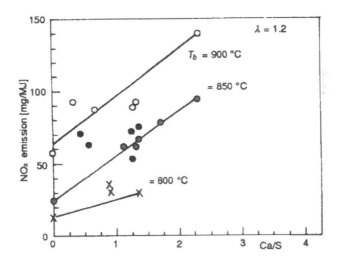

Figure 7.36. *Effect of $CaCO_3$ addition on NO_x emission in circulation fluidized bed combustion boiler from Leckner (Reproduced by kind permission of the ASME from [40])*

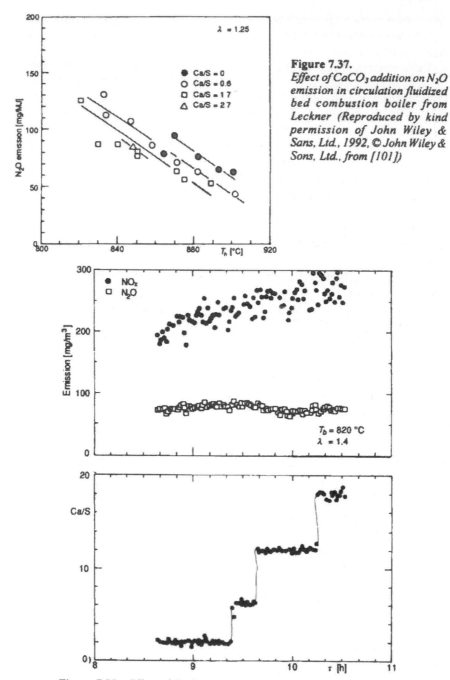

Figure 7.37.
Effect of CaCO₃ addition on N₂O emission in circulation fluidized bed combustion boiler from Leckner (Reproduced by kind permission of John Wiley & Sans, Ltd., 1992, © John Wiley & Sons, Ltd., from [101])

Figure 7.38. *Effect of Ca/S ratio on NOₓ and N₂O emissions during coal combustion in bubbling fluidized bed from Braun [87]*

7.5.5. Measures for reduction of NO_x and N_2O emission in fluidized bed combustion

A large number of operation parameters affect NO_x and N_2O emission in fluidized bed combustion and most frequently produce opposite behavior for these compounds, which make it difficult to find methods and strategies for emission reductions. The most convenient strategy would be one that simultaneously brought about reduction of both nitrogen oxide and nitrous oxide. The results presented in Section 7.5.4 show that this is usually impossible, because a reduction of NO_x emission is usually associated with an increased N_2O production.

Apart from choosing optimum values for the operating parameters, three possible strategies for NO_x and N_2O reduction have generally been considered:

— introduction of two-stage combustion,
— introduction of nitrogen compounds into the furnace to produce NO_x reduction, ammonium (NH_3) or urea, and
— combustion of natural gas at the furnace exit, to produce a strong temperature increase that will destroy N_2O.

Two-stage introduction of air for combustion and effects of sub-stoichiometric conditions in the bed to reduce NO_x emission and their affect on N_2O were discussed in the previous section. Since secondary air favorably affects combustion efficiency, this method can be considered as the simplest and possibly most convenient measure for simultaneous reduction of NO_x and N_2O emission.

The effect of reagents (NH_3) has been investigated in detail [55, 79, 85, 90, 105]. From reference [90], for NO_x emission control, the following conclusions can be drawn:

• It is easiest to introduce NH_3 into the furnace directly above the bed surface. Introduction with primary air or in the transportation air for coal feeding does produce good results.
• Excess air reduction is favorable during introduction of NH_3 for NO_x reduction.
• Introduction of NH_3 produces better results for overbed coal feeding.
• Introduction of NH_3 is better for combustion of high volatile coal.
• Bed temperature change appears to have no significant effects.
• At the site of NH_3 introduction, excess air of 2-4% is the most favorable.

Figure 7.39 shows a reduction of NO_x emission from 110 to about 20 mg/kJ with an increase of NH_3/NO_{out} molar ratio from 0 to 18, in a 16 MW$_{th}$ bubbling FBC boiler at Chalmers University in Göteborg [90]. Although introduction of ammonium into the furnace significantly reduces NO_x emission, a significant disadvantage was observed — an increase of carbon monoxide concentration from 20 to 1000 mg/kJ.

With a molar ratio $NH_3/NO_{out} = 3$, it is possible to achieve a reduction of NO_x emission by 50% without a significant increase of carbon monoxide concen-

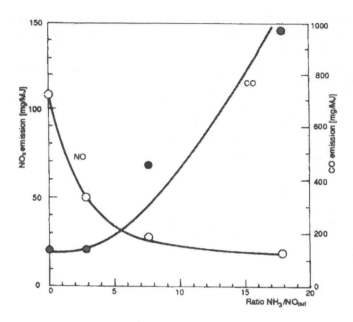

Figure 7.39. *Effect of ammonia addition during coal combustion in bubbling fluidized bed on NO$_x$ and CO emissions [90]*

tration. Further increase of this ratio leads to increased concentration of NH$_3$ in the flue gases exiting the furnace (i. e. NH$_3$ slip), in addition to increases in CO concentrations, and is therefore not recommended [90]. These investigations have also shown that introduction of ammonium in fluidized bed does not affect NO$_x$ emission. Introduction of ammonium with secondary air or at some increased distance from the bed surface only leads to increased NH$_3$ slip, which is highly undesirable given the toxic nature of this compound and its high cost. Reductions of NO$_x$ emission by 50% can be achieved by introduction of two-stage combustion. Similar data are cited in [105].

The results of comprehensive investigations in [85] are similar for NH$_3$ addition. With a molar ratio NH$_3$/NO$_{in}$ = 3, up to 80% of NO$_x$ emission reduction can be achieved in a bubbling fluidized bed. Greater reduction of NO$_x$ emission can be achieved by using urea. For a molar ratio H$_2$NCONH$_2$/NO$_{in}$ of 0.8, a reduction of NO$_x$ emission of 90% was seen. Changes of N$_2$O emission with introduction of ammonium and urea have also been observed. In both cases, a significant increase of N$_2$O emission was noted. With a molar ratio NH$_3$/NO$_{in}$ = 3, N$_2$O emission increases by almost 100%, from about 35 to about 70 ppm. With the introduction of urea, at a molar ratio H$_2$NCONH$_2$/NO$_{in}$ = 0.8, an increase of N$_2$O concentration occured from 40 to 80 ppm.

As a possible measure for reduction of N_2O emission without increasing NO_x emission, combustion of gaseous fuel in flue gases at furnace exit should be suggested, in order to increase the flue gas temperature [82]. This method of reduction N_2O emission, has so far only been demonstrated for a circulating fluidized bed boiler, but it proved highly successful. By combustion of about 10% of natural gas (in relation to the quantity of coal) at the exit from the furnace of an experimental 12 MW_{th} CFBC boiler at Chalmers University in Göteborg, a temperature rise of about 100 °C was achieved, with a resulting reduction of N_2O by about 40% for the same flue gas exit temperature (from 250 to 150 ppm). For a temperature rise of 200 °C, a further reduction to 30 ppm was produced.

Although the analyses of the possible routes for nitrogen oxide and nitrous oxide formation and reduction is highly complex, results presented in [102] show that it is possible by a choice of optimum regime parameters, bed temperature, fly ash recirculation degree and introduction of two-stage combustion, to design a boiler with bubbling fluidized bed combustion whose NO_x emission is below 100 ppm.

NO_x and N_2O production firing biomass. A discussion of NO_x and N_2O production from bubbling and circulating FBC boilers would not be complete without mentioning a rather suprising result. It has long been known that although biomass often has relatively low nitrogen contents when compared with coal [106], more of its fuel nitrogen is released as NO_x in FBC boilers than is the case for coal [107]. This is probably due to two factors, first any biomass char is very reactive so that the char loadings in a FBC will necessarily be lower, and second, the amount of fixed carbon or char produced from biomass is also rather low [106]. However, early measurements on a full-scale circulating boiler clearly showed that N_2O production was extremely low for biomass firing (only a few ppm) [108], and similar findings have been made for bubbling FBC firing biomass [109, 110]. When co-firing coal and wood, an interesting study by Leckner and Karlsson showed that N_2O production increased sharply with increasing coal addition (from virtually zero to typical FBC levels) whereas for NO_x it showed a much more curious behaviour and increased to a maximum at about 20-30% by weight of coal firing. The explanation for this behavior was that the higher absolute nitrogen levels in the coal were not being compensated at this level of co-firing by a resulting increase in bed char loading and hence NO_x was not being reduced effectively in the bed.

7.6.   Emission of solid particles in fluidized bed combustion

7.6.1.   Types and characteristics of solid combustion products from FBC boilers

The types and characteristics of solid combustion products from FBC boilers differ considerably from the solid particles formed in conventional pulverized coal combustion boilers. That is why the problem of removal and disposal of solid combustion products from FBC boilers is important. The specific nature of solids formed in FBC boilers must be taken into account in designing the boiler and in the choice and calculation for the capacity of the flue gas cleaning system and dust collectors.

For FBC boilers, the following three considrations much be noted.

(a) There is an excess of solid inert material in the fluidized bed. The height of the bed increases if particles larger than the chosen nominal bed particle size accumulate in the bed. This excess material must be removed from the bed either by means of an overflow system (rarely) or by draining through openings in the distribution plate. Through these draining openings the bed material, if fluidized, can flow freely out of the bed, whereafter they can be removed by a mechanical system such as described in Chapter 5.

Any excess material that remains in the bed consists of coal derived ash, tramp material fed with the coal, CaO and $CaSO_4$ and the initial inert bed material – e. g. silica sand. Particle sizes for this solid material is usually greater than 1mm, and depends on the selected nominal size of inert material (that is, the chosen fluidization velocity) and coal particle size distribution. Removing this material from the furnace and from the bottom of the bed and its disposal are normally not major problems.

(b) Solid particles separated in the cyclones can be returned to the furnace for reburning. Fly ash removed from flue gases by mechanical inertial separators or cyclones consists mainly of incompletely burnt char, ash, CaO and $CaSO_4$. In many cases, combustion efficiencies over 90% can be achieved only if the elutriated unburned particles in the fly ash from the boiler are recirculated. The quantity of solid material which is returned into the furnace to be "burned out" depends on its chemical composition and the particle size distribution, which in turn depend on the type and size distribution of coal, the fluidization velocity, the furnace height, and whether limestone is used or not, and on the chosen cyclone efficiency.

For a typical design and operating parameters for BFBC boilers, particles caught by the cyclones and returned for combustion are usually below 0.6 mm in size. More than 90% of them are smaller than 150 μm [111]. Only a part of this material is returned to the furnace, depending on the selected degree of recirculation, and the remaining ash is usually mixed with fly ash from the baghouse filters or electrostatic precipitators. Again the transportation and disposal of this material is not a major technical problem.

(c) Fly ash is very fine material that cannot be eliminated by multicyclones. It presents the greatest problem, both from an environmental and technical point of view. Baghouse filters or electrostatic precipitators must be used to remove fly ash from flue gases in order to meet permissible emission limits for particulates (typically around 50 mg/m^3) [112]. Both these types of dust collectors achieve satisfactory results. Typical fly ash particle size is below 5 µm [111, 113]. Fly ash contains mostly particles of ash, CaO and $CaSO_4$ and higher carbon loadings than seen with bed drain materials. Disposal of this fine material can pose a significant problem [111], as is the case for conventional pulverized coal combustion boilers.

The characteristics of fly ash, its removal and operational problems for the dust collectors used in bubbling FBC boilers will be discussed in more detail in this section.

The physical and chemical characteristics of fly ash in bubbling FBC boilers differ considerably from the characteristics of fly ash in pulverized coal combustion boilers, due to the different combustion conditions [113, 114].

Combustion in a fluidized bed takes place at temperatures 800-900 °C, which are considerably lower than typical ash softening or melting temperature. In pulverized coal combustion boilers, temperature in the furnace reaches 1200-1500 °C, and fly ash particles melt, and due to surface tension, acquire spherical shape. Particles of fly ash in FBC boilers do not melt, and are of irregular shape, with very high specific area, which results in their increased cohesiveness and bulk porosity. The chemical composition of fly ash in FBC boilers can also differ due to a limestone addition and the resulting larger quantity of $CaSO_4$, limestone and unused CaO. This is why fly ash from an FBC with sulphur capture is extremely alkaline. The use of multicyclones and inertial separators for particle recycling in the furnace also considerably reduces the size of particles that reach flue gas cleaning devices.

Investigations made in the U.S.A. [113, 114], clearly indicated differences between fly ash produced from FBC boilers and pulverized coal combustion boilers as shown in the Table 7.9.

Table 7.9. *Characteristics of fly ash from bubbling FBC and PC boilers*

	Bubbling FBC boilers	Pulverized coal combustion boilers
Mean diameter [µm]	3-5	5-8.5
Specific particle area [m²/g]	5-20	1-4
Bulk porosity [%]	82-84	58-75

The following two characteristics are significant for efficient flue gas cleaning:

(a) specific aerodynamic resistance of ash layer retained on baghouse filters, and
(b) electrostatic resistance of ash layer in electrostatic precipitators.

Both of these characteristics for FBC fly ash are less favorable than for pulverized coal combustion ash. Therefore, special attention must be paid to the design and use of flue gas cleaning devices for FBC boilers.

The specific aerodynamic resistance of the layer of fly ash particles trapped by baghouse filters, according to [113, 114], ranges up to 7000-12500 mm H_2O m/kg/s.

The electrostatic resistance of fly ash from FBC boilers is 10^{11}-10^{13} Ωcm, while from pulverized coal combustion boilers a typical value is 10^{10} Ωcm [115]. Electrostatic resistance increases as the Ca/S molar ratio rises when limestone is used for sulphur capture. When Ca/S increases from 0.45 to 1.72, ash resistance in an operating FBC boiler increases from 10^{12} to $1.5 \cdot 10^{13}$ Ωcm [112].

7.6.2. Experience with baghouse filters

Baghouse filters and electrostatic precipitators are equally successfully for flue gas cleaning in FBC boilers. Initially, the capacity of FBC boilers was small, so baghouse filters were used more frequently. In such small units, baghouse filters were cheaper than electrostatic precipitators. Due to the expected increased electrostatic resistance with limestone addition, electrostatic precipitators were significantly more expensive [112]. Nowadays, these two types of dust collectors are equally represented for bubbling FBC boilers, although high capacity boilers (100 MW_{th}) tend to use electrostatic precipitators especially in Europe [112–114].

In about 20 industrial operating boilers (up to 165 MW_{th}) the pressure drop across the baghouse filters (from inlet to the outlet) was 50-200 mm H_2O, despite regular dedusting with counter gas flow and a considerably lower specific flow rate than is usually used in gas cleaning in conventional boilers. Filters are designed with specific gas flow rate per square meter of filters 0.15-0.4 m^3/min./m^2, while in conventional boilers 0.5-0.8 m^3/min./m^2 is used.

The main cause of the increased pressure drop for much lower specific gas flow rate is that FBC fly ash forms a layer on the filter cloth whose specific resistance, as mentioned before, is much higher than in conventional boilers, despite the fact that the quantity of ash which is retained by the filter is lower (1.5-2 kg/m^2 compared to 2.5-5 kg/m^2). Due to the considerably smaller particles, the mean pore size of this layer is 4-8 times smaller (about 0.7 μm) than from conventional boilers, which causes much higher flow resistance despite a somewhat greater porosity of the layer (due to irregular shape of the particles) [113, 114].

Due to higher specific flow resistance of the fly ash layer in FBC boilers, baghouse filters for these boilers ought to be designed with significantly lower specific gas flow rate (0.5 m^3/min./m^2), if modern methods of efficient filter dedusting are not used. Modern filters in FBC boilers are designed with considerably higher specific flow rate 1-1.5 $m^3/min./m^2$, but also with a highly efficient method of pulsation dedusting of filters. Such design maintains the value of pressure drop of the filter 50-200 mmH_2O, due to the efficient "shaking" and cleaning process.

Flue gas temperature in front of the filter ranges from 120-200 °C. Glass fiber is increasingly more frequently used as it is a highly suitable material at such temperature and performs well given the specific chemical composition of ash, despite its shorter lifetime.

Although not explicitly noted, it seems that it is necessary to pay special attention in the case of combustion of high moisture fuels with limestone addition, due to a higher quantity of $CaSO_4$ and CaO in the fly ash, which can create problems in the baghouse filters. Problems may appear especially during start-up of the boiler, or during load following [116].

A general conclusion after significant investigations on such filters in the U.S.A. is that an adequate design of baghouse filters and a correct choice of design and operating parameters together with application of efficient dedusting (pulsating) enables a moderate pressure drop to be employed with a reasonable lifetime (2.5-4 years) and low particle emission from the stack. Data from investigations on industrial and utility FBC boilers published in 1991 indicate an extremely low concentration of particulates when baghouse filters are used, 2-15 mg/MJ [113].

7.6.3. Experience in the application of electrostatic precipitators

In the open literature there are very few data on the specific design and use of electrostatic precipitators for removal of fly ash from flue gases in FBC boilers [112, 115, 116]. However, available data indicate that despite the considerably higher electrostatic resistance of fly ash in FBC boilers and its different chemical composition, electrostatic precipitators used in FBC boilers, "operate much better than expected" [115].

Operation of electrostatic precipitators in three different 40 MW_{th} boilers indicated a very high separation efficiency, 99.5-99.97% [112]. Particle emissions were 10-30 mg/m^3.

Investigation of electrostatic precipitators in the TVA 20 MW_{th} pilot facility, despite the considerably higher electrostatic resistance of the fly ash (approximately $1 \cdot 10^{12}$ Ωcm), demonstrated that an electrostatic precipitator operating in conditions non-typical for conventional boilers (mean density of current about 60 nA/cm^2, voltage drop 20-50 kV/cm) achieved separation efficiency over 99% [115].

Thus, it is clear that both baghouse filters and electrostatic precipitators can be successfully applied for FBC boilers in order to meet appropriate environmental regulations for particulates. The actual choice of technology depends on technical and economical considerations.

Nomenclature

Ca/S	Ca (calcium in the limestone added to the bed) to S (sulphur added to the bed with coal) molar ratio
d	mean coal particle diameter, [mm] or [m]
d_p	mean limestone particle diameter, [mm] or[m]
H_b	bed height, [m]
NO_{in}	NO_x concentration before amonia addition, [ppm]
NO_{out}	NO_x concentration after amonia addition, [ppm]
r_c	recirculation ratio,
r_{pore}	mean pore radius in limestone particle, [Å]
r_s	primary to secondary air ratio,
T_h	bed temperature, [°C] or [K]
v_f	superficial velocity of fluidizing gas, [m/s]
V_{pore}	total volume of the pores, [cm³]
VM_a	volatile matter content in coal, as measured by proximate analysis, on "as received" basis, [%]
X_{CaO}	degree of sulphation

Greek symbols

η_s	degree of desulphurization (efficiency), ratio of SO_2 emission to maximum SO_2 emission calculated using sulphur content in coal, [%]
η_{CaO}	degree of limestone calcination, ratio of the CaO formed to the maximum possible amount and CaO calculated using amount of $CaCO_3$ present in limestone, [%]
η_{sk}	sulphation degree of limestone, ratio of the $CaSO_4$ formed to the amount of $CaCO_3$ present in limestone, [%]
λ	excess air
λ_b	excess air in bed for staged combustion with secondary air injected above the bed surface
τ	time, [s]

References

[1] Technical recommendation for clean air protection given by Yugoslav Society for Clean Air (in Serbian). Sarajevo (Yugoslavia): Yugoslav Society for Clean Air, 1986.

[2] A Hemenway, WA Williams, DA Huber. Effects of clean air act amendments of 1990 on the commercialization of fluidized bed technology. Proceedings of 11th International Conference on FBC, Montreal, 1991, Vol. 1, pp. 219-224.

[3] M Gavrilović, Lj Nesić, M Škundrić. Review of the clean air regulations over the world and recommendation for implementation of the clean air regulations in Yugoslavia (in Serbian). Proceedings of Workshop Influence of Thermal Power Plants in Vicinity of Belgrade on Air Quality. Belgrade: Serbian Electricity Board and Society of Engineers and Technicians of Belgrade, 1991, Vol. 1, pp. I-1-I-36.

[4] Z Kostić, B Repić, Lj Jovanović, D Dakić. Protection of the atmosphere from NO_x emission from thermal power plants (in Serbian). Report of the VINČA Institute of Nuclear Sciences, Belgrade, IBK-ITE-868, 1991.

[5] B Arsić, M Radovanović, S Oka. Choice of methodology for comparative analysis of limestone efficiency in sulphur capture during fluidized bed combustion (in Serbian). Report of the Institute of Nuclear Sciences Boris Kidrič, Vinča, Belgrade, IBK-ITE-650, 1987.

[6] M Radovanović. Sulphur retention in fluidized bed combustion (in Serbian). Report of the Institute of Nuclear Sciences Boris Kidrič, Vinča, Belgrade, IBK-ITE-595, 1986.

[7] B Paradiz. Contemporary problems in environmental protection in electricity production (in Slovenian). Proceedings of Symposium JUGEL Development of Electricity Production in Yugoslavia from 1991 till 2000, Ohrid (Yugoslavia), 1990. Belgrade: YUGEL, Vol. 2, pp. 377-387.

[8] D Kisić, Z Žbogar, D Djurdjević. Necessity of the ecological survey of thermal power plants operation (in Serbian). Proceedings of Symposium JUGEL Development of Electricity Production in Yugoslavia from 1991 till 2000, Ohrid (Yugoslavia), 1990. Belgrade: JUGEL, Vol. 2, pp. 393-408.

[9] J Rupar, M Vedenik-Novak. Ecological and other advantages of the district heating (in Slovenian). Proceedings of Symposium JUGEL Development of Electricity Production in Yugoslavia from 1991 till 2000, Ohrid (Yugoslavia), 1990. Belgrade: JUGEL, Vol. 2, pp. 409-416.

[10] D Ćemalović, A Čampara. VIDOS Technology and its application for control of the high SO_2 concentration in flue gases (in Serbian). Proceedings of Symposium JUGEL Development of Electricity Production in Yugoslavia from 1991 till 2000, Ohrid (Yugoslavia), 1990. Belgrade: JUGEL, Vol. 2, pp. 417-422.

[11] F Jevšek. Program of emission control in thermal power plants in Slovenia (in Serbian). Proceedings of Symposium JUGEL Development of Electricity Production in Yugoslavia from 1991 till 2000, Ohrid (Yugoslavia), 1990. Belgrade: YUGEL, Vol. 2, pp. 423-426.

[12] A Knežević. Strategy of the clean air control in thermal power plants in Yugoslavia (in Serbian). Proceedings of Symposium JUGEL Development of Electricity Production in Yugoslavia from 1991 till 2000, Ohrid (Yugoslavia), 1990. Belgrade: JUGEL, Vol. 2, pp. 673-680.

[13] M Škundrić, S Matić, D Milošević. Emission of harmful matter from thermal power plants and air quality around Belgrade (in Serbian). Proceedings of Workshop Influence of Thermal Power Plants in Vicinity of Belgrade on Air Quality. Belgrade: Serbian Electricity Board and Society of Engineers and Technicians of Belgrade, 1991, Vol. 1, 1-37-1-45.

[14] Dj Čobanović. Particle emission and operation of electrostatic precipitators in thermal power plants around Belgrade (in Serbian). Proceedings of Workshop Influence of Thermal Power Plants in Vicinity of Belgrade on Air Quality. Belgrade: Serbian Electricity Board and Society of Engineers and Technicians of Belgrade, 1991, Vol. 1, 1-46-1-57.

[15] B Jevtić. NO_x emission from thermal power plants around Belgrade. Proceedings of Workshop Influence of Thermal Power Plants in Vicinity of Belgrade on Air Qual-

ity. Belgrade: Serbian Electricity Board and Society of Engineers and Technicians of Belgrade, 1991, Vol. 1, pp. 1-58-1-67.

[16]   S Marković, Z Žbogar, D Kisić, D Ćarović. Following of operation of electrostatic precipitators in thermal power plant Nikola Tesla in Obrenovac (in Serbian). Proceedings of Symposium JUGEL Development of Electricity Production in Yugoslavia from 1991 till 2000, Ohrid (Yugoslavia), 1990. Belgrade: YUGEL, Vol. 1, pp. 1-68-1-75.

[17]   WW Hiskins, RJ Keeth, S Tavoulareas. Technical and economic comparison of circulating AFBC vs. pulverized coal plants. Proceedings of 10ᵗʰ International FBC Conference, San Francisco, 1989, Vol. 1, pp. 175-180.

[18]   RE Allen, et al. Comparison of a year 2000 atmospheric circulating fluidized bed and conventional coal-fired plant. Technical features and costs. Proceedings of 10ᵗʰ International FBC Conference, San Francisco, 1989, Vol. 1, pp. 511-518.

[19]   S Oka. Circulating fluidized bed boilers – State of art and experience in exploitation (in Serbian). Proceedings of Symposium JUGEL Development of Electricity Production in Yugoslavia from 1991 till 2000, Ohrid (Yugoslavia), 1990. Belgrade: YUGEL, Vol. 2, pp. 593-600.

[20]   D Wiegan. Technical and economic status of FBC in West-Germany. Presented at International Conference on Coal Combustion, Copenhagen, 1986.

[21]   R Kirchhoff, D Sill, HD Schilling. Reducing emissions from FBC by intrafluid technology – First results. Proceedings of 9ᵗʰ International Conference on FBC, Boston, 1987, Vol. 2, pp. 1054-1061.

[22]   F Verhoeff. The first two years operating experience with the AKZO 90 MW$_{th}$ FBC boiler in Holland. Presented at 20ᵗʰ IEA-AFBC Technical Meeting, Lisbon, 1990.

[23]   S Ikeda. Wakamatsu 50 MW atmospheric fluidized-bed combustion test results and EPDC's development schedule of FBC. Presented at 21ˢᵗ IEA-AFBC Technical Meeting, Belgrade, 1990.

[24]   Test results of 156 t/h bubbling type FBC for electric power utility. EPDC report. Presented at 22ⁿᵈ IEA-AFBC Technical Meeting, Montreal, 1991.

[25]   JT Tang, F Engstrom. Technical assessment on the Ahlstrom pyroflow circulating and conventional bubbling fluidized bed combustion systems. Proceedings of 9ᵗʰ International Conference on FBC, Boston, 1987, Vol. 1, pp. 38-53.

[26]   F Verhoeff. Design and operation of the 115 t/h FBC-Boiler for AKZO-Holland, Proceedings of 9ᵗʰ International Conference on FBC, Boston, 1987, Vol. 1, pp. 61-75.

[27]   Y Nakabayashi. Demonstration test program of the 50 MW$_e$ AFBC boiler in Japan. Proceedings of 9ᵗʰ International Conference on FBC, Boston, 1987, Vol. 1, pp. 177-184.

[28]   F Verhoeff, PHG van Heek. Two year operating experience with the AKZO 90 MW$_{th}$ coal-fired AFBC boiler in Holland. Proceedings of 10ᵗʰ International Conference on FBC, San Francisco, 1989, Vol. 1, pp. 289-296.

[29]   WJ Larva, et al. AFBC retrofit at Black Dog – Testing update. Proceedings of 10ᵗʰ International Conference on FBC, San Francisco, 1989, Vol. 2, pp. 729-738.

[30]   K Furuya. EPDC's fluidized bed combustion R&D&D: A progress report on Wakamatsu 50 MW demonstration test and the world's largest FBC retrofit project. Proceedings of 10ᵗʰ International Conference on FBC, San Francisco, 1989, Vol. 2, pp. 811-825.

[31] F Verhoeff. AKZO – 90 MW$_{th}$ SFBC boiler in Holland: Noteworthy results of a two-year demonstration program. Proceedings of 11th International Conference on FBC, Montreal, 1991, Vol. 1, pp. 251-261.

[32] B Imsdhal, et al. Montana-Dakota Utilities Co.: 25,000 hours of successful AFBC operation. Proceedings of 11th International Conference on FBC, Montreal, 1991, Vol. 1, pp. 263-270.

[33] R Carson, et al. TVA 160 MW$_e$ AFBC Demonstration Plant Process Performance. Proceedings of 11th International Conference on FBC, Montreal, 1991, Vol. 1, pp. 391-401.

[34] AM Manaker, et al. Project overview for the 160 MW AFBC demonstration plant at Tennessee Valley Authority's (TVA) Shawnee fossil plant reservation. Proceedings of 11th International Conference on FBC, Montreal, 1991, Vol. 1, pp. 507-513.

[35] J Stallings, et al. Environmental performance of utility-scale fluidized bed combustors. Proceedings of 11th International Conference on FBC, Montreal, 1991, Vol. 1, pp. 225-232.

[36] JW Regan, H Beisswenger. Utility applications of AFBC. Proceedings of 9th International Conference on FBC, Boston, 1987, Vol. 1, pp. 125-131.

[37] JW Wormgoor, et al. Enhanced environmental and economical performance of atmospheric fluidized bed boilers. Proceedings of 11th International Conference on FBC, Montreal, 1991, Vol. 2, pp. 665-676.

[38] S Kawada, et al. Fuel evaluation and operating experiences of Babcock Hitachi multiple fuel fluidized bed boiler in Japan. Proceedings of 9th International Conference on FBC, Boston, 1987, Vol. 1, pp. 405-414.

[39] MLG van Gasselt. "Which is the Best" atmospheric FBC or atmospheric CFBC? Presented at 18th IEA-AFBC Technical Meeting, Paducah (U.S.A.), 1989.

[40] B Leckner, LE Amand. Emissions from a circulating and a stationary fluidized bed boiler: A comparison. Proceedings of 9th International Conference on FBC, Boston, 1987, Vol. 2, pp. 891-897.

[41] UHC Bijvoet, JW Wormgoor, HHJ Tossaint. The characterization of coal and staged combustion in the TNO 4 MW$_{th}$ AFBB research facility. Proceedings of 10th International Conference on FBC, San Francisco, 1989, Vol. 2, pp. 667-673.

[42] A Fernandez, J Otero, A Cabanillas, G Morales. Characterization of Spanish high sulphur coals for FBC process. Proceedings of 10th International Conference on FBC, San Francisco, 1989, Vol. 1, pp. 93-98.

[43] D Keairns, et al. Sulphur emission control. In: SE Tung, GC Williams, eds. Atmospheric Fluidized-Bed Combustion: A Technical Source Book (Final report). MIT, Cambridge, and U.S. Department of Energy, 1987, DOE/MC/14536-2544 (DE88001042), pp. 7-1-7-99.

[44] FP Fennelly, et al. Environmental aspects of AFBC. In: SE Tung, GC Williams, eds. Atmospheric Fluidized-Bed Combustion: A Technical Source Book (Final report). MIT, Cambridge, and U.S. Department of Energy, 1987, DOE/MC/14536-2544 (DE88001042), pp. 11-1-11-94.

[45] B Arsić, M Radovanović, S Oka. Comparative analysis of efficiency of sulphur retention of Yugoslav limestones (in Serbian). Report of the Institute of Nuclear Sciences Boris Kidrič, Vinča, Belgrade, IBK-ITE-717, 1988.

[46] B Arsić, S Oka, M Radovanović. Characterization of Yugoslav limestones in a fluidized bed reactor. Presented at 4th Conference on Fluidized Bed Combustion, London, 1989, pp. I/17/1–I/17/9.

[47]   HT Kim, JM Stencel, JR Byrd. Limestone calcination and sulfation microstructure, porosity and kinetics under AFBC environments. Proceedings of 9th International Conference on FBC, 1987, Boston, Vol. 1, pp. 449-457.

[48]   RA Newby, DL Keairns. FBC removal – Do we know enough? Proceedings of 11th International Conference on FBC, 1991, Montreal, Vol. 1, pp. 65-71.

[49]   A Lyngfelt, B Leckner. The effect of reductive decomposition of $CaSO_4$ on sulphur capture in fluidized bed boilers. Proceedings of 10th International Conference on FBC, San Francisco, 1989, Vol. 2, pp. 675-684.

[50]   PFB Hansen, et al. Sulphur retention on limestone under fluidized bed combustion conditions – An experimental study. Proceedings of 11th International Conference on FBC, Montreal, 1991, Vol. 1, pp. 73-82.

[51]   JQ Zhang, et al. Evaluation of SO_2 emission from six fluidized bed combustors. Proceedings of 11th International Conference on FBC, Montreal, 1991, Vol. 2, pp. 639-648.

[52]   WVZ Khan, BM Gibbs. Simultaneous removal of NO_x and SO_2 by limestone and ammonia during unstaged and staged FBC. Proceedings of 11th International Conference on FBC, Montreal, 1991, Vol. 1, pp. 99-107.

[53]   M Horio, M Harada, H Moritomi. Current SO_x control status of FB boilers in Japan. Presented at 20th IEA-AFBC Technical Meeting, Lisbon, 1990.

[54]   BJ Zobeck, et al. Western U.S. coal performance in a pilot-scale fluidized bed combustor. Proceedings of 9th International Conference on FBC, Boston, 1987, Vol. 1, pp. 330-339.

[55]   M Mjornell, et al. Emission control with additives in CFB coal combustion. Proceedings of 11th International Conference on FBC, Montreal, 1991, Vol. 2, pp. 655-664.

[56]   EJ Anthony, et al. Pilot-scale trials on AFB combustion of a petroleum coke and a coal-water slurry. Proceedings of 10th International Conference on FBC, San Francisco, 1989, Vol. 2, pp. 653-660.

[57]   L Saroff, et al. A relationship between solids recycling and sulphur retention in fluidized bed combustors. Proceedings of 10th International Conference on FBC, 1989, San Francisco, Vol. 2, pp. 1003-1008.

[58]   GJ Snell, et al. Sorbent testing for AFBC applications. Proceedings of 10th International Conference on FBC, San Francisco, 1989, Vol. 1, pp. 351-357.

[59]   W Lin, MK Senary, CM van den Bleek. SO_2/NO_x emission in FBC of coal: Experimental validation of the DUT SURE-model with data from the Babcock and Wilcox 1-foot × 1-foot AFBC unit. Proceedings of 11th International Conference on FBC, Montreal, 1991, Vol. 2, pp. 649-654.

[60]   R Korbee, et al. A general approach to FBC sulfur retention modeling. Proceedings of 11th International Conference on FBC, Montreal, 1991, Vol. 2, pp. 907-916.

[61]   JQ Zhang, et al. Interpretation of FBC sulphur retention tests on Canadian coals. Proceedings of 10th International Conference on FBC, San Francisco, 1989, Vol. 1, pp. 367-373.

[62]   M Valk, EA Bramer, HHJ Tossaint. Optimal staged combustion conditions in a fluidized bed for simultaneous low NO_x and SO_2 emission levels. Proceedings of 10th International Conference on FBC, San Francisco, 1989, Vol. 2, pp. 995-1001.

[63] M Valk, EA Bramer, HHJ Toissant. Effect of staged combustion of coal on emission levels of NO_x and SO_2 in a fluidized bed. Proceedings of 9th International Conference on FBC, Boston, 1987, Vol. 2, pp. 784-792.

[64] JL Hodges, GD Jukkola, PP Kantesaria. Model prediction of combustion and sulfur capture processes in AFBC. Proceedings of 9th International Conference on FBC, Boston, 1987, Vol. 1, pp. 494-500.

[65] LS Barron, et al. AFBC pilot plant test of six Kentucky limestones. Proceedings of 9th International Conference on FBC, Boston, 1987, Vol. 1, pp. 474-480.

[66] AT Yeh, YY Lee, WE Genetti. Sulphur retention by mineral matter in lignite during fluidized bed combustion. Proceedings of 9th International Conference on FBC, Boston, 1987, Vol. 1, pp. 345-352.

[67] AP Raymant. Sulphur capture by coal ash and freeboard processes during fluidized bed combustion. Proceedings of 10th International Conference on FBC, San Francisco, 1989, Vol. 1, pp. 597-602.

[68] MK Senary, J Pirkey. Limestone characterization for AFBC applications. Proceedings of 10th International Conference on FBC, San Francisco, 1989, Vol. 1, pp. 341-350.

[69] CA Hamer. Evaluation of SO_2 sorbents in a fluidized bed reactor. Proceedings of 9th International Conference on FBC, Boston, 1987, Vol. 1, pp. 458-466.

[70] MF Couturier, HA Becker, RK Code. Sulphation characteristics of twelve canadian limestones. Proceedings of 9th Internationals Conference on FBC, Boston, 1987, Vol. 1, pp. 487-493.

[71] B Arsić, S Oka, M Radovanović. Characterization of limestones for SO_2 absorption in fluidized bed combustion. Presented at 5th Conference on Fluidized Bed Combustion, London, 1991, pp. 171-177.

[72] MZ Haji-Sulajman, AW Scaroni. Evaluation of the properties of natural sorbents for fluidized-bed combustion. Proceedings of 9th International Conference on FBC, Boston, 1987, Vol. 1, pp. 481-486.

[73] PT Daniell, HO Kono. A chemical reaction model of porous CaO particles and SO_2 gas, when the intergrain gas diffusion controls the overall rate. Proceedings of 9th International Conference on FBC, Boston, 1987, Vol. 1, pp. 467-473.

[74] D Celentano, et al. Review of methods for characterizing sorbents for AFBC. Proceedings of 9th International Conference on FBC, Boston, 1987, Vol. 1, pp. 501-510.

[75] B Arsić, S Oka, M Radovanović. Experimental investigation of the reactivity of Yugoslav limestones (in Serbian). Termotekhnika 3-4:187-186, 1990.

[76] B Arsić, S Oka, M Radovanović. Influence of limestone origin and bed temperature on SO_2 retention in fluidized bed (in Serbian). Report of the Institute of Nuclear Sciences Boris Kidrič, Vinča, Belgrade, IBK-ITE-791, 1989.

[77] U Stitsbergen, et al. Limestone addition and flue gas sampling system. In: M Radovanović, ed. Fluidized Bed Combustion. New York: Hemisphere Publ. Co., 1986, pp. 233-260.

[78] LE Amand, S Andersson. Emissions of nitrous oxide (N_2O) from fluidized bed boilers. Proceedings of 10th International Conference on FBC, San Francisco, 1989, Vol. 1, pp. 49-56.

[79] EA Bramer, M Valk. Nitrous oxide emissions from a fluidized bed combustor. Presented at 20th IEA-AFBC Technical Meeting, Lisbon, 1990.

[80] T Halgaard. Nitrous oxide from combustion. PhD dissertation, Technical University of Denmark, Lyngby (Denmark), 1991.

[81] V Langer. Modelling of N_2O in FB combustors. Presented at 20th IEA-AFBC Technical Meeting, Lisbon, 1990.

[82] L Gustavsson, B Leckner. Reduction of N_2O emissions from FB boilers through gas injection. Presented at 20th IEA-AFBC Technical Meeting, Lisbon, 1990.

[83] RA Brown, L Muzio. N_2O emissions from fluidized bed combustion. Proceedings of 11th International Conference on FBC, Montreal, 1991, Vol. 2, pp. 719-724.

[84] A Boemer, A Braun. Emissions of N_2O and NO from large scale CFB combustor. Proceedings of 11th International Conference on FBC, Montreal. 1991, Vol. 2, pp. 719-724.

[85] A Braun, et al. Emission of NO and N_2O from a 4 MW fluidized bed combustor with NO reduction. Proceedings of 11th International Conference on FBC, Montreal, 1991, Vol. 2, pp. 709-717.

[86] M Harada. N_2O emissions from FBC. Presented at 24th IEA-FBC Technical Meeting, Turku (Finland), 1992.

[87] A Braun. Emission of NO and N_2O from a 4 MW fluidized bed combustor. Presented at 21st IEA-AFBC Technical Meeting, Belgrade, 1990.

[88] LE Amand, B Leckner. Influence of air supply on the emissions of NO and N_2O from CFB. Presented at 24th IEA-FBC Technical Meeting, Turku (Finland), 1992.

[89] Y Suzuki, H Moritomi, N Kido. On the formation mechanism of N_2O during circulating fluidized bed combustion. Proceedings of 4th SCEJ Symposium on CFB, Japan, 1991, also presented at 24th IEA-FBC Technical Meeting, Turku (Finland), 1992.

[90] LE Amand, B Leckner. Ammonia addition into the freeboard of a fluidized bed boiler. Presented at 13th IEA-AFBC Technical Meeting, Liège (Belgium), 1986.

[91] JE Johnsson, LE Amand, B Leckner. Modeling of NO_x formation in CFBC boiler. Presented at 3rd International Conference on CFB, Nagoya (Japan), 1990.

[92] H Moritomi, Y Suzuki, N Kido, Y Ogisu. NO_x emission and reduction from CFB. Presented at 3rd International Conference on CFB, Nagoya (Japan), 1990.

[93] H Moritomi, Y Suzuki. Nitrous oxide formation in fluidized bed combustion conditions. Proceedings of 7th International Conference on Fluidization, Brisbein (Australia), 1992. United Engineering Foundation, pp. 495-507

[94] H Moritomi, Y Suzuki, N Kido, Y Ogisu. NO_x formation mechanism of circulating fluidized bed combustion. Proceedings of 11th International Conference on FBC, Montreal, 1991, Vol. 2, pp. 1005-1011.

[95] T Shimizu, et al. Emission control of NO_x and N_2 of bubbling fluidized bed combustor. Proceedings of 11th International Conference on FBC, Vol. 2, Montreal, 1991, pp. 695-700.

[96] S Andersson, LE Amand, B Leckner. N_2O emission from fluidized bed combustion. Presented at IEA AFBC Technical Meeting, Amsterdam, 1988.

[97] JE Johnsson. A NO_x module for the IEA-Model. Presented at 21st IEA-AFBC Technical Meeting, Belgrade, 1990.

[98]   T Hirama, K Takeuchi, M Horio. Nitric oxide emission from circulating fluidized-bed coal combustion. Proceedings of 9th International Conference on FBC, Boston, 1987, Vol. 2, pp. 898-905.

[99]   LE Amand, B Leckner. Emissions of nitrogen oxide from a CFB boiler – The influence of design parameters. Presented at 2nd International Conference on CFB, Compiegne (France), 1988.

[100] HA Becker, et al. Detailed gas and solids measurements in a pilot scale AFBC with results on gas mixing and nitrous oxide formation. Proceedings of 11th International Conference on FBC, Montreal, 1991, Vol.1, pp. 91-98.

[101] B Leckner. Optimization of emissions from fluidized bed boilers. International Journal of Energy Research Vol. 16, 351-363, 1992.

[102] T Hasegawa, et al. Application of AFBC to a very low NO_x coal fired industrial boiler. Proceedings of 10th International Conference on FBC, San Francisco, 1989, Vol. 2, pp. 897-904.

[103] M Mjornell, C Hallstrom, M Karlson, B Leckner. Emissions from a circulating fluidized bed boiler. Chalmers University, Götteborg (Sweden), 1989, II Report A 89-180.

[104] EA Bramer, M Valk. Nitrous oxide and nitric oxide emissions by fluidized bed combustion. Proceedings of 11th International Conference on FBC, Montreal, 1991, Vol. 2, pp. 701-707.

[105] TF Salam, SF Subtain, BM Gibbs. Reduction of NO_x by staged combustion combined with ammonia injection in a fluidized bed combustor: Influence of fluidizing velocity and excess air level. Proceedings of 10th International Conference on FBC, San Francisco, 1989, Vol. 1, pp. 69-76.

[106] DA Tillman. The Combustion of Solid Fuels and Wastes, London: Academic Press, 1991.

[107] B Leckner, M Karlsson. Gaseous emissions from circulating fluidized bed combustion of wood. Biomass and Bioenergy Vol. 4, 5:379-389, 1993.

[108] B Leckner, M Karlsson, M Mjornell, U Hagman. Emissions from a 165 MWth circulating fluidized bed boiler. J. of Inst. of Energy Vol. 65, 464:122-130, 1992.

[109] EJ Anthony, F Preto, BE Herb, JJP Lewnnard. The technical, environmental and economic feasiblity of recovering energy from paper mill residual fiber. Proceedings of 12th International Conference on FBC, San Diego, 1993, Vol. 1, pp. 239-247.

[110] SH Vayda, EO Jauhiainen, L Astrom. Operating experience of an ecoenergy bubbling bed combustion boiler burning paper mill sludges in combination with woodwaste and other fuels. Proceedings of 12th International Conference on FBC, San Diego, 1993, Vol.1, pp. 521-538.

[111] CE Bazzel, RG Mallory, MW Milligan. Disposal of multicyclone and baghouse catch at TVA's 20 MW AFBC pilot plant. Proceedings of 9th International Conference on FBC, Boston, 1987, Vol. 2, pp. 967-973.

[112] S Maartmann, K Bradburn. Further particular control experience after fluidized bed boilers. Proceedings of 9th Intrenational Conference on FBC, Boston, 1987, Vol. 2, pp. 974-976

[113] TJ Boyd, KM Cushing, VH Belba. Status of fabric filtration applied to fluidized bed combustion boilers. Proceedings of 11th International Conference on FBC, Montreal, 1991, Vol. 1, pp. 303-309.

[114] KM Chushing, et al. Fabric filtration – AFBC versus pulverized-coal combustion. Proceedings of 9th International Conference on FBC, Boston, 1987, Vol. 2, pp. 977-982.

[115] WR Carson, WC Nobles, MH Anderson, TJ Boyd. Electrostatic precipitator test program at TVA's 20 MW AFBC pilot plant. Proceedings of 9th International Conference on FBC, Boston, 1987, Vol. 2, pp. 977-982.

[116] D Wilson, et al. Auxiliary equipment. In: CE Tung, GC Williams, eds. Atmospheric Fluidized Bed Combustion. A Technical Source Book. MIT, Cambridge, and U.S. Department of Energy, 1987, DOE/MC/14536-2544, pp. 10-1-10-94.

Index